Protocols for Cytogenetic Mapping of Arthropod Genomes

Protocols for Cytogenetic Mapping of Arthropod Genomes

Edited by Igor V. Sharakhov

CRC Press
Taylor & Francis Group
Boca Raton London New York

CRC Press is an imprint of the
Taylor & Francis Group, an **informa** business

CRC Press
Taylor & Francis Group
6000 Broken Sound Parkway NW, Suite 300
Boca Raton, FL 33487-2742

First issued in paperback 2018

© 2015 by Taylor & Francis Group, LLC
CRC Press is an imprint of Taylor & Francis Group, an Informa business

No claim to original U.S. Government works

ISBN-13: 978-1-4665-9815-7 (hbk)
ISBN-13: 978-1-138-37487-4 (pbk)

Library of Congress Cataloging-in-Publication Data

Protocols for cytogenetic mapping of arthropod genomes / edited by Igor V. Sharakhov.
  p. ; cm.
  Includes bibliographical references and index.
  ISBN 978-1-4665-9815-7 (hardcover)
  I. Sharakhov, Igor V., editor.
  [DNLM: 1. Arthropods--genetics. 2. Chromosome Mapping. 3. Cytogenetic Analysis. 4. Genome. QU 550.5.C4]

  QL434.72
  592.8'615--dc23   2014022168

Visit the Taylor & Francis Web site at
http://www.taylorandfrancis.com

and the CRC Press Web site at
http://www.crcpress.com

Contents

Preface

Arthropods are the most abundant, diverse, and ubiquitous group in the animal kingdom. Originating about 500 million years ago, during the Cambrian period, over 1 million arthropod species are now playing a major role in terrestrial, aquatic, and marine ecosystems. Moreover, arthropods are important to worldwide agriculture, food safety, human health, and energy production. Besides their practical significance, various species represent excellent model systems for biological investigations of evolution, development, physiology, reproduction, and social interaction. For these reasons, arthropod genomics is receiving increasing attention from researchers around the globe. Ambitious projects to obtain whole-genome sequences for insect and related arthropod species, such as the i5K project (http://arthropodgenomes.org/wiki/i5K), have been initiated. However, genome assemblies obtained by next-generation sequencing can be highly fragmented. The level of assembly fragmentation depends on the levels of genetic polymorphism and the abundance of repetitive elements. Success of genomic analyses will be limited if researchers deal with numerous sequencing contigs rather than with chromosome-based genome assemblies. If a genome sequence is associated with real chromosomes, new types of analyses become possible. For example, association mapping, which links phenotypes to genotypes using historic linkage disequilibrium, requires chromosome mapping data. In addition, an interpretation of population genomics data depends on the chromosomal location of markers in the reference genome. Finally, a number of studies including rearrangement phylogeny, chromosome evolution, gene movements, sex-biased expression, epigenomic modifications, and chromatin interactions depend on the availability of chromosome-based genome assemblies. In addition to increasing the value of genome sequence data to the research community, chromosome mapping can potentially identify gaps, misassembled scaffolds, and different haplotypes within assemblies. Therefore, the development of high-resolution physical maps is an important framework for improving the quality of genome assembly, annotation, and analysis. The main reason for this book is to bring together the expertise of cytogeneticists working on diverse groups of arthropods including Diptera (tephritid fruit flies, hessian flies, tsetse flies, and mosquitoes), Coleoptera (beetles), Lepidoptera (silkmoths), Hymenoptera (parasitoid wasps), Hemiptera (aphids, bed bugs, and spittlebugs), Orthoptera (grasshoppers), and Ixodida (ticks). Because the included arthropod species have been studied cytogenetically, they can serve as model species in efforts to chromosomally map genomic sequences. Furthermore, the book intends to facilitate the exchange of cytogenetic expertise among entomologists working with various taxonomic groups, including species that have medical, veterinary, or agricultural importance. Each chapter demonstrates approaches to tissue dissection, chromosome preparation, fluorescence in situ hybridization, and imaging. This book can be viewed as the main source of information about detailed protocols for physical chromosome mapping and their applications for studying genome organization and evolution in arthropod species.

Igor V. Sharakhov
Virginia Polytechnic and State University

Acknowledgments

I thank each author for contributing the detailed cytogenetic protocols for this book. I also thank acquisitions editor Leong Li-Ming for inviting me to publish these protocols with CRC Press and Taylor & Francis.

Editor

Igor V. Sharakhov is an associate professor in the Department of Entomology at Virginia Polytechnic and State University (Virginia Tech). He was appointed to the Virginia Tech faculty as an assistant professor in 2004. He leads a research program focused on the comparative and evolutionary genomics of arthropods of medical importance. He earned his PhD in genetics in 1996 at the Institute of Cytology and Genetics in Novosibirsk, Russia, and his university diploma magna cum laude with a major in biology from Tomsk State University, Russia, in 1989. He joined the laboratory of Guiyun Yan at the Department of Biological Sciences at the State University of New York at Buffalo in 1999 as a research instructor, where he undertook research on molecular cytogenetics of African malaria mosquitoes. In 2001, he was appointed as a research associate in the laboratory of Nora Besansky at the Department of Biological Sciences of the University of Notre Dame, where he completed research on the comparative genome mapping of African malaria mosquitoes. He serves on the editorial boards of *PLoS ONE*, *Journal of Insect Science*, *Journal of Visualized Experiments (JoVE)*, and *Scientific World Journal*.

Contributors

Rajat Aggarwal
DowAgrosciences LLC
Indianapolis, Indiana

Serap Aksoy
Department of Epidemiology of Microbial
    Diseases
Yale School of Public Health
New Haven, Connecticut

Boris A. Anokhin
Zoological Institute of the Russian Academy
    of Sciences
Saint Petersburg, Russia

Antonios A. Augustinos
Department of Biology
University of Patras
Patras, Greece

and

Department of Environmental and Natural
    Resources Management
University of Patras
Agrinio, Greece

and

Insect Pest Control Laboratory
Joint Food and Agriculture Organization/
    International Atomic Energy Agency
    Division of Nuclear Techniques in Food and
    Agriculture
Vienna, Austria

Kostas Bourtzis
Insect Pest Control Laboratory
Joint Food and Agriculture Organization/
    International Atomic Energy Agency
    Division of Nuclear Techniques in Food and
    Agriculture
Vienna, Austria

Diogo C. Cabral-de-Mello
Department of Biology
Institute of Biosciences, Sao Paulo State
    University
Rio Claro, São Paulo, Brazil

Josefa Cabrero
Department of Genetics
Faculty of Sciences
University of Granada
Granada, Spain

Juan Pedro M. Camacho
Department of Genetics
Faculty of Sciences
University of Granada
Granada, Spain

Elena Drosopoulou
Department of Genetics, Development
    and Molecular Biology
School of Biology
Aristotle University
Thessaloniki, Greece

Marco Falchetto
Department of Biology and Biotechnology
University of Pavia
Pavia, Italy

Jürgen Gadau
School of Life Sciences
Arizona State University
Tempe, Arizona

Phillip George
Department of Entomology
Fralin Life Science Institute
Virginia Polytechnic and State University
Blacksburg, Virginia

Snejana Grozeva
Institute of Biodiversity and Ecosystem
    Research
Bulgarian Academy of Sciences
Sofia, Bulgaria

Monika Gulia-Nuss
Department of Entomology
Purdue University
West Lafayette, Indiana

Catherine A. Hill
Department of Entomology
Purdue University
West Lafayette, Indiana

Tatyana Karamysheva
Institute of Cytology and Genetics
Siberian Branch of the Russian Academy
    of Sciences
Novosibirsk, Russia

Valentina G. Kuznetsova
Zoological Institute of the Russian Academy
    of Sciences
Saint Petersburg, Russia

Maria Dolores López-León
Department of Genetics
Faculty of Sciences
University of Granada
Granada, Spain

Anna R. Malacrida
Department of Biology and Biotechnology
University of Pavia
Pavia, Italy

Mauro Mandrioli
Department of Life Sciences
University of Modena and Reggio Emilia
Modena, Italy

Gian Carlo Manicardi
Department of Life Sciences
University of Modena and Reggio Emilia
Reggio Emilia, Italy

Frantisek Marec
Laboratory of Molecular Cytogenetics
Institute of Entomology, Biology Centre ASCR
Ceske, Budejovice, Czech Republic

Anna Maryańska-Nadachowska
Institute of Systematics and Evolution of Animals
Polish Academy of Sciences
Kraków, Poland

Penelope Mavragani-Tsipidou
Department of Genetics, Development and
    Molecular Biology
School of Biology
Aristotle University
Thessaloniki, Greece

Jason M. Meyer
Department of Entomology
Purdue University
West Lafayette, Indiana

and

Department of Biotechnology
Monsanto Company
Chesterfield, Missouri

Michaela Neusser
Institute of Human genetics
Ludwig Maximilian University
München, Germany

Ashley Peery
Department of Entomology
Fralin Life Science Institute
Virginia Polytechnic and State University
Blacksburg, Virginia

Francisco J. Ruiz-Ruano
Department of Genetics
Faculty of Sciences
University of Granada
Granada, Spain

Karsten Rütten
Emil-Fischer-Gymnasium
Euskichen, Germany

Ken Sahara
Laboratory of Applied Entomology,
   Faculty of Agriculture
Iwate University
Morioka, Japan

Brandon J. Schemerhorn
United States Department of Agriculture—
   Agricultural Research Service
Department of Entomology
Purdue University
West Lafayette, Indiana

Francesca Scolari
Department of Biology and Biotechnology
University of Pavia
Pavia, Italy

Igor V. Sharakhov
Department of Entomology
Fralin Life Science Institute
Virginia Polytechnic and State University
Blacksburg, Virginia

Maria V. Sharakhova
Department of Entomology
Fralin Life Science Institute
Virginia Polytechnic and State University
Blacksburg, Virginia

Atashi Sharma
Department of Entomology
Fralin Life Science Institute
Virginia Polytechnic and State University
Blacksburg, Virginia

Jeff J. Stuart
Department of Entomology
Purdue University
West Lafayette, Indiana

Vladimir Timoshevskiy
Department of Entomology
Fralin Life Science Institute
Virginia Polytechnic and State University
Blacksburg, Virginia

Yuji Yasukochi
Insect Genome Research Unit
National Institute of Agrobiological
   Sciences
Tsukuba, Ibaraki, Japan

Atsuo Yoshido
Division of Biological Sciences and
   Center for Genome Dynamics
Faculty of Science
Hokkaido University
Sapporo, Japan

Antigone Zacharopoulou
Department of Biology
University of Patras
Patras, Greece

Tephritid Fruit Flies (Diptera)

Penelope Mavragani-Tsipidou,
Antigone Zacharopoulou,
Elena Drosopoulou,
Antonios A. Augustinos,
Kostas Bourtzis, and Frantisek Marec

CONTENTS

LIST OF ABBREVIATIONS

BCIP, 5-bromo-4-chloro-3-indolylphosphate p-toluidine salt
BSA, bovine serum albumin
B&W, black-and-white
CCD, charge-coupled device
DAB, Dimethylaminoazobenzene
DAPI, 4′,6-diamidino-2-phenylindole
dH_2O, distilled water
DIG, digoxigenin
EDTA, ethylenediaminetetraacetic acid
FISH, fluorescence in situ hybridization
GSS, genetic sexing strain
ID, imaginal disc
NBT, nitroblue tetrazolium chloride
PBS, phosphate buffered saline
PEN, polyethylene naphthalate
RT, room temperature
SDS, sodium dodecyl sulphate
SSC, saline sodium citrate

1.1   INTRODUCTION

1.1.1   Taxonomy and Importance of the Species

Fruit flies of the family Tephritidae with about 4450 species are classified into six subfamilies and 484 genera (Norrbom et al. 1999; Korneyev 2000; Systematic Entomology Laboratory 2004). The majority of species attack fruits and other plant crops worldwide. The five genera, *Anastrepha*, *Bactrocera*, *Ceratitis*, *Dacus*, and *Rhagoletis*, include the most destructive agricultural pests in the world and are therefore considered pests of global economic importance (Fletcher 1989; White and Elson-Harris 1992).

Bactrocera with 629 described species is the largest genus (Drew and Hancock 2000; Drew 2004). It includes about 40 species of economic importance, mainly found in Asia, Australia, and the Pacific (White and Elson-Harris

1992) with a sole representative in Europe, the olive fruit fly, *Bactrocera oleae*. *Dacus* is also a large genus (248 species), closely related to *Bactrocera* (Drew and Hancock 2000; Drew 2004). It inhabits the Afrotropical region and includes 11 species of economic importance (White and Elson-Harris 1992). The genus *Anastrepha* includes about 198 species, 15 of which are of economic importance (White and Elson-Harris 1992); fruit flies of this genus have a mainly neotropical distribution (McPheron et al. 2000; Norrbom et al. 2000). The Afrotropical genus *Ceratitis* includes more than 89 species (Barr and McPheron 2006). Eleven of them are important pests (White and Elson-Harris 1992) with the Mediterranean fruit fly, *Ceratitis capitata*, being the best-studied member of the whole family and a model system for genetic, molecular, and cytogenetic studies including the development of genetic sexing strains (GSSs) for sterile insect technique (SIT) applications (Robinson et al. 1999; Gariou-Papalexiou et al. 2002; Franz 2005). The genus *Rhagoletis*, finally, contains more than 60 species, mostly native to North America and only few of Eurasian origin (Smith et al. 2005). Seventeen species are considered pests of economic importance (White and Elson-Harris 1992).

The taxonomy of the Tephritidae family has undergone many revisions and more can be expected. This is due to the high diversity and plasticity of this family, as well as to the application of new techniques employed to resolve the status of species complexes, cryptic species, or cases of incipient speciation that are very common in the Tephritidae. Much attention has been given to deciphering of the *B. dorsalis*, *Ceratitis* FAR (*fasciventris-anonae-rosa*), and *Anastrepha fraterculus* species complexes, as well as the sympatric speciation through host shift and arthropod symbiosis in the genus *Rhagoletis*.

The challenge to reduce the damage caused by tephritid pests and as a consequence to increase food production led to the development of many pest management strategies. Although insecticides are still heavily used for the control of many tephritids, there are also environmental friendly control strategies, such as the SIT, for the application of which knowledge of the genetic structure of populations of pests is essential. Such methods have been gaining more and more attention in the recent years for use on an increasing number of species (Robinson et al. 1999; Gariou-Papalexiou et al. 2002; Franz 2005; Ant et al. 2012). It has become apparent that basic research has led to a much better understanding of the insect pests and how to control them effectively.

1.1.2 Karyotype and Genome Analysis

Cytogenetic studies on tephritid species are greatly facilitated by the existence, in all dipterans, of two basic forms of chromosomes with different morphology and function, "standard" chromosomes (mitotic and meiotic) and polytene chromosomes. Standard chromosomes are found in proliferating tissues, such as the larval nervous system, imaginal discs (IDs), ovaries, and testes, whereas polytene chromosomes suitable for microscopic analysis are found in the interphase nuclei of differentiated cells of certain larval tissues (e.g., salivary glands, Malpighian tubules, and fat body) as well as in the pupal trichogen cells. Cytology of mitotic and meiotic chromosomes is essential for identifying the number, relative size, structure, and rearrangements of the chromosomes of a given species. Basic questions of chromosome biology can be investigated using standard cytogenetic methods and banding techniques (e.g., Giemsa staining and C-banding). Moreover, with fluorescence in situ hybridization (FISH) and derived methods, such as whole chromosome painting, genomic in situ hybridization, and comparative genomic hybridization, targeted sequences can be mapped to the chromosomes, specific chromosomes or genomic regions can be characterized, and homologous chromosomal sections can be identified in different species. Polytene chromosomes of dipterans are used as excellent experimental material to study chromosome structure and function, and temporal gene activity and genomic organization, to assess phylogenetic relationships among closely related species, to distinguish members of a species complex group, and to provide a means for the construction of detailed genetic–cytogenetic maps through accurate mapping of genetic loci to the banding pattern (Zhimulev et al. 2004). Not only genes but also any nonrepetitive DNA sequence can be precisely localized by in situ hybridization.

The best studied Tephritidae species from the genetic and cytogenetic point of view are the following: *C. capitata*, *B. oleae*, *B. dorsalis*, *B. tryoni*, *B. cucurbitae*, *A. ludens*, *D. ciliatus*, *R. cerasi*, *R. cingulata*, and *R. completa*. The karyotypes of these species consist of six pairs of chromosomes ($2n = 12$), with the sex chromosomes forming pair no. 1 (Figure 1.1a through e). The autosomal complement has, thus, five pairs of chromosomes that are designated as 2–6 in order of descending size. In most of the above-mentioned species, the autosomes are metacentric or submetacentric, but in *A. ludens* acrocentric (Figure 1.1c) (Bedo 1986; Zacharopoulou 1987, 1990; Mavragani-Tsipidou et al. 1992; Zhao et al. 1998; Frias 2002; Kounatidis et al. 2008; Garcia-Martinez et al. 2009; Drosopoulou et al. 2010, 2011a,b; Zacharopoulou et al. 2011a,b).

Tephritid females are the homogametic (XX) and males the heterogametic sex (XY). Both sex chromosomes are largely heterochromatic and in most species are heteromorphic and easily distinguished. Even though a great variability in the length ratio of the sex chromosomes has been observed, the Y chromosome is, in most cases, the smallest and most intensely stained element of the mitotic set (Figure 1.1a through c) (Zacharopoulou 1987, 1990; Mavragani-Tsipidou et al. 1992; Zhao et al. 1998; Garcia-Martinez et al. 2009; Zacharopoulou et al. 2011a,b). An exception is *D. ciliatus*, where the X chromosome is the smallest (dot-like) element and Y is almost twice as big (Figure 1.1d) (Drosopoulou et al. 2011b). Distinguishing female and male karyotypes is difficult in the three *Rhagoletis* species, *R. cerasi*, *R. completa*, and *R. cingulata*, due to the similar length of their sex chromosomes (Figure 1.1e) (Kounatidis et al. 2008; Drosopoulou et al. 2010, 2011a).

In tephritids, as all dipterans, the most useful tissue for studying the function and structure of polytene chromosomes

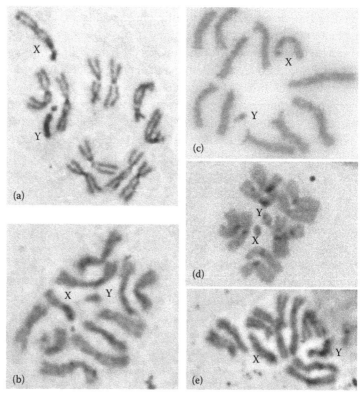

FIGURE 1.1 Mitotic karyotypes of five Tephritidae species: (a) *Ceratitis capitata*, Giemsa staining; (b) *Bactrocera dorsalis*, Giemsa staining; (c) *Anastrepha ludens*, Giemsa staining (From Garcia-Martinez V. et al., *Genome*, 52, 1–11, 2009); (d) *Dacus ciliatus*, C-banding; and (e) *Rhagoletis cerasi*, Giemsa staining. X and Y, sex chromosomes.

is that of the salivary gland cells. The analysis of the salivary polytene nuclei (Figure 1.2a) of tephritid species showed the existence of five long chromosomes (Figure 1.2b), which correspond to five autosomes of the mitotic complement. The polytene nuclei lack a typical chromocenter resulting in separation of chromosomes (Bedo 1986, 1987; Zacharopoulou 1987, 1990; Mavragani-Tsipidou et al. 1992; Zambetaki et al. 1995; Zhao et al. 1998; Kounatidis et al. 2008; Drosopoulou et al. 2010, 2011a,b; Zacharopoulou et al. 2011a,b). The sex chromosomes remain largely underreplicated and form a granular heterochromatic network spread in the space between the polytene arms (Figure 1.2a). However, in polytene nuclei of *C. capitata* pupal trichogen cells, a distinct sex chromosome body composed of a granular network together with a compact heterochromatic sphere associated with the nucleolus is found (Figure 1.3) (Bedo 1986). On the basis of observations in several Y–autosome translocation strains, it was suggested that the compact sphere represents the Y chromosome, while the granular network corresponds to the X chromosome (Figure 1.3) (Bedo 1987; Bedo and Zacharopoulou 1988; Kerremans et al. 1990, 1992; Zacharopoulou et al. 1991a,b; Franz et al. 1994; Cladera and Delprat 1995; Franz 2005). Further evidence for this hypothesis comes from in situ hybridization of rRNA gene clusters (Bedo and Webb 1989) and X-linked genes (ceratotoxins) in both mitotic and polytene chromosomes of *C. capitata* (Rosetto et al. 2000). Moreover, direct proof for the location of the X and Y chromosomes in the granular network and the heterochromatic compact sphere, respectively, was accomplished in *B. oleae* salivary gland polytene chromosomes using FISH with sex chromosome painting probes prepared by laser microdissection of the respective chromosomes from mitotic metaphases (Drosopoulou et al. 2012). Taken together,

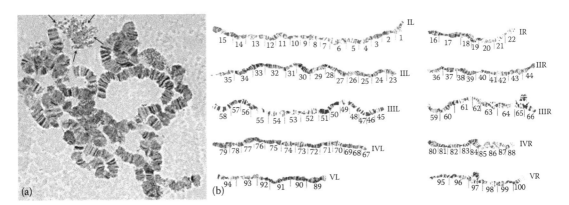

FIGURE 1.2 Salivary gland polytene chromosomes of *Bactrocera oleae*: (a) polytene nucleus and (b) polytene chromosome map. Arrows in (a) indicate the heterochromatic mass corresponding to the nonpolytenized sex chromosomes. (From Zambetaki, A. et al., *Genome.*, 38, 1070–1081, 1995.)

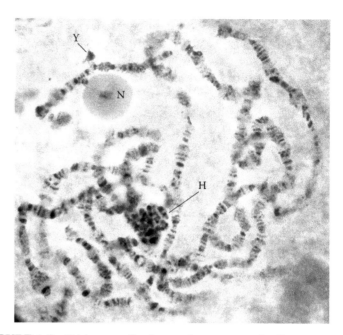

FIGURE 1.3 Trichogen cell polytene chromosomes of a *Ceratitis capitata* genetic sexing strain (GSS). H indicates the heterochromatic net representing the X chromosome, N indicates the nucleolus, and Y indicates the spherical body representing the Y chromosome. Both the Y chromosome and the nucleolus are attached to the autosome that is involved in the Y–autosome translocation at the position of the breakpoint.

these results suggest a conserved sex chromosome structure in polytene tissues of tephritid fruit flies.

The comparison of polytene chromosomes prepared from salivary glands and other tissues showed in some cases (e.g., fat body and Malpighian tubules) considerable similarities between banding patterns (Kerremans et al. 1990; Mavragani-Tsipidou et al. 1992; Zambetaki et al. 1995; Mavragani-Tsipidou 2002), while in others (e.g., pupal orbital trichogen cells in *C. capitata*), a completely different pattern was observed (Bedo 1986, 1987; Bedo and Zacharopoulou 1988; Zacharopoulou et al. 1991a; Semeshin et al. 1995). Even though the main reason for these differences is related to the differential activity of loci, differences in the structural organization of the chromosomes in these tissues cannot be excluded (Semeshin et al. 1995).

1.2 PROTOCOLS

1.2.1 Species Culture

1. Rearing on fruits
   The studied Tephritidae species (*C. capitata*, *B. oleae*, *B. tryoni*, *B. cucurbitae*, *B. dorsalis*, *A. ludens*, *D. ciliatus*,

R. cerasi, R. cingulata, and R. completa) can be reared in laboratory cages where they feed, grow, and oviposit on fruits of their preference. For this purpose, undamaged fruits are selected, washed (borax solution), and stored at 4°C. For each generation, a sufficient number of fruits is placed in the cages. After oviposition, the fruits are transferred to boxes lined with disinfected sand for pupation. Pupae are collected and used to grow the next generation. The disadvantages of this method are the quantity of fruits needed and the declining quality of the fruits after long storage (Roller 1989; Tsitsipis 1989; Tzanakakis 1989).

2 Rearing on artificial diet
   With the exception of *Rhagoletis* species, laboratory-adapted colonies and strains for the above-mentioned species are routinely maintained in insect cages (wood, plastic, or plexiglass) and reared on an artificial diet under specific conditions of temperature, humidity, and light–dark cycle. For cytogenetic analysis, larvae should be grown in uncrowded larval medium to provide mitotic and polytene chromosome materials of sufficient quality.

   Equipment, chemicals, and solutions needed for artificial diets are listed in alphabetical order. Suppliers and catalogue numbers are given only when specific products are used.

1.2.1.1 Equipment

1. Absorbent wipes
2. Cages made from wood, plastic, or plexiglas
3. Dry sand
4. Filter paper
5. Mixer
6. Nylon cloth
7. Petri dishes
8. Plastic bottles

1.2.1.2 Chemicals and Additives for Diet

1. Brewer's yeast
2. Carrot powder
3. Cholesterol
4. Dry yeast
5. Egg yolk
6. Ethanol
7. Hydrochloric acid (HCl) 37%
8. Methyl-*p*-hydroxybenzoate (Nipagin)
9. Olive oil
10. Paraffin mixture

11. Potassium sorbate
12. Propionic acid
13. Sodium benzoate
14. Soy hydrolyzate, enzymatic
15. Streptomycin
16. Sugar
17. Sucrose
18. Tween 80
19. Wheat germ
20. Wheat bran
21. Yeast hydrolyzate, enzymatic

1.2.1.3 Media and Mixtures Required

1. *B. dorsalis* larval medium: 28 g wheat bran, 7 g brewer's yeast, 13 g sugar, 0.28 g sodium benzoate, 1.7 mL 37% HCl and tap water in final volume 100 mL. Mix well. Store at 4°C.

2. *B. oleae* larval medium: 300 g cellulose powder, 30 g soy hydrolyzate, enzymatic, 75 g brewer's yeast, 20 g sucrose, 20 mL olive oil, 7.5 mL Tween 80, 2 g Nipagin, 0.5 g potassium sorbate, 30 mL 2 N HCl in 550 mL tap water. Mix well. Store at 4°C.

3. *B. oleae* adult diet: 40 g sucrose, 10 g yeast hydrolyzate, 3 g dried egg yolk, 25 mg streptomycin. Mix well. Store at 4°C.

4. *B. tryoni* larval medium: 12 g carrot powder, 6.7 g inactivated dry yeast, 4 mL 37% HCl, 2 g Nipagin in 83 mL tap water. Mix well. Store at 4°C.

5. *C. capitata* larval medium 1: 15.2 mL 37% HCl, 4 g sodium benzoate, 42 g yeast hydrolyzate, 115 g dehydrated carrot powder in 1 L tap water. Mix well. Adjust pH to 4.5. Store at 4°C.

6. *C. capitata* larval medium 2: 10 mL 14.5% HCl, 10 mL benzoate solution (12.5 g sodium benzoate in 71.2% ethanol), 30 g yeast hydrolyzate, 30 g sugar, 30 g soft paper in small pieces, 10 mL cholesterol solution (5.3% cholesterol in 25% ethanol) in 500 mL tap water. Mix well. Store at 4°C.

7. *D. ciliatus* adult diet: 10 g yeast hydrolyzate and 30 g sugar. Mix well. Store at 4°C.

8. Paraffin mixture: 45 g paraffin wax (melting point 52°C–54°C), 30 g paraffin wax (melting point 46°C–48°C), 7.5 g bee wax. The mixture is heated to 70°C–90°C and stored at room temperature (RT).

1.2.1.4 Culture

1. *Ceratitis capitata*

   Cultures of *C. capitata* are maintained in the laboratory at 25°C and 70% relative humidity with a 12-hour
   light–12-hour dark cycle. Adults are maintained in insect
   cages and fed on a mixture of sugar and yeast (1:2). The
   water is supplied with pieces of absorbent wipe dipped
   in boxes with water. Adults oviposit through mesh holes
   of a nylon cloth affixed to one side of the box. The eggs
   are collected in water and transferred to artificial larval
   medium in Petri dishes. Several types of larval medium
   are used for medfly; two of them are described in Section
   1.2.1.3. Sterilized sand is placed at the bottom of the rearing dishes, about 2 days before the larvae start jumping
   out of the larval medium to pupate in the sand.

2. *Bactrocera oleae*

   Cultures of *B. oleae* are maintained in the laboratory
   at 25°C and 70% relative humidity with a 12-hour
   light–12-hour dark cycle. The adults live in insect
   cages feeding on a solid diet containing sucrose/yeast
   hydrolysate/egg yolk (Tsitsipis and Kontos 1983). Water
   is continuously supplied using pieces of absorbent wipe
   dipped in plastic bottles filled with water. Adults oviposit on cones of oviposition substrates consisting of
   nylon cloth covered by a paraffin mixture (Tzanakakis
   1989). The eggs are collected from the paraffin cones
   by rinsing them off with water and incubated on filter
   paper in Petri dishes for 24–48 hours. The filter paper
   is soaked with 0.3% propionic acid (Manoukas and
   Mazomenos 1977). The eggs are transferred to plastic
   containers with artificial larval diet (Tzanakakis et al.
   1970) with a brush and left there until pupation. The
   pupae are transferred to Petri dishes and placed in cages
   for oviposition.

3. *B. dorsalis*, *B. cucurbitae*, and *Anastrepha ludens*

   Cultures of these three species are maintained in the laboratory at 25°C and 70% relative humidity with a 12-hour
   light–12-hour dark cycle. Adults are maintained in insect
   cages and fed on a mixture of yeast, wheat germ, and sugar
   (1:2:3). The water is supplied through pieces of absorbent
   wipe dipped in bottles with water. Adults oviposit through
   mesh holes of a nylon cloth affixed to one side of the box.
   The eggs are collected in water and transferred to artificial larval medium in Petri dishes.

4. *B. tryoni*

Wild-type laboratory-adapted strains of *B. tryoni* are reared at 25°C and 70% relative humidity with a 14-hour light–10-hour dark cycle. Adults are feed on a mixture of sucrose and yeast. Eggs are obtained by allowing flies to oviposit through a perforated wax film and transferred to the larval medium.

5. *Dacus ciliatus*

Laboratory colonized adults are maintained in insect cages and provided with a mixture of yeast hydrolyzate and sugar (1:3) (Jeyasankar et al. 2009; Zur et al. 2009). Water is continuously supplied using pieces of absorbent wipe dipped in plastic bottles filled with water. The colony, established from field-infested fruits and maintained under quarantine conditions (in the Agricultural Research Organization of Israel), is refreshed once a year with wild-collected flies. Zucchini fruits are used for female oviposition and larval development.

1.2.2   Mitotic Chromosomes: Karyotype Analysis

The required equipment and materials are listed in alphabetical order. Suppliers and catalogue numbers are given only when specific products are used.

1.2.2.1   Equipment

1. Hot plate
2. Microscope equipped with digital camera
3. Microscope slides and cover slips
4. Parafilm
5. Staining jar
6. Stereoscope
7. Thin needles
8. Well slides

1.2.2.2   Chemicals

1. Acetic acid, glacial (CH_3COOH)
2. Barium hydroxide [$Ba(OH)_2$]
3. Calcium chloride hydrate ($CaCl_2 \ H_2O$)
4. Chloroform ($CHCl_3$)
5. Colchicine
6. Disodium hydrogen phosphate ($Na_2HPO_4 \cdot 12H_2O$)
7. Ethanol
8. Giemsa

9. Hydrochloric acid (HCl) 37%
10. Methanol (CH$_3$OH)
11. Potassium chloride (KCl)
12. Potassium hydrogen phosphate (KH$_2$PO$_4$)
13. Sodium bicarbonate (NaHCO$_3$)
14. Sodium chloride (NaCl)
15. Sodium citrate (C$_6$H$_5$Na$_3$O$_7$·2H$_2$O)

1.2.2.3 Solutions Required

1. Acetic acid 60% in distilled H$_2$O (dH$_2$O).
2. Barium hydroxide [Ba(OH)$_2$] saturated solution.
3. Carnoy's fixative: ethanol/chloroform/acetic acid in 6:3:1 ratio.
4. Colchicine, 0.1 mM in Ringer's solution.
5. Fixative solution: methanol/acetic acid in 3:1 ratio.
6. Giemsa staining—optional solutions:
    a. Solution 1: 5% Giemsa in 0.1 M Sörensen phosphate buffer.
    b. Solution 2: 5% Giemsa in 0.1 M sodium phosphate buffer.
7. HCl 0.2 N in dH$_2$O.
8. Hypotonic solutions:
    a. Solution 1: 0.075 M KCl in dH$_2$O. Store at 4°C.
    b. Solution 2: 1% sodium citrate dehydrate in dH$_2$O. Store at 4°C.
9. Physiological solutions:
    a. Solution 1: 0.7% NaCl in dH$_2$O. Store at 4°C.
    b. Solution 2 (Insect Ringer's solution): 6.5 g NaCl, 0.14 g KCl, 0.12 g CaCl$_2$, 0.2 g NaHCO$_3$ in 1 L dH$_2$O. Adjust the pH to 6.8. Store at 4°C.
10. Sodium phosphate buffer 0.1 M, pH 6.8: 24.50 mL solution A (200 mM Na$_2$HPO$_4$), 25.50 mL solution B (200 mM NaH$_2$PO$_4$), and 50 mL dH$_2$O.
11. Sörensen phosphate buffer 0.1 M: 0.05 M Na$_2$HPO$_4$, 0.05 M KH$_2$PO$_4$ in dH$_2$O. Adjust pH to 6.8 and store at 4°C.
12. Saline sodium citrate (SSC) 20×: 88.2 g sodium citrate, 175.3 g NaCl in 1 L dH$_2$O. Adjust pH to 7.0. Autoclave.
13. SSC 2×: 1:10 dilution of 20× SSC in dH$_2$O.

1.2.2.4 Chromosome Preparations

Neural ganglia of third instar larvae or young pupae are the tissue of choice for preparations of mitotic metaphase chromosomes in all species of Tephritidae fruit flies (Figure 1.1).

1. Protocol 1. Spreading technique with or without colchicine The spreading technique using hot plate was originally developed for pachytene mapping in Lepidoptera (Traut 1976) and later adopted for preparation of mitotic and meiotic chromosomes from other insects and arthopods (Sahara et al. 1999). Then, a procedure described in Frydrychová and Marec (2002) was applied with minor modifications for *R. cerasi* (Kounatidis et al. 2008), *R. completa* (Drosopoulou et al. 2010), *R. cingulata* (Drosopoulou et al. 2011a), and *D. ciliatus* (Drosopoulou et al. 2011b).

Dissection. Dissection is performed in physiological solution.

Note 1. A third instar larva or one young pupa is transferred into a drop of Ringer's solution or physiological solution on a well slide. The brain is dissected under a stereoscope using two thin needles. The slide should be placed on a dark surface (if this is not provided with the stereoscope, black paper can be placed under the slide). Usually the brain ganglia remain attached to mouth parts together with the salivary glands and several IDs. Therefore, the ganglia must be separated and the unwanted material must be removed using the needles.

Colchicine treatment. The ganglia are incubated for 15–25 minutes in 0.1 mM colchicine. Colchicine incubation is used only when a large number of metaphase nuclei are needed. In most cases, this step is omitted because the ganglia have sufficient number of metaphases even without colchicine treatment.

Note 2. The ganglia are transferred into a drop of colchicine placed on a new slide and incubated for 15–25 minutes. It is best to use a well slide or a slide covered with parafilm. The slide should be placed on a dark surface.

Hypotonic pretreatment. The ganglia are incubated for 15–20 minutes in hypotonic solution 1 (0.075 M KCl).

Note 3. The ganglia are transferred into a drop of hypotonic solution placed on a new slide. It is better to use a well slide or a slide covered with parafilm. The incubation time in the hypotonic solution is very critical; sometimes it must be adjusted to obtain well-spread preparations. During the treatment, any remaining waste material should be removed. The slide should be placed on a dark surface.

Fixation. The material is fixed in freshly prepared Carnoy's fixative for 15–20 minutes.

Note 4. The material is transferred into a drop of freshly prepared Carnoy's fixative. Duration of fixation

(15–30 minutes in total) is critical and it must be optimized for every different species. To ensure complete removal of water at least three changes of the fixative solution are proposed. It is best to use a well slide or a slide covered with parafilm. The slide should be placed on a dark surface.

Maceration. The sample is homogenized in 60% acetic acid.

Note 5. The fixed ganglia are transferred to a small drop (~15 µL) of 60% acetic acid placed on a slide. Small drop helps to minimize lose of material. It is best to use a well slide. Under the stereoscope, the ganglia must be homogenized before spreading. The ganglia are first torn into fine pieces using a pair of dissecting needles and then the material is pipetted in and out of a micropipette tip (2–20 µL) or pushed through a syringe needle several times and released out. Finally, an additional 15 µL of 60% acetic acid are added to the homogeneous material. Duration of the maceration step should not exceed 2 minutes. The slide should be placed on a dark surface.

Spreading. The homogenized material is spread onto a slide and the preparations are air-dried.

Note 6. The material is transferred with a micropipette or a syringe onto a clean slide, which is placed on a warm hot plate (40°C–45°C) for drying. With the help of the micropipette, the drop is spread gently over an area of approximately 20 × 20 mm until the acetic acid is almost evaporated. The preparations can be used immediately or be stored in suitable containers at RT for a long period.

2. Protocol 2. Air-drying technique

This technique was described by Guest and Hsu (1973) and applied with minor modifications for *C. capitata* (Zacharopoulou et al. 1987, 1990), *B. oleae* (Mavragani-Tsipidou et al. 1992), *A. ludens* (Garcia-Martinez et al. 2009), *B. dorsalis* (Zacharopoulou et al. 2011a), *B. cucurbitae* (Zacharopoulou et al. 2011b), and *B. tryoni* (Zhao et al. 1998).

Dissection. Dissection of ganglia is performed in physiological solution (see Note 1).

Hypotonic pretreatment. The ganglia are incubated for 10–15 minutes in hypotonic solution 2 (1% sodium citrate), essentially as described in Note 3.

Fixation. The material is fixed in freshly prepared fixative (methanol/acetic acid, 3:1) for 3 minutes. During this step, the fixative is changed twice. We use a well slide or a slide covered with parafilm.

Maceration. The sample is homogenized in 60% acetic acid (see Note 5).

Spreading. The homogenized suspension is dripped onto a clean slide, which is placed on a hot plate, let to spread and then air-dried (see Note 6).

1.2.2.5   Chromosome Staining

1. Giemsa staining

   Staining. The preparations are stained in Giemsa solution for 5–10 minutes.

   Note 7. The air-dried preparations are immersed into a solution of 5% Giemsa in 0.1 M Sorensen phosphate buffer in slide staining racks and the slides are left there for 5–10 minutes. Duration of the Giemsa staining must be optimized with respect to the quality of the spreads. After staining, the slides are washed with tap water for about 30 seconds and are left in a vertical position to air-dry at RT. If the chromosomes are not sufficiently stained, the slide can be restained. If the chromosomes are overstained, the slide can be destained in water. The stain residues on the lower side of the slide are removed with alcohol before microscopic inspection.

2. C-banding

   The method was described by Canovai et al. (1994) for *C. capitata* and adapted by Selivon and Perondini (1997) for *Anastrepha* species.

   Chromosome aging. The air-dried preparations are "aged" either by heating for 3 hours at 60°C or by storing for about a week at 25°C.

   Hydrolization. Slides are immersed in 0.2 N HCl for 10 minutes and rinsed with dH_2O.

   Depurination. Slides are transferred to a saturated solution of barium hydroxide, at 50°C for 2 minutes.

   Rinsing. Slides are rinsed in acid water (dH_2O with a few drops of acetic acid).

   Denaturation. Slides are incubated in 2× SSC at 60°C for 30 minutes.

   Staining. Following abundant rinsing in dH_2O, the slides are stained with 5% Giemsa in 0.01 M phosphate buffer for about 30–40 minutes. The staining time is critical, and over staining must be avoided.

1.2.2.6   Imaging and Karyotype Construction

For the construction of karyotypes, images of mitotic metaphase complements are taken in a phase contrast or bright field microscope at 100× magnification using an optical or digital camera, and the best micrographs are selected. Following the labeling system used by Radu et al. (1975) for the medfly *C. capitata*, the sex chromosomes are labeled as the first pair of the mitotic karyotype, whereas the five autosomes from 2 to 6 in order of descending size.

1.2.3 Polytene Chromosomes: Banding Pattern Analysis—Chromosome Maps

The required equipment and materials are listed in alphabetical order. Suppliers and catalogue numbers are given only when specific products are used.

1.2.3.1 Equipment

1. Cover slips
2. Microscope slides
3. Nail polish
4. Parafilm
5. Phase contrast microscope equipped with digital camera
6. Staining jar
7. Stereoscope
8. Thin needles
9. Well slides

1.2.3.2 Chemicals

1. Acetic acid, glacial (CH_3COOH)
2. Calcium chloride hydrate ($CaCl_2 \ H_2O$)
3. Ethanol
4. Hydrochloric acid (HCl) 37%
5. Lactic acid [$CH_3CH(OH)COOH$]
6. Orcein powder (Gurr's Orcein, BDH, product No. 34210)
7. Potassium chloride (KCl)
8. Sodium bicarbonate ($NaHCO_3$)
9. Sodium chloride (NaCl)

1.2.3.3 Solutions Required

1. Acetic acid 45% in dH_2O.
2. Ethanol/acetic acid solution in 3:1 ratio.
3. HCl 1 N in dH_2O.
4. HCl 3 N in dH_2O.
5. Lactoacetic acid—optional solutions:
   a. Solution 1: 80% lactic acid/60% acetic acid in 1:1 ratio.
   b. Solution 2: lactic acid/dH_2O/acetic acid in 1:2:3 ratio.

6. Lactic acetic orcein—optional solutions:
   a. Solution 1: 1 g orcein powder, 45 mL acetic acid, 25 mL lactic acid, 30 mL dH$_2$O. Orcein is boiled in the solution for 45 minutes in a reflux apparatus. Filtration through filter paper is important before use. Store at RT or 4°C.
   b. Solution 2: 2 g of natural orcein in 100 mL of a mixture of equal parts of 85% lactic acid, glacial acetic acid, and dH$_2$O. Heat and filter while hot. Store at RT or 4°C.
7. Physiological solutions (Section 1.2.2.3).

1.2.3.4 Polytene Chromosome Preparations

Salivary glands of late third instar larvae or pupae (1–4 days old) are the tissue of choice for the preparation of polytene chromosomes in all species of Tephritidae fruit flies (Figure 1.2). In addition, in *C. capitata* the orbital trichogen cells from male pupae can be used (Figure 1.3). The quality of chromosome spreads depends greatly on the growth conditions of the larvae. For best preparations, it is important that the colonies are not crowded and the larvae reared at 18°C–20°C.

1. Protocol 1. Preparations from salivary glands
   This technique has been used for *B. oleae* (Mavragani-Tsipidou et al. 1992; Zambetaki et al. 1995), *R. cerasi* (Kounatidis et al. 2008), *R. completa* (Drosopoulou et al. 2010), *R. cingulata* (Drosopoulou et al. 2011a), and *D. ciliatus* (Drosopoulou et al. 2011b). The best preparations were obtained from pupae (1–4 days old).
   Dissection. Dissection of salivary glands is performed in physiological solution.
   Note 8. A pupa (or third instar larva) is transferred into a drop of Ringer's solution or physiological solution on a well slide. The glands are dissected under a stereoscope using two thin needles. The slide should be placed on a dark surface (if this is not available for the stereoscope, black paper can be placed under the slide). Usually the glands remain attached to the mouth parts and other larval structures. The glands must be freed of this unwanted material using a pair of needles. The two glands may remain attached or can be separated. Attached to the glands is also a layer of flat fat bodies. Do not try to remove the fat body from the glands at this step.
   Fixation 1. The glands are fixed in 45% acetic acid for approximately 2–3 minutes.
   Note 9. The material is transferred into a drop of freshly prepared fixative 45% acetic acid. We use a well slide or a slide covered with parafilm. The fat bodies must now be

removed from the glands using thin needles. Take care not to damage the glands. The slide should be placed on a dark background.

Hydrolyzation. The glands are hydrolyzed in 1 N HCl for approximately 2 minutes.

Note 10. The material is transferred into a drop of 1 N HCl. The glands become transparent.

Fixation 2. The glands are passed through lactoacetic acid solution 1 for approximately 6 seconds.

Note 11. The material is transferred into a drop of lactoacetic acid (80% lactic acid/60% acetic acid, 1:1). The glands are left in this solution only briefly because incubation for more than 30 seconds might destroy the glands. The glands become more transparent.

Staining. The material is stained in lactic acetic orcein staining solution 1 for 10–20 minutes.

Note 12. The glands are transferred into a drop of approximately 10 µL of staining solution, placed on a new parafilm-covered slide. The parafilm helps to avoid drying of the stain. On the same slide, many small drops of the stain can be placed, so that glands of different individuals can be simultaneously stained. Duration of staining must be adjusted to the conditions, the quality of the chromosomes, and the species used.

Washing. The stained glands are then washed in lactoacetic acid solution 1.

Note 13. Excess stain is removed by briefly washing the glands in a drop of lactoacetic acid before squashing. Each salivary gland can be used for one preparation. Alternatively, each salivary gland can be cut in two or three pieces and each piece is used for a separate preparation. In the latter case, the quality of the preparations tends to improve.

Squashing. The material is squashed in lactoacetic acid solution 1.

Note 14. Each salivary gland or a piece of the salivary gland is transferred into a drop of lactoacetic acid placed on a clean slide. A coverslip (18 × 18 mm) is placed on the material and the glands are presquashed by moving the coverslip very gently in a circular motion by a needle. Then, the slide is fold in filter paper and the preparation is squashed firmly by the thumb (coverslip must not move sideways). The quality of chromosome spreading is checked in a microscope. If the result is satisfactory, nail polish is used to seal the edges of the cover slip. The preparations are semipermanent; they can be analyzed immediately or stored in suitable containers at 4°C until use (for ~2–4 weeks).

2. Protocol 2. Preparations from salivary glands

This technique was used for *C. capitata* (Zacharopoulou et al. 1987, 1990), *B. tryoni* (Zhao et al. 1998), *A. ludens* (Garcia-Martinez et al. 2009), *B. dorsalis* (Zacharopoulou et al. 2011a), and *B. cucurbitae* (Zacharopoulou et al. 2011b).

Dissection. Dissection of salivary glands is performed in 45% acetic acid.

Note 15. A third instar larva is transferred into a drop of 45% acetic acid placed onto a well slide. The glands are quickly dissected under a stereoscope using two thin needles. To facilitate dissection the slide should be placed on a dark surface (see Note 8).

Hydrolyzation. The glands are fixed in 3 N HCl for 1 minute.

Note 16. The material is transferred into a drop of 3 N HCl placed on a well slide.

Fixation. The glands are fixed with lactoacetic acid solution 2 for 3–5 minutes.

Note 17. The material is transferred into a drop of lactoacetic acid (lactic acid/dH$_2$O/glacial acetic acid, 1:2:3) on a well slide until it becomes transparent.

Staining. The material is stained in lactic acetic orcein solution 2 for 5–7 minutes (see Note 12).

Washing. The material is washed in lactoacetic acid solution 2 (see Note 13).

Squashing. The material is squashed in lactoacetic acid solution 2 (see Note 14).

3. Preparations from orbital trichogen cells

Male 5–6 days old pupae (with orange eyes, but not bristles or cuticle pigmentation) are used. This technique was described for *C. capitata* by Bedo (1986, 1987) and used for the cytogenetic characterization of numerous Y chromosome–autosome translocation strains constructed for the development of GSSs of *C. capitata* (Franz 2005).

Pretreatment. The puparium and pupal cuticle are removed and the heads are fixed in ethanol/acetic acid (3:1) and stored at 4°C (for at least 24 hours), until used for slide preparation.

Dissection. Isolation of the specific cuticle region containing the trichogen cells is performed in ethanol.

Note 18. The head is transferred into a drop of 95% ethanol. A segment of the head cuticle containing the trichogen cells (there are two cells in each male pupa) of the spatulate superior orbital bristles, as well as a part of eye tissue from the dorsal part of the cuticle is isolated. Most of the internal tissues of the dissected segment are removed, leaving a thin layer of fat.

Fixation. Fixation is performed in 45% glacial acetic acid.

Note 19. The cuticle segment is transferred into a drop of 45% acetic acid on a clean slide. The fat layer is removed from the cuticle and macerated, whereas the clean cuticle is removed.

Staining. The material is stained in lactic acetic orcein solution 2 for 3–4 minutes.

Note 20. A drop of lactic acetic orcein is added to the preparation for 3–4 minutes. A coverslip (18 × 18 mm) is placed on the material and it is pressed very gently. Then, the preparation is checked at low magnification (10×) for the presence of the two trichogen cells and the quality of the chromosome spreads. If the result is not satisfactory, light thumb pressure is applied using several layers of blotting paper to absorb excess stain. Chromosome flattening is carefully monitored to avoid excess pressure, which might destroy their morphology. The transfer of slides to a hot plate (30°C–40°C) improves chromosome flattening. It is preferable to use a bright field microscope.

1.2.3.5 Imaging and Construction of Photographic Chromosome Maps

For the construction of photographic chromosome maps, images of polytene nuclei are taken with a phase contrast microscope at magnifications of 40×, 63×, and 100×. The 40× magnification is used for overall observation of different nuclei of the same individual and for the identification of different chromosome elements and structures of each nucleus (Figure 1.2a). Micrographs taken at magnifications of 63× and 100× are used for the construction of polytene maps (Figure 1.2b). It should be pointed out that the maps of each species must be constructed (1) from larvae or pupae of same age (because the banding pattern changes between different developmental stages, which may lead to inconsistencies as a consequence of differential gene expression) and (2) from nuclei of approximately the same chromosome size (because variations in chromosome polytenization among nuclei of the same fly are extremely high). For the best analysis of the polytene complements and the construction of photographic chromosome maps, the largest nuclei of each preparation of similar aged individuals with well-spread chromosomes are used. The polytene complement is divided into sections from 1 to 100. The longer arm of each chromosome is designated as left (L) and the shorter arm as right (R).

Two different numbering systems have been used for the polytene chromosome of Tephritidae. In the first system, chromosomes are numbered from 2 to 6 based on the correlation of mitotic and polytene chromosomes, which was achieved through the analysis of several Y–autosome and autosome–autosome translocation strains in *C. capitata* (Zacharopoulou 1990). For *B. tryoni*, *B. cucurbitae*, *B. dorsalis*, and *A. ludens*, the system of *C. capitata* was used based mainly on the banding pattern similarity of their polytene elements (Zhao et al. 1998; Garcia-Martinez et al. 2009; Zacharopoulou et al. 2011a,b). In the second system, chromosomes are numbered from I to V (see the maps of *B. oleae*, *D. ciliatus*, *R. cerasi*, *R. completa*, and *R. cingulata*) based on size and banding pattern similarities (Mavragani-Tsipidou et al. 1992; Zambetaki et al. 1995; Kounatidis et al. 2008; Drosopoulou et al. 2010, 2011a,b).

1.2.4   In Situ Hybridization on Polytene Chromosomes

The equipment, chemicals, and solutions needed for in situ hybridization on polytene chromosomes are listed separately for each step of the technique in alphabetical order. Suppliers and catalogue numbers are given only when specific products are used.

1.2.4.1   Equipment

1.2.4.1.1   Chromosome Preparations

 1. Coplin jars
 2. Cover slips 18 × 18 mm
 3. Insulated flask (Thermos bottle)
 4. Microscope slides
 5. Parafilm
 6. Phase contrast microscope equipped with digital camera
 7. Razor blade
 8. Stereoscope
 9. Thin needles
10. Well slides

1.2.4.1.2   Pretreatment and Denaturation of Chromosomes

 1. Slide staining racks
 2. Slide carriers
 3. Water bath

1.2.4.1.3   Hybridization

 1. Absorbent paper
 2. Centrifuge
 3. Coverslips 18 × 18 and 24 × 30 mm

4. Freezer (−20°C and −80°C)
5. Incubation oven
6. Plastic or glass boxes with cover
7. Slide staining racks
8. Slide carriers
9. Thermoblock
10. Water bath

1.2.4.1.4   Signal Detection

1. Coverslips 22 × 22 mm
2. Slide staining racks
3. Slide carriers

1.2.4.2   Chemicals and Reagents

1.2.4.2.1   Chromosome Preparations

1. Acetic acid, glacial (CH_3COOH)
2. Calcium chloride hydrate ($CaCl_2 \, H_2O$)
3. Ethanol
4. Lactic acid [$CH_3CH(OH)COOH$]
5. Liquid nitrogen
6. Potassium chloride (KCl)
7. Sodium bicarbonate ($NaHCO_3$)
8. Sodium chloride (NaCl)

1.2.4.2.2   Pretreatment and Denaturation of Chromosomes

1. Ethanol
2. Sodium chloride (NaCl)
3. Sodium citrate ($C_6H_5Na_3O_7$)
4. Sodium hydroxide (NaOH)

1.2.4.2.3   Hybridization

1. Formamide ($HCONH_2$) deionized
2. Sodium chloride (NaCl)
3. Sodium citrate ($C_6H_5Na_3O_7$)

1.2.4.2.4   Signal Detection

1. Detection of digoxigenin-labeled probes by anti-DIG-alkaline phosphatase conjugates and colorimetric substrates.
   a. Dig Nucleic acid Detection kit (Roche Diagnostics, Mannheim, Germany, Cat. No. 11 175 041 910). Alternatively: Anti-Digoxigenin-AP Fab fragments (Roche Diagnostics, Cat. No. 082 736

103), Blocking Reagent (Roche Diagnostics, Cat. No. 10 057 177 103), 5-Bromo-4-chloro-3-indolylphosphote p-toluidine salt (BCIP), Nitroblue tetrazolium chloride (NBT).

   b. Hydrochloric acid (HCl) 37%

   c. Magnesium chloride (MgCl$_2$)

   d. *N,N*-Dimethylformamide [(CH3)$_2$NC(O)H]

   e. Sodium chloride (NaCl)

   f. Tris-(hydroxymethyl)-aminomethane (Tris)

2. Detection of biotin-labeled probes with an avidin-/biotin-based peroxidase system and dimethylaminoazobenzene (DAB) substrate.

   a. DAB Substrate Kit for Peroxidase (Vector Laboratories, Inc., Burlingame, CA, Cat. No. SK-4100). The kit includes four bottles: buffer stock solution, DAB stock solution, H$_2$O$_2$ solution, nickel solution. Alternatively: DAB (Isopac, Sigma-Aldrich, St. Louis, MO), H$_2$O$_2$ 30%.

   b. Potassium chloride (KCl)

   c. Potassium dihydrogen phosphate (KH$_2$PO$_4$)

   d. Sodium chloride (NaCl)

   e. Sodium hydrogen phosphate (Na$_2$HPO$_4$)

   f. Triton X-100

   g. VECTASTAIN ABC kit Standard (Vector Laboratories, Cat. No. PK-6100). The kit includes two reagents: reagent A (Avidin DH solution) and reagent B (Biotinylated peroxidase).

1.2.4.3   Solutions Required

1.2.4.3.1   Chromosome Preparations

1. Acetic acid 45% in dH$_2$O
2. Lactoacetic acid—optional solutions
   a. Solution 1: lactic acid/dH$_2$O/acetic acid in 2:3:4.5 ratio
   b. Solution 2: lactic acid/dH$_2$O/acetic acid in 1:2:3 ratio
3. Physiological solutions (Section 1.2.2.3)

1.2.4.3.2   Pretreatment and Denaturation of Chromosomes

1. Ethanol 30%, 50%, 70%, and 96% in dH$_2$O.
2. NaOH 0.07 N: Dissolve 0.84 g NaOH pellets in 300 mL dH$_2$O. Prepare fresh.
3. SSC 20× (see Section 1.2.2.3).
4. SSC 2×: (see Section 1.2.2.3).

1.2.4.3.3   Hybridization

1. SSC 20× (see Section 1.2.2.3)
2. SSC 2× (see Section 1.2.2.3)

3. SSC 4×: 1:5 dilution of 20× SSC in dH_2O
4. Labeled probe (see Section 1.2.6)

1.2.4.3.4   Signal Detection

1. Detection of digoxigenin-labeled probes by anti-DIG-alkaline phosphatase conjugates and colorimetric substrates.
    a. Antibody solution: 1/500 dilution of Anti-Digoxygenin-AP Fab fragments in Buffer 2.
    b. BCIP solution 50 mg/mL in *N,N*-dimethylformamide.
    c. Buffer 1: 100 mM Tris-HCl (pH 7.5)/150 mM NaCl. Dilution 1/10 of Tris-HCl 1 M (pH 7.5) and 1/20 of NaCl 3 M in dH_2O.
    d. Buffer 2: 0.5% Blocking Reagent in Buffer 1.
    e. Buffer 3: 100 mM Tris-HCl (pH 9.5)/100 mM NaCl/50 mM $MgCl_2$. Dilution 1/10 of Tris-HCl 1 M (pH 9.5), 1/30 of NaCl 3 M, and 1/40 of $MgCl_2$ 2 M in dH_2O.
    f. Color solution: 35 µL BCIP solution and 35 µL NBT solution in 10 mL of Buffer 3.
    g. Giemsa staining solution (see Section 1.2.2.3).
    h. $MgCl_2$ 2 M in dH_2O.
    i. NaCl 3 M in dH_2O.
    j. NBT solution 100 mg/mL in 70% *N,N*-dimethylformamide.
    k. Tris-HCl 1 M (pH 7.5): dissolve 121.14 g in 800 mL dH_2O. Adjust the pH to 7.5 with HCl 37%. Add dH_2O to 1 L.
    l. Tris-HCl 1 M (pH 9.5): dissolve 121.14 g in 800 mL dH_2O. Adjust the pH to 9.5 with HCl 37%. Add dH_2O to 1 L.
2. Detection of biotin-labeled probes with an avidin-/biotin-based peroxidase system and DAB.
    a. Avidin/biotinylated enzyme solution: Add 16 µL avidin (kit component reagent A) and 16 µL biotinylated enzyme (kit component reagent B) into 1 mL 1× phosphate buffered saline (PBS), mix immediately and allow the reagent to stabilize for about 30 minutes before use. (The two reagents are included in the VECTASTAIN ABC Kit (Standard) PK-6100.)
    b. DAB substrate optional solutions.
        i. Solution 1: To 2.5 mL dH_2O, add 1 drop (~40 µL) of buffer solution, mix well, add 2 drops of DAB solution, mix well, and add 1 drop of H_2O_2 and mix well. For a gray-black stain add 1 drop of nickel solution and mix well. The reagents are included in the DAB substrate kit for peroxidase (Vector Laboratories SK-4100).

    ii. Solution 2: 100 μL DAB (Isopac, Sigma) solution (0.5 mg/mL in 1× PBS) and 2μL 30% H_2O_2 in 1 mL 1× PBS. Prepare fresh for each experiment.

c. Giemsa staining solution (see Section 1.2.2.3).

d. PBS 10×: 1.37 M NaCl, 0.012 M phosphate, 2.7 mM KCl, pH 7.4. Add 80 g NaCl, 2 g KCl, 14.4 g Na_2HPO_4 (or 18.1 g Na_2HPO_4 $2H_2O$ or 27.2 g Na_2HPO_4 $7H_2O$), 4 g KH_2PO_4 (monobasic anhydrous) to 800 mL dH_2O. Adjust pH to 7.4 with HCl. Add H_2O to 1 L. Autoclave for 20 minutes on liquid cycle. Store at RT.

e. PBS 1×:1:10 dilution of 10× PBS in H_2O.

f. Triton X-100 0.1%/PBS 1× in dH_2O: 1 ml Triton X-100 and 100 mL 10× PBS in 1 L dH_2O.

1.2.4.4 Chromosome Preparations

Dissection. Salivary glands are dissected in physiological solution (see Note 8).

Fixation. The glands are fixed in 45% acetic acid for approximately 30–60 seconds (see Note 9).

Squashing. The material is squashed in lactoacetic acid solution.

Note 21. Each salivary gland or piece of salivary gland is transferred to a drop of 10–15 μL lactoacetic acid solution 1 or 2 placed in the middle of a siliconized or ethanol cleaned coverslip (18 × 18 mm). The coverslip should be rest on a flat dark background. After the glands are fixed (2–3 minutes), a slide is laid onto the coverslip and turned over. The slide is placed on filter paper under a stereoscope and the material is spread by moving the coverslip very gently in a circular motion using a needle. The quality of chromosome spreads is checked under a phase contrast microscope. If the chromosomes are well spread, the slide is folded into a piece of filter paper, and the preparation is squashed (strongly) by thumb (coverslip must not move sideways). Satisfactory preparations are placed horizontally at −20°C to flatten overnight.

Coverslip removal. The slide is dipped into liquid nitrogen until bubbling stops, and the coverslip is immediately removed with a razor blade. Slides are dehydrated in absolute or 95% ethanol for 10 minutes, air-dried, and kept at RT or 4°C until use (maximum for 2–3 months).

1.2.4.5 Pretreatment and Denaturation of Chromosomes

1. Protocol 1

This technique has been used in *B. oleae* (Zambetaki et al. 1999; Augustinos et al. 2008; Tsoumani et al. 2011; Kakani et al. 2012), *D. ciliatus* (Drosopoulou et al. 2011b), *R. cerasi* (E. Drosopoulou, I. Nakou, P. Mavragani-Tsipidou, unpublished data) and *B. dorsalis*

(A. Augustinos, E. Drosopoulou, A. Gariou-Papalexiou, K. Bourtzis, P. Mavragani-Tsipidou, A. Zacharopoulou, unpublished data).

Predenaturation. Slides are incubated in preheated 2× SSC at 65°C for 30 minutes, then washed twice in 2× SSC at RT for 5 minutes, dehydrated in 70% (twice) and 96% (once) ethanol for 5 minutes and air-dried.

Denaturation. Preparations are incubated in 2× SSC at RT for 3 minutes and then in 0.07 N NaOH at RT for 90–120 seconds.

Note 22. The 0.07 N NaOH solution must be freshly prepared. Denaturation time varies slightly depending on species and the quality of the preparations.

Washes. Preparations are washed twice in 2× SSC at RT for 5 minutes.

Dehydration. Preparations are dehydrated in 70% (twice) and 96% (once) ethanol for 5 minutes and then air-dried. The hybridization procedure must follow within 4 hours.

2. Protocol 2

This technique was used for *C. capitata* and *B. tryoni* (Zacharopoulou et al. 1992; Scott et al. 1993; Banks et al. 1995; Kritikou 1997; Papadimitriou et al. 1998; Zhao et al. 1998; Verras et al. 1999, 2008; Rosetto et al. 2000; Michel et al. 2001; Theodoraki and Mintzas 2006; Kokolakis et al. 2008; Schetelig et al. 2009).

Rehydration. Preparations are rehydrated in 70%, 50%, and 30% ethanol and 2× SSC (2 minutes each).

Predenaturation. Slides are incubated in preheated 2× SSC at 65°C for 30 minutes.

Denaturation. Preparations are incubated in 2× SSC at RT for 2 minutes and 0.07 N NaOH at RT for 2 minutes.

Washes. Preparations are washed in 2× SSC at RT for 2 minutes.

Dehydration. Preparations are dehydrated in 30%, 50%, 70%, and 96% ethanol for 2 minutes and then air-dried. The hybridization procedure must follow within 4 hours.

1.2.4.6 Hybridization

Denaturation of the probe. The probe is denatured at 100°C for 5–10 minutes and the tubes are immediately transferred to ice.

Hybridization. A volume of 15 μL of the probe (20–100 ng) is placed in the center of the preparation and carefully covered with an 18 × 18-mm coverslip (avoid bubbles). Slides are placed in a moist chamber and incubated in an oven at the appropriate hybridization temperature for 16–20 hours.

Note 23. It is important that the preparations remain moist during hybridization. A "moist chamber" can be a plastic or glass box with an airtight cover. Absorbent paper moistened with 4× SSC is placed on the bottom of the box. Slides are placed horizontally in the box (avoid contact with wet paper) and the box is sealed. The hybridization temperature is selected based on the homology of the probe and on the hybridization solutions used (e.g., for homologous probes and aqueous hybridization solution, the temperature used is about 65°C–67°C, whereas for hybridization solutions containing 50% formamide it is approximately 40°C–42°C).

Washes. The coverslip is removed in 2× SSC and the preparations are washed for 3×20 minutes in preheated 2× SSC at 53°C for homologous probes, or 40°C for heterologous probes. Do not allow the slides to dry from this point on.

1.2.4.7 Signal Detection

The signal detection protocol depends on the label of the probe as well as on the detection system for each label. Here we describe protocols that have been successfully used for in situ hybridization on Tephritidae polytene chromosomes.

1. Protocol 1. Detection of digoxigenin-labeled probes by anti-DIG-alkaline phosphatase conjugates and colorimetric substrates

    Either the DIG DNA Detection kit (Roche) can be used or the reagents (Blocking Reagent, Anti-Digoxigenin-AP Fab fragments, BCIP, NBT) can be purchased separately (Section 1.2.4.2.4). The protocol has been used for *B. oleae* (Zambetaki et al. 1999; Augustinos et al. 2008; Tsoumani et al. 2011; Kakani et al. 2012), *D. ciliatus* (Drosopoulou et al. 2011b), *R. cerasi* (E. Drosopoulou, I. Nakou, P. Mavragani-Tsipidou, unpublished data) and *B. dorsalis* (A. Augustinos, E. Drosopoulou, A. Gariou-Papalexiou, K. Bourtzis, P. Mavragani-Tsipidou, A. Zacharopoulou, unpublished data).

    Blocking of unspecific antibody binding. The slides are briefly incubated in Buffer 1 and then in Buffer 2 for 30 minutes. Subsequently, they are washed in Buffer 1 for 1 minute.

    Antibody binding. A volume of 200 μL antibody solution is pipetted on each preparation and covered with a 22×22-mm cover slip. Slides are incubated at RT for 45–60 minutes.

    Washes. The coverslip is removed, and the slides are washed twice in Buffer 1 for 15 minutes.

    Color reaction. The slides are briefly placed in Buffer 3. One milliliter of color solution is applied to each preparation, and the slides are incubated horizontally for 30–60 minutes in the dark. Then the slides are washed with dH$_2$O to stop the color reaction.

Staining and observation. Preparations are stained in 5% Giemsa solution for 5–10 minutes. Hybridization signals are observed under a phase contrast microscope (Figure 1.4a).

Note 24. The preparations can be kept at 4°C for several weeks.

2. Protocol 2. Detection of biotin-labeled probes by a biotin-/ avidin-based peroxidase system using DAB

Either the DAB substrate kit for peroxidase (Vector Laboratories) can be used or the reagents (DAB Isopac-Sigma and 30% H$_2$O$_2$) can be purchased separately (see Section 1.2.4.2.4). This protocol has been used for *C. capitata* and *B. tryoni* (Zacharopoulou et al. 1992; Scott et al. 1993; Banks et al. 1995; Kritikou 1997; Papadimitriou et al. 1998; Zhao et al. 1998; Verras et al. 1999, 2008; Rosetto et al. 2000; Michel et al. 2001;

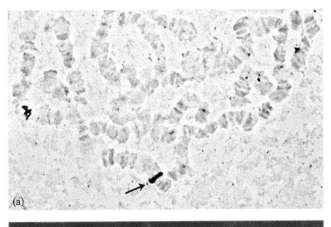

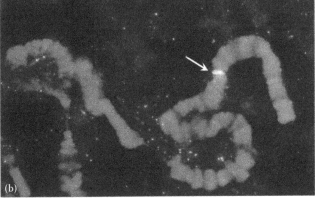

FIGURE 1.4   (See color insert.) (a) In situ hybridization on the polytene chromosomes of *Bactrocera oleae* using a homologous probe (*ovo* gene, cDNA clone). (b) Fluorescence in situ hybridization on the polytene chromosomes of a transgenic strain of *Ceratitis capitata* using as marker the DsRed. Arrows indicate hybridization signals.

Theodoraki and Mintzas 2006; Kokolakis et al. 2008; Schetelig et al. 2009; Gabrieli et al. 2010).

Washes. The slides are washed in 1× PBS twice for 5 minutes and in 0.1% Triton-X/1× PBS for 2 minutes and then kept in 1× PBS until the next step.

Antibody binding. A volume of 50 μL avidin/biotinylated enzyme solution is applied to each preparation and then covered with a 22 × 22-mm cover slip. Slides are incubated at RT for 60 minutes.

Washes. The coverslip is removed and the washing step is repeated.

Immunostaining. A volume of 50 μL DAB solution is placed on each preparation, covered with a coverslip and incubated for 15–20 minutes in the dark. The slides are washed twice with dH_2O and twice with 1× PBS. They are then kept in 1× PBS until staining.

Staining and observation. Preparations are stained in 5% Giemsa solution for 1 minute and hybridization signals are observed under a phase contrast microscope. Overstaining can be alleviated by dipping the slides into ethanol.

1.2.5 Fluorescence In Situ Hybridization on Mitotic and Polytene Chromosomes

The equipment, chemicals, and solutions needed for FISH on mitotic and polytene chromosomes are listed separately for each step of the technique in alphabetical order. Suppliers and catalogue numbers are given only when specific products are used.

1.2.5.1 Equipment

1.2.5.1.1 *Chromosome Preparations*

Equipment for mitotic and polytene chromosome preparations is listed in Sections 1.2.2.1 and 1.2.4.1.1, respectively.

1.2.5.1.2 *Pretreatment and Denaturation of Chromosomes*

1. Coplin jars
2. Cover slips
3. Forceps
4. Freezer (−20°C)
5. Oven
6. Plastic foil
7. Slide staining racks
8. Slide carriers
9. Water bath

1.2.5.1.3 Hybridization

1. Absorbent paper
2. Centrifuge
3. Coplin jars
4. Coverslips 18 × 18 and 24 × 32 mm
5. Forceps
6. Freezer (−20°C and −80°C)
7. Incubation oven
8. Plastic or glass boxes with a cover
9. Refrigerated centrifuge
10. Rubber cement
11. Slide staining racks
12. Slide carriers
13. Thermomix
14. Waterbath

1.2.5.1.4 Signal Detection

1. Centrifuge
2. Coplin jars
3. Coverslips 22 × 22 and 24 × 50 mm
4. Dark plastic boxes
5. Filters with 0.45 μm pore size
6. Forceps
7. Freezer (−20°C)
8. Nail polish
9. Slide storage boxes

1.2.5.1.5 Observation and Imaging

1. Color charge-coupled device (CCD) camera and appropriate software for capturing fluorescent images
2. Epifluorescence microscope equipped with standard fluorescence filter sets
3. Peltier cooled, black-and-white CCD camera and appropriate software for capturing fluorescent images
4. Software for pseudocoloring and processing images such as Adobe Photoshop

1.2.5.2 Chemicals and Reagents

1.2.5.2.1 Chromosome Preparations

Chemicals for mitotic and polytene chromosome preparations are listed in Sections 1.2.2.2 and 1.2.4.2.1, respectively.

1.2.5.2.2 Pretreatment and Denaturation of Chromosomes

1. Ethanol
2. Bovine serum albumin (BSA)
3. Ficoll (type 400)

4. Formamide ($HCONH_2$) deionized
5. Hydrochloric acid (HCl) 37%
6. Polyvinylpyrrolidone
7. Potassium chloride (KCl)
8. Potassium dihydrogen phosphate (KH_2PO_4)
9. Proteinase K
10. RNase A (DNase free)
11. Sodium chloride (NaCl)
12. Sodium citrate ($C_6H_5Na_3O_7$)
13. Sodium hydrogen phosphate (Na_2HPO_4)
14. Sodium hydroxide (NaOH)
15. Tween 20

1.2.5.2.3   Hybridization

1. Dextran sulfate
2. Ethanol
3. Formamide ($HCONH_2$) deionized
4. Salmon sperm DNA (sonicated)
5. Sodium chloride (NaCl)
6. Sodium citrate ($C_6H_5Na_3O_7$)
7. Sodium acetate ($C_2H_3NaO_2$)
8. Tween 20

1.2.5.2.4   Signal Detection

1. Detection of digoxigenin-labeled probes with fluoro-chrome-conjugated antibodies.
   a. DABCO (1,4-diazabicyclo(2.2.2)octane, Sigma-Aldrich, St. Louis, MO, Cat. No. D2522)
   b. Fluorescent Antibody Enhancer Set for DIG Detection (Roche Diagnostics, Mannheim, Germany, Cat. No. 1 768 506). The kit includes vial 1: Anti-DIG monoclonal antibody against digoxigenin, mouse IgG1 (12.5 µg in 500 µL 1× blocking solution); vial 2: Anti-mouse Ig-DIG, F(ab)$_2$ fragment (12.5 µg in 500 µL 1X blocking solution); vial 3: Anti-DIG-Fluorescein, Fab fragments (12.5 µg in 500 µL 1× blocking solution); vial 4: 10× blocking solution.
   c. Glycerol ($C_3H_8O_3$)
   d. Hydrochloric acid (HCl) 37%
   e. Propidium iodide ($C_{27}H_{34}I_2N_4$)
   f. Sodium chloride (NaCl)
   g. Sodium citrate ($C_6H_5Na_3O_7$)
   h. Tris-(hydroxymethyl)-aminomethane (Tris)
   i. Tween 20
2. Detection of biotin-labeled probes with the streptavidin–antistreptavidin system.

    a. BSA

    b. Biotinylated antistreptavidine (Vector Laboratories, Inc., Burlingame, CA, Cat. No. BA-0500)

    c. Cy3-streptavidine (Jackson ImmunoResearch Laboratories, Inc., West Grove, PA, Cat. No. 016-160-084)

    d. DABCO (1,4-diazabicyclo(2.2.2)octane, Sigma-Aldrich, Cat. No. D2522)

    e. DAPI (4',6-diamidino-2-phenylindole, Sigma-Aldrich, Cat. No. D9542)

    f. Glycerol

    g. Hydrochloric acid (HCl) 37%

    h. Kodak Photo-Flo solution

    i. Potassium chloride (KCl)

    j. Potassium dihydrogen phosphate (KH_2PO_4)

    k. Sodium chloride (NaCl)

    l. Sodium citrate ($C_6H_5Na_3O_7$)

    m. Sodium hydrogen phosphate (Na_2HPO_4)

    n. Triton X-100

    o. Tris-(hydroxymethyl)-aminomethane (Tris)

    p. Tween 20

1.2.5.3 Solutions Required

1.2.5.3.1 Chromosome Preparations

Solutions to be prepared for mitotic and polytene chromosome preparations are listed in Sections 1.2.2.3 and 1.2.4.3.1, respectively.

1.2.5.3.2 Pretreatment and Denaturation of Chromosomes

1. Denhardt's solution 50×: Ficoll (type 400) 1%, polyvinyl-pyrrolidone 1%, BSA 1% in dH_2O. Sterilize by filtration. Store in aliquots at −20°C.
2. Ethanol 70%, 80%, 96%, and 100% in dH_2O.
3. Formamide 70% in 2× SSC: 70 µL deionized formamide plus 10 µL 20× SSC plus 20 µL dH_2O.
4. HCl 10 mM in dH_2O.
5. NaOH 0.07 N (see Section 1.2.4.3.2).
6. PBS 10× (see Section 1.2.4.3.2).
7. PBS 1× (see Section 1.2.4.3.2).
8. Proteinase K 1 µg/µL in 1× PBS.
9. RNase A—optional.
    a. Solution 1: 100 µg/mL in 2× SSC.
    b. Solution 2: 50 µg/mL in dH_2O.
10. SSC 20× (see Section 1.2.2.3).
11. SSC 2× (see Section 1.2.2.3).
12. SSC 0.4×/0.1% Tween 20: 20 mL 20× SSC and 1 mL Tween 20 in 1 L dH_2O.

1.2.5.3.3   Hybridization

1. Dextran sulfate 20%: Add 0.5 g dextran sulfate to 0.5 mL 20× SSC and 2 mL dH_2O. Dissolve at 70°C, vortex, aliquot and store at −20°C.
2. Dextran sulfate 50%: Add 0.5 g dextran sulfate in 1 mL dH_2O. Dissolve at 70°C, vortex, aliquot and store at −20°C.
3. Ethanol 70% and 100% (ice cold) in dH_2O.
4. Formamide 50% in 2× SSC: 50 mL formamide, 10 mL 20× SSC, 40 mL dH_2O.
5. Master mix: 100 µL dextran sulfate 50%, 100 µL dH_2O, 100 µL 20× SSC, 50 µL salmon sperm DNA (10 mg/mL).
6. Salmon sperm DNA (10 mg/mL): Dilute sonicated salmon sperm DNA in dH_2O to a final concentration of 10 mg/mL. Aliquot and store at −20°C.
7. Sodium acetate 3 M in dH_2O.
8. SSC 20× (see Section 1.2.2.3).
9. SSC 4× (see Section 1.2.4.3.3).
10. SSC 2× (see Section 1.2.2.3).
11. SSC 0.1×: 1/200 dilution of 20× SSC in dH_2O.
12. SSC 4×/0.1% Tween 20: 200 mL 20× SSC and 1 mL Tween 20 in 1 L dH_2O.
13. SSC 0.4×/0.1% Tween 20 (see Section 1.2.5.3.2).

1.2.5.3.4   Signal Detection

1. Detection of digoxigenin-labeled probes with fluorochrome-conjugated antibodies.
   a. Antibody solution 1: 1:25 dilution of Anti-DIG monoclonal antibody against digoxigenin, mouse IgG1 (vial 1, Fluorescent Antibody Enhancer Set for DIG Detection) in 1× blocking solution.
   b. Antibody solution 2: 1:25 dilution of Anti-mouse Ig-DIG, F(ab)₂ fragment (vial 2, Fluorescent Antibody Enhancer Set for DIG Detection) in 1× blocking solution.
   c. Antibody solution 3: 1:25 dilution of Anti-DIG-Fluorescein, Fab fragment (vial 3, Fluorescent Antibody Enhancer Set for DIG Detection) in 1× blocking solution.
      Note 25. Aliquot the antibody solutions and store protected from light: working aliquot at 2°C–8°C, the remaining at −15°C to −25°C.
   d. Antifade-working solution: 0.233 g DABCO, 800 µL dH_2O; 200 µL 1 M Tris-HCl, pH 8.0; 9 mL glycerol;

filter through 0.45 μm filter and aliquot into 1.5-mL Eppendorf tubes; store at 4°C.

e. Blocking solution 1×: 1:10 dilution of blocking solution 10× (vial 4, Fluorescent Antibody Enhancer Set for DIG Detection) in 2× SSC. Store under sterile conditions at 2°C–8°C.

f. Propidium iodide staining solution: 5–10 ng/mL in dH$_2$O.

g. SSC 2× (see Section 1.2.2.3).

h. Wash solution: 0.2% Tween 20 in 2× SSC.

2. Detection of biotin-labeled probes with streptavidin–antistreptavidin system.

a. Antifade working solution (see Section 1.2.5.3.4).

b. Biotinylated antistreptavidine stock solution: reconstitute 0.5 mg in 1 mL dH$_2$O, add 0.6 mL glycerol, aliquot and store at –20°C.

c. Biotinylated antistreptavidine working solution: for 4 slides 10 μL antistreptavidine stock solution in 250 μL 2.5% BSA.

d. BSA 10% stock solution: Dissolve 1 g molecular biology grade BSA in 10 mL of dH$_2$O. Gently move the capped tube until the BSA has dissolved completely. Do not stir. Store in aliquots at –20°C.

e. BSA 2.5%: 1:4 dilution of 10% stock solution in dH$_2$O.

f. Cy3-streptavidine stock solution: reconstitute 1 mg in 0.6 mL dH$_2$O, add 0.6 mL glycerol, aliquot and store at –20°C.

g. Cy3-streptavidine working solution: for 4 slides 0.85 μL Cy3-streptavidine stock solution in 850 μL 2.5% BSA.

h. DAPI stock solution: 5 mg/mL in dH$_2$O; store at 4°C.

i. Kodak Photo Flo 1%/1× PBS: 1 mL Kodak Photo Flo and 10 mL 10× PBS in 89 ml dH$_2$O.

j. Kodak Photo Flo 1%: 1 mL of Kodak Photo Flo in 100 mL dH$_2$O.

k. PBS 10× (see Section 1.2.4.3.4).

l. SSC 20× (see Section 1.2.2.3).

m. SSC 4×/0.1% Tween 20 200 mL 20× SSC and 1 mL Tween 20 in 799 mL dH$_2$O

n. Tris-HCl, pH 8.0: dissolve 121.14 g in 800 mL dH$_2$O. Adjust the pH to 8.0 with HCl 37%. Add dH$_2$O to 1 L.

o. Triton X-100 1%/PBS 1×: 1 mL Triton X-100 and 10 mL 10× PBS in 89 mL dH$_2$O.

p. Triton X-100 1%/SSC 2×: 1 mL Triton X-100 and 10 mL 20× SSC in 89 mL dH$_2$O.

1.2.5.4   Chromosome Preparations

The procedure for preparations of mitotic chromosomes for FISH is identical to that for karyotype analysis as is described in Section 1.2.2.4. The procedure for preparations of polytene chromosomes for FISH is identical to that for in situ hybridization on polytene chromosomes as described in Section 1.2.4.4.

1.2.5.5   Pretreatment and Denaturation of Chromosomes

1. Protocol 1. Mitotic and polytene chromosomes
   This method is described by Fuková et al. (2005) and has been used for FISH in *B. oleae* mitotic and polytene chromosomes using an 18S rDNA and a satellite DNA probe (Drosopoulou et al. 2012; Tsoumani et al. 2013).

   Chromosome aging: dehydration. Slides with chromosome preparations are aged at −20°C for 2 hours or longer (after dehydration). After removal from the freezer, the slides are immediately immersed into cold 70% ethanol for 2 minutes. Then, they are passed through 80% and 100% ethanol, 30 seconds each, and air-dried.

   Note 26. Dehydrated slides can be stored at RT (for a few days) or at −20°C (for several weeks). Long-term storage at −20°C is not recommended because melting and refreezing can harm chromosome preparations badly. Alternatively, slides can be stored at −80°C for long periods.

   Cytoplasm removal (optional). Air-dried slides are incubated in 10 mM HCl for 10 minutes at 37°C in shaking water bath. Alternatively, slides are baked at 60°C for 2–4 hours.

   RNA removal. Slides are incubated in RNase A solution (100 μg/mL in 2× SSC) for 1 hour at 37°C.

   Note 27. RNase treatment serves to remove endogenous RNA and thus reduce noise in DNA–DNA hybridizations.

   Protein removal (optional). Slides are incubated in proteinase K solution (1 μg/μL in 1× PBS) at 37°C for 5 minutes and subsequently washed twice in 1× PBS at 37°C for 5 minutes.

   Blocking of unspecific hybridization. Slides are incubated in 5× Denhardt's solution at 37°C for 30 minutes and then drained on paper towel.

   Denaturation. Chromosomes are denatured in 70% formamide for 3.5–4 minutes at 68°C–72°C.

   Note 28. 70% formamide must be prepared shortly before use. To cover the samples, add 100 μL denaturation solution to the center of the preparation, put one (short)

edge of a coverslip or plastic foil (24 × 50 mm) in contact with the slide and lower gently, supporting with forceps. Several slides can be prepared quickly at RT and then placed on a metal plate in the prewarmed oven. After denaturation, shake off cover glass with sharp wrist movement and place slide immediately into ice-cold 70% ethanol (stored at –20°C) for 2 minutes.

Dehydration. Slides are dehydrated in 70% ethanol (ice-cold) for 2 minutes, then through 80% and 100% ethanol at RT for 30 seconds each, and air-dried.

2. Protocol 2. Mitotic chromosomes

This technique has been applied for *C. capitata* mitotic chromosomes (Willhoeft and Franz 1996a,b; A. Zacharopoulou and G. Franz, unpublished data).

Chromosome aging: dehydration. Slides are baked at 80°C for 2 hours. They are then dehydrated through 70%, 80%, and 100% ethanol at RT, 30 seconds each, and air-dried.

RNA removal. Slides are incubated in 25 µL RNase A solution (50 µg/mL) for 30 minutes at 37°C, and then washed in 0.4× SSC/0.1% Tween 20 and dehydrated as described earlier.

Note 29. This step is used especially for rDNA probes.

Denaturation. Chromosomes are denatured in 0.05 M NaOH for 60 seconds, transferred into 0.4× SSC/0.1% Tween 20 for 10 seconds.

Dehydration. Slides are dehydrated through 80% and 100% ethanol at RT, 30 seconds each, and air-dried.

3. Protocol 3. Polytene chromosomes

Pretreatment and denaturation of polytene chromosomes for FISH can be performed by the same procedure used for nonfluorescence in situ hybridization, as described in Section 1.4.2.2 (protocol 2). The protocol was used for *C. capitata* polytene chromosomes (Stratikopoulos et al. 2002, 2008; Krasteva et al. 2004).

1.2.5.6 Hybridization

1. Protocol 1. Mitotic and polytene chromosomes

Probe denaturation. Labeled probes are prepared and denatured as follows:

   a. Mix a labeled probe (~5–50 ng/slide) with blocking DNA (25 µg of unlabeled sonicated salmon sperm DNA).

   b. Add 1/10 volume 3 M NaOAc and 2.5 volumes of 100% ethanol.

   c. Precipitate at –80°C for 30–60 minutes.

    d. Centrifuge at 13,000 rpm, 4°C for 20 minutes.

    e. Discard supernatant, wash with about 400 μL prechilled 70% ethanol.

    f. Centrifuge again at 13,000 rpm, 4°C for 10 minutes.

    g. Remove supernatant with pipetting; air-dry for 3 minutes at 37°C.

Note 30. It is important to remove all ethanol before drying.

    h. Dissolve hybridization mixture in 5 μL prewarmed deionized formamide and incubate at 37°C for 30 minutes. Thermomix at about 300 rpm.

    i. Add 5 μL 20% dextran sulfate prewarmed to 37°C and vortex; if needed spin small droplets down.

    j. Denature probe at 90°C for 5 minutes.

    k. Chill immediately on ice; incubate at least for 3 minutes.

Hybridization. 10 μL of the probe is placed in the center of the preparation and carefully covered with a 24 × 32-mm coverslip (avoid bubbles). Edges are completely sealed with rubber cement. Slides are placed in a moist chamber (see Note 23) and are incubated at 37°C for 16–20 hours.

Washes. Washes are performed as follows:

    a. Peel off rubber cement with forceps.

    b. Dip slides briefly in 50% formamide in 2× SSC and remove cover slip.

    c. Wash 3 times in 50% formamide in 2× SSC at 46°C for 5 minutes.

    d. Wash 5 times in 2× SSC at 46°C for 2 minutes.

    e. Wash 3 times in 0.1× SSC at 62°C for 5 minutes.

    f. Wash in 0.1% Tween 20 in 4× SSC at RT for 5 minutes (or longer if needed).

2. Protocol 2. Mitotic chromosomes

Denaturation of the probe. Labeled probe is prepared and denatured as follows:

    a. Mix together labeled probe (~10 ng/slide) in 1.5 μL with 3.5 μL master mix and 5 μL deionized formamide.

    b. Heat to 80°C for 8 minutes and chill immediately on ice.

Hybridization. Hybridization is performed as described in Protocol 1.

Washes. Washes are performed as follows:

    a. Peel off rubber cement with forceps.

    b. Wash twice in 0.4× SSC/0.1% Tween 20 at RT for 5 minutes.

3. Protocol 3. Polytene chromosomes
The hybridization step can be performed with the same procedure used for nonfluorescence in situ hybridization as described in Section 1.2.4.6.

1.2.5.7 Signal Detection

Signal detection depends on the label of the probe, as well as on the detection system used for the label. Here, we describe protocols that have been successfully used in our labs for FISH on Tephritidae chromosomes.

1. Detection of digoxigenin-labeled probes with fluorochrome-conjugated antibodies
This protocol was used in *C. capitata* (Willhoeft and Franz 1996a,b; Stratikopoulos et al. 2002, 2008; Krasteva et al. 2004) using the fluorescent Antibody Enhancer Set for DIG Detection (Roche).
    Detection of hybridization signals. The signal detection is performed as follows:
    a. Remove coverslip by dipping the slides in 2× SSC, then keep the slides in 2× SSC at RT for 2–5 minutes.
    b. Incubate slides for 30 minutes in 1× blocking solution at RT and let the slides drain on paper.
    c. Add onto each slide 25 µL antibody solution 1, cover with a coverslip and incubate for 1 hour in a humid box at 37°C.
    d. Remove the coverslip, wash briefly three times in wash buffer at 37°C. Add 25 µL antibody solution 2 and incubate for 1 hour, as in step c.
    e. Repeat step d by adding 25 µL antibody solution 3 and incubate in dark.
    Staining. Staining with propidium iodide is performed as follows:
    a. Wash the slides thoroughly (three to four times for 5 minutes) in wash buffer at 37°C and incubate in propidium iodide staining solution for 2–5 minutes in dark.
    b. Wash the slides in dH$_2$O, put approximately 20 µL antifade solution and cover with a coverslip (24 × 30 mm).
    c. Store the slides in dark at approximately 4°C–8°C.
2. Detection of biotin-labeled probes with a streptavidin–antistreptavidin system
The following protocol was used on *B. oleae* mitotic and polytene chromosomes with an rRNA (Drosopoulou et al. 2012) and a satellite DNA probe (Tsoumani et al. 2013).

Detection of hybridization signals (step 1a). The first step of the signal detection is performed as follows:

   a. Pipet 450 µL 2.5% BSA on each slide, cover with a 24 × 50-mm coverslip and incubate at RT for 20 minutes in dark.

   b. Add 100 µL Cy3-streptavidine to each slide, cover with a 24 × 50-mm coverslip and incubate at RT for 30 minutes in dark (centrifuge Cy3-streptavidine before use at 13,000 rpm, 2 minutes)—from now work in moderately dark room!

   c. Wash three times in 0.1% Tween 20 in 4× SSC at 37°C for 3 minutes.

Detection of hybridization signals (step 1b). The second step of the signal detection is performed as follows:

   a. Pipet 450 µL 2.5% BSA onto each slide, cover with a 24 × 50-mm coverslip and incubate at RT for 10 minutes in dark.

   b. Add 50 µL antistreptavidine to each slide, cover with a 24 × 50-mm coverslip and incubate at 37°C for 20 minutes in dark (centrifuge antistreptavidine before use at 13,000 rpm, 2 minutes).

   c. Wash three times in 0.1% Tween 20 in 4× SSC at 37°C for 3 minutes.

   d. Place 450 µL 2.5% BSA on each slide, cover with a 24 × 50-mm coverslip and incubate at RT for 10 minutes in dark.

   e. Spot 100 µL Cy3-streptavidine on each slide, cover with a 24 × 50-mm coverslip and incubate at 37°C for 20 minutes in dark (centrifuge Cy3-streptavidine before use at 13,000 rpm, 2 minutes).

   f. Wash three times in 0.1% Tween 20 in 4× SSC at 37°C for 3 minutes.

Staining. DAPI staining is performed as follows:

   a. Wash in 1% Triton X-100 in 2× SSC at 25°C for 5 minutes.

   b. Stain in 0.5 µg/mL DAPI in 1% Triton X-100/1× PBS at 25°C for 15 minutes (polytene chromosomes) or 8 minutes (mitotic chromosomes).

   c. Wash in 1% Kodak Photo Flo/1× PBS at 25°C for 4 minutes.

   d. Short wash in 1% Kodak Photo Flo/H_2O at RT for 10 seconds.

   e. Let excess fluid drain off (do not let dry completely!).

   f. Mount in 30 µL antifade solution and cover with 24 × 40-mm cover slip.

   g. Press cover slip to squeeze out excess of antifade, avoiding bubbles.

   h. Seal cover slip with nail polish.

   i. Store at 4°C–8°C in dark.

1.2.5.8 Observation and Imaging

FISH preparations are observed in an epifluorescence microscope equipped with standard filter sets. Black-and-white (B&W) images are recorded with a cooled CCD camera and captured separately for each fluorescent dye. Then the images are pseudocolored, superimposed, and further processed with a graphics editing software such as Adobe Photoshop. Intensities of fluorescence for individual dyes can be aligned manually using software (Figure 1.5b through d). Alternatively, images for each fluorescent dye can be recorded with a cooled color CCD camera (but these are generally less sensitive than B&W cameras) and superimposed without necessity of pseudocoloring.

In some cases (e.g., when only one probe is used and the intensity of hybridization signals is high and comparable to counterstaining), FISH preparations can be observed in an epifluorescence microscope equipped with two-pass filters and a cooled color CCD camera. Then the camera captures simultaneously both fluorescence dyes in different colors (Figure 1.4b).

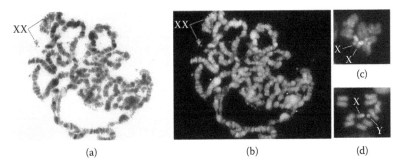

(a)   (b)   (d)

FIGURE 1.5 **(See color insert.)** Chromosome painting of sex chromosomes in *Bactrocera oleae*. Chromosomes were counterstained with DAPI (blue); hybridization signals of the X- or Y-chromosome-derived probes are red. (a) Phase contrast image of female polytene nucleus before hybridization. Lines indicate the granular network corresponding to the X chromosomes. (b) The same polytene nucleus as in (a) after fluorescence in situ hybridization with the X-painting probe. (c) Female metaphase showing blocks of strong hybridization signals of the X-painting probe on the X chromosomes. (d) Male metaphase showing strong hybridization signals of the Y-painting probe covering the entire Y chromosome. (From Drosopoulou, E. et al., *Genetica.*, 140, 169–180, 2012.)

1.2.6 Labeling of Probes

There are several methods and a variety of kits and products that can be used for DNA labeling. In the following paragraphs we describe products and procedures that have been successfully used and applied by the authors for in situ hybridization or FISH on Tephritidae chromosomes.

1.2.6.1 Equipment

1. Centrifuge
2. Freezer
3. Micropipettes
4. Thermoblock
5. Thermocycler
6. Waterbath

1.2.6.2 Chemicals and Reagents

The chemicals and solutions needed for DNA labeling are listed separately for each method in alphabetical order. Suppliers and catalogue numbers are given only when specific products are used.

1.2.6.2.1 *Labeling by Random Priming*

1. Biotin High Prime kit (Roche Diagnostics, Cat. No. 11 585 649 910)
2. Blocking reagent (Roche Diagnostics, Cat. No. 10 057 177 103)
3. DIG DNA Labeling Kit (Roche Diagnostics, Cat. No. 11 175 033 910) Alternatively: Random primer mix and Klenow enzyme
4. Ethanol
5. Ethylenediaminetetraacetic acid (EDTA)
6. Formamide ($HCONH_2$) deionized
7. Lithium chloride (LiCl)
8. Sodium chloride (NaCl)
9. Sodium citrate ($C_6H_5Na_3O_7$)
10. Sodium dodecyl sulfate (SDS)
11. Sodium lauroyl sarcosinate (Sarcosyl)

1.2.6.2.2 *Labeling by Nick Translation*

1. DIG Nick Translation MIX (Roche Diagnostics, Cat. No. 11 745 816 910)
2. EDTA
3. Formamide ($HCONH_2$) deionized
4. Sodium chloride (NaCl)
5. Sodium citrate ($C_6H_5Na_3O_7$)

1.2.6.2.3 Labeling through Polymerase Chain Reaction

1. Biotin-16-dUTP (Roche Diagnostics, Cat. No. 11 093 070 910)
2. Buffer for *Taq* polymerase
3. dNTPs (dATP, dCTP, dGTP, dTTP)
4. *Taq* polymerase
5. Primer F
6. Primer R

1.2.6.3 Solutions Required

1.2.6.3.1 Labeling by Random Priming

1. EDTA 0.2 M pH 8.0: Add 37.224 g EDTA in 350 mL dH$_2$O. To dissolve, slowly adjust pH to 8.0 with 10 M NaOH while mixing. After complete dissolving, add dH$_2$O to 500 mL. Autoclave, store at RT.
2. LiCl 4 M: dissolve 16.96 g LiCl in dH$_2$O. Autoclave, store at RT.
3. Ethanol 100% and 70% in dH$_2$O.
4. SSC 20× (see Section 1.2.2.3).
5. Hybridization buffer: 1/4 dilution of 20× SSC, 1% blocking reagent, 0.1% Sarcosyl, 0.02% SDS in dH$_2$O. Aliquot and store at −20°C.

1.2.6.3.2 Labeling by Nick Translation

1. SSC 20× (see Section 1.2.2.3).
2. EDTA 0.5 M pH 8.0: add 93.06 g EDTA in 350 mL dH$_2$O (see Section 1.2.6.3.1).

1.2.6.3.3 Labeling through Polymerase Chain Reaction

1. dNTP labeling mixture: Mix dATP, dCTP, dGTP to a final concentration of 1 mM each, dTTP to 0.65 mM and biotin-16-dUTP to 0.35 mM in sterile dH$_2$O. Aliquot and store at −20°C.
2. Primer working solution: dilute Primer to 20 μM in sterile dH$_2$O. Aliquot and store at −20°C.

1.2.6.4 Labeling by Random Priming

Random priming is a technique widely used for labeling linear or circular DNA. It produces labeled fragments of 200–1000 bp (Feinberg and Vogelstein 1983).

1. Random priming labeling with digoxigenin-dUTP
   This protocol has been used for generating digoxigenin-labeled probes used for in situ hybridization on polytene

chromosomes of *B. oleae* (Zambetaki et al. 1999; Augustinos et al. 2008; Tsoumani et al. 2011; Kakani et al. 2012), *D. ciliatus* (Drosopoulou et al. 2011b), and *R. cerasi* (E. Drosopoulou, I. Nakou, P. Mavragani-Tsipidou, unpublished data), as well as for FISH on mitotic chromosomes of *C. capitata* (Willhoeft and Franz 1996a,b). Either the DIG DNA Labeling Kit (Roche) can be used or the reagents (Random primers, dNTPs labeling mixture, Klenow enzyme) can be purchased separately (see Section 1.2.6.2.1).

DNA labeling. DNA labeling is performed as follows:
  a. Denature 100–200 ng DNA in 15 μL at 100°C for 10 minutes. Chill on ice.
  b. Add 2 μL hexanucleotide mixture.
  c. Add 2 μL dNTP labeling mixture.
  d. Add 2 U of Klenow.
  e. Mix well and incubate at 37°C overnight.
  f. Stop the reaction with 2 μL EDTA 0.2 M.

Precipitation of labeled DNA. Precipitation of labeled DNA is performed as follows:
  a. Add 2.5 μL 4M LiCl.
  b. Add 75 μL ice-cold 100% ethanol and incubate at −70°C for 30 minutes or −20°C for 2 hours.
  c. Centrifuge at 12,000*g* at 4°C for 20 minutes.
  d. Discard supernatant and add 100 μL ice-cold 70% ethanol.
  e. Centrifuge at 12,000*g* at 4°C for 10 minutes.
  f. Discard supernatant and dry the pellet.
  g. Resuspend labeled DNA in hybridization buffer. Store at −20°C.

Note 31. One labeling reaction is used for five to six preparations.

2. Random priming labeling with biotin16-dUTP
   This method has been used for in situ hybridization on polytene chromosomes of *C. capitata* and *B. tryoni* (Zacharopoulou et al. 1992; Zwiebel et al. 1995; Papadimitriou et al. 1998; Zhao et al. 1998; Verras et al. 1999, 2008; Michel et al. 2001; Gariou-Papalexiou et al. 2002; Schetelig et al. 2009) and *B. dorsalis* (A. Augustinos, E. Drosopoulou, A. Gariou-Papalexiou, K. Bourtzis, P. Mavragani-Tsipidou, A. Zacharopoulou, unpublished data). An updated protocol based on the availability of kits, as the Biotin-High Prime DNA labeling kit (Roche) is given.
   a. Denature 1 μg template DNA in 16 μL at 100°C for 10 minutes. Chill on ice.

b. Centrifuge briefly the denatured DNA and add 4 µL Biotin-High Prime mix.
c. Incubate at 37°C for 1–16 hours.
d. Stop the reaction by adding 2 µL 0.2 M EDTA.
e. Add 5 µL dH$_2$O, 25 µL 20× SSC, 50 µL deionizied formamide, and store it at −20°C until use.

1.2.6.5 Labeling by Nick Translation

Nick translation labeling with digoxigenin-dUTP. The described protocol uses the DIG-Nick Translation Mix (Roche) and has been used for in situ hybridization on polytene chromosomes of *C. capitata* (Stratikopoulos et al. 2002, 2008; Krasteva et al. 2004).

a. Add 1 µg template DNA to sterile ddH$_2$O (final volume of 16 µL).
b. Add 4 µL nick translation mix, mix well and centrifuge briefly.
c. Incubate for 90 minutes at 15°C.
d. Stop the reaction by adding 1 µL 0.5 M EDTA, pH 8.0 and heating to 65°C for 10 minutes.
e. Add 5 µL dH$_2$O, 25 µL 20× SSC, 50 µL deionized formamide, and keep it at −20°C until use.

1.2.6.6 Polymerase Chain Reaction Labeling

Polymerase chain reaction (PCR) labeling with biotin-11-dUTP. The protocol has been used for labeling rRNA and repetitive DNA probes for FISH on *B. oleae* polytene and mitotic chromosomes (Drosopoulou et al. 2012; Tsoumani et al. 2013).

a. Add 1-4 ng DNA to a PCR Eppendorf tube.
b. Add dH$_2$O up to 15 µL.
c. Add 2 µL Enzyme buffer.
d. Add 1.6 µL dNTPs labeling mixture.
e. Add 0.6 µL each primer.
f. Add 1 unit *Taq* Polymerase.
g. Spin at 15,000g for 5 seconds.
h. Incubate in thermocycler at the following program.
Initial denaturation: 94°C, 5 minutes.
Denaturation: 94°C, 30 seconds.
Annealing: 50°C–60°C (depending on the primers used), 20 seconds.
Elongation: 72°C, 1 min/kb.
Repeat the Denaturation, Annealing, and Elongation steps 34 times.
Final elongation: 72°C, 3 minutes.

i. Measure DNA concentration spectrophotometrically.
j. Store at −20°C.

1.2.7   Chromosome Painting

The equipment, chemicals, and solutions needed for chromosome painting on mitotic and polytene chromosomes are listed separately for each step of the technique in alphabetical order. Suppliers and catalogue numbers are given only when specific products are used.

1.2.7.1   Equipment

1.2.7.1.1   Chromosome Preparations for Microdissection

1. Hot plate
2. Membrane Slide 0.17 PEN (D) (Carl Zeiss Int., Cat. No. 415190-9061-000). Alternatively: Polyethylene naphthalate membrane (PEN membrane) slides freshly prepared in the lab using: glass slides 76 × 24 × 0.17 mm (custom made, e.g., by Menzel-Gläser, Braunschweig, Germany); polyethylene naphthalate (PEN) film, thickness 0.0013 mm (Goodfellow, Huntingdon, England, Cat. No. ES361010); 96%–100% ethanol
3. Staining jar
4. Thin needles
5. UV lamp
6. Well slides

1.2.7.1.2   Chromosome Microdissection

1. P.A.L.M. MicroLaser System (Carl Zeiss MicroImaging GmbH, Munich, Germany) for laser microdissection, equipped with a laser pressure catapulting system.
2. Stereoscopic microscope with diascopic stand.

1.2.7.1.3   Preparation and Labeling of Chromosome Painting Probes

1. Centrifuge
2. Freezer (−20°C and −80°C)
3. Hybridization oven
4. Micropipettes
5. Thermocycler
6. Water bath

1.2.7.1.4   Fluorescence In Situ Hybridization of Chromosome Painting Probes

Equipment for FISH is listed in Section 1.2.5.1.

1.2.7.2   Chemicals and Reagents

1.2.7.2.1   Chromosome Preparations for Microdissection

The chemicals needed for chromosome preparations are included in Section 1.2.2.2.

1.2.7.2.2   Preparation and Labeling of Chromosome Painting Probes

1. Buffer for *Taq* polymerase
2. DNase-free ultrapure water
3. dNTPs mixture (10 mM dATP, dCTP, dGTP, 8.5 mM dTTP)
4. GenomePlex® Single Cell Whole Genome Amplification Kit (Sigma-Aldrich, Cat. No. WGA4)
5. GenomePlex WGA Reamplification Kit (Sigma-Aldrich, Cat. No. WGA3)
6. Fluorochrome-labeled dUTP (1 mM)
7. *Taq* polymerase
8. Wizard SV Gel and PCR Clean-Up System (Promega, Madison, WI, Cat. No. A9281)

1.2.7.2.3   Fluorescence In Situ Hybridization with
*   Chromosome Painting Probes*

The chemicals needed for FISH with chromosome painting probes are included in Section 1.2.5.2.

1.2.7.3   Solutions Required

1.2.7.3.1   Chromosome Preparations for Microdissection

The solutions needed for chromosome preparations are included in Section 1.2.2.3.

1.2.7.3.2   Fluorescence In Situ Hybridization with
*   Chromosome Painting Probes*

The solutions needed for FISH with chromosome painting probes are included in Section 1.2.5.3.

The following methods were used for painting X and Y chromosomes in mitotic and polytene nuclei of *B. oleae* (Drosopoulou et al. 2012).

1.2.7.4   Chromosome Preparations for Microdissection

1.2.7.4.1   Preparation of PEN Membrane Slides

The researcher can either purchase commercially available PEN membrane slides, such as MembraneSlide 0.17 PEN (D),Carl Zeiss Int. (see Section 1.2.7.1.1), or prepare PEN slides as follows:

1. Place the PEN membrane on a sheet of paper with drawn rectangles 18 × 35 mm and cut to pieces together with the paper.

2. Put a drop of about 30 μL ethanol in the center of the glass slide and immediately place a piece of the PEN membrane (with paper up) on the layer of ethanol.
3. Remove the piece of paper and use it to expel excess of ethanol and smooth the membrane.
4. Close the edges of the membrane with nail polish to prevent evaporation of ethanol. Keep the PEN membrane slides in the fridge until use.
5. Sterilize 30 minutes under UV light shortly before use.

1.2.7.4.2   *Mitotic Chromosome Preparations on PEN Membrane Slides*

The procedure applied for preparations of mitotic chromosomes on PEN membrane slides is identical to those applied for karyotype analysis and is described in Section 1.2.2.4. Staining is performed as described in Section 1.2.2.5. Prepared slides can be kept in the fridge until microdissection, but no longer than a few weeks.

1.2.7.5   Chromosome Microdissection

Selected chromosomes are identified in an inverted microscope according to their specific features (usually morphology and size) and microdissected with the help of a P.A.L.M. MicroLaser System as described by Kubickova et al. (2002). Each microdissected chromosome is catapulted by a single laser pulse into the cap (containing 3 μL of mineral oil) of a PCR tube. DNA of microdissected samples, each containing 1–30 (usually 10–15) specific chromosomes is then used as a template for PCR amplification.

1.2.7.6   Preparation and Labeling of Chromosome Painting Probes

1.2.7.6.1   *Amplification of Microdissected Chromosomes*

For amplification of DNA of microdissected chromosomes, we use a WGA4 GenomePlex Single Cell Whole Genome Amplification Kit and a WGA3 GenomePlex WGA Reamplification Kit (Sigma-Aldrich), essentially following the manufacturer's instructions. Briefly, shortly before the amplification using the WGA4 kit, 9 μL DNase-free ultrapure water is added to each sample and the sample is spun down in a microcentrifuge at 2000g for 3 minutes. The amplified products are purified using a Wizard SV Gel and PCR Clean-Up System (Promega), reamplified by PCR using the WGA3 kit, and purified again. The finished products are stored in a deep freezer.

1.2.7.6.2   Labeling of Microdissected Chromosomes

Labeling of the amplified chromosomes is performed as follows:

a. Add 10–20 ng template DNA of the WGA3 reamplified product to a PCR tube.
b. Add dH$_2$O up to 56.5 μL.
c. Add 7.5 μL Master mix.
d. Add 3 μL dNTPs mixture.
e. Add 3 μL fluorochrome-labeled dUTP.
f. Add 5 μL WGA Polymerase.
g. Spin at 15,000×*g* for 5 seconds.
h. Incubate in thermocycler under the following conditions.
   Initial denaturation: 95°C, 3 minutes.
   Denaturation: 94°C, 15 seconds.
   Annealing and Elongation: 65°C, 5 minutes.
   Repeat the Denaturation, Annealing and Elongation steps 14 times.
i. Store at −20°C.

Note 32. One labeling reaction can be used for 15 preparations.

1.2.7.7   Fluorescence In Situ Hybridization of Chromosome Painting Probes

1.2.7.7.1   Chromosome Preparations

The procedure applied for preparations of mitotic and polytene chromosomes for FISH is described in Section 1.2.5.4.

1.2.7.7.2   Pretreatment and Denaturation of Chromosomes

Aging-Dehydration. Slides with chromosome preparations are aged at −20°C for 2 hours or more (after dehydration). After removal from the freezer, slides are immediately put into ice-cold 70% ethanol for 2 minutes. Then they are passed through 80% and 100% ethanol, 30 seconds each, and air-dried (see Note 26).

Denaturation. Chromosomes are denatured in 70% formamide for 3.5–4 minutes at 68°C–72°C (see Note 28).

Dehydration. Slides are dehydrated through 70%, 80%, and 100% ethanol at RT, 30 seconds each, and air-dried.

1.2.7.7.3   Hybridization

Preparation and denaturation of the probe. The procedure for the preparation and denaturation of the probe is described in Section 1.2.5.6, Protocol 1.

Hybridization. Place 10 μL probe in the center of the preparation and carefully cover with a 24 × 32-mm coverslip (avoid bubbles). Edges are completely sealed with rubber cement. Slides are placed in a moist chamber (see Note 23) and incubated at 37°C for 3 days.

Washing. Washing is performed as follows:

a. Peel off rubber cement with forceps.
b. Dip briefly in 1% Triton X-100 in 0.1× SSC and remove cover slip.
c. Wash once in 1% Triton X-100 in 0.1× SSC at 62°C for 5 minutes.
d. Wash once in 1% Triton X-100 in 2× SSC at RT for 2 minutes.

1.2.7.7.4   *Signal Detection*

After FISH with a fluorochrome-labeled probe no signal detection procedure is necessary. Immediately after washing, the slides can be stained with DAPI as described in Section 1.2.5.7.

1.2.7.7.5   *Observation and Imaging*

Observation and image capturing and processing is done as described in section 1.2.5.8 (Figure 1.5b through d).

1.2.8   Troubleshooting

The most important factor in obtaining good results in all the above described cytogenetic techniques is the use of high-quality preparations. Other significant factors are the use of clean materials, good quality chemicals, and, if relevant, freshly prepared solutions. Sufficient denaturation of chromosome preparations and well-prepared and sufficiently labeled molecular probes are crucial for successful hybridization and localization of target DNA sequences.

1.2.8.1   Mitotic Chromosome Preparations

1. Incubation time in the hypotonic solution is critical for adequate spreading of chromosomes and should be adjusted for each species examined.
2. Carnoy's fixative should be freshly prepared, that is, used within several hours after preparation of the solution. The tissue used for spreading should not be overfixed, because then it might be difficult to macerate the tissue and spread the chromosomes.

1.2.8.2   Polytene Chromosome Preparations

1. Incubation time in the fixation solution is critical and should be adjusted for each species examined.
2. The quality of spreads depends greatly on the growth conditions of the larvae. For best preparations, larvae are reared at 18°C–20°C in uncrowded colonies.

3. The use of Gurr's orcein (BDH, product No. 34210) is highly recommended, as it gives the best staining results in our laboratories. However, duration of staining should be adjusted for each species examined.

4. Arms of the polytene chromosomes should be spread but not broken. Therefore, spreading should be performed by gentle movements with occasional inspection under a microscope. Also, chromosome flattening should be carefully monitored to avoid excess pressure, which might destroy their morphology.

1.2.8.3 Probe Preparation and Labeling

1. Materials and reagents used for probe labeling should be carefully checked and protocols should be strictly followed.

2. Labeling of probes can be checked by dot blots.

3. Labeled probes should consist of DNA fragments of adequate length (optimally 200–500 bp). Long fragments could hinder access of the probe to target sequences of compact chromosomes; short fragments could be lost during precipitation of the probe. Therefore, it is useful to check the probe length before use by gel electrophoresis.

1.2.8.4 In Situ Hybridization and Fluorescence In Situ Hybridization

1. The use of positive controls is recommended for accuracy and optimization of procedures.

2. Denaturation of chromosomes should be carefully monitored. Longer denaturation or denaturation at a higher temperature could damage chromosome morphology and also increase noise on preparations after hybridization and signal detection. Conversely, insufficient denaturation may result in weak hybridization signals or even complete absence of signals.

3. Fluorochromes are sensitive to direct light. Therefore, all steps with fluorochrome-conjugated antibodies or fluorochrome-labeled probes and fluorochrome counterstaining of chromosome preparations should be carried out in moderate darkness.

4. Remnants of ethanol in hybridization mixture after precipitation may reduce solubility of the mixture and consequently result in lowered concentration of the probe and thus, in weak hybridization signals. Therefore, it is very important to remove all ethanol before drying of the mixture (we do it by repeated spinning of hybridization mixture and repeated discarding of the supernatant after last step of precipitation (i.e., after last centrifugation).

However, overdrying can also reduce solubility of hybridization mixture. Therefore, recommended drying time (about 3 minutes at 37°C) should be respected.

5. Caution should be taken during observation of FISH preparation and capturing images at high magnification because fluorescence fades and signal becomes weak.

1.3 DISCUSSION

1.3.1 Integration of Cytogenetic, Linkage, and Genome Maps

Significant progress has been achieved in the field of Tephritidae cytogenetics during the last three decades. Several studies on the mitotic chromosomes of tephritid species confirmed that the modal number of their karyotype is $2n = 12$ including one pair of sex chromosomes, XX and XY, and five pairs of autosomes (Figure 1.1) (Radu et al. 1975; Southern 1976; Bhatnagar et al. 1980; Singh and Gupta 1984; Bedo 1986, 1987; Zacharopoulou 1987, 1990; Mavragani-Tsipidou et al. 1992; Procunier and Smith 1993; Hunwattanakul and Baimai 1994; Canovai et al. 1994; Baimai et al. 1995, 1999, 2000; Zhao et al. 1998; Frias 2002; Rocha and Selivon 2002; Selivon et al. 2002, 2005a,b, 2007; Cevallos and Nation 2004; Goday et al. 2006; Kounatidis et al. 2008; Caceres et al. 2009; Garcia-Martinez et al. 2009; Drosopoulou et al. 2010, 2011a,b; Zacharopoulou et al. 2011a,b; Hernandez-Ortiz et al. 2012).

Moreover, polytene chromosome maps constructed from larval salivary glands for several tephritid species showed that their polytene complement consists of a total of five long chromosomes (Figure 1.2) (Singh and Gupta 1984; Bedo 1986, 1987; Zacharopoulou 1987, 1990; Mavragani-Tsipidou et al. 1992; Zhao et al. 1998; Kounatidis et al. 2008; Garcia-Martinez et al. 2009; Drosopoulou et al. 2010, 2011a,b; Zacharopoulou et al. 2011a,b). Polytene chromosome maps from other tissues showed considerable similarities among the banding patterns of salivary glands with those of fat body and Malpighian tubules (Kerremans et al. 1990; Zambetaki et al. 1995; Mavragani-Tsipidou 2002). Interestingly, the banding pattern of pupal trichogen cells in *C. capitata* was found to be so different than that of salivary glands (Figure 1.3) that even the chromosome tips cannot be matched (Bedo and Zacharopoulou 1988). However, analysis of autosomal breakpoints in several translocation lines allowed the two chromosome maps to be aligned (Zacharopoulou et al. 1991a), while their detailed comparative analysis was achieved by in situ hybridization (Kritikou 1997).

Cytogenetic analysis including both mitotic and polytene chromosomes clearly indicated that the five polytene elements found in the polytene complements correspond to the five autosomes of the mitotic ones. The sex chromosomes being largely heterochromatic are underreplicated in polytene tissues (Figure 1.5) (Bedo 1987; Bedo and Webb 1989; Zacharopoulou 1990; Rosseto et al. 2000; Drosopoulou et al. 2012). Moreover, the correspondence of *C. capitata* polytene chromosomes to the mitotic complement and to genetic linkage groups was achieved by the cytological analysis of chromosomal rearrangements (Zacharopoulou 1990).

1.3.2 Practical and Scientific Benefits of Genome Mapping

In situ mapping of DNA sequences permitted the precise localization of cloned DNA sequence of interest, including a number of genes, on the polytene chromosomes (Figure 1.4) and thus enabled the integration of molecular genetic and cytogenetic maps in *C. capitata* (Figure 1.6), *B. oleae*, and *B. tryoni* (Zacharopoulou et al. 1992; Scott et al. 1993; Banks et al. 1995; Papadimitriou et al. 1998; Zhao et al. 1998; Zambetaki et al. 1999; Gariou-Papalexiou et al. 2002; Stratikopoulos et al. 2002, 2008, 2009; Theodoraki and Mintzas 2006; Kokolakis et al. 2008; Augustinos et al. 2008; Verras et al. 2008; Tsoumani et al. 2011; Kakkani et al. 2012).

The localization of transgenes has facilitated insect germ-line transformation-based applications of *C. capitata* (Loukeris et al.

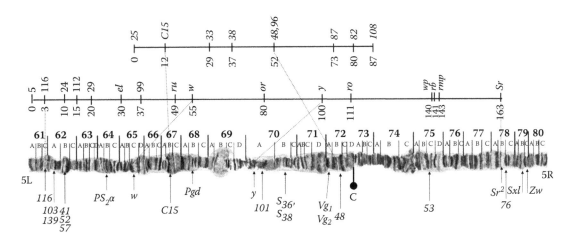

FIGURE 1.6 Integrated cytogenetic (bottom) and genetic maps (top) of *Ceratitis capitata* chromosome 5, including genes (letters in italics) and microsatellite clones (numbers in italics). Arrows indicate the cytological position of the markers, whereas dotted lines link the genetic loci to the respective cytological position on the polytene chromosome map. Numbers beneath the genetic maps indicate genetic distance. C indicates the centromere. (From Stratikopoulos, E. et al., *Mol. Genet. Genomics.*, 282, 283–306, 2009.)

1995; Handler et al. 1998; Michel et al. 2001; Krasteva et al. 2004; Schetelig et al. 2009). In addition, the parallel mapping on mitotic and polytene nuclei by in situ hybridization and/or FISH (Figure 1.5) permitted the correspondence of the sex chromosomes in the two complements (Bedo and Webb 1989; Rosetto et al. 2000; Drosopoulou et al. 2012).

The cloning, isolation, and characterization of genes and DNA sequences (Verras et al. 1999, 2008; Stratikopoulos et al. 2002, 2008, 2009; Drosopoulou et al. 2009; Tsoumani et al. 2011) has provided genetic and molecular information for the tephritid genome including their potential use as diagnostic tools for differentiating major agricultural pests. In addition, cytogenetics can support, by providing landmarks, the assembly and the completion of ongoing genome projects, such as those of *C. capitata*, *B. dorsalis*, and *B. tryoni*, which are large in size and contain high degree of repetitive elements.

Cytogenetics has played a catalytic role in unraveling sex determination in the medfly, *C. capitata*. Aneuploid $XX22^Y$ offspring generated by a male-linked translocation strain were males suggesting that the male determining factor is localized in the Y chromosome (Zapater and Robinson 1986). Using male translocation lines, Willhoeft and Franz (1996a) analyzed a series of Y chromosome deletions, derived through adjacent-1 segregation during meiosis of several male translocation strains, which was then used to map the male-determining factor (M, maleness factor) using FISH. The M factor was localized to the long arm of the Y chromosome, in a region encompassing approximately 15% of the entire Y chromosome. This finding was of paramount importance and resulted to the development of stable GSSs strains based on Y–autosome translocation lines (Franz 2005). The isolation of induced inversions helped to improve the stability of the medfly GSSs by significantly reducing recombination events (Franz 2005) and also permitted the construction of the first balancer chromosome, which is an important genetic tool for the manipulation of laboratory strains (Gourzi et al. 2000). The availability of stable GSSs greatly improved the SIT in *C. capitata*, the model organism of tephritid fruit flies, at two levels: (1) economy in production (no females reared, sterilized, and released) and (2) efficiency in action (no assortative mating, fruit damage, and trapped females) (Robinson et al. 1999; Gariou-Papalexiou et al. 2002; Franz 2005). At this moment, the new generation of medfly GSS, VIENNA 8, is being reared in mass-rearing facilities and used for SIT applications for medfly population control in all continents (Guatemala, Mexico, United States, Brazil, Peru, Chile, Argentina, South Africa, Spain, Israel, and Australia).

1.3.3 Chromosome Organization and Evolution

Polytene chromosome maps facilitated classical, molecular, and evolutionary genetic research; provided information for chromatin and genome organization; and offered a rapid way of inferring phylogenetic relationships among species as well as a "snapshot" of their current/ongoing chromosomal evolution (Zhao et al. 1998; Gariou-Papalexiou et al. 2002; Mavragani-Tsipidou 2002; Drosopoulou et al. 2011b; Tsoumani et al. 2011). Through these studies, it was revealed that species differentiation was based on paracentric inversions and/or transpositions. Interestingly, only one pericentric inversion was detected in comparisons between *C. capitata* and the rest tephritid species analyzed so far (Figure 1.7).

The in situ hybridization technique has also provided clear molecular evidence for the homology between the *Drosophila melanogaster* X chromosome and a specific autosome of tephritid species (chromosome 5 in *C. capitata* and its homologue autosome in other tephritid species (Zacharopoulou et al. 1992; Zhao et al. 1998; Zambetaki et al. 1999). It is worth noting that a remarkable conservation of linkage groups was observed between *Drosophila* and the analyzed tephritid species, which supports the concept that the major chromosomal elements retain their identity not only among closely but also between distantly related Diptera (Zacharopoulou et al. 1992; Zhao et al. 1998; Zambetaki et al. 1999; Gariou-Papalexiou 2002; Mavragani-Tsipidou 2002; Tsoumani et al. 2011).

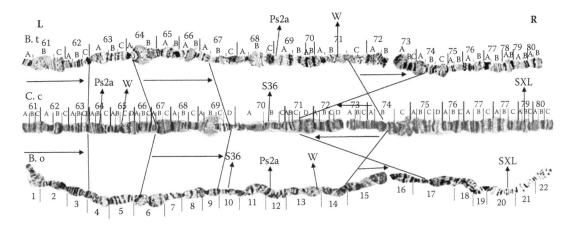

FIGURE 1.7 Comparison of a polytene chromosome element in three tephritid species, *Bactrocera tryoni* (B. t), *Ceratitis capitata* (C. c), and *B. oleae* (B. o), based on banding pattern similarities and hybridization sites (vertical arrows) of the *C. capitata* gene probes *white* (*w*), *PS2a*, *S36*, and *SexLethal* (*SXL*). L and R indicate the left and the right chromosome arm, respectively. Lines connecting the chromosomes indicate sections with similar banding pattern and horizontal arrows show the relative orientation between them. A pericentric inversion is evident among *C. capitata* and the two *Bactrocera* species. The hybridization sites of the *white* and *PS2a* genes support a transposition of the respective chromosome region among *C. capitata* and the two *Bactrocera* species. (From Zambetaki, A. et al., *Genome.*, 38, 1070–1081, 1995; Zhao, J.T et al., *Genome.*, 41, 510–526, 1998; Gariou-Papalexiou, A. et al., *Genetica.*, 116, 59–71, 2002.)

The development of enriched cytogenetic maps can also detect "early" signs of speciation and provide evidence for the presence of cryptic species or species complexes as has recently been reported for the genera *Anastrepha* and *Bactrocera* (Baimai et al. 1995, 1999, 2000; Frias 2002; Rocha and Selivon 2002; Selivon et al. 2002, 2005a,b, 2007; Cevallos and Nation 2004; Goday et al. 2006; Caceres et al. 2009).

In conclusion, advances in the field of cytogenetics have allowed the development of tools of both basic and applied scientific importance for tephritid species, including the integration of cytogenetic, linkage, and genome maps, unraveling sex determination, improvement of sterile insect technique, as well as elucidating chromosome organization during Diptera evolution and incipient speciation phenomena.

ACKNOWLEDGMENTS

The authors would like to thank the Joint FAO/IAEA Division of Nuclear Techniques in Food and Agriculture for the financial support provided through various Coordinated Research Programs on the development of tools and strategies for the control of tephritid pest species.

REFERENCES

Ant, T., M. Koukidou, P. Rempoulakis, H. F. Gong, A. Economopoulos, J. Vontas, and L. Alphey. 2012. Control of the olive fruit fly using genetics-enhanced sterile insect technique. *BMC Biol* 10:51.

Augustinos, A. A., E. E. Stratikopoulos, E. Drosopoulou, E. G. Kakani, P. Mavragani-Tsipidou, A. Zacharopoulou, and K. D. Mathiopoulos. 2008. Isolation and characterization of microsatellite markers from the olive fly, *Bactrocera oleae*, and their cross-species amplification in the Tephritidae family. *BMC Genomics* 9:618.

Baimai, V., J. Phinchongsakuldit, and C. Sumrandee. 2000. Cytological evidence for a complex of species within the taxon *Bactrocera tau* (Diptera: Tephritidae) in Thailand. *Biol J Linn Soc* 69:399–409.

Baimai, V., J. Phinchongsakuldit, and S. Tigvattananont. 1999. Metaphase karyotypes of fruit flies of Thailand. IV. Evidence for six new species of the *Bactrocera dorsalis* complex. *Cytologia* 64:371–377.

Baimai, V., W. Trinachartvanit, S. Tigvattananont, P. J. Grote, R. Poramarcom, and U. Kijchalao. 1995. Metaphase karyotypes of fruit flies of Thailand. I. Five sibling species of the *Bactrocera dorsalis* complex (Diptera: Tephritidae). *Genome* 38:1015–1022.

Banks, G. K., A. S. Robinson, J. Kwiatowski, F. J. Ayala, M. J. Scott, and D. Kritikou. 1995. A second superoxide dismutase gene in the medfly *Ceratitis capitata*. *Genetics* 140:697–702.

Barr, N. B. and B. A. McPheron. 2006. Molecular phylogenetics of the genus *Ceratitis* (Diptera: Tephritidae). *Mol Phylogenet Evol* 38:216–230.

Bedo, D. G. 1986. Polytene and mitotic chromosome analysis in *Ceratitis capitata* (Diptera; Tephritidae). *Canadian J Genet Cytol* 28:180–188.

Bedo, D. G. 1987. Polytene chromosome mapping in *Ceratitis capitata* (Diptera: Tephritidae). *Genome* 29:598–611.

Bedo, D. G. and G. C. Webb. 1989. Conservation of nucleolar structure in polytene tissue of *Ceratitis capitata* (Diptera: Tephritidae). *Chromosoma* 98:443–449.

Bedo, D. G. and A. Zacharopoulou. 1988. Inter-tissue variability of the polytene chromosome banding. *Trends Genet* 4:90–91.

Bhatnagar, S., D. Kaul, and R. Chaturvedi. 1980. Chromosomal studies in three species of the genus *Dacus* (Trypetidae: Diptera). *Genetica* 54:11–15.

Caceres, C., D. F. Segura, M. T. Vera, V. Wornoaypon, J. L. Cladera, P. Teal, P. Sapountzis, K. Bourtzis, A. Zacharopoulou, and A. S. Robinson. 2009. Incipient speciation revealed in *Anastrepha fraterculus* (Diptera; Tephritidae) by studies on mating compatibility, sex pheromones, hybridization, and cytology. *Biol J Linnean Soc* 97:152–165.

Canovai, R., B. Caterini, L. Contadini, and L. Galleni. 1994. Karyology of the medfly *Ceratitis capitata* (Wied.) mitotic complement: ASG bands. *Caryology* 47:241–247.

Cevallos, V. E. and J. L. Nation. 2004. Chromosomes of the Caribbean fruit fly (Diptera: Tephritidae). *Fla Entomol* 87:361–364.

Cladera, J. L. and A. Delprat. 1995. Genetic and cytological mapping of a "Y-2" translocation in the Mediterranean fruit fly *C. capitata*. *Genome* 38:1091–1097.

Drew, R. A. I. 2004. Biogeography and speciation in the Dacini (Diptera: Tephritidae: Dacinae). *Bishop Mus Bull Entomol* 12:165–178.

Drew, R. A. I. and D. L. Hancock. 2000. Phylogeny of the tribe Dacini (Dacinae) based on morphological, distributional, and biological data. In *Fruit Flies (Tephritidae): Phylogeny and Evolution of Behavior*, Edited by M. Aluja and A. L. Norrbom, pp. 491–504. Boca Raton, FL: CRC Press.

Drosopoulou, E., A. A. Augustinos, I. Nakou, K. Koeppler, I. Kounatidis, H. Vogt, N. T. Papadopoulos, K. Bourtzis, and P. Mavragani-Tsipidou. 2011a. Genetic and cytogenetic analysis of the American cherry fruit fly, *Rhagoletis cingulata* (Diptera: Tephritidae) *Genetica* 139:1449–1464.

Drosopoulou, E., A. Chrysopoulou, V. Nikita, and P. Mavragani-Tsipidou. 2009. The heat shock 70 genes of the olive pest *Bactrocera oleae*: Genomic organization and molecular characterization of a transcription unit and its proximal promoter region. *Genome* 52:210–214.

Drosopoulou, E., K. Koeppler, I. Kounatidis, I. Nakou, N. T. Papadopoulos, K. Bourtzis, and P. Mavragani-Tsipidou. 2010. Genetic and cytogenetic analysis of the Walnut-Husk fly (Diptera: Tephritidae). *Ann Entomol Soc Am* 103:1003–1011.

Drosopoulou, E., I. Nakou, J. Šíchová, S. Kubíčková, F. Marec, and P. Mavragani-Tsipidou. 2012. Sex chromosomes and associated rDNA form a hetero chromatic network in the polytene nuclei of *Bactrocera oleae* (Diptera: Tephritidae). *Genetica* 140:169–180.

Drosopoulou, E., D. Nestel, I. Nakou, I. Kounatidis, N. T. Papadopoulos, K. Bourtzis, and P. Mavragani-Tsipidou. 2011b. Cytogenetic analysis of the Ethiopian fruit fly *Dacus ciliatus* (Diptera: Tephritidae). *Genetica* 139:723–732.

Feinberg, A. P. and B. Vogelstein. 1983. A technique for radiolabeling DNA restriction endonuclease fragments to high specific activity. *Anal Biochem* 132:6–13.

Fletcher, B. S. 1989. Life history strategies of tephritid fruit flies. In *Fruit flies, Their Biology, Natural Enemies and Control*, Edited by A. S. Robinson and G. Hooper, pp. 195–208. Amsterdam, The Netherlands: Elsevier.

Franz, G. 2005. Genetic sexing strains in Mediterranean fruit fly, an example for other species amenable to large scale rearing for the sterile insect technique. In *Sterile Insect Technique: Principles and Practice in Area Wide Integrated Pest Management,* Edited by V. A. Dyck, J. Hendrichs, and A. S. Robinson, pp. 427–452. The Netherlands: Springer.

Franz, G., E. Genscheva, and P. Kerremans. 1994. Improved stability of sex-separation strains of the Medfly, *Ceratitis capitata*. *Genome* 37:72–82.

Frias, D. 2002. Importance of larval morphology and heterochromatic variation in the identification and evolution of sibling species in the genus *Rhagoletis* (Diptera: Tephritidae) in Chile. In *Proceedings of the 6th International Symposium on Fruit Flies of Economic Importance (May 6–10, 2002, Stellenbosch)*, Edited by B. N. Barnes, pp. 267–276. Irene, South Africa: Isteg Scientific Publications.

Frydrychová, R. and F. Marec. 2002. Repeated losses of TTAGG telomere repeats in evolution of beetles (Coleoptera). *Genetica* 115:179–187.

Fuková, I., P. Nguyen, and F. Marec. 2005. Codling moth cytogenetics: Karyotype, chromosomal location of rDNA and molecular differentiation of sex chromosomes. *Genome* 48:1083–1092.

Gabrieli, P., A. Falaguerra, P. Siciliano, L. M. Gomulski, F. Scolari, A. Zacharopoulou, G. Franz, A. R. Malacrida, and G. Gasperi. 2010. Sex and the single embryo: Early development in the Mediterranean fruit fly, *Ceratitis capitata*. *BMC Dev Biol* 10:12.

Garcia-Martinez, V., E. Hernandez-Ortiz, C. S. Zepeta-Cisneros, A. S. Robinson, A. Zacharopoulou, and G. Franz. 2009. Mitotic and polytene analysis in the Mexican fruit fly, *Anastrepha ludens* (Loew) (Diptera: Tephritidae). *Genome* 52:1–11.

Gariou-Papalexiou, A., P. Gourzi, A. Delprat, D. Kritikou, K. Rapti, B. Chrysanthakopoulou, A. Mintzas, and A. Zacharopoulou. 2002. Polytene chromosomes as tools in the genetic analysis of the Mediterranean fruit fly, *Ceratitis capitata*. *Genetica* 116:59–71.

Goday, C., D. Selivon, A. L. P. Perondini, P. G. Greciano, and M. F. Ruiz. 2006. Cytological characterization of sex chromosomes and ribosomal DNA location in *Anastrepha* species. (Diptera: Tepritidae). *Cytogen Genom Res* 114:70–76.

Gourzi, P., D. Gubb, Y. Livadaras, C. Caceres, G. Franz, C. Savakis, and A. Zacharopoulou. 2000. The construction of the first balancer chromosome for the Mediterranean fruit fly, *Ceratitis capitata*. *Mol Genet Genom* 1–2:127–136.

Guest, W. C. and T. C. Hsu. 1973. A new technique for preparing *Drosophila* neuroblast chromosomes. *Drosoph Inf Serv* 50:193.

Handler, A. M., S. D. McCombs, M. I. Frazer, and S. H. Saul. 1998. The lepi-dopteran transposon vector, *piggyBac*, mediate germ-line transformation in the Mediterranean fruit fly. *Proc Natl Acad Sci* USA 95:7520–7525.

Hernandez-Ortiz, V., A. F. Bartolucci, P. Morales-Valles, D. Frias, and D. Selivon 2012. Cryptic species of the *Anastrepha fraterculus* complex (Diptera: Tephritidae): A multivariate approach for the recognition of South American morphotypes. *Ann Entomol Soc Am* 105:305–318.

Hunwattanakul, N. and V. Baimai. 1994. Mitotic karyotype of four species of fruit flies (*Bactrocera*) in Thailand. *Kasetsart J Nat Sci* 28:142–148.

Jeyasankar, A., D. Nestel, D. Dragushich, E. Nemny-Lavy, L. Anshelevich, A. Zada, and V. Soroker. 2009. Identification of host attractants for the Ethiopian fruit fly, *Dacus ciliatus* (Loew). *J Chem Ecol* 35:542–551.

Kakani, E. G., M. Trakala, E. Drosopoulou, P. Mavragani-Tsipidou, and K. D. Mathiopoulos. 2012. Genomic structure, organization and local-ization of the acetylcholinesterase locus of the olive fruit fly, *Bactrocera oleae*. *B Entomol Res* 103:36–47.

Kerremans, P., K. Bourtzis, and A. Zacharopoulou. 1990. Cytogenetic analysis of three genetic sexing strains of *Ceratitis capitata*. *Theor Appl Genet* 80:177–182.

Kerremans, P., E. Genscheva, and G. Franz. 1992. Genetic and cytogenetic anal-ysis of Y-autosome translocations in the Mediterranean fruit fly *Ceratitis capitata*. *Genome* 35:264–272.

Kokolakis, G., M. Tatari, A. Zacharopoulou, and A. C. Mintzas. 2008. The *hsp27*gene of the Mediterranean fruit fly, *Ceratitis capitata*: Structural characterization, regulation and developmental expression. *Insect Mol Biol* 17:699–671.

Korneyev, V. A. 2000. Phylogenetic relationships among higher groups of Tephritidae. In *Fruit Flies (Tephritidae): Phylogeny and Evolution of Behaviour*, Edited by M. Aluja and A. L. Norrbom, pp. 73–113. Boca Raton, FL: CRC Press.

Kounatidis, I., N. Papadopoulos, K. Bourtzis, and P. Mavragani-Tsipidou. 2008. Genetic and cytogenetic analysis of the fruit fly *Rhagoletis cerasi* (Diptera: Tephritidae). *Genome* 51:479–491.

Krasteva, R., A. M. Handler, A. Zacharopoulou, C. Caceres, and G. Franz. 2004. Generation and initial analyses of transgenic medfly strains. Paper presented at the 5th Meeting of the Working Group on Fruit Flies of the Western Hemisphere (May16–21). Florida.

Kritikou, D. 1997. Cytological mapping of genes and anonymous DNA clones-Molecular Analysis of the *hsp70* gene family of the Mediterranean fruit fly *Ceratitis capitata*. PhD diss., University of Patras, Rio Patras, Greece.

Kubickova, S., H. Cernohorska, P. Musilova, and J. Rubes. 2002. The use of laser microdissection for the preparation of chromosome-specific painting probes in farm animals. *Chromosome Res* 10:571–577.

Loukeris, T. C., I. Livadaras, B. Arca, S. Zabalou, and C. Savakis. 1995. Gene transfer into the medfly, *Ceratitis capitata*, with a *Drosophila hydei* transposable element. *Science* 270:2002–2005.

Manoukas, A. G. and B. Mazomenos. 1977. Effect of antimicrobials upon eggs and larva of *Dacus oleae* (Diptera: Tephritidae) and the use of propionates for larval diet preservation. *Annales de Zoologie Ecologie Animale* 9:277–285.

Mavragani-Tsipidou, P. 2002. Genetic and cytogenetic analysis of *Bactrocera oleae* (*Dacus oleae*) (Diptera: Tephritidae). *Genetica* 116:45–57.

Mavragani-Tsipidou, P., G. Karamanlidou, A. Zacharopoulou, S. Koliais, and C. Kastritsis. 1992. Mitotic and polytene chromosome analysis in *Dacus oleae* (Diptera: Tephritidae). *Genome* 35:373–378.

McPheron, B. A., H. Y. Han, J. G. Silva, and A. L. Norrbom. 2000. Phylogeny of the genera *Anastrepha* and *Toxotrypana* (Trypetinae: Toxotrypanini) based upon 16S rRNA mitochondrial DNA sequences. In *Fruit Flies (Tephritidae): Phylogeny and Evolution of Behaviour*, Edited by M. Aluja and A. L. Norrbom, pp. 343–361. Boca Raton, FL: CRC Press.

Michel, K., A. Stamenova, A. C. Pinkerton, G. Franz, A. S. Robinson, A. Gariou-Papalexiou, A. Zacharopoulou, D. A. O'Brochta, and P. W. Atkinson. 2001. Hermes-mediated germ-line transformation of the Mediterranean fruit fly *Ceratitis capitata*. *Insect Mol Biol* 10:155–162.

Norrbom, A. L., L. E. Carroll, F. C. Thompson, I. M. White, and A. Feinberg. 1999. Systematic database of names. In *Fruit Fly Expert Identification System and Systematic Information Database: A Resource for Identification and Information on Fruit Flies and Maggots, with Information on Their Classification, Distribution and Documentation*, Edited by F. C. Thompson, Myia 9, vii + 524 pp. and Diptera Data Dissemination Disk, pp. 65–251. Leiden, The Netherlands: Backhuys Publications for the North American Dipterists' Society.

Norrbom, A. L., R. A. Zucchi, and V. Hernández-Ortiz. 2000. Phylogeny of the genera *Anastrepha* and *Toxtrypana* (Trypetinae: Toxotrypanini) based on morphology. In *Fruit Flies (Tephritidae): Phylogeny and Evolution of Behaviour*, Edited by M. Aluja and A. L. Norrbom, pp. 299–342. Boca Raton, FL: CRC Press.

Papadimitriou, E., D. Kritikou, M. Mavroidis, A. Zacharopoulou, and A. C. Mintzas. 1998. The heat shock 70 gene family in the Mediterranean fruit fly Ceratitis capitata. *Insect Mol Biol* 7:279–290.

Procunier, W. S. and J. J. Smith. 1993. Localization of ribosomal DNA in *Rhagoletis pomonella* (Diptera: Tephritidae) by in situ hybridization. *Insect Mol Biol* 2:163–174.

Radu, M., Y. Rossler, and Y. Koltin. 1975. The chromosomes of the Mediterranean fruit fly *Ceratitis capitata* (Wied): Karyotype and chromosomal organization. *Cytologia (Tokyo)* 40:823–828.

Robinson, A. S., G. Franz, and K. Fisher. 1999. Genetic sexing strains in the medfly, *Ceratitis capitata*: Development, mass rearing and field application. *Trends Entomol* 2:81–104.

Rocha, L. S. and D. Selivon. 2002. Studies on highly repetitive DNA in cryptic species of the *Anastrepha fraterculus* complex (Diptera: Tephritidae). In *Proceedings of the 6th International Symposium on Fruit Flies of Economic Importance (May 6–10, 2002, Stellenbosch)*, Edited by B. N. Barnes, pp. 415–418. Irene, South Africa: Isteg Scientific Publications.

Roller, E. F. 1989. Small-scale rearing: *Rhagoletis* spp. In *Fruit Flies: Their Biology, Natural Enemies and Control*, Edited by A. S. Robinson and G. H. S. Hooper, pp. 119–127. Amsterdam, The Netherlands: Elsevier.

Rosetto, M., T. de Filippis, M. Mandrioli, A. Zacharopoulou, P. Gourzi, and A. G. O. Manetti. 2000. Ceratotoxins, female-specific X-linked genes from the medfly *Ceratitis capitata*. *Genome* 43:707–711.

Sahara, K., F. Marec, and W. Traut. 1999. TTAGG telomeric repeats in chromosomes of some insects and other arthropods. *Chromosome Res* 7:449–460.

Schetelig, M. F., C. Caceres, A. Zacharopoulou, G. Franz, and E. A. Wimmer. 2009. Conditional embryonic lethality to improve the sterile insect technique in *Ceratitis capitata* (Diptera: Tephritidae). *BMC Biol* 7:4.

Scott, M. J., D. Kritikou, and A. S. Robinson. 1993. Isolation of cDNAs encoding 6-phosphogluconate dehydrogenase and glucose-6-phosphate dehydrogenase from the Mediterranean fruit fly *Ceratitis capitata*. *Insect Mol Biol* 1:213–222.

Selivon, D. and A. L. P. Perondini. 1997. Evaluation of techniques for C and ASG banding of the mitotic chromosomes of *Anastrepha* species (Diptera, Tephritidae). *Brazilian J Genet* 20:651–653.

Selivon, D., A. L. P. Perondini, and J. Morgante. 2005a. A genetic–morphological characterization of the two cryptic species of the *Anastrpeha fraterculus* complex. *Ann Entomol Soc Am* 98:367–381.

Selivon, D., A. L. P. Perondini, and L. S. Rocha. 2005b. Systematics, morphology and physiology—Karyotype characterization of *Anastrepha* fruit flies (Diptera: Tephritidae). *Neotrop Entomol* 34:273–279.

Selivon, D., F. M. Sipula, L. S. Rocha, and A. L. P. Perondini. 2007. Karyotype relationships among *Anastrepha bistrigata*, *A. striata*, and *A. serpentina* (Diptera: Tephritidae). *Genet Mol Biol* 30:1082–1088.

Selivon, D., C. Vretos, L. Fontes, and A. L. P. Perondini. 2002. New variant forms in the *Anastrepha fraterculus* complex (Diptera: Tephritidae). In: *Proceedings of the 6th International Symposium on Fruit Flies of Economic Importance (May 6–10, 2002, Stellenbosch)*, Edited by B. N. Barnes, pp. 253–258. Irene, South Africa: Isteg Scientific Publications.

Semeshin, V. F., D. Kritikou, A. Zacharopoulou, and I. F. Zhimulev. 1995. Electron microscope investigation of polytene chromosomes in the Mediterranean fruit fly *Ceratitis capitata*. *Genome* 38:652–660.

Singh, O. P. and J. P. Gupta. 1984. Studies on mitotic and salivary chromosomes of *Dacus cucurbitae* Coquilett (Diptera, Tephritidae). *Genetica* 62:217–221.

Smith, J. J., M. Jaycox, M. R. B. Smith-Caldas, and G. L. Bush. 2005. Analysis of mitochondrial DNA and morphological characters in the subtribe *Carpomyina* (Diptera: Tephritidae). *Israel J Entomol* 35–36: 317–340.

Southern, D. I. 1976. Cytogenetic observations on *Ceratitis capitata*. *Experientia* 32:20–22.

Stratikopoulos, E. E., A. A. Augustinos, A. Gariou-Papalexiou, A. Zacharopoulou, and K. D. Mathiopoulos. 2002. Identification and partial characterization of a new *Ceratitis capitata*-specific 44-bp pericentromeric repeat. *Chromosome Res* 10:287–295.

Stratikopoulos, E. E., A. A. Augustinos, I. D. Pavlopoulos, K. P. Economou, A. Mintzas, K. D. Mathiopoulos, and A. Zacharopoulou. 2009. Isolation and characterization of microsatellite markers from the Mediterranean fruit fly, *Ceratitis capitata*: Cross-species amplification in other Tephritidae species reveals a varying degree of transferability. *Mol Genet Genomics* 282:283–306.

Stratikopoulos, E. E., A. A. Augustinos, Y. G. Petalas, M. N. Vrahatis, A. Mintzas, K. D. Mathiopoulos, and A. Zacharopoulou. 2008. An integrated genetic and cytogenetic map for the Mediterranean fruit fly, *Ceratitis capitata*, based on microsatellite and morphological markers. *Genetica* 133:147–157.

Systematic Entomology Laboratory, ARS, USDA. 2004. The Diptera Site. Fruit fly (Diptera: Tephritidae) classification and diversity. http://www.sel.barc.usda.gov/diptera/tephriti/TephClas.htm

Theodoraki, M. A. and A. C. Mintzas. 2006. cDNA cloning, heat shock regulation and developmental expression of the *hsp83* gene in the Mediterranean fruit fly *Ceratitis capitata*. *Insect Mol Biol* 15:839–852.

Traut, W. 1976. Pachytene mapping in the female silkworm, *Bombyx mori* L. (Lepidoptera). *Chromosoma* 58:275–284.

Tsitsipis, J. A. 1989. Nutrition: Requirements. In *Fruit Flies: Their Biology, Natural Enemies and Control*, Edited by A. S. Robinson and G. H. S. Hooper, pp. 101–116. Amsterdam, The Netherlands: Elsevier.

Tsitsipis, J. A. and A. Kontos. 1983. Improved solid adult diet for the olive fruit fly *Dacus oleae* (Diptera: Tephritidae). *Entomologia Hellenica* 1:24–29.

Tsoumani, K. T., A. A. Augustinos, E. G. Kakani, E. Drosopoulou, P. Mavragani-Tsipidou, and K. D. Mathiopoulos. 2011. Isolation, annotation and applications of expressed sequence tags from the olive fly, *Bactrocera oleae*. *Mol Genet Genomics* 285:33–45.

Tsoumani, K. T., E. Drosopoulou, P. Mavragani-Tsipidou, and K. D. Mathiopoulos. 2013. Molecular characterization and chromosomal distribution of a species-specific transcribed centromeric satellite repeat from the olive fruit fly, *Bactrocera oleae*. *PLoS ONE* 8:e79393.

Tzanakakis, M. 1989. Small-scale rearing: *Dacus oleae* (Gmelin). In *Fruit Flies: Their Biology, Natural Enemies and Control*, Edited by A. S. Robinson and G. H. S. Hooper, pp. 105–118. Amsterdam, The Netherlands: Elsevier.

Tzanakakis, M., A. Economopoulos, and J. A. Tsitsipis. 1970. Rearing and nutrition of the olive fruit fly. I. Improved larval diet and simple containers. *J Econ Entomol* 63:317–318.

Verras, M., P. Gourzi, K. Kalosaka, A. Zacharopoulou, and A. C. Mintzas. 2008. cDNA cloning, characterization, and developmental expression of the 20S proteasome α5 subunit in the Mediterranean fruit fly *Ceratitis capitata*. *Arch Insect Biochem Physiol* 67:120–129.

Verras, M., M. Mavroidis, G. Kokolakis, P. Gourzi, A. Zacharopoulou, and A. C. Mintzas. 1999. Cloning and characterization of CcEcR, an ecdysone receptor homolog from the Mediterranean fruit fly *Ceratitis capitata*. *Eur J Biochem* 265:798–808.

White, I. M. and M. M. Elson-Harris. 1992. *Fruit Flies of Economic Significance: Their Identification and Bionomics*. Wallingford, CT: CAB International Publications.

Willhoeft, U. and G. Franz. 1996a. Identification of the sex-determining region of the *Ceratitis capitata* Y chromosome by deletion mapping. *Genetics* 144:737–745.

Willhoeft, U. and G. Franz. 1996b. Comparison of the mitotic karyotypes of *Ceratitis capitata*, *Ceratitis rosa* and *Trirhithrum coffeae* (Diptera: Tephritidae) by C-banding and FISH. *Genome* 39:884–889.

Zacharopoulou, A. 1987. Cytogenetic analysis of mitotic and salivary gland chromosomes in the medfly *Ceratitis capitata*. *Genome* 29:67–71.

Zacharopoulou, A. 1990. Polytene chromosome maps in the medfly *Ceratitis capitata*. *Genome* 33:184–197.

Zacharopoulou, A., A. A. Augustinos, W. A. Sayed, A. S. Robinson, and G. Franz. 2011a. Mitotic and polytene chromosome analysis of the oriental fruit fly, *Bactrocera dorsalis* (Hendel) (Diptera: Tephritidae). *Genetica* 139:79–90.

Zacharopoulou, A., K. Bourtzis, and P. Kerremans. 1991a. A comparison of polytene chromosomes in salivary glands and orbital bristle trichogen cells in *Ceratitis capitata*. *Genome* 34:215–219.

Zacharopoulou, A., M. Frisardi, C. Savakis, A. S. Robinson, P. Tolias, M. Konsolaki, K. Komitopoulou, and F. C. Kafatos. 1992. The genome of the Mediterranean fruit fly *Ceratitis capitata*: Localization of molecular markers by in situ hybridization to salivary gland polytene chromosomes. *Chromosoma* 101:448–455.

Zacharopoulou, A., E. Riva, A. Malacrida, and G. Gasperi. 1991b. Cytogenetic characterization of a genetic sexing strain of *Ceratitis capitata*. *Genome* 34:606–611.

Zacharopoulou, A., W. A. Sayed, A. A. Augustinos, F. Yesmin, A. S. Robinson, and G. Franz. 2011b. Analysis of mitotic and polytene chromosomes and photographic polytene chromosome maps in *Bactrocera cucurbitae* (Diptera: Tephritidae). *Ann Entomol Soc Am* 104:306–318.

Zambetaki, A., K. Kleanthous, and P. Mavragani-Tsipidou. 1995. Cytogenetic analysis of Malphigian tubule and salivary gland polytene chromosomes of *Bactrocera oleae* (*Dacus oleae*) (Diptera: Tephritidae). *Genome* 38:1070–1081.

Zambetaki, A., A. Zacharopoulou, Z. G. Scouras, and P. Mavragani-Tsipidou. 1999. The genome of the olive fruit fly *Bactrocera oleae*: Localization of molecular markers by in situ hybridization to salivary gland polytene chromosomes. *Genome* 42:744–751.

Zapater, M. and A. S. Robinson. 1986. Sex chromosome aneuploidy in a male-translocation in *Ceratitis capitata*. *Can J Genet Cytol* 28:161–167.

Zhao, J. T., M. Frommer, J. A. Sved, and A. Zacharopoulou. 1998. Mitotic and polytene analyses in the Queensland fruit fly, *Bactrocera tryoni* (Diptera: Tephritidae). *Genome* 41:510–526.

Zhimulev, I. F., E. S. Belayaeva, V. F. Semeshin, D. E. Koryakov, S. A. Demakov, O. V. Demakova, G. V. Pokholkova, and E. N. Andreyeva. 2004. Polytene chromosomes: 70 Years of genetic research. *Internat Rev Cytol* 241:203–275.

Zur, T., E. Nemny-Lavy, N. T. Papadopoulos, and D. Nestel. 2009. Social interactions regulate resource utilization in a Tephritidae fruit fly. *J Internat Physiol* 55:890–897.

Zwiebel, L. J., G. Saccone, A. Zacharopoulou, N. J. Besansky, G. Favia, F. H. Collins, C. Louis, and F. C. Kafatos. 1995. The white gene of *Ceratitis capitata*: A phenotypic marker for germline transformation. *Science* 720:2005–2008.

2

Hessian Flies (Diptera)

Jeff J. Stuart, Rajat Aggarwal, and
Brandon J. Schemerhorn

CONTENTS

LIST OF ABBREVIATIONS

AFRI, Agricultural Food and Research Initiative
BSA, bovine serum albumin
DAPI, 4′,6-diamidino-2-phenylindole
dNTP, the four deoxyribonucleotide triphosphates: dATP, dCTP, dGTP and dTTP

EDTA, ethylenediaminetetraacetic acid
FISH, fluorescence in situ hybridization
NIB, nuclei isolation buffer
NIFA, National Institute of Food and Agriculture
PBS, phosphate buffered saline
R, resistance
SDS, sodium dodecyl sulfate
SSC, saline-sodium citrate buffer
STE, sodium chloride-Tris-EDTA buffer
TE, tris EDTA buffer
TMN1, tris magnesium chloride sodium chloride buffer 1

2.1   INTRODUCTION

The Hessian fly (*Mayetiola destructor*) is an important insect pest of wheat (*Triticum* spp.) (Harris et al. 2003). It has been a persistent problem in the United States since it was first discovered along the Atlantic coast just after the American Revolutionary War (Pauly 2002). It is now present nearly everywhere where wheat is grown, but poses its greatest threat to agriculture in northern Africa, Southwest Asia, and the United States (Harris et al. 2003).

The small first and second instar larvae of the insect feed on or near meristematic tissues causing abnormal growth in wheat plants (Harris et al. 2006). The growth of plants attacked in the seedling stage is permanently stunted. The stems of these plants never recover and fail to produce wheat seed, although tillering sometimes allows these plants to compensate for the damage done to the main stem (Stuart et al. 2012). Plants attacked at the jointing stage are permanently weakened and produce fewer and smaller seeds (Hatchett et al. 1987).

The most effective and economical method of Hessian fly control is the planting of Hessian fly–resistant cultivars (Ratcliffe and Hatchett 1997). These plants carry single dominant resistance (*R*) genes that trigger a resistance reaction in the plant, which kills first instar larvae as they attempt to feed. Over 30 different *R* genes have been discovered (Sardesai et al. 2005). Unfortunately, only a handful of these are truly effective where Hessian fly is the greatest problem because of the evolution of Hessian fly biotypes (genotypes) that are capable of living (are virulent to) cultivars carrying most *R* genes. Hessian fly biotype evolution has been the major focus of genetic investigations of the insect (Stuart et al. 2012). Ongoing investigations are discovering the genes and mutations that allow the insect to overcome the resistance conferred by several different *R* genes.

The Hessian fly is a member of one of the largest families within the Diptera, the Cecidomyiidae (gall midges) (Roskam 2005). Cytological investigations of the Cecidomyiidae date back to the early 1900s. The earliest investigations focused on understanding the anomalous parthenogenic form of reproduction (paedogenesis) found in the most primitive species within the group, in which female larvae give rise to offspring without maturing to an adult stage (White 1973; Kloc 2008; Stuart et al. 2012). These and later investigations focused on the unusual segregation of chromosomes during both oogenesis and spermatogenesis and the discovery of polytene chromosomes in the salivary glands of certain species. For the most recent and comprehensive review of these investigations the reader should refer to Matuszewski (1982).

Metcalfe (1935) performed the first investigation of Hessian fly cytology. She recognized the unusual number of chromosomes and the elimination of chromosomes during critical developmental stages, but her conclusions regarding the details of these events were incorrect. Bantock (1961) clarified the timing of chromosome elimination during Hessian fly embryogenesis and more properly established the germ line and somatic chromosome numbers. He was the first to show that the eliminated chromosomes are essential for Hessian fly (gall midge) fertility (Bantock 1970). The first genetic experiments were performed at about the same time (Gallun and Hatchett 1969; Hatchett and Gallun 1970). These were the first to demonstrate the existence of a gene-for-gene relationship (Flor 1956) between the Hessian fly and wheat (Triticum spp.). The discovery of polytene chromosomes in the salivary glands of the Hessian fly (Figure 2.1) and the relationship

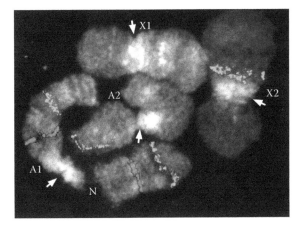

FIGURE 2.1 **(See color insert.)** The Hessian fly salivary gland polytene chromosomes (8 S chromosomes). Shown is an example of in situ hybridization of Hessian fly polytene chromosomes A1, A2, X1, and X2. As in most dipteran genomes, the Hessian fly diploid chromosome number ($2n = 8$) is low and the homologs are often paired in diploid polytene nuclei, as they are here. Four biotin-labeled BAC clones (green) and two digoxigenin-labeled BAC clones (red) are visible on the chromosomes. The position of the nucleolus (N) on chromosome A1 is indicated. Centromeric heterochromatin is visible as brighter staining DNA near constrictions (arrows) that correspond to the chromosome centromeres.

between chromosome behavior and sex determination were not established until nearly two decades later (Stuart and Hatchett 1987, 1988a, b; Stuart et al. 2012). (Gallun and Hatchett 1969; Hatchett and Gallun 1970; Stuart and Hatchett 1987, 1988a,b).

2.2 HESSIAN FLY GENOME ORGANIZATION

As we have already alluded, the chromosome cycle of the Hessian fly, like other gall midges, is complicated and unusual (Stuart and Hatchett 1988a,b). Embryos receive a set of four chromosomes from their father. For historical and pragmatic reasons (White 1950, 1973), these four chromosomes are called the "S chromosomes" and are said to be composed of two autosomes (A1 and A2) and two X chromosomes (X1 and X2). Together these chromosomes make up one-half of the chromosomes that are present in somatic tissues. Embryos receive a second set of S chromosomes from their mothers. In addition, they receive 28–32 "E chromosomes." The embryo, therefore, begins its life with a diploid number of S chromosomes and a full complement of E chromosomes. During the early, pre-blastulation, embryonic nuclear divisions, the E chromosomes are eliminated from the cells that give rise to all somatic tissues, but they are retained in the germ line (Bantock 1970). Cytological data indicate that the E chromosomes are extra copies of the S chromosomes (Stuart, J. J., B. J. Schemerhorn, and Y. M. Crane, unpublished data). If this is correct, the Hessian fly has a haploid number of $n = 4$ (equivalent to the haploid S chromosome number and the number of chromosomes in the sperm) and the germ line is decaploid ($10n = 40$), although, the precise numbers of chromosomes present in the germ line varies between individuals. The mechanisms that retain the integrity of the strictly maternally inherited E chromosomes are unknown.

Sex is determined during the embryonic chromosome elimination events (Stuart and Hatchett 1991; Benatti et al. 2010). Males arise when the paternally derived X1 and X2 chromosomes are eliminated with the E chromosomes during embryogenesis, leaving the somatic cells diploid for the autosomes, but haploid for the X chromosomes (A1A2X1X2/A1A2OO). Females arise when both copies of the X1 and X2 chromosomes are retained in the soma (A1A2X1X2/A1A2X1X2). Maternal genotype controls whether the X1 and X2 chromosomes are eliminated during embryogenesis. Thus, female Hessian flies produce unisexual families and are usually either female-producers or male-producers. Interestingly, female-producers are heterozygous for a small A1 inversion that male-producers always lack (Benatti et al. 2010), and Hessian fly populations are composed of three sexual forms: female-producing females (Z/W'; where Z and W' represent the noninverted and inverted A1 chromosomes, respectively), male-producing females (Z/Z), and males (Z/Z). Matings between female-producing females (Z/W') and males (Z/Z) produce female-producing (Z/W') and male-producing (Z/Z) females in equal proportions. Matings between male-producing females (Z/Z) and males (Z/Z) produce only males.

Oogenesis is unusual in that the S chromosomes pair and form chiasma, whereas the E chromosomes remain decondensed (Stuart and Hatchett 1988a). The S chromosomes then segregate "normally," so that only a haploid set are present in each ova. Spermatogenesis is also unusual; the S chromosomes do not pair and only the maternally inherited S chromosomes segregate into the primary spermatocyte. The primary spermatocytes then undergo a mitotic division to form the secondary spermatocytes, whereas the remainder of the chromatin is retained in a residual cell. Two sperms are therefore formed from each spermatogonial cell, and all sperms carry only the maternally derived S chromosomes.

2.3 HESSIAN FLY AS AN EXPERIMENTAL MODEL

The life cycle and the unusual chromosome cycle of the Hessian fly offer a few advantages for genetic experimentation (Stuart et al. 2012), and make the Hessian fly a genetic model for studies of plant–insect interactions and plant gall formation. The Hessian fly can be reared as families of single females on small caged pots (8–10 cm in diameter) containing wheat seedlings. Thus, unlike most gall midges, Hessian flies can be easily reared in a small space. They are also easily reared at room temperature (18°C–24°C) in the greenhouse or the laboratory where the life cycle from egg to adult is only 28–32 days. Third instar Hessian fly larvae enter a diapause when placed at 4°C for 10 days, and they can be conveniently maintained in this diapause for several months. Simply bringing the larvae back into a constant room temperature breaks this diapause. Genetic experiments are preferably conducted in growth chambers where environmental conditions are maintained at a constant temperature (20°C ± 2°C) and light regimen (12:12 hours, light:dark). Because the families that develop on the plants in these pots are typically of only one sex, virgin females are easily collected for experimental matings from these cages. Genes on the X1 and X2 chromosomes, which together compose nearly 40% of the genome, are more easily isolated and mapped using haploid males.

2.4 REVIEW OF GENETIC AND PHYSICAL MAPS

Two additional factors make the Hessian fly a good genetic model for plant–insect interactions: a small genome (158 Mb) and polytene chromosomes in the larval salivary glands (Stuart et al. 2012). These have allowed the construction of deep bacterial artificial chromosome (BAC) libraries and the ability to determine the relative positions of DNA fragments directly on the chromosomes (Behura et al. 2004; Stuart et al. 2008). This has permitted two complementary molecular approaches to gene mapping in the Hessian fly genome: (1) begin by genetically mapping markers and then determining their positions on the chromosomes (Behura et al. 2004) or (2) begin by physically positioning large DNA fragments on the chromosomes and then genetically map markers developed from those sequences later (Aggarwal et al. 2009). Approximately 60% of the Hessian fly genome has been positioned on the chromosomes using the later method in combination with contigs consisting of BAC clones. The ability to develop markers within this portion of the genome will soon be significantly improved with the complete sequencing and assembly of the Hessian fly genome.

A major improvement in the ability to tie Hessian fly contigs together was the application of Fiber–fluorescent in situ hybridization (FISH) (Fransz et al. 1996; Heng and Tsui 1998; Cheung et al. 2001). This procedure applies FISH to extended DNA fibers (Florijn et al. 1995; Fransz et al. 1996). The targeted fibers can be derived from nuclear DNA or cloned fragments such as BACs. The procedure can detect small genomic rearrangements, measure with high resolution the lengths of contiguous DNA fragments, and physically order DNA sequences that are on the same chromosomes. In the Hessian fly, the technique complements both molecular genetic mapping and physical mapping of DNA sequences to the polytene chromosomes (Lobo et al. 2006).

2.5 CHROMOSOME PREPARATIONS

The following preparations prepare chromosomes that can be observed without staining using phase contrast optics. They can also be stained with Giemsa or orcein and observed using standard light microscopy, or stained with a fluorescent dye, such as 4,6-diamidino-2-phenylindole (DAPI), and observed using a fluorescent microscope. These preparations are also suitable for FISH.

2.5.1   Mitotic S Chromosomes

1. Dissect cerebral ganglia from 7- to 9-day-old second instar larvae in Ringer's solution (6.50 g NaCl, 0.42 g KCl, 0.27 g $CaCl_2$ dissolved in 1 L sterile distilled water) under a stereomicroscope using 21-gauge syringe needles.
2. Use the needles to transfer the dissected ganglia to ~50 μL hypotonic solution (0.1 g colchicine, 0.8 g NaCl, 0.02 g $CaCl_2$, 0.02 g KCl, 0.02 g $NaHC_3$ dissolved in 200 mL sterile distilled water) in a 1.5-mL Eppendorf tube and allow the cells to sit in this solution for 20 minutes.
3. Raise the volume in the tube to ~1 mL with freshly prepared fixative (ethanol:glacial acetic acid, 3:1) and allow the cells to fix in this solution for 10 minutes.
4. Gently spin the tube in a microcentrifuge to move the ganglia to the bottom of the tube.
5. Remove the fixative, and add 1 mL fresh fixative and allow the ganglia to sit in this solution for another 10 minutes.
6. Gently spin the tube in a microcentrifuge to move the ganglia to the bottom of the tube. Remove as much fixative as possible without allowing the ganglia to dry.
7. Add 50% glacial acetic acid to the tube to disperse the cells. Use 10–20 mL dilute acid per ganglia.
8. Pipette small drops of the solution onto clean glass microscope slides and allow the drops to dry in a clean environment at room temperature.

2.5.2   Mitotic S + E Chromosomes

1. Dissect gonads from 12- to 15-day-old third instar larvae in Ringer's solution under a stereomicroscope using 21-gauge syringe needles.
2. Proceed from steps 2 through 6 as described in Section 2.5.1.
3. Add 50% glacial acetic acid to the tube to disperse the cells. Use 5–15 mL dilute acid per gonad.
4. Pipette small drops of the solution onto clean glass microscope slides and allow the drops to dry in a clean environment at room temperature.

2.5.3   Salivary Gland Polytene (S) Chromosomes

1. Dissect salivary glands from early to mid-staged second instar larvae (8–10 days post egg hatch) in 45% acetic acid in a well slide under a stereomicroscope using 21-gauge syringe needles. Glands from several larvae

should be collected for each slide. Note: the polytene X chromosomes of male larvae have a decondensed morphology that prevents one from distinguishing the regions of the chromosomes. We collect the larvae from families of individual females and sex the entire family by a chromosome preparation with a few larvae collected from that family. We then continue making preparations only from the all-female families.

2. Wash the salivary gland material by extracting the dissecting solution and replacing it with clean 45% acetic acid.

3. Place a single drop (~12 μL) of lactic–acetic acid (1 part lactic acid, 2 parts double distilled water [ddH$_2$O], and 3 parts glacial acetic acid) in the middle of a clean glass microscope slide. Transfer the salivary gland cells to this drop using a 10-μL pipette. Transfer as little solution with the glands as possible.

4. Allow the glands to sit in the lactic–acetic acid for about 5 minutes. Then lower an 18 × 18-mm coverslip over the drop on the slide taking care to prevent air bubbles from forming under the coverslip.

5. Carefully tap the edges of the slide against a piece of paper towel on the bench top. This knocks the chromosomes free of the nuclear membrane and the cytoplasm.

6. Check the condition of the chromosomes periodically under phase contrast microscope (10 × and 40 ×), but continue tapping until suitable spreading of the chromosomes is obtained.

7. Place the slide upside down on a piece of bibulous paper or paper towel. Place another piece of absorbent paper over the bottom of the slide. Place a large rubber stopper (size 10 works well) upside down directly over the position of the coverslip on the slide. Using the ball of your hand, bear down on the rubber stopper with your weight to flatten the chromosomes. Take care not to move the coverslip—this will roll the chromosomes.

8. To flatten the chromosomes further, place the slide, coverslip down, between two paper towels sandwiched between two glass plates and leave a weight (brick) on top of the upper glass slide for at least 6 hours. If the slides are left for too long, the preparation may dry up. However, slides can be left to flatten overnight if the "sandwich" is performed at 4°C.

9. Freeze the coverslip to the slide in liquid nitrogen or on a block of dry ice. Immediately flick off the coverslip with a razor blade and plunge the slide into 95% ethanol. Leave in 95% ethanol for at least 1 hour; it will not hurt to leave the slide overnight.

10. Air-dry in a dust-free atmosphere. The preparation is now stable and can be stored indefinitely, preferable at 4°C, in a dry, dust-free atmosphere.

2.5.4 Spermatogonial Preparations

1. Identify all-male families of Hessian fly larvae by making polytene chromosome preparations as described in Section 2.5.3.
2. Collect 12- to 18-day-old male third instar larvae (16–22 days post egg hatch).
3. Dissect testes in Ringer's solution using 21-gauge syringe needles under a stereoscope.
4. Transfer the testes to a 1.5-mL Eppendorf tube containing ~50 μL hypotonic solution (0.8 g NaCl, 0.02 g CaCl$_2$, 0.02 g KCl, and 0.02 g NaHC$_3$ dissolved in 200 mL sterile distilled water) and allow the cells to sit in this solution for 5 minutes. Note: this hypotonic solution contains no colchicine.
5. Proceed from steps 3 through 6 as described in Section 2.5.1.
6. Add 50% glacial acetic acid to the tube to disperse the cells. Use 5–15 mL dilute acid per testis.
7. Pipette small drops of the solution onto clean glass microscope slides and allow the drops to dry in a clean environment at room temperature.

2.6 FLUORESCENCE IN SITU HYBRIDIZATION

2.6.1 Preparing Slides for Hybridization

1. Heat slides prepared as described earlier for 30 minutes in 2 × saline sodium citrate (SSC) at 65°C.
2. Wash slides for 2 minutes in 2 × SSC at room temperature.
3. Dehydrate in ethanol series of 5-minute washes: 70%, 90%, 95%, and 100%.
4. Allow the slides to air-dry.
5. Denature chromosomes by incubating in freshly prepared 0.07 N NaOH for 3 minutes (0.7 g in 250 mL ddH$_2$O or 0.35 mL 10 N NaOH in 50 mL ddH$_2$O).
6. Wash slides in 2 × SSC for 5 minutes.
7. Dehydrate in ethanol series of 5-minute washes: 70%, 90%, 95%, and 100% ethanol.
8. Air-dry thoroughly. At this point, the slides are ready to hybridize.

2.6.2   Probe Preparation Using Nick Translation

1. Preparation of BAC clone DNA—PSI Clone BAC DNA Kit for 3–5 mL cultures (Cat. no. PP-120; Princeton Separations, Adelphia, NJ)
2. Nick translation reactions for biotin probes. Suggested chemicals: Nick Translation Kit (Cat. No. 42803; Enzo Life Sciences, Ann Arbor, MI)

10× polymerase buffer	5 μL
0.5 mM dNTP	5 μL
biotin-dUTP mix	5 μL
1:5 DNase I	1 μL
DNA polymerase I	1 μL
DNA template (~1 μg)	10 μL
ddH$_2$O	23 μL
Total	50 μL

3. Nick translation reactions for digoxygenin probes:

Digoxygenin nick translation mix (Roche Diagnostics, Indianapolis, Indiana)	4 μL
DNA template (~1 μg) and ddH$_2$O	16 μL
Total	20 μL

4. Incubate reactions at 15°C for 2 hours.
5. Stop reactions by adding 1/10th volume of 0.2 M EDTA.
6. Clear reactions with QIAquick Nucleotide Removal Kit (Qiagen, Valencia, CA) and elute DNA with 50 μL elution buffer.
7. Use 1–2 μL probe (20–40 ng) in the hybridization mixture described in Section 2.6.3.

2.6.3   Hybridization

1. Probe mixture per slide: Dry each probe DNA (20–40 ng each) and dissolve in 1μL ddH$_2$O.

Hybridization mixture	
Probe DNA (20–40 ng)	x μL
10 mg/mL salmon sperm DNA	1 μL
20 × SSC	1 μL
Deionized formamide	5 μL
50% dextran sulfate	2 μL
Total	10 μL

2. Denature probe mixture at 80°C–90°C for 5–15 minutes.

3. Place the mixture on ice.
4. Place 10 µL probe mixture on slide and cover with 22 × 22-mm coverslip. Seal with rubber cement.
5. Place slide in a humid box at 37°C overnight.

2.6.4   Detection

1. Gently remove the rubber cement.
2. Place sides into 2 × SSC at room temperature to allow coverslip to fall away.
3. Wash in 2 × SSC at room temperature for 5 minutes.
4. Wash in 2 × SSC at 42°C for 10 minutes.
5. Wash in 2 × SSC at room temperature for 5 minutes.
6. Wash in 1 × phosphate buffered saline (PBS) at room temperature for 5 minutes.
7. Incubate with 100 µL antibody mix under a 22 × 44-mm coverslip for 30 minutes at 37°C.

Antibody mix per slide	
5× Antibody buffer (5× PBS, 5% BSA)	20 µL
Rhodamin-conjugated antidigoxigenin (Roche)	1 µL
Alexa Fluor488 conjugated antibiotin	1 µL
ddH$_2$O	78 µL
Total	100 µL

8. Wash three times with 1× PBS at room temperature for 5 minutes each time.
9. Add 12 µL Vectashield mounting medium with DAPI counterstain (Vector Labs, Burlingame, CA), cover with a 22 × 40-mm coverslip and gently squash.
10. Take digital images using a compound microscope with ultraviolet optics, a mounted digital camera, and a compatible computer running MetaMorph (Universal Imaging, West Chester, PA) imaging software.

2.6.5   Solutions

PBS: 130 mM NaCl, 7 mM Na$_2$HPO$_4$-2H$_2$O, 3 mM NaH$_2$PO$_4$-2H$_2$O

TMN1: 100 mM Tris-HCl (pH 7.5), 100 mM NaCl, 2 mM MgCl$_2$, 0.05% Triton X-100

2% BSA in TMN1, Triton X-100: Dissolve 2 g BSA/100 mL TMN1, Triton X-100 buffer

2 × SSC: Dissolve 17.53 g NaCl and 8.82 g sodium citrate in 1 L distilled water.

2.7   FIBER-FISH

The fiber-FISH procedure described by Jackson et al. (1998) was adapted for Hessian fly. Here, we have divided the protocol into four major steps: isolation of insect nuclei, preparation of DNA fibers, probe hybridization, and signal or probe detection.

2.7.1   Isolation of Hessian Fly Nuclei

1. To prepare nuclei for Fiber-FISH, take 2–5 second and third instar larvae and ground to a fine powder in liquid nitrogen with a precooled mortar and pestle. We observed that the quality of DNA fibers, which largely depends on the quality of nuclei, is better when using early larval instar tissue as compared to the adults.

2. Transfer powder to a 50-mL centrifuge tube and add 10 mL chilled nuclei isolation buffer (NIB) (10 mM Tris-HCl pH 9.5, 10 mM EDTA, 100 mM KCl, 0.5 M sucrose, 4 mM spermidine, 1 mM spermine, and 0.1% 2-mercaptoethanol). Stock solution of NIB can be prepared and stored at 4°C. 2-Mercaptoethanol should be added fresh just before use and should not be included in the stock. Mix very gently to break up any clumps that might have formed. Place the tube in an ice bucket on a gentle shaker and incubate for 5 minutes.

3. Filter the solution through a series of progressively smaller nylon meshes, beginning with a 250-μm mesh and proceeding through a 149-μm, a 49-μm, and finally a 20-μm mesh (Small Parts Inc., Miami Lakes, FL), into ice-cold 50-mL centrifuge tubes using a chilled funnel.

4. Add 1mL NIB containing 10% (v/v) Triton X-100 and gently mix the filtrate. Centrifuge at 2000×g for 10 minutes at 4°C. Decant the supernatant. If the resulting pellet is very small, further cleaning steps (5 through 7) can be skipped.

5. Resuspend the nuclei pellet in 10 mL NIB (with 2-mercaptoethanol).

6. Filter through 49-and 30-μm nylon meshes sequentially, as in step 3.

7. Gently mix the filtrate with 1 mL NIB containing 10% Triton X-100, and centrifuge at 2000×g for 10 minutes at 4°C.

8. Decant the supernatant and resuspend the pellet in 1–5 mL solution containing 1:1 NIB (without 2-mercaptoethanol and Triton X-100):glycerol. Store at −20°C for at least 24 hours for nuclei and cellular debris to settle at bottom of the tube. Nuclei will generally be settled as a thin layer on top of the debris.

2.7.2 Preparation of Extended DNA Fibers

1. Carefully pipette 1–5 µL of prepared nuclei suspension into 80 µL NIB (without 2-mercaptoethanol and Triton X-100) in an Eppendorf tube to dilute the glycerol. Depending on the concentration of the suspension, this volume can be used for making one to five DNA fiber slides. Gently mix the solution and centrifuge at 3000×g for 5 minutes. Carefully remove the supernatant with a pipette.
2. Resuspend the pellet in 2.5 µL PBS (10 mM sodium phosphate, pH 7.0; 140 mM NaCl) per slide.
3. Pipette 2.5 µL suspension in a line across one end of a clean poly-L-lysine glass microscope slide (Sigma-Aldrich, St. Louis, MO) and allow to air-dry until the solution appears sticky (5–10 minutes). Overdrying the suspension would prevent efficient lysis of the nuclei. Poly-L-lysine-treated slides helps in better adhesion of the DNA molecule.
4. Pipette 8 µL Sodium dodecyl sulfate, Tris, EDTA (STE) lysis buffer (0.5% SDS, 5 mM EDTA, 100 mM Tris, pH 7.0) on top of the nuclear suspension and incubate for 4 minutes at room temperature.
5. Slowly drag the solution down the surface of the slide with the edge of a clean coverslip held just above the slide's surface and without touching the glass slide. Air-dry for 10 minutes at room temperature.
6. Fix in fresh 3:1 100% ethanol:glacial acetic acid for 2 minutes using Coplin jar.
7. Bake the slide at 60°C for 30 minutes. Slides are best when used immediately but can be stored for several weeks.

2.7.3 Probe Labeling and Hybridization

1. DNA probe preparation and hybridization procedures are the same as in the regular FISH protocol; labeling DNA with either biotin- or digoxigenin-conjugated dUTP by nick translation.
2. Apply 20 µL hybridization solution (10% dextran solution, 2 × SSC, 20 µg Herring sperm DNA, and 50% formaldehyde) to each slide, cover with a 22 × 40-mm coverslip and seal with rubber cement.
3. After the cement dries, place the slide on a heated surface at 80°C for 5 minutes. Place the slide in a prewarmed humid chamber and hybridize overnight at 37°C.

2.7.4   Probe Detection

Detection procedure for fiber-FISH signals differs only slightly from regular FISH. As compared to single-layer detection in most regular FISH protocols, we recommend using three layers of antibodies for the detection of biotin-labeled probes to amplify the green signal. Detection of digoxigenin-labeled probes can be performed with two or three layers of antibodies. Each antibody layer hybridization step requires 100 µL hybridization solution covered with 22 × 40-mm coverslip and incubation in a 37°C humid chamber for at least 30 minutes and followed by three washes in 1 × TNT (0.1 M Tris-HCl, 0.15 M NaCl, 0.05% Tween 20, pH 7.5) for 5 minutes each.

ACKNOWLEDGMENTS

This work was supported by a NIFA grant 2010-03741 to Jeff J. Stuart. The authors recognize the technical support provided by Sue Cambron and advice from Dr. Scott A. Jackson, Dr. Ming-Shun Chen, and Dr. Marion O. Harris.

REFERENCES

Aggarwal, R., T. Benatti, N. Gill, C. Zhao, M.-S. Chen, B. Schemerhorn, J. P. Fellers, and J. J. Stuart. 2009. A BAC-based physical map of the Hessian fly genome anchored to polytene chromosomes. *BMC Genomics* **10**:293.

Bantock, C. 1961. Chromosome elimination in cecidomyidae. *Nature* **190**:466–467.

Bantock, C. R. 1970. Experiments on chromosome elimination in the gall midge, *Mayetiola destructor*. *J. Embryol. Exp. Morph.* **24**:257–286.

Behura, S. K., F. H. Valicente, S. D. Rider Jr., M. Shun-Chen, S. Jackson, and J. J. Stuart. 2004. A physically anchored genetic map and linkage to aviurlence reveals recombination suppression over the proximal region of Hessian fly chromosome A2. *Genetics* **167**:343–355.

Benatti, T., F. H. Valicente, R. Aggarwal, C. Zhao, J. G. Walling, M.-S. Chen, S. E. Cambron, B. J. Schemerhorn, and J. J. Stuart. 2010. A neo-sex chromosome that drives postzygotic sex determination in the Hessian fly (*Mayetiola destructor*). *Genetics* **184**:769–777.

Cheung, V., N. Nowak, W. Jang, I. R. Kirsch, S. Zhao, X.-N. Chen, T. S. Furey et al. 2001. Integration of cytogenetic landmarks into the draft sequence of the human genome. *Nature* **409**:953–958.

Flor, H. H. 1956. The complementary genic systems in flax and flax rust. *Adv. Genet.* **8**:29–54.

Florijn, R. J., L. A. J. Bonden, H. Vrolijk, J. Wiegant, J.-W. Vaandrager, F. Bass, J. T. d. Dunnen, H. J. Tanke, G.-J. B. van Ommen, and A. K. Raap. 1995. High-resolution DNA Fiber-FISH for genomic DNA mapping and colour bar-coding of large genes. *Hum. Mol. Genet.* **4**:831–836.

Fransz, P. F., C. Alonso-Blanco, T. B. Liharska, A. J. M. Peeters, P. Zabel, and J. H. de Jong. 1996. High-resolution physical mapping in *Arabidopsis thaliana* and tomato by fluorescence in situ hybridization to extended DNA fibres. *Plant J.* **9**:421–430.

Gallun, R. L. and J. H. Hatchett. 1969. Genetic evidence of elimination of chromosomes in the Hessian fly. *Ann. Entomol. Soc. Am.* **62**:1095–1101.

Harris, M. O., T. P. Freeman, O. Rohfritsch, K. G. Anderson, S. A. Payne, and J. A. Moore. 2006. Virulent Hessian fly (Diptera: Cecidomyiidae) larvae induce a nutritive tissue during compatible interactions with wheat. *Ann. Entomol. Soc. Am.* **99**:305–316.

Harris, M. O., J. J. Stuart, M. Mohan, S. Nair, R. J. Lamb, and O. Rohfritsch. 2003. Grasses and gall midges: Plant defense and insect adaptation. *Annu. Rev. Entomol.* **48**:549–577.

Hatchett, J. H. and R. L. Gallun. 1970. Genetics of the ability of the Hessian fly, *Mayetiola destructor*, to survive on wheats having different genes for resistance. *Ann. Entomol. Soc. Am.* **63**:1400–1407.

Hatchett, J. H., K. J. Starks, and J. A. Webster. 1987. Insect and mite pests of wheat. In *Wheat and Wheat Improvement*, E. G. Heyne (ed.). American Society of Agronomy, Madison, WI.

Heng, H. H. and L. C. Tsui. 1998. High resolution free chromatin/DNA fiber fluorescent in situ hybridization. *J. Chromatogr. A.* **806**:219–229.

Jackson S. A., M. L. Wang, H. M. Goodman, J. Jiang. 1998. Application of fiber-FISH in physical mapping of Arabidopsis thaliana. *Genome* **41**:566–72.

Kloc, M. 2008. Basic science B.D. (before *Drosophila*): cytology at Warsaw University (Poland). *Int. J. Dev. Biol.* **52**:115–119.

Lobo, N. F., S. K. Behura, R. Aggarwal, M.-S. Chen, C. A. Hill, F. H. Collins, and J. J. Stuart. 2006. Genomic analysis of a 1 Mb region near the telomere of Hessian fly chromosome X2 and avirulence gene *vH13*. *BMC Genomics* **7**:7.

Matuszewski, B. 1982. Diptera I. Cecidomyiidae. In *Animal Cytogenetics, Volume 3, Insecta*, B. John (ed.), pp 1–140. John Gebrüder Burtraeger, Berlin, Germany.

Metcalfe, M. E. 1935. The germ cell-cycle of *Phytophaga destructor* Say. *Quart. J. Microscopical Soc.* **77**:585–606.

Pauly, P. J. 2002. Fighting the Hessian fly: American and British responses to insect invasion, 1776–1789. *Environ. Hist.* **7**:485–507.

Ratcliffe, R. H. and J. H. Hatchett. 1997. Biology and genetics of the Hessian fly and resistance in wheat. In *New Developments in Entomology*, K. Bondari (ed.), pp. 47–56. Research Signpost, Scientific Information Guild, Trivandurm, India.

Roskam, H. C. 2005. Phylogeny of gall midges (Cecidomyiidae). In *Biology, Ecology, and Evolution of Gall-Inducing Arthropods*, A. Raman, C. W. Schaefer, and T. M. Withers (eds.), pp. 307–309. Science Publishers, Enfield, CT.

Sardesai, N., J. A. Nemacheck, S. Subramanyam, and C. E. Williams. 2005. Identification and mapping of *H32*, a new wheat gene conferring resistance to Hessian fly. *Theor. Appl. Genet.* **111**:1167–1173.

Stuart, J. J., M.-S. Chen, and M. Harris. 2008. Hessian fly. In *Genome Mapping and Genomics in Arthropods*, W. Hunter and C. Kole (eds.), pp. 93–102. Springer-Verlag, Berlin, Germany.

Stuart, J. J., M. S. Chen, R. Shukle, and M. O. Harris. 2012. Gall midges (Hessian flies) as plant pathogens. *Annu. Rev. Phytopathol.* **50**:339–357.

Stuart, J. J. and J. H. Hatchett. 1987. Morphogenesis and cytology of the salivary gland of the Hessian fly, *Mayetiola destructor* (Diptera: Cecidomyiidae). *Ann. Entomol. Soc. Am.* **80**:475–482.

Stuart, J. J. and J. H. Hatchett. 1988a. Cytogenetics of the Hessian fly: II. Inheritance and behavior of somatic and germ-line-limited chromosomes. *J. Hered.* **79**:190–199.

Stuart, J. J. and J. H. Hatchett. 1988b. Cytogenetics of the Hessian fly: I. Mitotic karyotype analysis and polytene chromosome correlations. *J. Hered.* **79**:184–189.

Stuart, J. J. and J. H. Hatchett. 1991. Genetics of sex determination in the Hessian fly, *Mayetiola destructor*. *J. Hered.* **82**:43–52.

White, M. J. D. 1950. *Cytological Studies on Gall Midges (Cecidomyidae)*, pp. 5–80. University of Texas Publication, Austin, TX.

White, M. J. D. 1973. *Animal Cytology and Evolution*. Ed. 3. Cambridge University Press, London.

3

Tsetse Flies (Diptera)

Anna R. Malacrida, Francesca Scolari,
Marco Falchetto, and Serap Aksoy

CONTENTS

LIST OF ABBREVIATIONS

AAT, animal African trypanosomiasis
CAD, carbamoyl-phosphate synthetase 2, aspartate transcar-
   bamylase, dihydroorotase)
CCD, charge-coupled device
COI, cytochrome oxidase I
DAPI, 4′,6-diamidino-2-phenylindole

EDTA, ethylenediaminetetraacetic acid
FACS, fluorescence-activated cell sorting
HAT, human African trypanosomiasis
PBS, phosphate buffered saline
SSC, saline sodium citrate

3.1 INTRODUCTION

3.1.1 Taxonomy

Glossina is the only genus in the family Glossinidae (Brues et al. 1954; Pollock 1971). Glossinidae are placed in the superfamily Hippoboscoidea, earlier proposed as Glossiniodea (Hennig 1971), and pertain to the dipteran group Calyptratae (Nirmala et al. 2001). Molecular analyses (Gooding et al. 1991) provided support for the monophyly of the three groups, that is, *morsitans*, *palpalis*, and *fusca*. The phylogeny of the Hippoboscoidea, including seven *Glossina* species, was estimated using two mitochondrial (cytochrome oxidase I [COI] and 16S rRNA) and two nuclear (Carbamoyl-phosphate synthetase 2, Aspartate transcarbamylase, Dihydroorotase [CAD] and 28S rDNA) markers (Petersen et al. 2007), confirming the monophyly of the Glossinidae. Gas chromatographic analysis of cuticular alkenes–derived phenetic relationships between 26 species and subspecies (Carlson et al. 1993), again supporting the three groups. However, species considered to belong to the *fusca* group (i.e., *Glossina longipennis*, *G. medicorum*, and *G. nigrofusca nigrofusca*) appeared to be mixed within the *palpalis* clade, possibly because of convergent environmental adaptation as occurring in other disease vectors (Maingon et al. 2003). The taxonomic position of other species is also uncertain (Gooding and Krafsur 2005). On the basis of classical taxonomy using characters of the male genitalia, *G. austeni* was placed in the *morsitans* group. However, female genital characters are shared with the *fusca* group and their ecology is very similar to that of the *palpalis* group (Gooding and Krafsur 2005). Enzyme analysis placed *G. austeni* as a sister group of *morsitans*, and DNA sequence data indicate that *G. austeni* is more closely related to the subspecies of the *G. morsitans* than to the species of the *palpalis* subgroup (Gooding et al. 1991). In addition to the uncertainties surrounding the validity of the subgeneric groupings, there are a number of taxa of uncertain taxonomic status at the species/subspecies level. Within the *palpalis* group, there are five taxa originally accorded subspecific status by Machado (1954). Even within these subspecies there is evidence for possible cryptic species (Gooding et al. 2004). Similarly within the *morsitans* subgroup there are three subspecific forms within the nominal taxon (Machado 1970), although Krafsur and Endsley (2006) used microsatellite data to argue the elevation of the three subspecies of *G. morsitans* (*G. morsitans morsitans*, *G. morsitans submorsitans*, and *G. morsitans centralis*) to specific status.

3.1.2 Importance of the Species

Tsetse flies include 33 species and subspecies, two of them being restricted to sub-Saharan Africa (Gooding and Krafsur 2005). The adults of both sexes are strictly hematophagous and these species are important vectors of two debilitating diseases human African trypanosomiasis (HAT), sleeping sickness, and animal African trypanosomiasis (AAT), nagana (Mattioli et al. 2004). Tsetse flies and trypanosomiasis render vast areas of agricultural land unexploitable, especially during the rainy seasons. Although, probably all tsetse species are capable of transmitting pathogenic trypanosomes, only a few (including *G. morsitans morsitans*, *G. morsitans centralis*, *G. pallidipes*, *G. palpalis palpalis*, *G. fuscipes fuscipes*, and *G. tachinoides*) are major vectors of trypanosomes. It is estimated by the World Health Organization that there are currently 10,000–45,000 cases of HAT with 60 million people at risk in 36 countries covering approximately 40% of Africa (almost 10 million square kilometer). After a devastating epidemic in the early twentieth century when a million people died of HAT, the disease almost disappeared from Africa by the 1960s. In the 1990s, another epidemic killed tens of thousands of people. Sustainable management of such diseases poses a formidable challenge to endemic countries and health ministries, which are faced with limited infrastructure and financing. If the decline in the reported HAT cases triggers African governments to abandon their local control efforts, and for funding agencies to relax their disease research priorities, it is certain that epidemics will continue to flare-up in the near future as has happened in the recent past (Aksoy 2011). At present the limited amount of tsetse control conducted is reliant on wide-scale insecticide use involving cattle pour-ons, aerial spraying, or targets (Allsopp 2001). Sterile insect release programs are also proposed for the later stages of control campaigns (Vreysen et al. 2000). These anti-vector measures are reliant on accurate identification of vector species (Gooding and Krafsur 2005). Their relevant medical and economic importance stimulates considerable research into trypanosomes, but the toolbox for disease control is still limited with neither vaccines nor effective and affordable drugs accessible in the near future. Nowadays, thanks to the genomic/transcriptomic/proteomic resources becoming increasingly available, new and/or improved control tools are being developed.

3.1.3   Karyotype and Genome Size

The basic mitotic karyotype of *Glossina* species include two pairs of autosomes (L1, L2) and one pair of sex chromosomes (X, Y), whereas the existence of heterochromatic supernumerary chromosomes (S or B) has been observed in a number of species (Southern 1980). Indeed, the close similarity between tsetse B and Y chromosomes in meiotic behavior and C-band pattern led Amos and Dover (1981) to propose that their B chromosomes originated from the Y. They proposed that the heterochromatic state of the Y and B chromosomes could be a reflection in part of their once common origin, as proposed by Southern and Pell (1973), and in part of the subsequent acquisition of satellite DNA sequences. They also suggest that the B chromosomes have arisen from Y in two stages (Amos and Dover 1981). A recent study has identified the presence of *Wolbachia* fragments inserted into the *G. m. morsitans* Y and B chromosomes, further supporting a possible common evolutionary origin of B and Y (Brelsfoard et al. 2014). Carvalho et al. (2009) do not exclude the alternative evolutionary scenario of Y originating from B. The *Glossina* genome has been recently sequenced and published (International *Glossina* Genome Initiative 2014). The genomic data, together with studies aimed at uncovering the ancestral state of sex chromosomes in related taxa, might soon clarify this key biological question.

The number of supernumerary or B chromosomes varies in different taxa (0–8 in *Glossina sensu stricto*, 8–12 in *Machadomyia*, and 12–22 in *Austenina*) (Gooding 1985; Willhoeft 1997). Functional genes on B chromosomes have not been demonstrated yet. Linkage groups (i.e., genes that are on the same chromosome) are established through standard "three-point-cross" experiments, and the number of linkage groups in a taxon equals the number of chromosomes that have functional genes. The demonstration of only three linkage groups in *G. (N.) p. palpalis* (Gooding and Rolseth 1995), *G. (G.) m. morsitans* (Gooding and Rolseth 1992), and *G. (G.) m. submorsitans* (Gooding and Challoner 1999) is consistent with the cytological information.

The tsetse polytene chromosome maps have been constructed for *G. m. submorsitans* (Gariou-Papalexiou et al. 2002), *G. austeni, G. pallidipes*, and *G. m. morsitans* (Gariou-Papalexiou et al. 2007). The polytene nuclei contain three long polytene elements representing the X chromosomes and the two autosomes (L1, L2), whereas the Y and B chromosomes are not polytenized due to their heterochromatic nature (Gariou-Papalexion et al. 2007). The homology of chromosomal elements between all four species was achieved by comparison of the respective banding patterns. The telomeric and subtelomeric regions were found to be identical in all species. The pericentromeric regions were found to be similar in the X chromosome and the left arm of L1 chromosome (L1L) but different in L2 chromosome and the right arm of L1 chromosome (L1R). The L2 chromosome differs by a pericentric inversion fixed in *G. pallidipes*, *G. m. morsitans*, and *G. m. submorsitans*. The two *morsitans* subspecies appeared to be homosequential and differ only by two paracentric inversions on XL and L2L arm. Interestingly, the relative position of specific chromosome regions was different due to chromosome inversions established during their phylogeny. However, there are regions with no apparent homology between the species, possibly due to the significant intrachromosomal rearrangements occurred following the species divergence.

The genome size of several *Glossina* species (*G. m. morsitans*, *G. pallidipes*, *G. p. palpalis*, *G. fuscipes*) has been investigated using a FACSCalibur flow cytometer (Aksoy et al. 2005). These results indicate that the genome size varies from 500 to 600 Mb in size, and is about 1.5 times the size of the *Drosophila virilis* genome. Interestingly, reassociation kinetics analysis aimed at determining the genome size of *G. p. palpalis* predicted a much larger size estimate of over 7000 Mb. The 35% of the *G. p. palpalis* genome corresponds to foldback DNA, indicating the presence of a large heterochromatic region.

3.1.4 Genome Sequencing Project

The International *Glossina* Genome Initiative (Aksoy et al. 2005) produced the annotated *G. m. morsitans* genome that has been recently published (International Glossina Genome Initiative 2014). Eight satellite papers on genomic and functional biology findings, which reflect the unique biology of this disease vector, have also been published (Tsetse Biology Collection in PLoS NTDs). The assembled *G. m. morsitans* genome is 366 Mb, represented by 13,807 scaffolds with an N50 of 120 Kb, a mean size of 27 Kb and a maximum size of 25.4 Mb. This is roughly twice the size of the *D. melanogaster* genome. Analysis of gene synteny between *Drosophila* and *Glossina* shows that roughly equal amounts of sequence (53 and 54 Mb, respectively) are syntenic in total. However, the actual size of syntenic blocks containing orthologous gene sequences is larger on average (~2.25 times larger) in *Glossina*. The larger syntenic regions in *Glossina* may be attributed to larger introns and an increase in the size of intergenic sequences as a result of possible transposon activity and/or repetitive sequence expansions. The *Glossina* genome is estimated to contain 12,220

protein-encoding genes based on automated and manual anno-
tations. Although this number is slightly less than *Drosophila*,
the average gene size in *Glossina* is almost double that of
Drosophila. The number of exons and their average size is
roughly equivalent in both fly species, but the average intron
size in *Glossina* appears roughly twice that of *Drosophila*. In
total, 9172 (74%) of *Glossina* genes (from 8374 orthologous
clusters) were found to have a diperan ortholog; 2803 genes
(23%) had no ortholog/paralog, and 482 (4%) had a unique
duplication/paralog. The analysis of genes in orthologous gene
clusters across the Diptera shows that 94% (7867/8374) of clusters containing a *Glossina* gene also contained an ortholog
with *Drosophila*.

3.2 PROTOCOLS

3.2.1 Equipment

- Axioplan Zeiss epifluorescence microscope fitted with
  Olympus D70 CCD camera
- Stereomicroscope Leica
- Millipore Milli-Q Plus water purification system
- Techne heat block
- PBI Alfa-10-Plus autoclave
- Scotsman AF80 Ice machine
- Sartorius Extend precision balance

3.2.2 Species Culture

Tsetse flies are reared at $24 \pm 1°C$ with 50%–55% relative humidity, and fed with defibrinated bovine blood every 48 hours using an
artificial membrane system (Moloo 1971).

3.2.3 Mitotic Chromosomes

Mitotic chromosome spreads can be obtained from the brains of
freshly deposited larvae. Larval nerve ganglia are incubated on
a slide in 100 µL 1% sodium citrate for 10 minutes at room temperature, and sodium citrate is replaced with methanol–acetic acid
(3:1 solution) for 4 minutes. The tissue is disrupted by pipetting in
100 µL 60% acetic acid for fixation and the material is spread onto
slides that are heated on a hot plate at 70°C until the acetic acid
evaporated. After dehydration in 80% ethanol, slides are stored at
−20°C for at least 2 weeks.

3.2.3.1 Fluorescent In Situ Hybridization on Mitotic Chromosomes

3.2.3.1.1 Probe Preparation

DNA is extracted from *G. m. morsitans* and polymerase chain reaction (PCR) products obtained are loaded on a 2% agarose gel. Gel-extracted PCR products corresponding to designated gene are cloned into the pCR® 2.1 TOPO® vector (Invitrogen, Carlsbad, CA). Plasmids are transferred into competent cells of *Escherichia coli*. Positive insert–containing colonies are selected and at least three clones per sample are sequenced. One-microgram-sequenced plasmid DNA is labeled using the Biotin High Prime kit (Roche, Basel, Switzerland), as explained in Section 3.2.3.1.2, and stocked at –20°C until use.

3.2.3.1.2 Probe Labeling

Day 1

1. Boil 1 μg DNA (vector or PCR) diluted in 16 μL ddH$_2$O for 10 minutes.
2. Put it in ice for 10 minutes.
3. Add 4 μL of labeling reagent (Biotin High Prime, 11585649910, Roche).
4. Leave it overnight (20–24 hours) at 37°C.

Day 2

5. Stop the reaction by adding 2 μL 0.2 M EDTA (ethylene-diaminetetraacetic acid).
6. Bring the reaction to 65°C for 10 minutes.
7. Add 5 μL H$_2$O, 25 μL 20× SSC (saline sodium citrate), 50 μL deionized formamide.
8. Keep it at –20°C until use.
9. Before use, when the chromosome slides are ready to use, boil the probe 10 minutes and put the Eppendorf tube in ice.

Chromosome slide preparation

10. Put the chromosome slides at around 70°C–80°C for 2 hours.
11. Dehydrate the chromosome on the slides with each ethanol concentration (30%, 50%, 70%, 95%) for 2 minutes.
12. Completely dry the slides in air.
13. Denaturate the chromosome in 0.07 M NaOH at room temperature for 2 minutes.
14. Wash the slides in 0.4× SSC and 0.1% Tween 20 for 10 seconds.
15. Dehydrate the chromosomes on the slides in each ethanol concentration (30%, 50%, 70%, 95%) for 2 minutes.
16. Completely dry it in air.

Hybridization

17. Boil the prepared probe for 10 minutes, spin and put it in ice.

18. Prepare a paper sheet soaked with 2× SSC and cover the bottom of the slide box to maintain the box humid.
19. Put 25 μL of the probe on the slide and put a cover slip on each slide (do not fix the cover slide, it has to be removed).
20. Put the slides horizontally in the box (not vertically) and fix them at a border using Scotch tape. Close the box and seal it using tape.
21. Put it at 37°C overnight (20–24 hours).
    Washing steps and antibody incubation

Day 3

22. Remove the cover slip from the slide in 2× SSC.
23. Wash twice the slide in 0.4× SSC and 0.1 Tween 20 for 5 minutes each.
24. Wash each slide in 1× PBS (phosphate buffered saline).

In the meantime, prepare the anti-biotin antisera: add 16 μL solution A and 16 μL solution B in 1 mL 1× PBS (Vectastain Elite ABC kit standard, PK-6100)

The slides must always be wet.

25. Put 100 μL of the antibody on the slide.
26. Put a large cover slip on the slide (24 × 40 mm).
27. Put the slides in the slide box with paper sheet soaked with 2× SSC and close it.
28. Leave it at room temperature for 1 hour.
29. Remove the cover slip in 2× PBS.
30. Wash twice each slide in PBS for 5 minutes.
31. Wash in PBS + 0.1% Triton for 2 minutes.
32. Leave the slides in PBS.
33. Add 3 μL of Milli-Q Plus water to 200 μL of Amplification buffer (stock solution).
34. Add 5 μL of the stock solution to 495 μL of Amplification buffer (working solution).
35. Add the Tyramide to the working solution and then put 100 μL of this mix on each slide.
36. Put the cover slip (24 × 40 mm) on the slides. The slides must be immediately put into the box to protect them from light.
37. Incubate the box at room temperature for half an hour.
38. Wash twice with Milli-Q Plus water, twice with PBS and leave in PBS until use (to maintain the humidity).
39. Use 4',6-diamidino-2-phenylindole (DAPI) VECTASHIELD solution (Vector Laboratories Inc., Burlingame, CA) or Hoechst stain reagent to mount the slides.
40. Chromosomes analyzed under an epifluorescence Zeiss Axioplan microscope equipped with an Olympus DP70 CCD camera.

3.2.4   Polytene Chromosomes

3.2.4.1   Preparation of Polytene Chromosome Spreads

Polytene chromosome preparations (Gariou-Papalexiou et al. 2002) are made from trichogen cells associated with apical scutellar bristles from pharate adults. Dissections are performed under a stereoscopic microscope in 45% v/v glacial acetic acid.

After removing the first cuticular layer, the scutellum can be isolated and transferred to a drop of 45% v/v glacial acetic acid on a clean coverslip. Then, the second cuticular layer can be removed, and the remaining tissue can be manipulated by a pair of needles and a drop of glacial acetic acid:H_2O:lactic acid (3:2:1) is added. One to two minutes later, a drop of lactic–acetic orcein is added and after 1–2 minutes of staining the tissue can be picked up by placing a clean slide over the coverslip. The slide is then inverted and the coverslip gently moved to spread the polytene chromosomes out of the nuclei. Finally, the slide is lightly blotted to remove the excess staining solution. Gentle tapping and squashing are applied to assess the suitability of the chromosomes for adequate spreading and separation.

3.2.4.1.1   *Construction of Photographic Chromosomes Maps*

Well-spread chromosomes with a clear banding pattern can be photographed under 100 × magnification. To construct chromosome maps, each individual arm is assembled from selected pictures using the Adobe Photoshop CC.

3.3   DISCUSSION

3.3.1   Integration of Cytogenetic, Linkage, and Genome Maps

Cytogenetic maps are able to show the positions of genetically mapped markers on the chromosomes, in relation to cytological landmarks including centromeres, telomeres, heterochromatin, and nucleolar organizer regions. The continuous availability of increasing number of sequenced insect genomes opens new possibilities to integrate cytogenetic and genomic data in comparative and evolutionary studies, but will also provide a unique insight into genome organization in the context of the chromosomes.

High-resolution cytogenetic maps can provide important biological information on the genomic organization, permitting the identification of conserved synteny shared by distinct genomes, and also identify potential mistakes in genome assembly. Moreover, several studies demonstrated that physical mapping has the potential to orient genomic supercontigs (Timoshevskiy et al.

2013). Thus, cytogenetic maps incorporating genetic, cytological, and physical data can significantly contribute to the improvement of sequence assembly, by confirming the positions of markers on the linkage groups, and also helping to evaluate the size of the putative remaining gaps.

3.3.2   Practical and Scientific Benefits of Genome Mapping

The availability of either an annotated genome or chromosomal maps will be important not only for the study of genome evolution but also for applied purposes. As such, this field of knowledge will contribute to significant advances not only in insect comparative field-based studies but also in biomedical and biological research. Availability of *G. m. morsitans* and other tsetse genomes will allow the identification of evolutionarily conserved genes/syntenies, the investigation of evolutionary divergence and mechanisms underlying key biological processes associated with the unique biology of tsetse (feeding, digestion, excretion, and reproduction) and the comparison with other insect with alternative life histories.

Genome mapping of multiple tsetse species will also provide a framework to illuminate the genetic basis of vectorial capacity. Specific knowledge on *Glossina* evolution and vector competence will be obtained from genome comparisons with *Stomoxys, Musca,* and the available *Drosophila* genomes, in addition to those from other dipteran species (mosquitoes and sand flies) currently under investigation.

Knowledge on genes related to host trypanosome resistance mechanisms can be immediately used to generate refractory strains of tsetse to be used in sterile insect technique programs to increase the efficacy of their application in human disease endemic areas (Aksoy et al. 2001).

Availability of mapped high-resolution molecular markers (i.e., microsatellites, single nucleotide polymorphisms, etc.) will allow improved population genomic analyses to investigate population genetic structuring (Beadell et al. 2010; Solano et al. 2010). These markers offer a unique opportunity to undertake genomic scan analyses to understand tsetse's vector competence traits. In addition, when applied to tsetse control, it will be possible to explore tsetse biological traits including multiple mating, sperm use mechanisms, paternity skew, important prerequisites for the feasibility, success, and sustainability of eradication campaigns in the target African areas.

3.3.3 Chromosome Organization and Evolution

The availability of integrated cytogenetic maps will reveal the accumulation of transposable elements and their mobility, but will also localize repetitive DNA chromosomal markers that are very useful for studying species evolution, supernumerary chromosomes, sex chromosomes, and for the identification of chromosomal rearrangements. In particular, sex chromosomes offer a unique possibility to study a wide range of aspects of genome evolution, including the degeneration of nonrecombining genomic portions, the selective pressures that lead to degeneration, and the timescales over which such changes take place. Gene shuffling in response to differential sexual selection, different rates of adaptation depending on the dominance of mutations, and epigenetic modifications affecting chromatin structure are additional topics that can be addressed using sex chromosomes.

Moreover, the evaluation of the transcriptional activities of the *Wolbachia* genes found to be inserted into the *G. m. morsitans* genome (Brelsfoard et al. 2014) will provide novel insights into how host–symbiont gene transfers could contribute to evolution/coevolution.

REFERENCES

Aksoy S (2011). Sleeping sickness elimination in sight: Time to celebrate and reflect, but not relax. *PLoS Negl Trop Dis*; 5:1–3.

Aksoy S, Berriman M, Hall N, Hattori M, Hide W, Lehane MJ (2005). A case for a *Glossina* genome project. *Trends Parasitol*; 21(3):107–111.

Aksoy S, Maudlin I, Dale C, Robinson AS, O'Neill SL (2001). Prospects for control of African trypanosomiasis by tsetse vector manipulation. *Trends Parasitol*; 17:29–35.

Allsopp R (2001). Options for vector control against trypanosomiasis in Africa. *Trends Parasitol*; 17(1):15–19.

Amos A, Dover G (1981). The distribution of repetitive DNAs between regular and supernumerary chromosomes in species of *Glossina* (Tsetse): A two-step process in the origin of supernumeraries. *Chromosoma*; 81:673–690.

Beadell JS, Hyseni C, Abila PP, Azabo R, Enyaru JC, Ouma JO, Mohammed YO, Okedi LM, Aksoy S, Caccone A (2010). Phylogeography and population structure of Glossina fuscipes fuscipes in Uganda: Implications for control of tsetse. *PLoS Negl Trop Dis*; 4(3):e636.

Brelsfoard C, Tsiamis G, Falchetto M, Gomulski LM, Telleria E, Alam U, Doudoumis E et al. (2014) Presence of extensive *Wolbachia* symbiont insertions discovered in the genome of its host *Glossina morsitans morsitans*. *PLoS Negl Trop Dis* 8(4):e2728.

Brues CT, Melander AL, Carpenter FM (1954). Classification of insects. Keys to the living and extinct families of insects, and to the living families of other terrestrial arthropods. *Bull Mus Comp Zool*; 108:1–917.

Carlson DA, Milstrey SK, Narang SK (1993). Genetic classification of the tsetse flies using cuticular hydrocarbons. *Bull Entomol Soc*; 83:507–515.

Carvalho AB, Koerich KL, Clark AG (2009). Origin and evolution of Y chromosomes: *Drosophila* tales. *Trends Genet*; 25:270–277.

Gariou-Papalexiou A, Yannopoulos G, Robinson AS, Zacharopoulou A (2007). Polytene chromosome maps in four species of tsetse flies *Glossina austeni, G. pallidipes, G. morsitans morsitans* and *G. m. submorsitans* (Diptera: Glossinidae): A comparative analysis. *Genetica*; 129(3):243–251.

Gariou-Papalexiou A, Yannopoulos G, Zacharopoulou A, Gooding RH (2002). Photographic polytene chromosome maps for *Glossina morsitans submorsitans* (Diptera: Glossinidae): Cytogenetic analysis of a colony with sex-ratio distortion. *Genome*; 45(5):871–880.

Gooding RH (1985). Electrophoretic and hybridization comparison of *Glossina morsitans morsitans, G. m. centralis* and *G. m. submorsitans* (Diptera: Glossinidae). *Can J Zool*; 63:2694–2702.

Gooding RH, Challoner CM (1999). Genetics of the tsetse fly, *Glossina morsitans submorsitans* Newstead (Diptera: Glossinidae): Further mapping of linkage groups I, II and III. *Can J Zool*; 77:1309–1313.

Gooding RH, Krafsur ES (2005). Tsetse genetics: Contributions to biology, systematics, and control of tsetse flies. *Annu Rev Entomol*; 50, 101–123.

Gooding RH, Moloo SK, Rolseth BM (1991). Genetic variation in *Glossina brevipalpis, G. longipennis,* and *G. pallidipes*, and the phonetic relationships of *Glossina* species. *Med Vet Entomol*; 5, 165–173.

Gooding RH, Rolseth BM (1992). Genetics of *Glossina morsitans morsitans* (Diptera: Glossinidae). XIV. Map locations of the loci for phosphoglucomutase and glucose-6-phosphate isomerase. *Genome*; 35:699–701.

Gooding RH, Rolseth BM (1995). Genetics of *Glossina palpalis palpalis*: Designation of linkage groups and the mapping of eight biochemical and visible marker genes. *Genome*; 38:833–837.

Gooding RH, Solano P, Ravel S (2004). X-chromosome mapping experiments suggest occurrence of cryptic species in the tsetse fly *Glossina palpalis palpalis*. *Can J Zool*; 82:1902–1909.

Hennig W (1971). Insektfossilien aus der unteren Kreide. III. Empidiformia ("Microphorinae") aus der unteren Kreide und aus dem Baltischen Bernstein; ein Vertreter der Cyclorrapha aus der unteren Kreide. Stuttg. Beitr. Naturkd.; 232:1–28.

International *Glossina* Genome Initiative. Genome sequence of the tsetse fly (*Glossina morsitans*): Vector of African trypanosomiasis (2014). *Science*; 344:380–386.

Krafsur ES, Endsley MA (2006). Shared microsatellite loci in *Glossina morsitans sensu lato* (Diptera: Glossinidae). *J Med Entomol*; 43:640–642.

Machado ADB (1954). Révision systématique des glossines du groupe palpalis (Diptera). Museo do Dundo, Companhia de Diamantes de Angola Lisboa, 22.

Machado ADB (1970). Les races géographiques de *Glossina morsitans*. In: *Tsetse Fly Breeding under Laboratory Conditions and Its Practical Application*, ed. JF de Azevedo, pp. 471–486. First International Symposium April 11–23, 1969, Lisbon, Portugal: Junta de Investigacoes do Ultramar.

Maingon RD, Ward RD, Hamilton JG, Noyes HA, Souza N, Kemp SJ, Watts PC (2003). Genetic identification of two sibling species of *Lutzomyia longipalpis* (Diptera: Psychodidae) that produce distinct male sex pheromones in Sobral, Ceara State, Brazil. *Mol Ecol*; 12:1879–1894.

Mattioli RC, Feldmann U, Hendrickx G, Wint W, Jannin J, Slingenbergh J (2004). Tsetse and trypanosomiasis intervention policies supporting sustainable animal-agricultural development. *Food Agr Environ*; 2:310–314.

Moloo SK (1971). An artificial feeding technique for *Glossina*. *Parasitology*; 63(3):507–512.

Nirmala X, Hypsa V, Zurovec M (2001). Molecular phylogeny of Calyptratae (Diptera: Brachycera): The evolution of 18S and 16S ribosomal rDNAs in higher dipterans and their use in phylogenetic inference. *Insect Mol Biol*; 10:475–485.

Petersen FT, Meier R, Kutty SN, Wiegmann BM (2007). The phylogeny and evolution of host choice in the Hippoboscoidea (Diptera) as reconstructed using four molecular markers. *Mol Phylogenet Evol*; 45:111–122.

Pollock JN (1971). Origin of the tsetse flies: A new theory. *J Entomol (B)*; 40:101–109.

Solano P, Kaba D, Ravel S, Dyer NA, Sall B, Vreysen MJ, Seck MT et al. (2010). Population genetics as a tool to select tsetse control strategies: Suppression or eradication of *Glossina palpalis gambiensis* in the Niayes of Senegal. *PLoS Negl Trop Dis*; 4(5):e692.

Southern DI (1980). Chromosome diversity in tsetse flies. In: *Insect Cytogenetics*, eds. RL Blackman, GM Hewitt, M Ashburner, Symposium of the Royal Entomological Society of London; 10:225–43. Oxford: Blackwell Scientific.

Southern DI, Pell PE (1973). Chromosome relationships and meiotic mechanisms of certain morsitans group tsetse flies and their hybrids. *Chromosoma*; 44:319–334.

Timoshevskiy VA, Severson DW, deBruyn BS, Black WC, Sharakhov IV, Sharakhova MV. (2013). An integrated linkage, chromosome, and genome map for the yellow fever mosquito *Aedes aegypti*. *PLoS Negl Trop Dis*; 7(2):e2052.

Vreysen MJB, Saleh KM, Ali MY, Abdulla AM, Zhu Z-R, Juma KG, Dyck VA, Msangi AR, Mkonyi PA, Feldmann HU (2000). *Glossina austeni* (Diptera: Glossinidae) eradicated on the island of Unguja, Zanzibar, using the sterile insect technique. *J Econ Entomol*; 93:123–135.

Willhoeft U (1997). Fluorescence in situ hybridization of ribosomal DNA to mitotic chromosomes of tsetse flies (Diptera: Glossinidae: *Glossina*). *Chromosome Res*; 5:262–267.

4

Mosquitoes (Diptera)

Maria V. Sharakhova, Phillip George,
Vladimir Timoshevskiy, Atashi Sharma,
Ashley Peery, and Igor V. Sharakhov

CONTENTS

LIST OF ABBREVIATIONS

2D, two-dimensional
3D, three-dimensional
AFLP, amplified fragment-length polymorphism
BAC, bacterial artificial chromosome
BSA, bovine serum albumin
cDNA, complementary DNA
cM, centimorgan
DAPI, 4',6-diamidino-2-phenylindole
DDT, dichlorodiphenyltrichloroethane
DEN2, dengue virus 2
DHF, dengue hemorrhagic fever
DMSO, dimethyl-sulfoxid
DSS, dengue shock syndrome
EDTA, ethylenediaminetetraacetic acid
EGTA, ethylene glycol tetraacetic acid
FISH, fluorescent in situ hybridization
IB12, 12 consecutive generations of single pair inbreeding
ID, imaginal disc
IGS, intergenic spacer
iMap, integrated map

JHB, Johannesburg
Kb, kilobases
LCM, laser capture microdissection
LVP, Liverpool
MAR, matrix-associated region
MEB, midgut escape barrier
MIB, midgut infection barrier
MR4, Malaria Research and Reference Reagent Resource Center
Mb, megabases
NaAC, sodium acetate
NBF, neutral buffered formalin
NHGRI, National Human Genome Research Institute
NIAID, National Institute of Allergy and Infectious Diseases
PBS, phosphate buffered saline
PCR, polymerase chain reaction
PEST, Pink Eye Standard
PET, polyethylene terephthalate
PIPES, piperazine-1,4-bis(2-ethanesulfonic acid)
QTL, quantitative trait locus (loci)
RAPD, random amplified polymorphism DNA
rDNA, ribosomal DNA
RFLP, restriction fragment length polymorphism
rpm, revolutions per minute
RT, room temperature
SD, segmental duplication
SNP, single nucleotide polymorphism
SSC, saline sodium citrate
SSCP, single-strand conformation polymorphism
TE, transposable element
TIGR, The Institute for Genomic Research
WGA, whole genome amplification
WHO, World Health Organization

4.1 INTRODUCTION

Mosquito-borne infectious diseases have a devastating impact on human health, and they pose unacceptable risks to public welfare (Tolle 2009). These diseases collectively account for more than a million human deaths per year (World Malaria Report 2010). Mosquito control has been a successful approach for disease elimination, especially in developed counties. Because of economic and practical reasons, vector control in tropical countries mainly relies on the use of synthetic insecticides (Takken and Knols 2009). However, this strategy is often jeopardized by the rapid spread of insecticide multiresistance in major mosquito vector species.

Moreover, mosquito control becomes inefficient if all vector species and populations are not targeted. Genomics is now offering an opportunity to explore novel strategies for vector-based disease control. Thousands of genes can be investigated as potential intervention targets. However, the full realization of the genome sequencing projects will not happen until the majority of sequencing scaffolds are assembled and anchored onto chromosomes. Fragmented, unmapped sequences create serious problems for genomic analyses because unidentified gaps and misassemblies cause incorrect or incomplete annotation of genomic sequences. Therefore, the development of high-quality reference genome assemblies with the help of cytogenetic mapping becomes a priority. In addition, taxonomic and population studies by whole-genome resequencing of wild mosquitoes heavily rely on chromosome-based reference assemblies.

All mosquitoes belong to the family Culicidae (order Diptera), which is divided into three subfamilies: Anophelinae, Toxorhynchitinae, and Culicinae (Knight and Stone 1977; Knight 1978). Anophelinae consists of three genera: *Chagasia* (four species in the Neotropical region), *Bironella* (nine species in the Australasian region), and *Anopheles* (about 500 species distributed throughout the world). Toxorhynchitinae includes only 65 species that do not consume blood. Subfamily Culicinae is further subdivided into tribes Culicini and Aedini. Various groups of mosquitoes, excluding Toxorhynchitinae, are responsible for the transmission of different pathogens including malaria parasites, transmitted by subfamily *Anophelinae*; dengue, yellow fever, and other arboviruses, mostly transmitted by aedine mosquitoes; and filarial worms and encephalitis viruses, which can be transmitted by both subfamilies of mosquitoes (Tolle 2009).

This chapter describes protocols developed for three species: the African malaria mosquito *Anopheles gambiae* Giles, 1902 (Anophelinae); the yellow fever mosquito *Aedes aegypti* Linnaeus, 1762 (Culicinae); and the southern house mosquito *Culex quinquefasciatus* Say, 1823 (Culicinae). These three species represent "a big tree" of mosquitoes with sequenced genomes. The divergence time between Anophelinae and Culicinae is about 145–200 million years, whereas the *Aedes* and *Culex* lineages split around 52–54 million years ago (Krzywinski et al. 2001, 2006).

4.1.1 *Anopheles gambiae* as a Disease Vector

Genus *Anopheles* is subdivided into six subgenera: *Anopheles, Cellia, Kerteszia, Lophopodomyia, Nyssorhynchus*, and *Stethomyia*. More than 500 recognized species of genus *Anopheles* inhabit every continent except Antarctica. The rich biodiversity of malaria mosquitoes has direct epidemiological implications. Of the approximately 500 anopheline species, no more than 30–50 significantly contribute to malaria transmission. Malaria remains a leading cause of human death and illness, having caused 225 million disease cases and about 800,000 deaths annually (World Malaria Report 2010). It is still not completely clear why some species or populations are efficient malaria vectors while others are of no medical importance. Therefore, understanding the genetic mechanisms of adaptation and speciation in malaria mosquitoes has not only a theoretical interest for evolutionary biology but also a practical application for vector control. The main African malaria vector *Anopheles gambiae* belongs to the subgenus *Cellia* and is a member of the *Anopheles gambiae* species complex. This complex consists of sibling malaria mosquito species with remarkably different geographic distributions, ecological adaptations, and host-seeking behaviors. *Anopheles gambiae* and *Anopheles arabiensis* Patton, 1905 are the two major vectors of malaria in Africa; they are both anthropophilic and can breed in temporal fresh water pools or human-made reservoirs. *Anopheles gambiae* occupies more humid areas, whereas *Anopheles arabiensis* dominates in arid savannas and steppes. *Anopheles gambiae* is further differentiated into two partly reproductively isolated incipient species named "M" and "S" forms first described on the basis of the virtual absence of hybrid genotypes between form-specific haplotypes found in ribosomal DNA-linked markers (Gentile et al. 2001). The "S" form is widely distributed, and the "M" form is restricted to West and Central Africa (Favia et al. 2003; della Torre et al. 2005). A recent study has proposed to elevate the taxonomic status of the "M" form to species level with the new name *Anopheles coluzzii* (Coetzee et al. 2013). *Anopheles merus* Donitz, 1902 and *Anopheles melas* Theobald, 1903 breed in brackish water. On the other hand, the habitat of *Anopheles bwambae* White, 1985 is restricted to mineral water breeding sites. These three species are relatively minor malaria vectors with narrow geographic distribution (Coluzzi et al. 1979). *Anopheles quadriannulatus* Theobald, 1911 (formerly *Anopheles quadriannulatus* A) and *Anopheles amharicus* Hunt, Coetzee, and Fettene, 1998 (formerly *Anopheles quadriannulatus* B) are fresh water breeders, zoophilic, and, although to some degree susceptible to *Plasmodium* infections, are not natural vectors of malaria (Takken et al. 1999; Coluzzi et al. 2002; Habtewold et al. 2008).

4.1.2   Cytogenetics of *Anopheles gambiae*

Cytogenetics studies have been essential in understanding the taxonomic and population complexity of the *Anopheles gambiae* complex. The major malaria vector *Anopheles gambiae* has karyotype

$2n = 6$. The chromosome complement of *Anopheles* consists of the
X chromosome, Y chromosome, and four autosomal arms: 2R, 2L,
3R, and 3L. Anopheline mosquitoes, like some other Diptera, have
giant polytene chromosomes that undergo endoreplication and can
be found in various tissues (Zhimulev 1996). These chromosomes
have different levels of compaction that appear as light and dark
bands, diffuse puffs, and heterochromatic regions. Banding patterns are mostly consistent within a species and, in some cases,
somewhat consistent between closely related species. Drawn and
photo chromosomal maps have been developed for about 50 species from the genus *Anopheles* (Sharakhov and Sharakhova 2008).
The first draft of a drawn cytogenetic map of polytene chromosomes from larval salivary glands of *Anopheles gambiae* was produced in 1956 (Frizzi and Holstein 1956). The first high-quality
drawn cytogenetic maps for species of the *Anopheles gambiae*
complex were developed for chromosomes from salivary glands
of larvae by Coluzzi and coauthors (Coluzzi and Montalenti 1966;
Coluzzi and Sabatini 1967). Importantly, they described several polymorphic inversions as well as fixed inversions between
Anopheles gambiae and *Anopheles arabiensis*. Soon after that,
Coluzzi (1968) discovered polytene chromosomes of a better
quality in ovarian nurse cells of *Anopheles gambiae* females, and
his group later created a new computationally enhanced drawn
map using these chromosomes (Coluzzi et al. 2002). Advances
in microphoto techniques led researchers to use photo images for
cytogenetic mapping. The first photomaps developed for various
species of malaria mosquitoes had numbered divisions but did not
have lettered subdivisions (Coluzzi et al. 1970; Kitzmiller et al.
1974; Mahmood and Sakai 1985; Kaiser and Seawright 1987). The
most recent cytogenetic photomap for *Anopheles gambiae* was
created by using a high-pressure squash technique that increases
overall band clarity (George et al. 2010).

From 1968 to the present time, all taxonomic and population cytogenetic studies on *Anopheles gambiae* have been performed using
cytogenetic maps developed for chromosomes of adult half-gravid
females. Cytogenetic analysis of chromosomal inversions helped to
identify species within the *Anopheles gambiae* complex (Coluzzi
and Sabatini 1967, 1968, 1969) and led to the discovery of chromosomal forms within *Anopheles gambiae s.s.* (Bryan et al. 1982;
Coluzzi et al. 1985). Members of the *Anopheles gambiae* complex
carry 10 fixed inversions that can be used for a phylogeny reconstruction (Coluzzi et al. 2002). Five fixed inversions are present on
the X chromosome, three inversions are found on the 2R arm, and
one is found on each of the 2L and 3L arms (Coluzzi et al. 2002).
The nonvectors in the complex, *Anopheles quadriannulatus* and
Anopheles amharicus, had been traditionally considered the closest

species to the ancestral lineage, because they have a large number of hosts, feed on animal blood, tolerate temperate climates, exhibit disjunctive distribution, and possess a "standard" karyotype (Coluzzi and Sabatini 1968, 1969; Coluzzi et al. 1979, 2002). A more recent study, however, obtained and analyzed the breakpoint sequences of fixed overlapping inversions 2Ro and 2Rp in the *Anopheles merus—Anopheles gambiae* clade and homologous sequences in *Anopheles stephensi*, *Aedes aegypti,* and *Culex quinquefasciatus* (Kamali et al. 2012). This work demonstrated that all studied out-group species had gene arrangements identical to that in the 2Ro breakpoints of *Anopheles merus* and in the 2R+^{p} breakpoints of *Anopheles gambiae*. Thus, sequencing, physical chromosome mapping, and bioinformatic analysis identified the 2Ro and 2R+^{p} arrangements in several out-group species indicating that these arrangements are ancestral. Because 2Ro and 2R+^{p} uniquely characterize the *Anopheles gambiae—Anopheles merus* clade, these two species have the least chromosomal differences from the ancestral species of the complex as compared to other members (Kamali et al. 2012). This methodology can be used for rooting chromosomal phylogenies in other complexes of sibling species.

Of the seven members of the *Anopheles gambiae* complex, *Anopheles arabiensis* and *Anopheles gambiae s.s.* are the only species, which have a highly polymorphic chromosome 2 and a continent-wide distribution in arid sub-Saharan Africa. Mosquito species with little or no chromosomal polymorphisms tend to occupy smaller and wetter geographic regions (Coluzzi et al. 2002). The 2Rb, 2Rbc, 2Rcu, 2Ru, 2Rd, and 2La inversions of *Anopheles gambiae* are frequent in arid Sahel Savanna mosquitoes and almost absent in those in humid equatorial Africa, strongly suggesting that these inversions confer adaptive fitness to the drier environment (Coluzzi et al. 1979, 2002; Toure et al. 1998; Powell et al. 1999). The variations in rates of water loss and in thermotolerance are associated with alternative arrangements of the 2La inversion in *Anopheles gambiae* (Gray et al. 2009; Rocca et al. 2009). Thus, the adaptive flexibility provided by this chromosomal polymorphism has probably allowed *Anopheles gambiae* to exploit a very broad range of climatic conditions, an important factor underlying the wide distribution and abundance of this species across Africa as well as its status as primary malaria vector. In addition, the inverted arrangements have been preferentially associated with indoor biting and resting behaviors (Coluzzi et al. 1979; Powell et al. 1999). Therefore, chromosomal inversions could influence epidemiologically important traits of *Anopheles gambiae*, such as its geographic distribution, its probability of vector–human contact, and the likelihood of vector exposure to insecticide-treated walls and bed nets.

4.1.3 Genome Assembly of *Anopheles gambiae*

The African malaria mosquito *Anopheles gambiae*, because of its epidemiological importance, was the first disease vector sequenced (Holt et al. 2002). Plasmid and bacterial artificial chromosome (BAC) DNA libraries were constructed from size-selected DNA fragments of the *Anopheles gambiae* PEST (Pink Eye Standard) strain. The PEST strain was chosen for genome sequencing for two major reasons. First, it had all fixed, standard chromosomal arrangements that could facilitate the genome assembly. Second, it had a sex-linked pink eye mutation that could be used as an indicator of cross-colony contamination. The genome was sequenced using plasmid libraries containing inserts of 2.5, 10, and 50 kilobases (Kb). The final sequencing data set had approximately equal coverage from male and female mosquitoes. The total size of the first genome assembly was 278 megabases (Mb), and the assembly had 10.2-fold coverage (Holt et al. 2002). The size of the *Anopheles gambiae* genome has been predicted by the C_0t analysis to be 260 Mb (Besansky and Powell 1992). The PEST strain was a hybrid between the M and S forms; therefore, the larger size of the PEST assembly was likely due to overrepresentation of incorrectly assembled haplotype scaffolds (Holt et al. 2002). For this first draft, 91% of the genome was organized into 303 scaffolds, and scaffolds constituting about 84% of the genome have been assigned to chromosomes by physical mapping of approximately 2000 BAC clones. In an effort to improve the assembly, the major scaffolds were reordered into a new golden path file by additional physical mapping of BAC clones and by bioinformatic analysis of haplotype scaffolds. The resulting AgamP3 assembly has a total of 80 scaffolds assigned to and ordered on the chromosome arms X, 2R, 2L, 3R, and 3L. The size of this new AgamP3 assembly is approximately 264 Mb (Sharakhova et al. 2007). Independent draft genome assemblies were generated for M and S forms of *Anopheles gambiae* based on approximately 2.7 million Sanger whole-genome shotgun reads (Lawniczak et al. 2010). Lower average read coverage (~6× in M and S forms vs. ~10× in PEST) contributed to assembly gaps in M and S scaffolds.

To develop a better understanding of genetic determinants of vectorial capacity and with support from vector biologists and members of the malaria community, the National Human Genome Research Institute and the National Institute of Allergy and Infectious Diseases have funded the sequencing of the genomes and transcriptomes of 16 *Anopheles* species (Neafsey et al. 2013). The genome sequences for 16 *Anopheles* species are now available (https://olive.broadinstitute.org/comparisons/anopheles.3). These species include *Anopheles albimanus* C. R. G. Wiedemann, 1820; *Anopheles arabiensis, Anopheles atroparvus* Van Thiel, 1927; *Anopheles christyi* Newstead & Carter, 1911; *Anopheles culicifacies* Giles, 1901; *Anopheles dirus* Peyton & Harrison, 1979; *Anopheles epiroticus* Linton & Harbach, 2005; *Anopheles farauti* Laveran, 1902; *Anopheles funestus* Giles, 1900; *Anopheles maculatus* Theobald, 1901; *Anopheles melas, Anopheles merus, Anopheles minimus* Theobald, 1901; *Anopheles quadrian-nulatus, Anopheles sinensis* Wiedemann, 1828; and *Anopheles stephensi* Liston, 1901. These draft genome assemblies consist of multiple unmapped and unoriented genomic scaffolds. Physical mapping by fluorescent in situ hybridization (FISH) is becoming instrumental in creating chromosome-based genome assemblies for these species and for improving the S form *Anopheles gambiae* genome assembly. Improvement of the S form assembly is necessary as VectorBase is transitioning the reference genome assembly from PEST to the S form.

4.1.4 *Aedes aegypti* as a Disease Vector

Aedes aegypti is recognized as a principal vector of dengue and yellow fever viruses (Tolle 2009). These two diseases have a significant worldwide impact on human health. Dengue virus (family Flaviviridae) occurs as four serotypes that are biologically transmitted between humans. This virus causes a nonspecific febrile illness termed dengue fever, which is the most widespread and significant arboviral disease in the world. It also is the etiological agent of dengue hemorrhagic fever and dengue shock syndrome, severe and sometimes fatal forms of the disease. Dengue fever is considered the most important vector-borne arbo-viral disease of the twenty-first century (Gubler 2012). The disease is a threat to 3.6 billion people and has an annual incidence of 230 million cases of infection resulting in 21,000 deaths per year. Since the 1950s, the incidence of dengue fever has expanded globally. The World Health Organization estimated a 30-fold increase in the incidence of dengue infections over the past 50 years (WHO 2009). The disease became endemic in 100 countries in Africa, West Asia, and America (Halasa et al. 2012) and is a growing threat to the United States (Morens and Fauci 2008). In addition to dengue, yellow fever, a devastating disease of the nineteenth century in North America and Europe, still affects up to 600 million lives and remains responsible for about 30,000 deaths annually (Gould and Solomon 2008). The disease is currently endemic in 32 countries in Africa and 13 in South America. Despite all control campaigns, *Aedes aegypti* has expanded its range to most subtropical and tropical regions during the last several decades; it is extremely well adapted to humans, prefers to feed on humans, and breeds in urban areas (Barrett and Higgs 2007).

Aedes aegypti is a convenient model system for experimental laboratory research. This species can be easily colonized and is highly tolerant to inbreeding (Severson 2008). Unlike *Anopheles* eggs, *Aedes aegypti* eggs are resistant to desiccation and can be stored in a dry place for several months. As a result of these advantages, genetic (linkage) mapping conducted on *Aedes aegypti* was very successful. The genetic mapping was originally inspired from the study of the inheritance of dichlorodiphenyltrichloroethane resistance as a single dominant trait (Coker 1958). A similar mechanism of inheritance, as a single gene or a single block of chromosome material, was demonstrated as the mechanism for sex determination in this species (McClelland 1962). In addition, 28 of 87 morphological mutations described for *Aedes aegypti* were mapped to the three linkage groups corresponding to the three chromosomes of this mosquito (Craig and Hickey 1967). The linkage map was extended by additional mapping of physiological and enzyme loci (Munstermann and Craig 1979). The classical linkage map included about 70 loci of morphological mutants, insecticide resistance, and isozyme markers (Munstermann 1990).

The possibility of using DNA molecular markers opened a new era in genetic mapping of traits that confer vector competence. The first molecular-marker-based linkage map for *Aedes aegypti* was constructed using restriction fragment length polymorphism (RFLP) of complementary DNA (cDNA) clones (Severson et al. 1993). This map included 50 DNA markers and covered 134 centimorgan (cM) across the three linkage groups. Thereafter, polymerase chain reaction (PCR) was used to generate a map based on random amplified polymorphic DNA (RAPD) loci, which consisted of 96 RAPD loci covering 168 cM (Antolin et al. 1996). Linkage maps based on single-strand conformation polymorphism (SSCP) and single nucleotide polymorphism (SNP) markers were also constructed. A composite map for RFLP, SSCP, and SNP markers incorporated 146 loci and covered 205 cM (Severson 2008). Later, an additional map using amplified fragment length polymorphism was also developed for 148 loci and covered about 180 cM of the genome (Sun et al. 2006). Finally, the genetic map of *Aedes aegypti* was extended by incorporating microsatellite loci (Chambers et al. 2007). The linkage map was used as a tool to localize several quantitative trait loci (QTL) related to pathogen transmission: the filarial nematode *Brugia malayi* (Severson et al. 1994), the avian malaria parasite *Plasmodium gallinaceum* (Severson et al. 1995; Sun et al. 2006), and dengue virus (Bosio et al. 2000; Gomez-Machorro et al. 2004). Among all mosquitoes, the linkage map developed for *Aedes aegypti* is the most densely populated.

4.1.5   Cytogenetics of *Aedes aegypti*

Detailed cytogenetic mapping in *Aedes aegypti* and other Culicinae is difficult because of the lack of high-quality, easily spreadable polytene chromosomes (Sharma et al. 1978; Campos et al. 2003b). The use of salivary glands as a source of polytene chromosomes yielded only 0.5% of chromosomal preparations useful for cytogenetic studies (Campos et al. 2003b). Fortunately, mitotic chromosomes in this mosquito are large and easily identifiable. The average size of the biggest metaphase chromosome in *Aedes aegypti* is 7.7 µm (Brown et al. 1995), which is bigger than the average size of human metaphase chromosomes and is comparable with the size of human chromosomes at prometaphase (Daniel 1985). Most of the classical cytogenetic studies in *Aedes aegypti* have been performed on mitotic or meiotic chromosomes from larval brain or male testis (Rai 1963; Newton et al. 1974; Motara and Rai 1977). These studies demonstrated that *Aedes aegypti* has a karyotype that includes three pairs of chromosomes. These chromosomes were originally designated as chromosomes I, II, and III in the order of increasing size (Rai 1963). Later, chromosomes were renamed as chromosomes 1, 2, and 3 in accordance with the *Aedes aegypti* linkage groups (McDonald and Rai 1970). Chromosome 1 was described as the shortest metacentric chromosome; chromosome 2 as the longest and also as a metacentric chromosome; and chromosome 3 as a medium-length submetacentric chromosome with the secondary constriction on the longer arm. However, more precise measuring of the centromeric indices made on spermatagonial metaphase chromosomes has indicated that all *Aedes aegypti* chromosomes fall into the category of metacentric chromosomes according to the standard classification (Levan et al. 1964; Motara et al. 1985).

Unlike anophelines, the pair of sex chromosomes is homomorphic in all culicine mosquitoes, including *Aedes aegypti* (Rai 1963). The sex determination alleles were linked to chromosome 1 and described as *Mm* in males and *mm* in females (McClelland 1962). The precise measuring of the sex chromosomes in males and females has indicated that the female chromosome 1 is slightly longer (Motara et al. 1985). The C-banding technique has also demonstrated differences between male and female sex chromosomes in *Aedes aegypti* (Motara and Rai 1977). Typically, females have pericentromeric and additional distinct intercalary bands on chromosome 1, both of which are absent on one of the sex chromosomes in males. The C-banding pattern has been found to be variable in different strains of *Aedes aegypti*. For example, an intercalary band can be present on the male chromosome in some strains, and intercalary C-bands may differ in size in females (Motara and Rai 1978; Wallace and Newton 1987). A silver staining technique (Wallace and Newton 1987) and in situ hybridization of 18S and 28S ribosomal genes (Kumar and Rai 1990) demonstrated the location of ribosomal locus on both sex chromosomes of *Aedes aegypti*.

4.1.6   Genome Assembly of *Aedes aegypti*

The genome of *Aedes aegypti* was sequenced in 2005 by the Broad Institute and The Institute for Genomic Research using a substrain of the Liverpool (LVP) strain. The LVP strain originated from West Africa and has been maintained at the Liverpool School of Tropical Medicine since 1936. This strain was selected for susceptibility to the filarial worm parasite *Brugia malayi*. A LVP substrain was derived after 12 consecutive generations of single-pair inbreeding (IB12) to decrease the level of heterozygosity. This LVP IB12 substrain was used for the genome project. The *Aedes aegypti* genome is among the largest within the Culicidae family (Nene et al. 2007). The draft genome assembly consists of 1376 Mb, which is approximately 5 times larger than the *Anopheles gambiae* genome (Holt et al. 2002). This 8× coverage genome assembly consists of 4758 supercontigs, with a contig N50 size of 82 Kb and supercontig N50 size of 1500 Kb. The *Aedes aegypti* genome has an extremely repetitive nature: about half of the genome consists of transposable elements (TEs). The genome shows "short-period interspersion," which means that, in general, approximately 1–2 Kb fragments of unique sequences alternate with approximately 0.2–4 Kb fragments of repetitive DNA (Severson 2008). When the genome sequence of *Aedes aegypti* was published (Nene et al. 2007), only approximately 31% of the *Aedes aegypti* genome was assigned to chromosomes mostly based on linkage mapping.

4.1.7   *Culex quinquefasciatus* as a Disease Vector

Culex quinquefasciatus (the southern house mosquito) is a principal vector of the lymphatic filarial worm and of encephalitis viruses, including West Nile virus. *Culex quinquefasciatus* belongs to the *Culex pipiens* complex that includes seven sibling species. Compared with *Anopheles gambiae* and *Aedes aegypti*, *Culex quinquefasciatus* has several unique epidemiologically important traits. A recent review highlighted the importance of *Culex quinquefasciatus* genome sequence for vector biology and argued that the "*Culex* genome sequence is more than just another genome for comparative genomics" (Reddy et al. 2012). Among the three groups of mosquitoes, *Culex* is the most diverse and widespread genus in both temperate and tropical areas

(Arensburger et al. 2010). This species transmits a much greater variety of viral, protozoan, and nematode human infections than do *Aedes* or *Anopheles* (Tolle 2009). In contrast to most other vectors, species from genus *Culex* have highly opportunistic host choice, which includes human, mammals, and birds. Another peculiar feature of *Culex* is that it may lay the first batch of eggs without a blood meal, and can use both polluted and nonpolluted water reservoirs as larval habitats.

Genetic mapping efforts in *Culex* attempted to chromosomally localize markers linked to traits related to vectorial capacity. The first linkage map for *Culex pipiens* Linnaeus, 1758 was published in 1999 (Mori et al. 1999). This map originally consisted of 21 cDNA markers and covered 7.1, 80.4, and 78.3 cM on chromosomes 1, 2, and 3, respectively. The sex determination locus was genetically mapped to the smallest chromosome 1 in *Culex pipiens* (Mori et al. 1999). The most recently developed genetic map of *Culex pipiens* complex includes 80 microsatellite markers (Hickner et al. 2010, 2013). In addition, QTL related to reproductive diapauses have been genetically mapped (Mori et al. 2007). Resistance of *Culex quinquefasciatus* and *Culex pipiens* to various insecticides has been extensively studied, and these studies revealed interesting patterns of mosquito adaptation in response to natural selection (Labbe et al. 2007, 2009).

4.1.8   Cytogenetics of *Culex quinquefasciatus*

Polytene chromosomes in *Culex* have low levels of polytenization, produce multiple ectopic contacts, and are almost unspreadable for slide preparations. It has been estimated that the percentage of preparations with recognizable polytene chromosomes is approximately 30% in salivary glands of the Johannesburg (JHB) strain of *Culex quinquefasciatus* (McAbee et al. 2007). Several attempts to create a cytogenetic photomap using *Culex* polytene chromosomes have been made. The Malpighian tubule chromosome map for *Culex pipiens* (Zambetaki et al. 1998) and *Culex quinquefasciatus* (Campos et al. 2003a) and, more recently, the salivary gland chromosome map for *Culex quinquefasciatus* (McAbee et al. 2007) have been developed. However, correspondence of arms and regions among these maps and the original drawn map (Dennhofer 1968) is uncertain. Almost no similarities between landmarks of different chromosome maps have been found (McAbee et al. 2007). Only two genes (esterase and odorant-binding protein) have been mapped to the polytene chromosomes of *Culex quinquefasciatus* (Perez-Requejo et al. 1985; McAbee et al. 2007). Several polymorphic inversions have been found on chromosomes 2 and 3 in laboratory strains of *Culex quinquefasciatus* (McAbee et al.

2007) and *Culex pipiens* (Tewfik and Barr 1976). However, no research has been done to link the inversion polymorphism with ecological adaptations in *Culex*.

Mitotic chromosomes of *Culex* could be more suitable for physical mapping, because they have a more reproducible morphology than polytene chromosomes. The *Culex pipiens* mitotic chromosomes have been described as three pairs of metacentric chromosomes and have been originally numbered in order of increasing size as chromosomes 1, 2, and 3 (Rai 1963). Two genes, 18S and 28S ribosomal DNA (rDNA), have been physically mapped to the mitotic chromosome 1 of *Culex pipiens* (Santos et al. 1985).

4.1.9   Genome Assembly of *Culex quinquefasciatus*

The genome of *Culex quinquefasciatus* was the third sequenced mosquito genome after the genomes of the malaria vector, *Anopheles gambiae,* and the yellow fever mosquito, *Aedes aegypti*. The JHB strain was the source of genomic DNA for the *Culex quinquefasciatus* genome project. The Broad Institute and the J. Craig Venter Institute produced 579 Mb of the genome assembly at 8× shotgun coverage. Although *Culex quinquefasciatus* has an intermediate genome size compared with *Aedes aegypti* and *Anopheles gambiae*, the total number of protein-coding genes in *Culex* (18,883) is 22% and 52% higher than that in *Aedes aegypti* and *Anopheles gambiae*, respectively. These expansions are primarily due to multiple gene-family expansions that include olfactory and gustatory receptors, salivary gland genes, and genes associated with xenobiotic detoxification (Bartholomay et al. 2010). However, the presence of different haplotypes for the same gene may also contribute to the overrepresentation of gene families, if genes are located in small supercontigs not mapped to chromosomes. The total assembly size (579 Mb) differs from an estimate based on reassociation kinetics (540 Mb) (Rao and Rai 1990), suggesting the possibility of the presence of alternative haplotype supercontigs in the assembly. Moreover, the *Culex quinquefasciatus* genome assembly is highly fragmented. The resulting assembly contains 48,671 contig fragments with an N50 contig size of 28.55 Kb, which were assembled into 3,171 supercontig sequences with an N50 size of 486.76 Kb. Contig sizes range from 201 to 11,094 bp, and supercontig sizes range from 1,197 to 3,873,010 bp. On the basis of sequence analysis, genomic supercontigs have been linked to the genetic map (Hickner et al. 2013). The linkage map covers 12% of the genome and is the most representative map of the *Culex quinquefasciatus* genome.

4.2   PROTOCOLS

4.2.1   Fluorescent In Situ Hybridization with Polytene Chromosomes of Malaria Mosquitoes

The cytogenetic advantage of the *Anopheles* species is the availability of high-quality polytene chromosome maps that allow for careful verification of new computational approaches for assembly, ordering, and orienting sequencing scaffolds. FISH is a method that uses fluorescently labeled DNA probes for mapping the position of a genetic element on polytene chromosomes. A DNA probe is prepared by incorporating Cy3 or Cy5 labeled nucleotides into DNA by nick translation or a random-primed labeling method. This protocol has been used to map genes (Sharakhova et al. 2010b, 2013; Bridi et al. 2013) and microsatellite markers (Kamali et al. 2011; Peery et al. 2011) on polytene chromosomes from ovarian nurse cells and salivary glands of malaria mosquitoes. Detailed physical genome mapping performed on polytene chromosomes has the potential to link DNA sequences to specific chromosomal structures such as heterochromatin (Sharakhova et al. 2010a). This method also allows comparative cytogenetic studies (Sharakhova et al. 2011a,c) and reconstruction of species phylogenies (Xia et al. 2008; Kamali et al. 2012). Here, we demonstrate both standard and innovative approaches to chromosome preparation, FISH, and imaging (George et al. 2012), which can be used for the physical map development. The innovative approaches include a high-pressure chromosome preparation and automated in situ hybridization and imaging. The scheme of the high-pressure chromosome preparation is shown in Figure 4.1. The chromosome preparation step involves the process of squashing and flattening chromosomes using a Dremel tool and mechanical vice, as well as chromosome visualization using a phase contrast microscope. The next step illustrates the hybridization of a fluorescently labeled probe to target DNA on the chromosome slide preparations using an automated slide staining system, the use of an automated scanning microscope for visualizing and mapping the probes after the FISH experiment, and the placing and orienting of genomic scaffolds on the chromosomes (Figure 4.2).

4.2.1.1   Materials for Fluorescent In Situ Hybridization with Polytene Chromosomes

1. MZ6 Leica stereo microscope (Leica, Cat. No. VA-OM-E194-354)
2. Olympus CX41 Phase Microscope (Olympus, Cat. No. CX41RF-5)
3. ACCORD™ PLUS Automated Scanning System (BioView, Billerica, Massachusetts, Cat. No. BV-5000-ACCP)

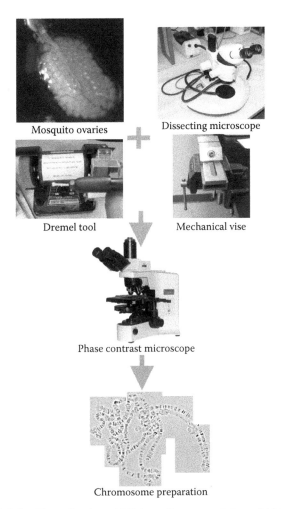

Mosquito ovaries

Dissecting microscope

Dremel tool

Mechanical vise

Phase contrast microscope

Chromosome preparation

FIGURE 4.1 **(See color insert.)** Schematic representation of high-pressure chromosome preparation. Mosquito ovaries are shown at the correct stage of development. (From George, P. et al., *J. Vis. Exp.*, (64), e4007 10.3791/4007, 2012).

4. Tin-coated rapid-vice precision ground square parallel to within 0.00025 (Avenger Gold Toolmaker, Cat. No. MTC-200-1)
5. Torque wrench (Craftsman, Cat. No. 44593)
6. Dremel 200 rotary tool with a Flex-Shaft attachment, 3″ Multipurpose Mini Bench Grinder (with rate limiter) (Rand)
7. ThermoBrite™ Slide Denaturation/Hybridization System (Abbott Molecular, Cat. No. 30-144110)
8. Xmatrx™ Automated Slide Staining System (Abbott Molecular, Cat. No. 08L46-001)
9. Specially designed 200-μL plastic tips (beaded plastic edges) for Dremel tool (Pipetman, Cat. No. F171300)
10. Dissecting needles (Fine Science Tools, Cat. No. 10130-10)

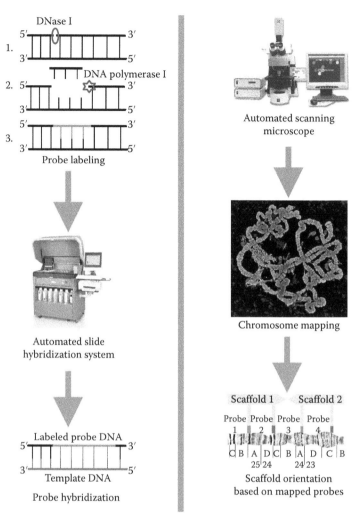

FIGURE 4.2 (**See color insert.**) A scheme representing automated fluorescent in situ hybridization, slide scanning, and chromosome mapping of genomic scaffolds. (From George, P. et al., *J. Vis. Exp.,* (64), e4007 10.3791/4007, 2012.)

11. Needle holders (Fine Science Tools, Cat. No. 26018-17)
12. 75 × 25-mm double frosted micro slides (Corning, Cat. No. 2949-75×25)
13. 22 × 22-mm microscope coverslips (Fisher Scientific, Cat. No. 12-544-10)
14. 18 × 18-mm microscope coverslips (Fisher Scientific, Cat. No. 12-553-402)
15. 25-mm barrier slides (Abbott Molecular, Cat. No. XT108-SL)
16. 25-mm coverslips (Abbott Molecular, Cat. No. XT122-90X)
17. Random Primed DNA Labeling kit (Roche, Cat. No. 11 004 760 001)
18. 10% NBF (neutral buffered formalin) (Sigma-Aldrich, Cat. No. HT501128)

19. 10× PBS (phosphate buffered saline) (Invitrogen, Cat. No. P5493)
20. 99% formamide (Fisher Scientific, BP227500)
21. Dextran sulfate sodium salt (Sigma-Aldrich, Cat. No. D8906)
22. 20× SSC (saline sodium citrate) buffer (Invitrogen, Cat. No. AM9765)
23. 50× Denhardt's solution (Sigma-Aldrich, Cat. No. D2532)
24. Sodium phosphate (Sigma-Aldrich, Cat. No. S3264)
25. 1 mM YOYO-1 iodide (491/509) solution in dimethyl-sulfoxid (DMSO) (Invitrogen, Cat. No. Y3601)
26. ProLong® Gold antifade reagent (Invitrogen, Cat. No. P36930)
27. dATP, dCTP, dGTP, dTTP (Fermentas, Cat. No. R0141, R0151, R0161, R0171)
28. Cy3-dUTP, Cy5-dUTP (GE Healthcare, PA53022, PA55022)
29. BSA (bovine serum albumin) (Sigma-Aldrich, Cat. No. A3294)
30. DNA polymerase I (Fermentas, Cat. No. EP0041)
31. DNase I (Fermentas, Cat. No. EN0521)
32. Acetic acid (Fisher Scientific, Cat. No. A491-212)
33. Methanol (Fisher Scientific, Cat. No. A412-4)
34. Propionic acid (Sigma-Aldrich, Cat. No. 402907)
35. Alcohol 200 Proof (Decon Laboratories, Cat. No. 2701)

4.2.1.2 Dissection of Ovaries from Half-Gravid *Anopheles* Females

1. Give blood to recently emerged (48–72 hours old) females using a live animal, and provide a container with water for them to lay eggs. Container of water should be provided 48 hours after blood feeding, and eggs will typically be laid approximately 72 hours post blood feed. After the mosquitoes lay eggs, feed them a second time. It is important that the mosquitoes lay eggs at least once before feeding them for dissection. If they do not lay eggs after the first blood feeding, feed them again until they do and then feed them for dissection (allow at least one gonotrophic cycle including oviposition).
2. After feeding mosquitoes 2nd or 3rd time, let them develop ovaries for 25 hours at 26°C and 80% humidity.
3. Prepare fresh modified Carnoy's solution: methanol:glacial acetic acid (3:1). Use only high-quality methanol and high-quality glacial acetic acid from trusted brands. Make aliquots of the solution by 0.5–1 mL in 1.5–2 mL tubes.

4. Immobilize mosquitoes by putting a cage into a −20°C freezer for 1 minute. Collect half-gravid females and place them on ice in a Petri dish before dissection. Keep mosquitoes alive.

5. Choose half-gravid females for dissection. At this stage, the light area occupied by the developing ovaries goes from the sixth dorsal to fourth ventral abdominal segments. The rest of the abdomen looks dark because of blood meal (Figure 4.3).

6. Dissect half-gravid females under a dissection microscope using sharp forceps or needles. Make a cut between third and fourth segments from the end of abdomen. Pull off the three segments with attached ovaries from an abdomen on a dry dust-free slide. Remove all blood and extra tissue because they contain water.

7. Check the stage of ovary development. They must be at Christophers' III stage (Clements 1992). At the underdeveloped stage, follicles are very small; a transparent area with nurse cells within follicles has not been developed. At the correct stage, follicles have an oval shape; a transparent area with nurse cells within follicles has a round shape. At the overdeveloped stage, follicles have an elongated, slightly bent oval shape (almost a banana shape) (Figure 4.4). *Note: Do not let ovaries dry out! Inspection of ovaries in the developmental stage must be done quickly, and dissected ovaries should be put in Carnoy's solution as soon as possible.*

8. Put ovaries from five females into 500 μL of modified Carnoy's solution in a 1.5-mL Eppendorf tube, and keep them at room temperature (RT) for 24 hours. Transfer ovaries to −20°C for a long-term storage.

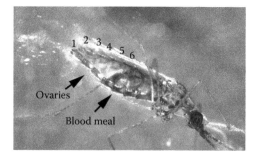

FIGURE 4.3 (See color insert.) A half-gravid *Anopheles gambiae* female at the correct stage for dissection. Arrows show the light area occupied by the developing ovaries and the dark area with a blood meal. The numbers indicate dorsal abdominal segments.

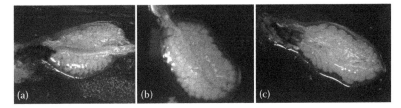

FIGURE 4.4 (**See color insert.**) Stages of ovarian development in *Anopheles*.
(a) Ovaries at the Christophers' II stage. (b) Ovaries at the Christophers' III stage
(the correct stage for chromosome preparations). (c) Ovaries at the Christophers'
IV stage.

4.2.1.3 Standard Protocol for Polytene Chromosome Preparation

1. To make the chromosome preparations, take one ovary
   out of the vials with a pair of forceps (or a transfer pipet)
   and place it into a drop of Carnoy's solution on microscope slide. After carefully removing tissues, trachea,
   and blood, quickly separate follicles from one ovary into
   2–4 pieces. Up to four preparations can be made from
   one ovary. *Note: while dissecting, the ovaries should
   never be allowed to dry. Continue adding drops of
   Carnoy's solution when needed to prevent drying of the
   ovaries.*

2. On four to eight microscope slides, add each piece of the
   divided ovary to one drop of 50% propionic acid on a separate slide. Let the pieces of ovary rest in propionic acid
   for about 5 minutes until follicles become clear and swell
   to about twice their original size.
   *Note: The number of slides depends on how many pieces
   each ovary is divided into.*

3. For each slide, use a dissecting microscope to separate the
   cleared follicles from each other and any other tissue or
   debris on the slide. Remove tissue and debris by wiping it
   away with a piece of paper towel, and apply a fresh drop
   of 50% propionic acid to the separated follicles.

4. Cover ovaries with a dust-free coverslip, and leave them
   for about 5 minutes. After 5 minutes, place a piece of filter
   paper over the coverslip and hold the four edges still with
   fingers. Gently tap it with a pencil eraser to release polytene chromosomes from the follicles.

5. Examine the banding pattern and spread of polytene chromosomes using a phase contrast microscope.

6. Place slides with good chromosomal preparations in a
   humid chamber with 4× SSC in the bottom of the chamber,
   at 60°C for 15–20 minutes. After heating, put slides at 4°C

overnight or until immersing the slides in liquid nitrogen. Heating can be done on the Thermobrite machine with the absorbent strips soaked in distilled water. *Note: Slides can dry out if left at 4°C for extended periods. Leaving them at 4°C for longer than overnight is not recommended.*

7. While holding one corner of the slide with forceps or a gloved hand, dip chromosome preparation into the liquid nitrogen so that the coverslip is completely immersed. Hold slide in liquid nitrogen until the bubbling stops (usually 10–15 seconds). Take it out of the liquid nitrogen, and immediately remove the coverslip with a razor blade from one corner. It sometimes helps to put the slide on a flat surface when trying to remove the coverslip. Put slide in a slide jar with prechilled 50% ethanol (−20°C), and keep at 4°C for at least 2 hours.

8. Dehydrate the preparations in slide jar with an ethanol series of 70% and 90% for 5 minutes each at 4°C and then with 100% ethanol for 5 minutes at RT. Air-dry, and keep slides in slide box until ready for use in FISH. *Note: Slides that are kept protected from dust and debris can be used for FISH at least within a year after the preparations are made.*

4.2.1.4 High-Pressure Polytene Chromosome Preparation

1. Prepare modified Carnoy's solution with ethanol (100% ethanol:glacial acetic acid, 3:1) and 50% propionic acid just before making slides. Place one pair of ovaries on a dust-free slide in one drop of modified Carnoy's solution. Split ovaries into approximately six sections with dissecting needles, and place them into drops of 50% propionic acid on clean slides under a dissection MZ6 Leica stereo microscope. Use a separate slide for each section.

2. Separate follicles using dissecting needles, and wipe out remaining tissue with filter paper or paper towel under a dissection microscope. Add a new drop of 50% propionic acid to the follicles, and allow them to sit for 3–5 minutes at RT. Place a coverslip on top of the 50% propionic acid droplet. Let the slide stand for approximately 1 minute.

3. Wrap the slide with filter paper and plastic. Using a Dremel 200 rotary tool with a Flex-Shaft attachment and soft plastic tip set between 3000 and 5000 revolutions per minute (rpm), express the follicles by swirling the tip in circles and lightly pressing the coverslip to evenly spread the nuclei. This step should take approximately 1 minute. Check the spread quality with an Olympus CX41 Phase Microscope using a 20× objective.

4. Prepare a sandwich by placing an additional coverslip next to the coverslip covering the chromosomes, and cover them with a second microscope slide. This reduces the chance of crushing the slide in the vice. Wrap the slides with cellophane or alternatively, the plastic sheet that comes in the glass slide box also works well. Wrap the plastic-covered slide sandwich in a filter paper to hold the layers in place and to protect the slide from scratching due to the vice.

5. Apply pressure to the slides via the mechanical vice. A pressure of 85- to 120-inch-lb is sufficient and is achieved by using a torque wrench. This step is necessary for flattening the chromosomes as much as possible.

6. Remove the second microscope slide and the additional coverslip. Heat the slide with the coverslip covering the chromosomes to 55°C on a slide denaturation/hybridization system for 10–15 minutes to further flatten the chromosomes. Dip the slide in liquid nitrogen for at least 15 seconds, and when bubbling has stopped, quickly remove the coverslip with a razor blade. Immediately place slides into cold 50% ethanol for 5 minutes. Dehydrate slides in 70%, 90%, and 100% ethanol for 5 minutes each. Air-dry slides.

4.2.1.5 Random Primer DNA Labeling

Note: Random Primer labeling is used for fragments shorter than 1 Kb.

1. If using PCR products as probe, purify the PCR product from an agarose gel or from the PCR reaction using a QIAquick® Gel Extraction Kit or QIAquick® PCR purification Kit. Similar kits that remove excess nucleotides can also be used. However, when using a kit, dissolve the DNA in double-distilled water instead of the elution buffer suggested in the final step.

2. Add 25 ng template DNA into double-distilled water to a final volume of 13.5 μL in a microcentrifuge tube.

3. Denature the DNA by heating in a boiling water bath for 10 minutes at 95°C and chilling quickly in an ice bath.

4. Add the following to the freshly denatured probes on ice.
   1 μL 1.0 mM dGTP.
   1 μL 1.0 mM dCTP.
   1 μL 1.0 mM dATP.
   2 μL Reaction mixture (vial 6 from Roche Random Primed DNA Labeling kit).
   1 μL Klenow enzyme.
   0.5 μL 1.0 mM Cy3- or Cy5-dUTP.
   Mix and centrifuge briefly.

5. Incubate for 1–20 hours (overnight) at 37°C.

6. Terminate the reaction by adding 1 μL 0.5 M ethylenediami-netetraacetic acid (EDTA) to the reaction. Add 1/10 volume 3 M sodium acetate (NaAC) and 2.5 volume 100% ethanol. And mix by inverting the tubes. Keep at −80°C or −20°C for at least 1 hour or until probes are needed for hybridization. If necessary, probes can be left in the freezer for long-term storage.

4.2.1.6   Nick Translation DNA Labeling

Note: Nick translation labeling is used for fragments longer than 1 Kb (1–150 Kb).

1. Prepare the following reaction mixture on ice.
   5 μL 10× buffer for DNA polymerase I.
   5 μL 10× dNTP mixture: 0.5 mM ATP, CTP, GTP, and 0.15 mM dTTP.
   4 μL DNase I freshly diluted to 0.02 units/μL.
   1 μL DNA polymerase I.
   1 μg Template DNA.
   1 μL Cy3- or Cy5-dUTP.
   5 μL BSA diluted to 0.5 mg/mL.
   Add water to final volume 50 μL.
   Note: This protocol can be scaled down by ½ to accommodate 500 ng of template DNA. Final concentrations of DNase I and DNA polymerase I have to be optimized based on factors including initial size of template DNA and reaction time. Larger template size generally requires more DNase I. The reaction time influences the final fragment size, because DNA polymerase I starts losing its activity earlier than DNase I.

2. Incubate the mix at 15°C for 1–3 hours.

3. Run 5–10 μL reaction mixture on an agarose gel to determine the size of digested fragments. Fragments should be 100–600 bp for best hybridization results. If fragments are still larger than this, incubate at 15°C for additional time.

4. Terminate the reaction by adding 1 μL 0.5 M EDTA to the reaction. Add 1/10 volume of 3 M NaAC and 2.5 volume of 100% ethanol. And mix by inverting the tubes. Keep at −80°C or −20°C for at least 1 hour or until probes are needed for hybridization. If necessary, probes can be left in the freezer for long-term storage. *Note: Fluorescently labeled probes should be protected from light! In the steps following, even where it is not explicitly stated, make efforts to protect probes from light.*

4.2.1.7 Standard Fluorescent In Situ Hybridization Protocol for Polytene Chromosomes

1. Perform chromosome fixation if slides are more than 2 months old. If slides are less than 2 months old, go directly to step 2. Fix slides in 1:3 glacial acetic acid:methanol at RT for 10 minutes and air-dry. Dehydrate slides in 100% ethanol for 10 minutes and air-dry again.

2. Immerse slides in 1× PBS for 20 minutes at RT.

3. Incubate slides at RT in 4% formalin for 1 minute. After incubation in formalin, slide should be immediately immersed in 50% ethanol. Drying the slide before dehydration in 100% ethanol damages the chromosome preparation. *Note: Mix 5 mL 10× PBS, 20 mL 10% NBF, and 25 mL water. Formalin is hazardous and should be handled carefully. Avoid breathing gas or dust during preparation, wear gloves and other personal protective equipment when handling. Formalin solution should not be dumped down drains.*

4. Dehydrate the slides through an ethanol series of 70%, 90%, and 100% for 5 minutes each at RT. Air-dry the slides.

5. Centrifuge the tubes of labeled probes at 20,817×g for 10 minutes. Carefully remove the supernatant and vacufuge the tubes for 20 minutes to dry pellets.

6. Dissolve dry probes in hybridization buffer (60% formamide, 10% dextran sulfate, 1.2× SSC) prewarmed to 37°C. The amount of hybridization buffer used to dissolve depends on the total amount of DNA probe you are dissolving. Dissolve 1 μg DNA probe in 20–40 μL warmed hybridization buffer.

7. In a clean microcentrifuge tube, combine at least 250 ng each of one Cy5-labeled and one Cy3-labeled probes. In situ hybridization on polytene chromosomes is efficient if at least 500 ng DNA is hybridized on the slide. Vortex and centrifuge the tube of combined probes briefly.

8. Transfer the above prepared solution of combined probes to a chromosome preparation slide and cover with a 22 × 22-mm coverslip. Remove any large air bubbles with gentle pressure.

9. Seal edges of coverslip with rubber cement. Denature the target and probe DNA by placing the slides on the Thermobrite machine at 75°C–90°C for 5–10 minutes. Thermobrite machine does not need to be humid.

10. Transfer the slides to prewarmed humid chambers with 4× SSC or water in the bottom of the chambers, and incubate at 39°C for interspecies (e.g., an *Anopheles gambiae*

probe with *Anopheles stephensi* chromosomes) or 42°C
for intraspecies hybridization. Allow the probes to hybridize for 3–18 hours (usually overnight). *Note: because there
are fluorescently labeled probes on the slide(s), humid
chambers should be impermeable to light. It is advisable
to cover the humid chamber with aluminum foil to minimize light impact on the probes.*

11. Carefully remove rubber cement with forceps and coverslip. In a slide jar covered with aluminum foil, wash the
slides with 1× SSC at 39°C after interspecies or 0.2× SSC
at 42°C after intraspecies hybridization for 20 minutes in
50 mL without agitation.

12. Wash the slides with 1× SSC after interspecies or 0.2×
SSC after intraspecies hybridization at RT for 20 minutes
in 50 mL without agitation.

13. Dilute fluorescent dye YOYO-1 100 times in 1× PBS to
make a stock solution. Mix 10 μL 100× diluted YOYO-1
with 90 μL 1× PBS for each slide that you want to stain.
The working solution of YOYO-1 is 1000× diluted relative to original concentration.

14. After washing in SSC for 20 minutes at RT, rinse slide in
1× PBS, and add 100 μL YOYO-1 in PBS on each slide.
Cover with parafilm. Leave at RT for 10 minutes inside a
slide box or somewhere dark.

15. Rinse in 1× PBS and add 10-μL ProLong Gold antifade
reagent, place coverslip on slide, and blot out bubbles.
Keep in the slide box at 4°C. Detect the signals using a
confocal or fluorescence microscope, and map them to the
polytene chromosomes of *Anopheles* mosquitoes.

4.2.1.8 Fluorescent In Situ Hybridization Using an Automated Slide Staining System

1. Put slides and reagents into the Xmatrx automated slide
staining system, and start the program to run the following
steps. Apply 800 μL 1× PBS for 20 minutes. Blow slides
with air. Perform formalin fixation by applying 450 μL
4% formalin in 1× PBS for 1 minute followed by washes
with 100% ethanol for 1 second twice and for 2 minutes
once. Blow slides with air.

2. Heat slides at 45°C for 2 minutes to avoid bubbling when
applying the probes. Apply 20 μL DNA probes, add drops
of mineral oil to avoid evaporation of a hybridization solution (2× SSC, 100 mM sodium phosphate, 1× Denhardt's
solution, and 10% dextran sulfate in formamide), and
place a coverslip on top. Denature chromosomes and
DNA probes by heating slides at 90°C for 10 minutes.

3. For hybridization, incubate slides at 42°C for 14 hours with coverslips on.

4. For stringency washes, heat slides at 42°C for 2 minutes, remove coverslips, wash slides in 2× SSC for 1 second 4 times. Blow slides with air. Apply 800 μL 0.4× SSC at 42°C for 10 minutes 2 times. Wash slides in 2× SSC at 25°C for 10 minutes.

5. Perform chromosome staining by applying 50 μL 1 μM YOYO-1 in 1× PBS. Apply drops of mineral oil to avoid evaporation of the staining solution, put on coverslips, and incubate at 25°C for 10 minutes. Remove coverslips, and wash slides in 2× SSC for 1 second 4 times. Blow slides with air. Apply 15 μL ProLong Gold antifade reagent. Put on coverslips.

4.2.1.9 Slide Reading Using an Automated Fluorescent Imaging System

Slide reading can be done using a fluorescent or confocal microscope. This procedure is usually done manually. However, the ACCORD PLUS automated scanning system frees researcher's time, and facilitates capturing and managing image acquisition. Instructions begin after turning on in this order: Olympus U-RFL-T power supply for a halogen bulb, computer Dell precision T3500, microscope Olympus BX61 with a connected camera Olympus U-CMAD3. A detailed video about this procedure has been published by George et al. (2012).

1. For setting up 10× Pre-Scan, open the Duet software in the ACCORD PLUS automated scanning system. Click "Online" button. Enter new Case ID and assign a Slide ID. Click the dot labeled "BF." This is the bright-field option. Set a scan choice to "10× Pre-scan." Use "2,500× circle," "10,000× circle," or "rectangle." Click "Set&run" for 10× Pre-Scan.

2. Click the "OK" button to run 10× Pre-Scan. Follow the prompts to adjust the scanning properly. Click "Finish" to start the scan. Press the "Main" tab to go back to the main screen. Click "Offline" button. Find the Case ID and Slide ID that was assigned and click "Offline Scan." In the black box at the top left area, click on an arrow, and select "10× Pre-scan." Using the arrows (< ‖ Δ > a.k.a. "back," "pause," "play," "forward" buttons), go through the scanned images.

3. After finding an image of interest, double click on the screen at the middle of the target region, and press "Snap." This will target an image for capture later on. After

selecting all targets, click "Classify." Select "10× Pre-scan." Right click an image, and get the chance to classify the images. Select "Polytene."

4. Set up 40× Pre-Scan in the Duet software. Click "Main" to go back to the main menu. Select "Online" again. Select a slide. Change "BF" to "FL." Change the task name to "Revisit-X40-RG." Change the section right below the last setting to "Revisit-ALL." Click "Set&Run." Press "OK." Follow the prompts again to set up the automation. Click the "Start matching views" button to match 10× and 40× images. Click "Finish" to start the scan. Once done, click "Classify," and look at the images.

4.2.1.10   Representative Results of Fluorescent In Situ Hybridization with Polytene Chromosomes

Figure 4.5 demonstrates a preparation of ovarian nurse cell poly-tene chromosomes from females of *Anopheles gambiae* made using the high-pressure technique. This method does not damage or change most of the chromosome structure.

The high-pressure procedure flattens spatially bent chromosomes and, thus, reveals hidden fine bands that are not seen on regular preparations (Figure 4.6).

The probes used in this protocol are genomic BAC DNA clones obtained from an ND-TAM BAC library generated from *Anopheles gambiae* PEST strain DNA. The genomic DNA for this library was extracted from newly hatched first instar larvae of both sexes. Figure 4.7 shows the results of FISH using BAC clones hybridized to polytene chromosomes of *Anopheles gambiae*. This procedure was performed using the Xmatrx automated slide staining system.

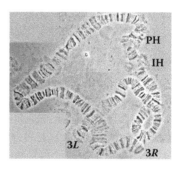

FIGURE 4.5   A spread of polytene chromosome 3 of *Anopheles gambiae* obtained using the high-pressure technique. 3L and 3R mark left and right arms at their telomeres. PH and IH indicate pericentric and intercalary heterochromatin, respectively. (From George, P. et al., *J. Vis. Exp.,* (64), e4007 10.3791/4007, 2012.)

(a)

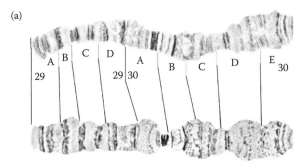

3R Standard (top) vs. high-pressure (bottom)

(b)

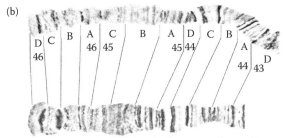

3L Standard (top) vs. high-pressure (bottom)

FIGURE 4.6 Comparison of *Anopheles gambiae* polytene chromosomes prepared using the standard (top) and high-pressure (bottom) techniques. The images cover subdivisions (a) 29A–30E of arm 3R and (b) 43D–46D of arm 3L. (From George, P. et al., *J. Vis. Exp.*, (64), e4007 10.3791/4007, 2012.)

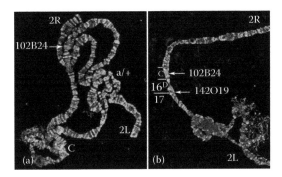

FIGURE 4.7 (**See color insert.**) Fluorescent in situ hybridization (FISH) of bacterial artificial chromosome (BAC) clones to polytene chromosomes of *Anopheles gambiae*. (a) Hybridization of 102B24 (red signal) with the 2R arm. (b) Dual-color FISH of 102B24 (red signal) and 142O19 (blue signal) to subdivisions 16C and 16D of the stretched 2R arm, respectively. Arrows indicate signals of hybridization of BAC clones labeled with Cy3 (red) and Cy5 (blue). (c) The centromeric region. a/+ shows the heterozygote 2La inversion. Chromosomes were counterstained with the fluorophore YOYO-1. (From George, P. et al., *J. Vis. Exp.*, (64), e4007 10.3791/4007, 2012.)

Using a cytogenetic photomap for *Anopheles gambiae* (George et al. 2010), two BAC clones, 102B24 (GenBank: BH372701, BH372694) and 142O19 (GenBank: BH368703, BH368698), were localized in subdivisions 16C and 16D of the 2R arm, respectively. However, the BLAST search against the *Anopheles gambiae* PEST strain AgamP3 assembly identified the sequences homologous to 102B24 and 142O19 in subdivisions 16B and 17A of the 2R arm, respectively (Table 4.1). Therefore, the correspondence between the genomic coordinates and the cytogenetic subdivisions can now be adjusted according to our mapping data. The BLAST search against the *Anopheles gambiae* M-form and S-form genome assemblies found that the two BAC clones are located in different contigs, but in the same scaffolds within each form (Table 4.1).

The identified contigs can now be associated with specific chromosomal locations. Moreover, the identified scaffolds can now be properly oriented within the cytological subdivisions 16CD. Interestingly, the distances between the 102B24 and 142O19 sequences in the scaffolds are 1,892,981, 1,658,391, and 1,688,426 bp in the PEST strain, M-form, and S-form genome assemblies, respectively. Because, the PEST strain is a hybrid between the M and S forms, this difference is likely due to incorrect assembly of the PEST genome. The detailed physical mapping performed on high-pressure chromosome preparations has the potential to link DNA sequences to specific chromosomal structures, such as bands, interbands, puffs, centromeres, telomeres, and heterochromatin, thus creating chromosome-based genome assemblies.

4.2.1.11   Troubleshooting of Fluorescent In Situ Hybridization with Polytene Chromosomes

Improper squashing can lead to insufficiently spread chromosomes, which can cause problems when trying to determine probe locations after FISH. Therefore, the most critical step of the high-pressure procedure is the proper squashing of ovarian nurse cells isolated at Christophers' III stage of ovarian development (Clements 1992). If the chromosomes are over-squashed, they can be become broken or elongated to the point where banding patterns are lost. Production of multiple slides should lead to a consistency when attempting to squash the slides using the Dremel tool, which will increase overall slide production efficiency. The high-pressure technique was first developed for freshly isolated salivary glands of *Drosophila melanogaster* (Novikov et al. 2007). However, ovaries of mosquitoes are routinely preserved in modified Carnoy's fixative solution (3 methanol:1 glacial acetic acid by volume) before they are used for chromosomal preparations. Therefore, we modified the existing high-pressure protocol to make it suitable for the fixed ovarian nurse cell polytene chromosomes of the

TABLE 4.1

BLAST Results of the BAC-End Sequences against the Genome Assemblies of *Anopheles gambiae*

BAC (Accession)	PEST-Strain AgamP3 Coordinates	M-Form Contigs Coordinates	M-Form Scaffolds Coordinates	S-Form Contigs Coordinates	S-Form Scaffolds Coordinates
102B24 (BH372694)	2R:AAAB01008844_26 (16B), 43,681,342–43,681,874	ABKP02024753.1, 32,513–33,045	EQ090167.1, 1,858,377–1,858,909	ABKQ01012332.1, 4,299–4,832	EQ099711.1, 215,891–216,424
102B24 (BH372701)	2R:AAAB01008844_28 (16B), 43,788,213–43,788,969	ABKP02024751.1, 24,816–25,575	EQ090167.1, 1,772,683–1,773,442	ABKQ01012335.1, 27,782–28,540	EQ099711.1, 305,514–306,272
142O19 (BH368698)	2R:AAAB01008805_5 (17A), 45,428,580–45,429,264	ABKP02024701.1, 2,499–3,185	EQ090167.1, 315,806–316,492	ABKQ01012376.1, 49,242–49,942	EQ099711.1, 1,791,625–1,792,325
142O19, (BH368703)	2R:AAAB01008805_8 (17A), 45,560,099–45,574,323	ABKP02022017.1, 30,971–32,469	EQ090167.1, 200,518–202,016	ABKQ01012379.1, 27,760–28,508	EQ099711.1, 1,903,569–1,904,317

malaria mosquito *Anopheles gambiae*. Because the high pressure is applied using a precision vice possessing a highly parallel work surface of the entire slide, it takes significantly less time to prepare the chromosome squash than using a standard tapping technique with a pencil's eraser.

It is important to sufficiently soak follicles in 50% propionic acid and heating the slide. Both of these steps are essential in helping to flatten the chromosomes. If they are neglected, chromosomes can appear shiny after dehydration, which potentially leads to an over-abundance of background that can be mistaken for a signal in FISH. Although regular squash preparations are sufficient for many pur-poses including FISH (Scriven et al. 2011) and immunostaining (Cai et al. 2010), the high-pressure method not only lowers potential vari-ance from one slide to the next but also increases overall chromo-some quality, leading to higher detail when mapping chromosomes.

Slide breakage and overstretching of the chromosomes are among the major limitations to the high-pressure method. Slide breakage caused by placing too much stress on the slide via the vice is possible, but is limited by using the pressures denoted in the article. For overstretching of the chromosomes, if the Dremel tool is applied to the slide for an extended period, there is the possibil-ity that the chromosomes can become too stretched out and lose resolution. Applying the tool for a brief period, checking under the microscope, and applying more time with the Dremel tool if the chromosomes are insufficiently spread can remedy this problem.

An automated FISH system can significantly increase the throughput. A standard FISH protocol includes several washing and incubation steps, which are usually 5–20 minutes long and require almost full attention of a researcher for the whole dura-tion of the experiment. Moreover, the number of slides that can be handled manually is usually limited to a few slides in a given experiment. In contrast, an automated slide staining system per-forms all steps (including washing, incubation, probe application, denaturation, hybridization, coverslip application, and removal) automatically. This frees up to 6–8 working hours for a researcher in a FISH experiment. The Xmatrx system is capable of process-ing up to 40 assays on single preparation slides and up to 80 assays on dual preparation slides simultaneously. Among limitations of this system is that it is not efficient for a small number of FISH experiments, as it requires preparing large volumes of solutions. In addition, the FISH protocol programmed in the system may require modifications and adjustments for new applications.

FISH results can be viewed using a regular fluorescent micro-scope or a slide scanning system, which is an automated version of a fluorescent microscope. Automated stage moving and a simple microscope control panel free a researcher's time and make operating the microscope extremely simple. For instance, the software in the ACCORD PLUS scanning system allows for easy capturing of multiple channels of fluorescence, and easy to manage image acquisition. An example of this is the inclusion of z-stack capturing, rather than taking a single image; the software captures a configurable z-stack of images to ensure that at least one image remains in focus. Although this system is made more for multiple image acquisition (a lot of cells on a single slide), it still makes finding chromosomes on the slide much easier than navigating through the slide manually.

4.2.2 Fluorescent In Situ Hybridization with Mitotic Chromosomes of Mosquitoes

Among the three mosquito genera, namely *Anopheles*, *Aedes,* and *Culex,* physical genome mapping techniques were well established only for *Anopheles*, whose members possess readable polytene chromosomes. For the genera of *Aedes* and *Culex*, however, cytogenetic mapping remains challenging because of the poor quality of polytene chromosomes. Here we present a universal protocol for obtaining high-quality preparations of mitotic chromosomes and an optimized FISH protocol for all three genera of mosquitoes. We developed a simple, robust technique for obtaining high-quality mitotic chromosome preparations from imaginal discs (IDs) of fourth instar larvae, which can be used for all three genera of mosquitoes (Timoshevskiy et al. 2012). A standard FISH protocol (Garimberti and Tosi 2010) is optimized for using BAC clones of genomic DNA as a probe on mitotic chromosomes of *Aedes aegypti and Culex quinquefasciatus,* and for using an intergenic spacer (IGS) region of rDNA as a probe on *Anopheles gambiae* mitotic chromosomes. A scheme showing the FISH protocol including DNA probe preparation, hybridization, and signal visualization is shown in Figure 4.8.

4.2.2.1 Materials for Fluorescent In Situ Hybridization with Mitotic Chromosomes

1. Olympus CX41 Phase Microscope (Olympus, Cat. No. CX41RF-5)
2. Olympus BX61 fluorescent microscope (Olympus, Cat. No. BX61)
3. ThermoBrite Slide Denaturation/Hybridization System (Abbott Molecular, Cat. No. 30-144110)
4. MZ6 Leica stereo microscope (Leica, Cat. No. VA-OM-E194-354)
5. Dissecting needles (Fine Science Tools, Cat. No. 10130-10)
6. Needle holders (Fine Science Tools, Cat. No. 26018-17)
7. 75 × 25-mm double frosted micro slides (Corning, Cat. No. 2949-75×25)

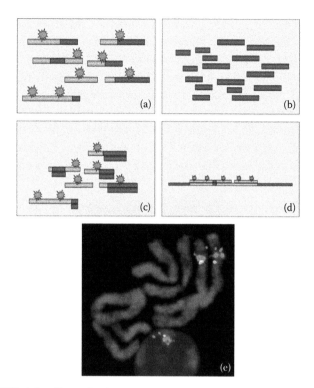

FIGURE 4.8 (**See color insert.**) A schematic representation of the fluorescent in situ hybridization (FISH) procedure. (a) Preparation of fluorescently labeled DNA probe. (b) Preparation of unlabeled repetitive DNA fraction. (c) Blocking unspecific hybridization of the probe with unlabeled repetitive DNA fraction. (d) Hybridization of fluorescently labeled DNA probe with chromosomes. (e) Visualization of FISH signals on mitotic chromosomes. (From Timoshevskiy, V.A. et al., *J. Vis. Exp.*, (67), e4215 10.3791/4215, 2012.)

8. 22 × 22-mm microscope coverslips (Fisher Scientific, Cat. No. 12-544-10)

9. 18 × 18-mm microscope coverslips (Fisher Scientific, Cat. No. 12-553-402)

10. Dissecting scissors (Fine Science Tools, Cat. No. 15000-03)

11. Parafilm (Fisher Scientific, Cat. No. 13-374-10)

12. Rubber Cement (Fisher Scientific, Cat. No. 50-949-105)

13. Qiagen Blood and Cell Culture Maxikit (Qiagen, Cat. No. 13362)

14. Qiagen Large Construct Kit (Qiagen, Cat. No. 12462)

15. Random Primed DNA Labeling kit (Roche, Cat. No. 11 004 760 001)

16. 10× PBS (Invitrogen, Cat. No. P5493)

17. 10% NBF (Sigma-Aldrich, Cat. No. HT501128)

18. 99% Formamide (Fisher Scientific, BP227500)

19. Dextran sulfate sodium salt (Sigma-Aldrich, Cat. No. D8906)

20. 20× SSC buffer (Invitrogen, Cat. No. AM9765)
21. Sodium phosphate (Sigma-Aldrich, Cat. No. S3264)
22. Sodium citrate dehydrate (Fisher Scientific, Cat. No. S279-500)
23. Sodium acetate trihydrate (Fisher Scientific, Cat. No. BP334-500)
24. 50× Denhardt's solution (Sigma-Aldrich, Cat. No. D2532)
25. 1 mM YOYO-1 iodide (491/509) solution in DMSO (Invitrogen, Cat. No. Y3601)
26. ProLong Gold antifade reagent (Invitrogen, Cat. No. P36930)
27. dATP, dCTP, dGTP, dTTP (Fermentas, Cat. No. R0141, R0151, R0161, R0171)
28. Cy3-dUTP, Cy5-dUTP (GE Healthcare, PA53022, PA55022)
29. BSA (Sigma-Aldrich, Cat. No. A3294)
30. DNA polymerase I (Fermentas, Cat. No. EP0041)
31. DNase I (Fermentas, Cat. No. EN0521)
32. Acetic acid (Fisher Scientific, Cat. No. A491-212)
33. Methanol (Fisher Scientific, Cat. No. A412-4)
34. Propionic acid (Sigma-Aldrich, Cat. No. 402907)
35. Alcohol 200 Proof (Decon Laboratories, Cat. No. 2701)
36. Hydrochloric acid (Fisher Scientific, Cat. No. A144-500)
37. Potassium chloride (Fisher Scientific, Cat. No. BP366-500)
38. EDTA (Fisher Scientific, Cat. No. S311-500)
39. Tris base (Fisher Scientific, Cat. No. BP152-1)
40. BSA (Sigma-Aldrich, Cat. No. A3294)
41. S1 Nuclease (Fermentas, Cat. No. EN0321)
42. RNase (Sigma-Aldrich, Cat. No. 9001-99-4)
43. Taq DNA polymerase (Invitrogen, Cat. No. 18038-042)
44. Pepsin (USB, Cat. No. 9001-75-6)
45. Salmon sperm DNA (Sigma-Aldrich, Cat. No. D7656)
46. Nonidet-P40 (NP40) (US Biological, Cat. No. NC9375914)

4.2.2.2 Mitotic Chromosome Preparation

1. Hatch mosquito eggs at 28°C, and after 2–3 days, transfer 2nd or 3rd instar larvae to a chamber kept at a lower temperature. Use 16°C for *Aedes aegypti* and *Culex quinquefasciatus* and 22°C for *Anopheles gambiae*. It is important to choose a larva at the correct stage (Figure 4.9).

2. Place fourth instar larvae on ice for several minutes to immobilize them.

3. Transfer a larva to a slide with a drop of cold hypotonic solution (0.5% sodium citrate or 0.075 M potassium chloride), and place it under the stereo microscope.

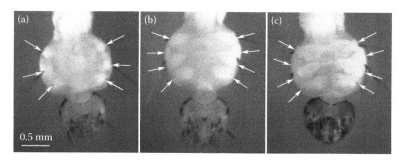

FIGURE 4.9   Stages of the imaginal disk (ID) development in fourth instar larva. (a) An early "round shape" stage. (b) An intermediate "oval shape" stage—optimal for the chromosome preparation. (c) A late stage—inappropriate for chromosome preparations. The positions of IDs are indicated by arrows on the ventral side of the larval thorax. (From Timoshevskiy, V.A. et al., *J. Vis. Exp.,* (67), e4215 10.3791/4215, 2012.)

4. Select larva with oval IDs for further dissection (Figure 4.10).
5. Decapitate larva, and cut the cuticle from the ventral side of the larval thorax using dissecting scissors (Figure 4.10a). Make an additional cut in second or third abdominal segment to dissect the gut from the larva. The directions of the cuts are shown by arrows.
6. Open the cuticle, and remove the gut and fat body from the larva. Remove the hypotonic solution from the slide using filter paper, and add a fresh drop of hypotonic solution directly to the IDs within the larva (Figure 4.10b). Keep the larva in hypotonic solution for 10 minutes at RT. In the case of *Anopheles gambiae*, dissect out the IDs from the larva (they are different in color than surrounding tissue), and place them in hypotonic solution on a fresh slide.
7. Remove hypotonic solution using filter paper, and apply Carnoy's solution (ethanol:acetic acid in 3:1 ratio). After adding fixative solution, IDs immediately turn white and become easily visible under the microscope (Figure 4.10c).
8. Using dissecting needles, remove IDs from the larva (Figure 4.10d), and transfer them to a drop of 50% propionic acid. Remove any other tissues, such as the gut and fat body, from the slide. Cover IDs with an unsiliconized 22 × 22 coverslip, and keep at RT for 5–10 minutes. In the case of *Anopheles gambiae*, keep the previously dissected IDs in 50% propionic acid for about 2 minutes before covering with an unsiliconized 22 × 22 coverslip.
9. Cover the slide with filter paper, and squash the tissue by tapping the eraser of a pencil on the perimeter of the coverslip.

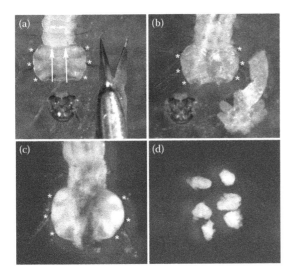

FIGURE 4.10   Steps of imaginal disk (ID) dissection. (a) Decapitated larva (the direction of cuts are indicated by arrows). (b) Larvae with dissected gut under hypotonic solution treatment (IDs swell and become almost invisible). (c) Larva after Carnoy's solution application (IDs become white and clearly visible). (d) Dissected IDs in Carnoy's solution. Positions of IDs in larva are indicated by asterisks. (From Timoshevskiy, V.A. et al., *J. Vis. Exp.,* (67), e4215 10.3791/4215, 2012.)

10. Briefly analyze the quality of the slide using the phase contrast microscope at 100× or 200× magnification (Figure 4.11). Preparations with >50 chromosome spreads can be considered suitable for FISH.
11. Dip and hold the slide in liquid nitrogen until it stops bubbling. Remove the coverslip from the slide using a razor blade, and immediately transfer the slide to a container of 50% ethanol chilled at −20°C. Store at 4°C for at least 2 hours for the best dehydration result (if necessary, slides can be stored at this step from several minutes to several days).
12. Dehydrate slides in a series of ethanol (70%, 80%, 100%) at 4°C for 5 minutes each, and air-dry at RT. Store dry slides at −20°C before using them for FISH.

4.2.2.3   Extraction of Repetitive DNA Fractions

Performing FISH of a BAC clone DNA probe on chromosomes from *Aedes aegypti* and *Culex quinquefasciatus* requires using unlabeled repetitive DNA fractions to block unspecific hybridization of the DNA repeats to the chromosomes. The reassociation of single-strand DNA fragmented into pieces of several hundred base pairs follows a C_0t curve, where C_0 is the initial concentration of single-stranded DNA and t the reannealing time. DNA fractions with C_0t values equal to 10^{-4}–10^{-1} or 10^0–10^2 are considered as highly and moderately repetitive, respectively.

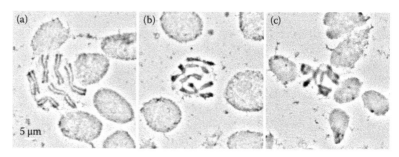

FIGURE 4.11 Different qualities of the chromosome spreads. (a) A perfect chromosome spread—round shape of the cells demonstrates sufficient treatment of the imaginal disks in hypotonic solution. (b) A perfect hypotonic treatment—chromosomes are slightly undersquashed. (c) A poor chromosome spread—the result of insufficient hypotonic treatment is indicated by oval shape of the cells. (From Timoshevskiy, V.A. et al., *J. Vis. Exp.*, (67), e4215 10.3791/4215, 2012.)

1. Extract 400–500 µg genomic DNA from entire adult mosquitos using a Qiagen Blood and Cell Culture Maxikit, and prepare a solution of 100–1000 ng/µL DNA in 1.2× SSC.
2. Denature DNA by placing a safe-lock tube with genomic DNA into a heating block prewarmed to 120°C for 2 minutes. High temperature helps to range DNA into 200–500 bp fragments.
3. Depending on the DNA concentration, reassociate DNA by placing the tube at 60°C for 15–150 minutes to obtain C_0t DNA fractions up to C_0t 3 (Table 4.2).
4. Place the tube with DNA on ice for 2 minutes.
5. Transfer the DNA to 42°C, add preheated 10× S1 nuclease buffer and S1 nuclease to a final concentration of 100 U per 1 mg of DNA, and incubate for 1 hour.
6. Precipitate DNA by adding 0.1 volume 3 M sodium acetate and 1 volume isopropanol at RT.
7. Centrifuge at 14,000 rpm for 20 minutes at 4°C.
8. Wash DNA in 70% ethanol, and centrifuge again at 14,000 rpm for 10 minutes at 4°C. Air-dry and dissolve DNA pellet in Tris-EDTA buffer.
9. Measure the DNA concentration, and visualize by gel electrophoresis. Usually, the final quantity of repetitive DNA fractions for *Aedes aegypti* represents 35%–50% of the original DNA amount.

4.2.2.4 Bacterial Artificial Chromosome Clone DNA Labeling Using Nick Translation

1. Extract BAC clone DNA from the BAC library using a Qiagen Large Construct Kit.
2. Prepare a reaction mixture for nick translation labeling on ice with a final volume of 50 µL: 1 µg isolated BAC clone DNA; 0.05 mM each of unlabeled dATP, dCTP, and

Table 4.2

DNA Concentration and Reannealing Times for Preparation of C_0t 2 and C_0t 3 Fractions

	DNA Concentration ($\mu g/\mu L$)	Reannealing Time (minutes)
C_0t 2	0.1	100
	0.3	33
	0.5	20
	0.7	14
	0.9	11
	1	10
C_0t 3	0.1	150
	0.3	50
	0.5	30
	0.7	21
	0.9	17
	1	15

dGTP and 0.015 mM dTTP; 1 μL Cy3-dUTP (or another fluorochrome); 0.05 mg/mL BSA; 5 μL 10× nick translation buffer; 20 U DNA polymerase I; and 0.0012 U DNase.

3. Incubate at 15°C for 2.5 hours.
4. Stop reaction by adding 1 μL 0.5 M EDTA.
5. Store probe at −20°C in a dark place.

4.2.2.5   Intergenic Spacer Ribosomal DNA Labeling Using Polymerase Chain Reaction

1. Prepare a reaction mixture on ice with a final volume of 50 μL: 200 ng genomic DNA; 0.05 mM each of unlabeled dATP, dCTP, and dGTP; 0.015 mM dTTP; 1 μL Cy3-dUTP (or another fluorochrome); 5 μL 10× PCR-buffer; 50 pmol of forward; UN (GTGTGCCCCTTCCTCGATGT) and reverse; GA (CTGGTTTGGTCGGCACGTTT) primers for IGS amplification; and 10 U Taq DNA polymerase (Scott et al. 1993).
2. Perform PCR reaction using standard PCR parameters for IGS amplification: 95°C/5 minutes × 1 cycle; (95°C/30 seconds, 50°C/30 seconds, 72°C/30 seconds) × 30 cycles; 72°C/5 minutes × 1 cycle; and 4°C hold (Scott et al. 1993).
3. Store probe at −20°C in a dark place.

4.2.2.6   Fluorescent In Situ Hybridization Using Bacterial Artificial Chromosome DNA Clones

1. Incubate slides in 2× SSC for 30 minutes at 37°C. Removing residual 2× SSC solution by placing a slide's edge to a paper towel and air-dry.

2. Put slides in a jar with 0.01% pepsin and 0.037% HCl solution, and incubate for 5 minutes at 37°C. Wash slides in 1× PBS for 5 minutes at RT.

3. Fix chromosome preparation in a jar with 1% formalin in 1× PBS prepared from 10% neutral-buffered formalin for 10 minutes at RT. Wash slides in 1× PBS for 5 minutes at RT. Dehydrate slides in series of 70%, 80%, and 100% ethanol for 5 minutes each at RT, and air-dry preparations at 37°C.

4. Denature slides in a jar with prewarmed 70% formamide for 2 minutes at 72°C. Dehydrate slides in series of cold (−20°C) 70%, 80%, and 100% ethanol for 5 minutes each, and air-dry at 37°C.

5. Prepare hybridization mixture: 5 μL labeled probe DNA from Step 3, 10 μL C_0t DNA fractions from Step 2 with final concentration of 0.5 ng/μL and 5 μL 1 μg/μL sonicated salmon sperm DNA.

6. Precipitate DNA by adding 0.1 volume 3 M sodium acetate and 2.5 volumes ethanol. Keep at −20°C for 1–3 hours.

7. Centrifuge at 14,000 rpm at 4°C for 20 minutes, remove the ethanol, and air-dry the pellet at RT. Thoroughly dissolve the pellet in 10 μL hybridization buffer: 60% formamide, 20% dextran sulfate, 2× SSC. Denature hybridization mixture for 5 minutes at 75°C, and immediately put on ice for 1 minute.

8. Prehybridize mixture at 37°C for 30 minutes to prevent unspecific hybridization of repetitive DNA to the chromosomes.

9. Place 10 μL hybridization mixture on the slide and cover with a 22 × 22-mm coverslip. Any air bubbles should be removed with gentle pressure to the coverslip using the tip of a pair of forceps. Glue coverslip around the perimeter using rubber cement. If using multiple probes, add 3–5 μL of each probe in a PCR tube to total volume10 μL and mix gently before applying on the coverslip.

10. Perform overnight hybridization in a humid chamber at 37°C.

11. Remove rubber cement and coverslip from the slide. Wash slide for 2 minutes in prewarmed Solution 1 (0.4× SSC, 0.3% Nonidet-P40) at 73°C. Wash slides in Solution 2 (2× SSC, 0.1% Nonidet-P40) for 5 minutes at RT.

12. Counterstain slide using 0.001 mM YOYO-1 in 1× PBS for 10 minutes in humid chamber at RT. Mount in a small amount of Prolong Gold antifade reagent with a coverslip.

13. Analyze preparations under a fluorescent microscope using appropriate filter sets at 1000× magnification.

4.2.2.7   Fluorescent In Situ Hybridization Using Intergenic Spacer Ribosomal DNA Probes

1. Incubate slides in 2× SSC for 30 minutes at 37°C. Dehydrate slides in series of 70%, 80%, and 100% ethanol for 5 minutes each at RT, and air-dry.

2. Incubate chromosome preparations in 0.1 mg/mL RNase solution under parafilm for 30 minutes at 37°C. Wash twice in 2× SSC for 5 minutes each at 37°C.

3. Put slides in a jar with 0.01% pepsin and 0.037% HCl solution, and incubate for 5 minutes at 37°C. Wash slides in 1× PBS for 5 minutes at RT.

4. Fix chromosome preparation in a jar with 1% formalin in 1× PBS prepared from 10% neutral-buffered formalin for 10 minutes at RT. Wash slides in 1× PBS for 5 minutes at RT. Dehydrate slides in series of 70%, 80%, and 100% ethanol for 5 minutes each at RT, and air-dry preparations at 37°C.

5. Prepare hybridization mixture: 5 μL labeled probe DNA from Step 3, and 5 μL 1 μg/μL sonicated salmon sperm DNA. Precipitate DNA by adding 0.1 volume 3 M sodium acetate and 2.5 volumes ethanol. Keep at −20°C for 1–3 hours. Centrifuge at 14,000 rpm at 4°C for 20 minutes, remove the ethanol, and air-dry the pellet at RT.

6. Thoroughly dissolve the pellet in 10 μL hybridization buffer: 50% formamide, 20% dextran sulfate, 2× SSC. Place 10 μL hybridization mixture on the slide and cover with a 22 × 22-mm coverslip. Prevent bubble formation by using gentle pressure on the coverslip.

7. Denature the probe and chromosome DNA simultaneously using a heating block at 72°C for 5 minutes. Glue coverslip around the perimeter using rubber cement.

8. Perform overnight hybridization in a humid chamber at 37°C.

9. Remove rubber cement and coverslip from the slide. Wash slide for 2 minutes in prewarmed Solution 1 (0.4× SSC, 0.3% Nonidet-P40) at 73°C. Wash slides in Solution 2 (2× SSC, 0.1% Nonidet-P40) for 5 minutes at RT.

10. Counterstain slide using 0.001 mM YOYO-1 in 1× PBS for 10 minutes in humid chamber at RT. Mount in a small amount of Prolong Gold antifade reagent with a coverslip. Alternatively, counterstain using 4′,6-diamidino-2-phenylindole (DAPI) with Prolong. Add a drop on selected

area, cover with coverslip, and store in dark for few hours or overnight before visualization under a fluorescent microscope.

11. Analyze preparations under a fluorescent microscope using appropriate filter sets at 1000× magnification.

4.2.2.8 Representative Results of Fluorescent In Situ Hybridization with Mitotic Chromosomes

Mosquito IDs can be found in each segment of the larval thorax. Depending on the position, IDs transform into different tissues at the adult stage of the insect. The IDs, which are used for the chromosome preparation in this protocol, develop into legs at the adult stage of the mosquito. These IDs are located at the ventral side of the larval thorax and are clearly visible through the cuticle under the microscope (Figure 4.9). At the early fourth instar larval stage, IDs have a round shape (Figure 4.9a). The largest numbers of mitosis, approximately 175 in one ID (Sharakhova et al. 2011b), are accumulated at a later "oval-shaped" stage (Figure 4.9b), which is considered the optimal stage for slide preparation. At this time, the intermediate ID splits into two: one transforms into a leg and another one transforms into a wing. We prefer using the large leg IDs at the "oval-shaped" stage for chromosome slide preparation. Figure 4.9c represents IDs at the latest stage of fourth instar larva development. At this stage, the IDs are already developed into legs and wings, and contain a significant amount of differentiated tissues and a low number of mitoses. This stage of ID development should be avoided for chromosome slide preparation. We also recommend rearing mosquito larvae at low temperatures: 16°C for *Aedes* and *Culex* and 22°C for *Anopheles*. This helps to increase the amount of mitosis in IDs (Sharakhova et al. 2011b).

The dissection of ID from the thorax of a fourth instar larva is shown in Figure 4.10. Because the cuticle of a live insect is hard to dissect, we recommend using dissecting scissors instead of the needles commonly used for larva preparation. The most crucial procedure for obtaining high-quality chromosome preparation is the hypotonic solution treatment. For best results, we remove the gut and fat body from the larval thorax before this treatment. Swelling of the ID cells during this procedure helps to spread chromosomes on a slide (Figure 4.11). The appropriate quality of the hypotonic solution treatment can be easily recognized by the round shape of cells in the preparations (Figure 4.11a and b). Cells with an oval shape indicate insufficient hypotonic solution treatment (Figure 4.11c). To be selected for FISH, chromosome preparation should contain at least 50 high-quality chromosome spreads. Normally, approximately 90% of the slides prepared using this protocol have sufficient quality for FISH (Sharakhova et al. 2011b).

Two slightly different FISH protocols are presented here: an advanced protocol for FISH using genomic BAC clone DNA probe on mitotic chromosomes of *Aedes* and *Culex* and a simple FISH protocol for IGS rDNA probe on mitotic chromosomes of *Anopheles*. The genomes of *Aedes* and *Culex* are highly repetitive because of the overrepresentation of TEs (Nene et al. 2007; Arensburger et al. 2010). Thus, performing FISH, which uses genomic BAC clone DNA as a probe, requires adding unlabeled repetitive DNA fractions to the probe to block unspecific hybridization of the DNA repeats to chromosomes. For the extraction of the repetitive DNA fractions, genomic DNA is denatured at 120°C for 2 minutes. Boiling DNA at a high temperature also helps to obtain DNA in fragments of 200–500 bp. DNA is allowed to reassociate after this treatment. The highly repetitive DNA fragments tend to find their mate for reassociation faster than DNA with unique sequences does. As a result, the reassociation of DNA follows a $C_0 \times t$ curve where C_0 is the initial concentration of single-stranded DNA, and t is the reannealing time. DNA fractions with C_0t values equal to 10^{-4}–10^{-1} or 10^0–10^2 are considered highly and moderately repetitive, respectively. The time of reassociation for different C_0t DNA fractions can be calculated using the formula $t = C_0tX \times 4.98/C_0$, where t is the time of incubation, C_0tX is C_0t fraction ($C_0t1 = 1$, $C_0t2 = 2$, etc.), and C_0 initial DNA concentration in µg/µL (Trifonov et al. 2009) (Table 4.2). After reassociation, the single-stranded DNA is digested using S1 nuclease. We prefer using all C_0t DNA fractions up to C_0t3 together instead of the commonly used C_0t1 DNA fraction. These C_0t fractions include some of the moderately repetitive DNA sequences and together usually represent 35%–50% of the original amount of the genomic DNA in *Aedes aegypti*. The correct proportion between labeled DNA probe and unlabeled C_0t DNA fraction depends on the repetitive DNA component in each particular BAC clone. On average, we use 1:20 probe to C_0t DNA fraction proportion for obtaining an acceptable signals/background ratio of the FISH result. Prehybridization of the DNA probe with C_0t DNA fractions in a tube for 30 minutes before the actual hybridization on the slide also helps to reduce background. Labeling, hybridization itself, and washing in this protocol are performed using standard conditions (Garimberti and Tosi 2010).

Figure 4.12 illustrates the results obtained by hybridization of various probes with mitotic chromosomes of three different species of mosquitoes. The FISH results of two differently labeled BAC clone DNA probes on mitotic chromosomes of *Aedes aegypti* and *Culex quinquefasciatus* are shown in Figure 4.12a and b, respectively. The BAC clone DNA probes produce strong

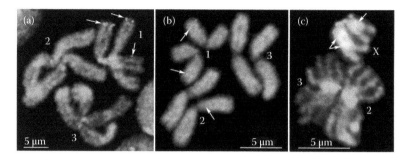

FIGURE 4.12 **(See color insert.)** Examples of fluorescent in situ hybridization (FISH) result with mitotic chromosomes. (a) FISH of bacterial artificial chromosome (BAC) clones with chromosomes of *Aedes aegypti*. (b) FISH of BAC clones with chromosomes of *Culex quinquefasciatus*. (c) FISH of intergenic spacer rDNA with chromosomes of *Anopheles gambiae*. 1, 2, and 3 are numbers of chromosomes; X—female sex chromosome in *Anopheles gambiae*. (From Timoshevskiy, V.A. et al., *J. Vis. Exp.*, (67), e4215 10.3791/4215, 2012.)

signals in a single position on the chromosomes. A simple version of the FISH protocol is designed for hybridization of IGS rDNA probe on mitotic chromosomes of *Anopheles*. Ribosomal genes in *Anopheles* are represented as a polymorphic cluster of genes located on sex chromosomes (Collins et al. 1987). A DNA probe in this protocol is labeled using standard PCR reaction by adding fluorescently labeled Cy3 or Cy5 dNTPs. Because blocking unspecific hybridization of repetitive DNA in euchromatin is not needed, all steps related to using C_0t DNA fractions are omitted. Instead, chromosome preparations are pretreated with RNase for preventing hybridization of the IGS rDNA probe to the nucleolus. Chromosomes and the DNA probe are denatured simultaneously by heating the slide together with a probe in a hybridization system at 75°C for 5 minutes. Hybridization and washing in this protocol are also performed using standard conditions for FISH (Garimberti and Tosi 2010). Figure 4.12c demonstrates the polymorphism of the IGS rDNA hybridization between two X chromosomes.

Figure 4.12 depicts chromosomes counterstained with YOYO-1 iodide. This dye produces the best banding patterns on *Aedes aegypti* chromosomes (Sharakhova et al. 2011b). Alternatively, other fluorescent dyes, such as DAPI (Figure 4.8e) or propidium iodide, can be used for the chromosome counterstaining. For suppressing photobleaching of the slides, we use Prolong Gold antifade mounting medium. This reagent has good signal preservation abilities and also can be easily removed from the slide by rinsing in 1× PBS if it is necessary to use the same slide for several hybridizations.

4.2.2.9 Troubleshooting of Fluorescent In Situ Hybridization with Mitotic Chromosomes

The protocol presented here is a simple, robust technique for obtaining high-quality chromosome preparations from IDs of fourth instar larvae (Sharakhova et al. 2011b). This method allows a high number of chromosomes to be obtained on one slide and can be used for all species of mosquitoes. However, some difficulties can limit the quality of chromosomal preparations and the final FISH results. Before slide preparation, it is important to choose larvae at the correct stage of development. The large number of chromosomes and high-quality preparation can be obtained from the larvae with IDs that are oval in shape. Choosing an earlier stage with round IDs will result in a low number of chromosomes. On the other hand, obtaining chromosome preparations from the later stage of the IDs development (when they start transforming into legs and wings) will cause the lower number of mitotic cells and poor spreading of the chromosomes. At the chromosome preparation step, the most critical procedure is treating the tissue with a hypotonic solution. Treatment of a tissue with 0.5% sodium citrate or 0.075 M potassium chloride causes swelling of IDs and simplifies releasing and spreading of chromosomes from the cells. To ensure that this procedure works correctly, the tissue of the IDs must interact with the solution. All additional tissues such as gut or fat body have to be removed from the larval thorax before the treatment. Overdeveloped IDs can also cause poor penetration of hypotonic solution into the cells. Growing larva at suggested lower temperatures would also help enhance the number of mitotic plates on a slide.

The FISH part of the protocol also has some critical steps. Using rDNA as a probe for FISH requires treatment of the preparation with RNase to reduce the background associated with hybridization of the probe to the nucleolus. Using BAC clones as a probe for FISH is especially challenging because of the necessity of using unlabeled repetitive DNA fractions to block unspecific hybridization to chromosomes. The correct fraction (C_0t 1–C_0t 10), the proportion of repetitive DNA to the probe, and the amount of time allowed for prehybridization of repetitive fractions with the probe can vary depending on the mosquito species and the quality of the BAC clone DNA. For *Aedes aegypti*, the best fraction to use is C_0t 3, the proportion of the probe to the fraction is 1:20, and the timing of prehybridization of the probe with repetitive DNA fractions is approximately 30 minutes. Extensive background on the chromosomes after FISH can be reduced by increasing the amount of C_0t DNA fraction up to 1:40 in probe/C_0t DNA fraction ratio. Mapping BAC clones from the heterochromatin of *Anopheles* will require using

C_0t 5–C_0t 10 fractions and increasing the time of preincubation up to 1 hour. However, these parameters have to be empirically selected for the specific purposes of the FISH performed.

4.2.3   Chromosome Painting in Malaria Mosquitoes

Chromosome painting is a useful method for studying organization and evolution of the karyotype. We developed an approach to isolate and amplify specific regions of interest from single polytene chromosomes that are subsequently used for FISH (George et al. 2014). The procedure shows how to efficiently isolate a euchromatic segment from a single polytene chromosome arm, amplify the DNA, and use it in downstream 2D (two-dimensional) and 3D (three-dimensional) FISH applications in malaria mosquitoes. First, we apply laser capture microdissection (LCM) to isolate and extract a single chromosome arm from specially prepared membrane slides. Second, whole genome amplification (WGA) is used to amplify the DNA from the microdissected material. Third, we hybridize the amplified DNA in FISH experiments to polytene squash preparations, metaphase and interphase chromosome slides, as well as 3D ovarian whole-mount samples. This procedure has been done to successfully paint a majority of the euchromatin in chromosomal arms of *Anopheles gambiae*. The overall flow-through of the protocol is described in Figure 4.13.

4.2.3.1   Materials for Chromosome Painting

 1. PALM MicroBeam LCM System (Zeiss)
 2. Olympus CX41 Phase Microscope (Olympus, Cat. No. CX41RF-5)
 3. Olympus BX61 fluorescent microscope (Olympus, Cat. No. BX61)
 4. ThermoBrite Slide Denaturation/Hybridization System (Abbott Molecular, Cat. No. 30-144110)
 5. MZ6 Leica stereo microscope (Leica, Cat. No. VA-OM-E194-354)
 6. Vacufuge vacuum concentrator (Eppendorf, Cat. No. 022820001)
 7. Spectroline Microprocessor-Controlled UV Crosslinker XL-1000 (Fisher Scientific, Cat. No. 11-992-89)
 8. Thermomixer (Eppendorf, Cat. No. 022670000)
 9. Thermo Scientific NanoDrop (Fisher Scientific, Cat. No. ND-2000)
10. Dissecting needles (Fine Science Tools, Cat. No. 10130-10)
11. Needle holders (Fine Science Tools, Cat. No. 26018-17)

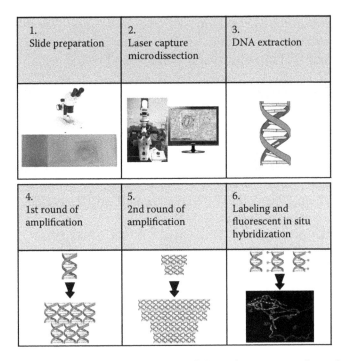

1. Slide preparation	2. Laser capture microdissection	3. DNA extraction
4. 1st round of amplification	5. 2nd round of amplification	6. Labeling and fluorescent in situ hybridization

FIGURE 4.13 (See color insert.) Schematic representation of the experimental procedures toward the preparation of chromosome paints. (From George, P. et al., *J. Vis. Exp.*, (83), e51173, 2014.)

12. Membrane slides 1.0 polyethylene terephthalate (PET) (Zeiss, Cat. No. 415190-9051-000)
13. AdhesiveCap 500 clear (Zeiss, Cat. No. 415190-9211-000)
14. 75 × 25 double frosted micro slides (Corning, Cat. No. 2949-75 × 25)
15. 22 × 22-mm microscope coverslips (Fisher Scientific, Cat. No. 12-544-10)
16. 18 × 18-mm microscope coverslips (Fisher Scientific, Cat. No. 12-553-402)
17. QIAamp DNA Micro Kit (50) (Qiagen, Cat. No. 56304)
18. Genomic DNA Clean and Concentrator Kit (Zymo Research, Cat. No. D4010)
19. REPLI-g Single Cell Kit (Qiagen, Cat. No. 150343)
20. GenomePlex Single Cell Kit (WGA4) (Sigma-Aldrich, Cat. No. WGA4-10RXN)
21. GenomePlex WGA Reamplification Kit (WGA3) (Sigma-Aldrich, Cat. No. WGA3-50RXN)
22. Rubber Cement (Fisher Scientific, Cat. No. 50-949-105)
23. Random Primed DNA Labeling kit (Roche, Cat. No. 11 004 760 001)
24. 10× PBS (Invitrogen, Cat. No. P5493)

25. 99% Formamide (Fisher Scientific, BP227500)
26. Dextran sulfate sodium salt (Sigma-Aldrich, Cat. No. D8906)
27. 20× SSC buffer (Invitrogen, Cat. No. AM9765)
28. 50× Denhardt's solution (Sigma-Aldrich, Cat. No. D2532)
29. 1 mM YOYO-1 iodide (491/509) solution in DMSO (Invitrogen, Cat. No. Y3601)
30. ProLong Gold antifade reagent (Invitrogen, Cat. No. P36930)
31. dATP, dCTP, dGTP, dTTP (Fermentas, Cat. No. R0141, R0151, R0161, R0171)
32. Cy3-dUTP, Cy5-dUTP (GE Healthcare, PA53022, PA55022)
33. BSA (Sigma-Aldrich, Cat. No. A3294)
34. DNA polymerase I (Fermentas, Cat. No. EP0041)
35. DNase I (Fermentas, Cat. No. EN0521)
36. Acetic acid (Fisher Scientific, Cat. No. A491-212)
37. Methanol (Fisher Scientific, Cat. No. A412-4)
38. Propionic acid (Sigma-Aldrich, Cat. No. 402907)
39. Alcohol 200 Proof (Decon Laboratories, Cat. No. 2701)
40. Hydrochloric acid (Fisher Scientific, Cat. No. A144-500)
41. Sodium citrate dehydrate (Fisher Scientific, Cat. No. S279-500)
42. Sodium acetate trihydrate (Fisher Scientific, Cat. No. BP334-500)
43. Sodium chloride (NaCl) (Fisher Scientific, Cat. No. BP3581)
44. Sodium phosphate (Sigma-Aldrich, Cat. No. S3264)
45. Potassium chloride (Fisher Scientific, Cat. No. BP366-500)
46. EDTA (Fisher Scientific, Cat. No. S311-500)
47. Tris base (Fisher Scientific, Cat. No. BP152-1)
48. BSA (Sigma-Aldrich, Cat. No. A3294)
49. RNase (Sigma-Aldrich, Cat. No. 9001-99-4)
50. Taq DNA polymerase (Invitrogen, Cat. No. 18038-042)
51. Buffer tablets "Gurr" (Life Technologies, Cat. No. 10582-013)
52. KaryoMAX Giemsa Stain (Life Technologies, Cat. No. 10092-013)
53. Spermidine (Sigma-Aldrich, Cat. No. S0266-1G)
54. Ethylene glycol tetraacetic acid (EGTA) (Sigma-Aldrich, Cat. No. E0396-25G)
55. Piperazine-1,4-bis(2-ethanesulfonic acid) (PIPES) (Sigma-Aldrich, Cat. No. P6757-25G)
56. Digitonin (Sigma-Aldrich, Cat. No. D141-100MG)
57. Triton-X100 (Fisher Scientific, Cat, No. BP151-100)
58. 37% Paraformaldehyde (Fisher Scientific, Cat, No. F79-500)

4.2.3.2 Polytene Chromosome Preparation for Laser Capture Microdissection

1. Dissect half-gravid *Anopheles* females at 25 hours after blood feeding. Fix ovaries from approximately five females into 500 μL fresh-modified Carnoy's solution (100% methanol:glacial acetic acid, 3:1) at RT for 24 hours. Transfer ovaries to −20°C for a long-term storage.

2. Prepare Carnoy's solution (100% ethanol:glacial acetic acid, 3:1) and 50% propionic acid just before making chromosome slides.

3. Place one pair of ovaries into one drop of Carnoy's solution on a Zeiss 1.0 PET membrane slide. Depending on the size, split ovaries into approximately two to four sections with dissecting needles, and place them into a drop of 50% propionic acid on clean slides under a dissection microscope.

4. Separate follicles, and remove remaining tissue using paper towel under a dissection stereo microscope. Add a new drop 50% propionic acid to the follicles, and allow them to sit for 3–5 minutes at RT.

5. Place a siliconized coverslip on top of the droplet. Let the slide stand for approximately 1 minute.

6. Cover the slide with an absorbent material (filter paper is used for this method), and while using the eraser side of a pencil, apply a generous amount of pressure to the coverslip by tapping on it repeatedly with the eraser.

7. Heat the slide to 60°C on a slide denaturation/hybridization system for 15–20 minutes to aid in flattening the polytene chromosomes. Place slides into a humid chamber at 4°C overnight to allow the acid to further flatten chromosomes.

8. Place slides into cold 50% ethanol for 10 minutes. Gently remove the coverslip, and replace in cold 50% ethanol for 10 more minutes.

9. Dehydrate slides in 70%, 90%, and 100% ethanol for 5 minutes each. Air-dry slides.

10. Prepare a solution of Gurr buffer solution by adding a single buffer tablet to 1 L distilled water. Autoclave.

11. Prepare the Giemsa solution by adding 1 mL Giemsa staining solution to 50 mL Gurr buffer.

12. Place air-dried slides into Giemsa solution for 10 minutes, and wash three times in 1× PBS. Air-dry slides again in a controlled sterile climate to avoid contamination.

4.2.3.3 Laser Capture Microdissection of a Single Polytene Chromosome Arm

This section details the use of the PALM Robo software, which comes with the PALM MicroBeam LCM system.

1. Clean the microscope with 100% ethanol. Sterilize gloves and tubes with an ultraviolet (UV) light in a UV-crosslinker.
2. Power up the PALM MicroBeam LCM system, and turn on laser. Open the laser dissection suite and PALM Robo, and configure the "Power" and "Focus" settings as necessary.
3. Search for polytene chromosome arm of interest.
4. Using the "Pencil" tool, outline the selected region.
5. Open the "Elements window" from the menu bar.
6. Select the "Drawn element," and ensure that you have selected "Cut."
7. Install the adhesive cap tube into the holder, place above the slide, leaving a small gap <1 mm in size, and start the laser cut.
8. Place the "Catapult selection" within the cut site, leaving space between the edge and chromosome.
9. Select "LPC" from the drop-down option, and begin catapulting.
10. Check to ensure sample was catapulted into cap by pressing the "Eye" icon.

4.2.3.4 Purification of DNA from a Single Chromosome Arm

Follow the instructions of the QIAamp DNA Micro Kit to release and purify the collected DNA. Step 1 was modified to accommodate the inverted tube.

1. Add 15 µL buffer ATL and 10 µL proteinase K to the inverted tube (inside the cap), and incubate at 56°C for 3 hours.
2. Add 25 µL buffer ATL, 50 µL buffer AL, and 1 µL carrier RNA; mix. Add 50 µL 100% ethanol; mix.
3. Transfer lysate to QIAamp column; centrifuge. Wash by adding 500 µL buffer AW1; centrifuge. Place column into new collection tube, add 500 µL buffer AW2; centrifuge. Place column into a new tube; centrifuge to remove excess liquid.
4. Place column into a 1.5-µL microcentrifuge tube, and add 20 µL water to elute; centrifuge.
5. Evaporate freshly eluted DNA down to a final volume of 9 µL using a vacufuge.

4.2.3.5 DNA Amplification and Probe Preparation via GenomePlex Whole Genome Amplification

1. Follow the GenomePlex Single Cell WGA4 Kit protocol to produce the first batch of amplified DNA.
   a. Add freshly prepared Proteinase K solution to the 9 μL sample; mix. Incubate DNA at 50°C for 1 hour, then heat to 99°C for 4 minutes. Keep on ice.
   b. Add 2 μL 1× Single Cell Library Preparation Buffer, and 1 μL Library Stabilization Solution; mix. Heat sample to 95°C for 2 minutes. Cool on ice and centrifuge.
   c. Add 1 μL Library Preparation Enzyme; mix and centrifuge. Incubate as follows.

$$\frac{16°C}{20'} + \frac{24°C}{20'} + \frac{37°C}{20'} + \frac{75°C}{20'} + \frac{4°C}{\infty}$$

   d. Add 7.5 μL 10× Amplification Master Mix, 48.5 μL water, 5.0 μL WGA DNA polymerase; mix and centrifuge.
   e. Thermocycle as follows.

$$\frac{95°C}{3'} + \left(\frac{94°C}{30"} + \frac{65°C}{5'}\right)^{25} + \frac{4°C}{\infty}$$

2. Purify DNA using the Genomic DNA Clean & Concentrator Kit. The protocol is as follows.
   a. Add 5:1 DNA-binding buffer:DNA sample (specifically for genomic DNA of less than 2 Kb. If sample DNA is greater than 2 Kb, use a 2:1 ratio), and transfer to provided spin column. Centrifuge.
   b. Add 200 μL DNA Wash Buffer and centrifuge. Repeat wash step.
   c. Add 50 μL water, and elute the DNA into a new 1.5 mL tube.

3. Reamplify sample DNA using the GenomePlex WGA3 Reamplification Kit as follows.
   a. Add 10 μL DNA to PCR tube (the kit recommends 10 ng total DNA) with 49.5 μL water, 7.5 μL 10× Amplification Master Mix, 3.0 μL 10 mM DNTP mix, and 5.0 μL WGA DNA polymerase. Mix and centrifuge.
   b. Use the following profile for the reaction.

$$\frac{95°C}{3'} + \left(\frac{94°C}{15"} + \frac{65°C}{5'}\right)^{14} + \frac{4°C}{\infty}$$

   c. Store DNA at −20°C.

4. Label DNA for FISH using the GenomePlex WGA3 Reamplification Kit as follows.

   a. Create a master mix from the GenomePlex WGA3 Reamplification Kit by adding 10 μL DNA to PCR tube with 49.5 μL water, 7.5 μL 10× Amplification Master Mix, 3.0 μL 1 mM dNTP mix (1 mM dATP, dCTP, dGTP, 0.3 μL 1 mM dTTP—if using labeled dUTP), 1 μL 25 nM labeled dUTP, and 5.0 μL WGA DNA polymerase.

   b. Use the following profile for the reaction.

$$\frac{95°C}{3'} + \left(\frac{94°C}{15''} + \frac{65°C}{5'}\right)^{14} + \frac{4°C}{\infty}$$

   c. Ethanol precipitate the labeled probe by adding 1/10 the final reaction volume (7.5 μL for 75 μL reaction) of 3 M sodium acetate pH 5.2 and 2–3 volumes of 100% ethanol. Chill DNA sample at −80°C for at least 30 minutes.

   d. Centrifuge sample at 4°C for 10 minutes to create labeled pellet, remove supernatant, and air-dry pellet.

   e. Create hybridization buffer as follows.
      0.2 g Dextran sulfate.
      1200 μL Deionized formamide.
      580 μL H₂O.
      120 μL 20× SSC.

   f. Add 40 μL hybridization buffer to air-dried pellet.

4.2.3.6   DNA Amplification and Probe Preparation via REPLI-g Whole Genome Amplification

1. Follow the REPLI-g Single Cell WGA Kit protocol to produce the amplified DNA.

   a. Prepare buffer D2 (3 μL 1 M DTT + 33 μL buffer DLB).

   b. Mix 4 μL purified microdissected material with 3 μL buffer D2. Flick tube to mix.

   c. Incubate for 10 minutes at 65°C. Add 3 μL Stop solution; mix.

   d. Add 9 μL H₂O, 29 μL REPLI-g Reaction Buffer, and 2 μL REPLI-g DNA polymerase to the sample.

   e. Incubate at 30°C for 8 hours. Inactivate DNA polymerase by heating to 65°C for 3 minutes. Store DNA at −20°C.

2. Follow the nick translation protocol to label REPLI-g amplified DNA.

   a. Prepare the following labeling mix.

      1 µg Amplified DNA.

      5 µL 10× DNA polymerase buffer.

      5 µL 10× dNTP.

      5 µL 1× BSA.

      1 µL 1 mM Labeled dNTP.

      4 µL 1 U/µL DNase I.

      1 µL 10 U/µL DNA polymerase I.

      H_2O to 50 µL.

   b. Incubate at 15°C for 2 hours. Add 2 µL 0.5 M EDTA to stop reaction. Check DNA fragment size by running on gel.

   c. Follow step 4, c through f, from Section 4.2.3.5 to precipitate and solubilize pellet.

4.2.3.7 Three-Dimensional Fluorescent In Situ Hybridization with Whole-Mount Ovaries

Note: For FISH on squash preparations of polytene and mitotic chromosomes, please refer to the detailed protocols in Sections 4.2.1 and 4.2.2.

1. Prepare the following Buffer A mix.

   60 mM KCl.

   15 mM NaCl.

   0.5 mM Spermidine.

   0.15 mM Spermine.

   2 mM EDTA.

   0.5 mM EGTA.

   15 mM PIPES.

2. Prepare slides for nuclear visualization by adding a layer of nail polish in a square pattern to match the size of the coverslips. This creates a raised surface to prevent squashing nuclei when placing a coverslip onto a slide.

3. Dissect fresh ovaries from Christopher's Stage III females, and keep in a solution of 150–250 µL Buffer A with 0.5% digitonin. Run larger dissection needle over follicles (in tube with Buffer A with 0.5% digitonin) to destroy follicular membrane.

4. Vortex for 5–10 minutes to further disturb follicles. Scrape down any large follicular pieces, and centrifuge tube for 30 seconds at lowest setting of approximately 500 rpm. Transfer supernatant to a new 2-mL Eppendorf tube, and add 100 µL Buffer A. Repeat step 6.4 between 5 and 7 times until the visible tissue is broken into small particles.

5. Spin both tubes for 10 minutes at 2,000 rpm. Discard supernatant in both tubes. *Note: Both tubes will be used for making final nuclear visualization slides. The tube with collected supernatant should contain primarily extracted nuclei, whereas the original tube with tissue will contain a mixture of tissue and nuclei embedded in nurse cells.* Add 200 µL Buffer A—0.1% Triton, and incubate overnight at 4°C. Centrifuge 5 minutes at 10,000 rpm (10,621×g) and remove supernatant.

6. Add 200 µL 4% paraformaldehyde in PBS. Incubate in thermomixer for 30 minutes, mixing at 450 rpm. Centrifuge 5 minutes at 5000 rpm, and remove supernatant. Wash with Buffer A with 0.1% Triton for 5 minutes, mixing at 450 rpm in thermomixer. Centrifuge 5 minutes at 5000 rpm (2655*g*), and remove supernatant.

7. Add the labeled DNA probe (prewarmed at 37°C) to the tube. Denature at 95°C in thermomixer, mixing at 450 rpm for 10 minutes. Continue denaturation at 80°C for 15 minutes with continued mixing. Incubate at 37°C in thermomixer, with 450 rpm mixing overnight. Centrifuge 5 minutes at 5000 rpm (2655*g*). Remove supernatant.

8. Wash with Buffer A with 0.1% Triton for 5 minutes. Centrifuge 5 minutes at 5000 rpm. Repeat 2 times. Apply drop of Prolong antifade with DAPI.

9. Pipet out nuclei/DAPI solution carefully (avoiding bubbles), apply to slide, and cover with coverslip.

4.2.3.8   Representative Results of Chromosome Painting

The overall flow-through of the chromosome painting protocol is illustrated in Figure 4.13. The user initially starts by microdissecting chromosome samples from membrane slides. Microdissected material is extracted and purified. The purified DNA is then amplified, reamplified, labeled, and then used for FISH to label chromosome spreads.

Figure 4.14 shows that the LCM protocol can be broken down into three overall steps: (1) finding chromosomes of interest and preparing the region for cutting (Figure 4.14a), (2) cutting and catapulting the chromosome region of interest via laser (Figure 4.14b), and (3) checking to determine if the sample is actually catapulted into adhesive cap (Figure 4.14c).

Figure 4.15 shows the results of gel electrophoresis ran for the GenomePlex and REPLI-g kits, as well as quantification of the samples by Nanodrop. GenomePlex and REPLI-g single-cell WGA kits used in this protocol differ greatly in resulting product size as well as overall yield.

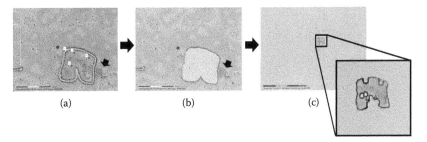

(a)   (b)   (c)

FIGURE 4.14 (**See color insert.**) The major steps in chromosome microdissection. (a) Laser-assisted cutting of the chromosomal region of interest through the membrane. (b) The membrane with a hole after the catapulting is performed. (c) The view of the catapulted piece of the membrane with a chromosomal segment in it attached to the adhesive cap. The arrow indicates the heterochromatin of the X chromosome that remained on the slide. The asterisk shows a piece of another chromosome that remained on the slide. (From George, P. et al., *J. Vis. Exp.*, (83), e51173, 2014.)

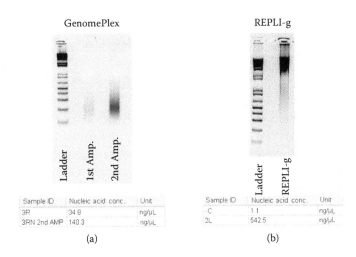

FIGURE 4.15 Agarose gel images showing DNA after whole genome amplification (WGA). (a) Low molecular weight (200–500 bps) DNA from arm 3R after using the WGA4 and WGA3 GenomePlex kits. (b) High molecular weight (10–20 Kb) DNA from arm 2L after REPLI-g amplification. The 100 bps ladder is shown in the left lanes. The tables below gel images show DNA concentrations measured by NanoDrop. (From George, P. et al., *J. Vis. Exp.*, (83), e51173, 2014.)

Chromosome painting probes are produced from the microdissected material. Figure 4.16 demonstrates FISH of five probes generated from microdissected material on polytene chromosomes of *Anopheles gambiae*. Four autosomal arms were labeled with three fluorophores using the WGA3 kit: the 3R chromosome is labeled in green (fluorescein), the 3L chromosome is in a mixture of red (Cy3) and yellow (Cy5), the 2R chromosome is in yellow (Cy5), and the 2L chromosome is in red (Cy3). The X chromosome was labeled in orange (Cy3) using nick translation of the REPLI-g material in a separate experiment.

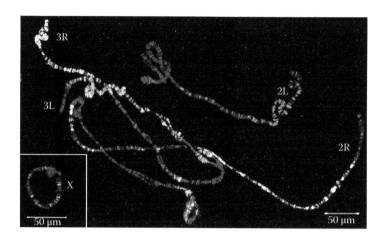

FIGURE 4.16 (**See color insert.**) Painting of polytene chromosomes from ovarian nurse cells of *Anopheles gambiae* using four probes generated from microdissected material. The X chromosome is labeled in orange (Cy3) by nick translation of the REPLI-g material. The 2R arm is labeled in yellow (Cy5); the 2L arm is in red (Cy3); the 3R arm is labeled in green (fluorescein); the 3L arm is labeled in a mixture of red (Cy3) and yellow (Cy5). Autosomes are labeled with the WGA3 amplification kit. Chromatin is stained in blue (DAPI). Chromosome names are placed near telomeric regions. (From George, P. et al., *J. Vis. Exp.*, (83), e51173, 2014.)

This procedure allowed us to establish the correspondence between euchromatic portions of polytene and mitotic chromosome arms. The same chromosome paints have been hybridized to interphase, prophase, prometaphase, and metaphase chromosomes of *Anopheles gambiae* (Figure 4.17).

To visualize the 3D organization of a single polytene chromosome arm in the cell nucleus, whole-mount FISH was performed on *Anopheles gambiae* Sua strain ovarian nurse cells. We successfully visualized a 2R-painting probe on polytene chromosomes in whole-mount ovarian nurse cells (Figure 4.18). Distinct chromosome arm territories are clearly seen in nuclei with interphase (Figure 4.17d).

4.2.3.9 Troubleshooting of Chromosome Painting

To successfully amplify DNA and prepare a painting probe from microdissected polytene chromosomes, several experimental steps must be carefully performed. The protocol employs the use of LCM, a method that increases overall efficiency and reduces exposure to foreign DNA by removing the interaction of physical tools with the sample. However, the amplification of foreign DNA is still the greatest potential pitfall of this experiment. Thus, during the entire process, it is essential to keep the samples protected from contamination. Throughout the slide preparation and microdissection phases, it is essential to ethanol wash dissection needles, slides, coverslips, and the workspace. It is critical to UV treat needles, slides, and coverslips prior to microdissection.

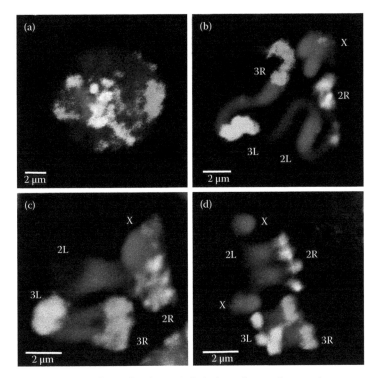

FIGURE 4.17 (**See color insert.**) Painting of nonpolytene chromosomes from larval imaginal disks (IDs) of *Anopheles gambiae*. (a) Interphase nucleus. (b) Prophase chromosomes. (c) Prometaphase chromosomes. (d) Metaphase chromosomes. Three probes were generated from microdissected material labeled by WGA3. The 2R arm is labeled in green (fluorescein); the 2L arm is unlabeled; the 3R arm is in pink, a mixture of red (Cy3) and orange (Cy5); the 3L arm is labeled in orange (Cy5). The X chromosome has a red label corresponding to the 18S rDNA probe. Chromatin is stained in blue (DAPI). Brightly stained regions of chromosomes correspond to the heterochromatin. (From George, P. et al., *J. Vis. Exp.*, (83), e51173, 2014.)

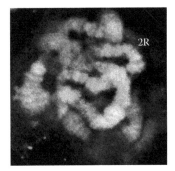

FIGURE 4.18 (**See color insert.**) Whole-mount three-dimensional fluorescent in situ hybridization performed on *Anopheles gambiae* ovarian nurse cells. The probe is labeled in Cy3 (depicted in red) and was made from a microdissected 2R chromosome arm. Chromatin is stained with DAPI and is depicted by cyan pseudo-coloring. (From George, P. et al., *J. Vis. Exp.*, (83), e51173, 2014.)

Spreading the chromosomes on the membrane of the dissection slides can be difficult. It is important to use more of the ovary (half to a full pair of ovaries) when making these slides to provide a greater chance of finding a well-spread nucleus. It is also recommended to use fresh tissue when making slides, as amplification rates appear to drop as tissues age (Frumkin et al. 2008) and chromosome spreading becomes more challenging. The preparation of slides for microdissection is the most time-consuming step in this protocol. Chromosomes must be well spread to avoid accidental acquisition of unwanted material. Giemsa staining allows the user to check spread quality with a phase contrast microscope prior to microdissection.

The described protocol allows extracting and analyzing chromosomal fragments ranging in size from small regions of interest to a majority of the arms. It is possible to obtain DNA from morphologically distinct regions such as inversions, specific euchromatic bands, and interbands, as well as telomeric, centromeric, and intercalary heterochromatin. The user can apply the generated painting probes to examining aberrant chromosomes, studying interspecies homology at particular loci, or characterizing spatial organization of chromosomes in an intact 3D nucleus. For developing chromosome paints, we selected euchromatic segments to avoid hybridization of repetitive DNA with multiple chromosomal regions. As a result, we obtained arm-specific painting without using a competitor, such as total genomic DNA or C_0t 1 DNA fraction.

Two different genomic DNA amplification kits, GenomePlex WGA and REPLI-g, have been chosen for this protocol based on reviews that compared the efficiency and dropout rate of multiple amplification kits (Hockner et al. 2009; Treff et al. 2011). Both kits performed the best among the available methods in both dropout rate (GenomePlex had a 12.5% rate compared to REPLI-g's 37.5%) and percentage of amplified markers (GenomePlex had a 45.24% amplification rate vs. REPLI-g's 30.0%) (Hockner et al. 2009). The GenomePlex kit also provided a higher quantity of DNA, and thus made it a better candidate for multiple downstream techniques. A reamplification kit for the GenomePlex system is also available, allowing for further amplification of DNA. However, it is important to note that amplification is not perfect. The possibility remains that successful amplification can introduce errors or has a bias toward specific loci in the target DNA. It is important to consider the final fragment size of the available genome amplification methods. GenomePlex fragmentation results in a library with fragments ranging from 200 to 500 bp, whereas the REPLI-g kit produces fragments approximately 10–20 Kb in size. The intended downstream application of this protocol is FISH, thus making the GenomePlex kit a more feasible option, as it provided the desired fragment size and the ability to label DNA fragments directly through WGA. However, nick translation labeling reaction must be used for long DNA molecules produced by the REPLI-g amplification.

Polytene chromosomes provide approximately 512 copies of a single DNA sequence and 1024 copies of two homologous DNA sequences. Therefore, it was possible to adapt this protocol to successfully amplify DNA from a single polytene chromosome arm. Although pooling of multiple chromosomes is possible using our method, the increased likelihood of sample contamination emphasizes the importance of beginning the experiment with as few chromosomes as possible. If polytene chromosomes are not available, our protocol could be adapted for use in mitotic chromosomes. It may be necessary, however, to pool 10–15 mitotic chromosomes prior to amplification for successful FISH (Drosopoulou et al. 2012). Amplification bias is lower with high quantities of starting template DNA (Raghunathan et al. 2005). Thus, pooling mitotic chromosomes will help to increase overall quality of DNA product.

4.3   DISCUSSION

4.3.1   Advances in Physical Mapping and Genome Analysis of *Anopheles gambiae*

The presence of readable polytene chromosomes in *Anopheles* species provides a unique opportunity for creating highly finished reference genome assemblies. Cytogenetic physical mapping has been crucial for developing the draft genome assembly of the *Anopheles gambiae* PEST strain (Holt et al. 2002). The positions of approximately 2000 BAC clones were assigned to the chromosomal locations. Additional cytogenetic mapping has improved the quality of the *Anopheles gambiae* PEST reference genome assembly (AgamP3) and identified potential haplotypes that belong to the M and S molecular forms (Sharakhova et al. 2007). This work improved the original version of the *Anopheles gambiae* PEST reference genome assembly and resulted in approximately 84.5% genome placement to the chromosomes (Figure 4.19).

The most significant improvement in the new *Anopheles gambiae* assembly is 24 scaffolds (8.64 Mb) located to pericentromeric regions. However, this improvement has not resulted in the complete assembly of the pericentromeric regions. Although the current AgamP3 genome assembly still has several physical gaps and 42 Mb of unmapped sequences, it is the best mosquito genome

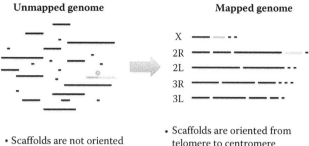

FIGURE 4.19 A scheme that explains the advantage of a chromosome-based genome assembly. Sequencing scaffolds of a mosquito genome assembly were mapped to five chromosome arms. Star indicates a misassembled scaffold in the unmapped genome. The misassembly is corrected in the mapped genome. Small scaffolds in the mapped genome correspond to pericentromeric heterochromation. Scaffolds assigned to the unknown chromosome and to the Y chromosome are not shown.

assembly available so far and the only chromosome-based mosquito genome assembly. For example, it has successfully served as a reference for newly sequenced individual genomes from natural populations of *Anopheles gambiae* (Cheng et al. 2012).

To establish the link between the chromosomal regions and the genome sequence, a study attached genome coordinates, based on 302 markers of BAC, cDNA clones, and PCR-amplified gene fragments, to the chromosomal bands and interbands at approximately 0.5–1 Mb interval (George et al. 2010). Because heterochromatic regions were not sufficiently covered with markers, additional physical mapping of PCR-amplified gene fragments near heterochromatin–euchromatin boundaries has been performed (George et al. 2010; Sharakhova et al. 2010a). The sizes of the mapped pericentric heterochromatin have been determined as the following: 4.4 Mb of the X chromosome, 2.6 Mb of the 2R arm, 2.4 Mb of the 2L arm, 1 Mb of the 3R arm, and 1.8 Mb of the 3L arm. In addition, physical mapping has identified three large regions of intercalary heterochromatin in *Anopheles gambiae*. These regions of intercalary heterochromatin are morphologically different: 0.7-Mb and 0.8-Mb regions of 2L and 3L are diffuse, whereas a 2.9-Mb region of 3R is compact heterochromatin. Because the *Anopheles gambiae* genome assembly successfully captured not only the euchromatin but also a significant portion of the heterochromatin, comparative analysis of chromatin types was possible. It has been shown that heterochromatin and euchromatin differ in gene density and the coverage of retroelements and segmental duplications (SDs). Gene ontology (GO) analysis revealed that heterochromatin is enriched in genes with DNA-binding and regulatory activities. The pericentric heterochromatin had the highest coverage of retroelements and tandem repeats, whereas intercalary heterochromatin was enriched with SDs (Sharakhova et al. 2010a).

The availability of the *Anopheles gambiae* genome sequence (Holt et al. 2002) and the physical maps for *Anopheles funestus* (Sharakhov et al. 2002) and *Anopheles stephensi* (Xia et al. 2010) enabled a fresh perspective to be gained on the relationships between the genomic landscape and evolutionary rates. Comparative mapping among *Anopheles gambiae, Anopheles funestus*, and *Anopheles stephensi* established arm homologies among these species; found no evidence for interarm transposition events, pericentric inversions, or partial-arm translocations; and confirmed that whole-arm translocations and paracentric inversions are the common rearrangements among species in the subgenus *Cellia* (Xia et al. 2010). The number of inversions between the species has been calculated using the Nadeau and Taylor method (Nadeau and Taylor 1984) and the Genome Rearrangements In Man and Mouse program (Tesler 2002). The rate of genome rearrangement in the subgenus *Cellia* has been found to be 0.003–0.005 inversions per 1 Mb per million years per lineage (Xia et al. 2010). Comparative cytogenetic studies performed on malaria mosquitoes provided some of the most obvious examples of the nonuniform inversion distribution among chromosomal arms (Xia et al. 2010; Coluzzi et al. 2002). These analyses have revealed that the X chromosome had the highest rate of inversion fixation and that the 2R arm evolved faster than other autosomes (Xia et al. 2010). Another study demonstrated a striking contrast among chromosome arms in the length of conserved segments: small conserved blocks (<1 Mb) are located on arm 2R, and large conserved blocks (up to 6–8 Mb) are located on arms 3R and 3L (Sharakhova et al. 2011c).

Of 10 inversions fixed among species of the *Anopheles gambiae* complex, 5 have been found on the X chromosome and 3 on the 2R arm (Coluzzi et al. 2002). Only a few polymorphic inversions have been found on the X chromosome in species from the *Anopheles gambiae* complex (Coluzzi et al. 2002). A comparison of the physical maps of *Anopheles gambiae, Anopheles funestus*, and *Anopheles stephensi* has also demonstrated an excess of fixed inversions, as compared to a deficit of polymorphic inversions on the X chromosome (Xia et al. 2010). The contrasting pattern of inversion polymorphism and inversion fixation on the X chromosome suggests that different forces govern sex chromosome and autosome evolution. This phenomenon, if confirmed by whole-genome analyses of multiple species, could indicate that polymorphic inversions on the X chromosome are underdominant, as was theoretically predicted earlier (Charlesworth et al. 1987).

In contrast to the X chromosome, the 2R and 2L arms of *Anopheles gambiae* and their homologous arms in *Anopheles stephensi* and *Anopheles funestus* harbor polymorphic inversions associated with ecological adaptations (Mahmood and Sakai 1984; Costantini et al. 1999; Coluzzi et al. 2002). Adaptive alleles or allelic combinations can be maintained within a polymorphic inversion by suppressing recombination between the loci (Kirkpatrick and Barton 2006). It has been predicted that chromosomal arms rich in polymorphic inversions would have higher gene densities (Krimbas and Powell 1992). Indeed, the 2R, 2L arms of *Anopheles gambiae* had the highest gene densities, whereas the polymorphic inversion-poor X chromosome had the lowest gene density (Xia et al. 2010). These observations highlight the fundamental differences between the evolutionary dynamics of the sex chromosome and autosomes. The rapid generation and fixation of inversions without maintenance of a stable inversion polymorphism are achieving the high rate of sex chromosome evolution. In contrast, the high rate of the autosomal evolution results from the high level of inversion polymorphism maintained by selection acting on gene-rich chromosomal arms. The polymorphic inversions 2Rb, 2Rbc, 2Rcu, 2Ru, 2Rd, and 2La of *Anopheles gambiae* are associated with adaptation of mosquitoes to dry environments (Coluzzi et al. 2002). The cuticle seems to play a major role in desiccation resistance of embryo and adult mosquitoes (Gray et al. 2009). These observations suggest a possibility that genes involved in the cuticle development may be disproportionally clustered on the 2R and 2L arms. A study of GO terms provides evidence that 2L is, indeed, enriched with genes involved in the structural integrity of a cuticle, whereas the 2R arm has overrepresentation of genes involved in cellular response to stress (e.g., temperature and humidity) and in building membrane parts (Xia et al. 2010). These data support the role of natural selection in maintaining polymorphic inversions associated with ecological adaptations. A recent study found that the several 2R inversions in *Anopheles gambiae*, *Anopheles stephensi*, and *Anopheles funestus* nonrandomly share common genes (Figure 4.20). This nonrandom distribution of markers is not the result of preservation of ancestral gene order. The gene orders have been extensively reshuffled within independently originated polymorphic inversions. Thus, it is likely that natural selection favors adaptive gene combinations within polymorphic inversions on 2R when distantly related species are exposed to similar environmental pressures.

Evolution of gene order likely has species- and chromosome-specific facilitators or inhibitors. The major consequences of unequal rates of karyotype evolution are differential plasticity of species and an increased role of certain chromosomes in

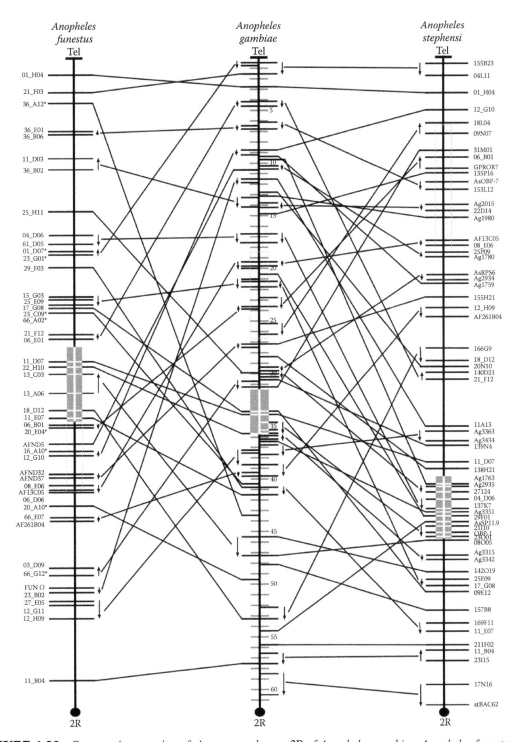

FIGURE 4.20 Comparative mapping of chromosomal arms 2R of *Anopheles gambiae*, *Anopheles funestus*, and *Anopheles stephensi*. Arrows denote oriented conserved gene orders. The darkly shaded boxes indicate positions of polymorphic inversions 2Rt of *Anopheles funestus*, 2Ru of *Anopheles gambiae*, and 2Rf of *Anopheles stephensi*. The lightly shaded boxes indicate positions of polymorphic inversions 2Rd/2Rh of *Anopheles funestus*, 2Rb of *Anopheles gambiae*, and 2Re of *Anopheles stephensi*. The centromere regions are shown by black circles at the end of the arms. (From Sharakhova, M.V. et al., *BMC Evol. Biol.*, 11(1), 91, 2011c.)

adaptation and evolution. What are the factors that constrain or promote chromosomal rearrangements? Contrary to expectation, the TE density in the *Anopheles gambiae* genome was found to be lowest on the 2R arm (Holt et al. 2002); thus, it is not clear whether the molecular content could be associated with inversion polymorphism and fixation rates. A recent study has shown that fragility of certain regions rather than functional constraints plays the main role in nonuniform distribution of inversions in *Drosophila* chromosomes (von Grotthuss et al. 2010). However, the molecular determinants of the fragile breakage have not yet been determined. If a nonrandom origin of inversions can be attributed to unequal density of repetitive DNA among chromosome arms, higher densities of break-causing elements on faster evolving arms can be predicted.

A Bayesian statistical model and procedure has been applied to studying differences between arms in molecular features, such as DNA-mediated TEs, RNA-mediated TEs, SDs, micro- and minisatellites, satellites, matrix-associated regions (MARs), and genes (Xia et al. 2010). The X chromosome had the highest density of TEs and the highest coverage of microsatellites, minisatellites, and satellites. The 2R arm had the highest density of genes and regions involved in SDs, but had the lowest densities of TEs and the lowest coverage of minisatellites and MARs (Figure 4.21).

Simple repeats have been shown to play a role in the formation of hairpin and cruciform structures, which can cause double-strand DNA breaks and rearrangements (Lobachev et al. 2007). SDs have been implicated in inversion generation in mosquitoes (Coulibaly et al. 2007) and are considered as a marker of genome fragility (Bailey et al. 2004). Because of the paucity of pericentric inversions and partial-arm translocations in mosquito evolution, the genome landscapes and evolutionary histories of individual arms are different. A strong association exists between the genome landscape characteristics and the rates of chromosomal evolution. A unique combination of various classes of genes and repetitive DNA in each arm, rather than a single type of repetitive element, is likely responsible for arm-specific rates of rearrangements. It is important to perform the genomic analyses considering individual chromosomal arms and using sequences physically mapped to the chromosomes.

4.3.2   Advances in Physical Mapping and Genome Analysis of *Aedes aegypti*

Physical mapping of the *Aedes aegypti* genome is difficult because of the poor development of polytene chromosomes and

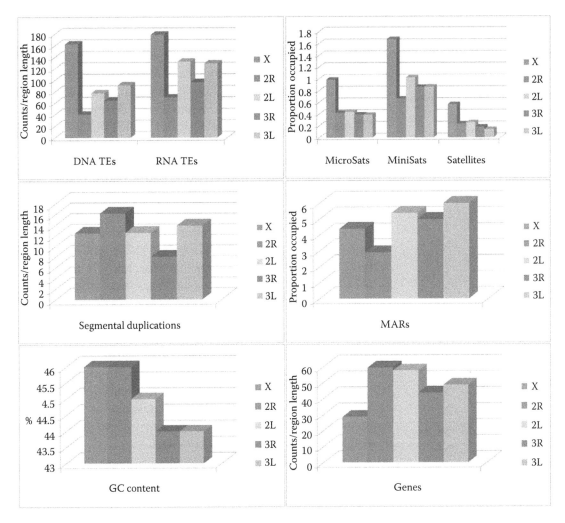

FIGURE 4.21 (**See color insert.**) Median values of density and coverage of molecular features in chromosomes of *Anopheles gambiae*. Counts per 1 Mb are given for DNA transposable elements (DNA TEs), RNA TEs, regions involved in segmental duplications (SDs), and genes. Percentage of region length occupied per 1 Mb is indicated for microsatellites, minisatellites, satellites, and matrix-associated regions (MARs). (From Xia, A. et al., *PLoS ONE*, 5(5), e10592, 2010.)

the abundance of repetitive elements in the genome. A FISH technique was first developed using mitotic chromosomes from the ATC-10 cell line of *Aedes aegypti*, resulting in direct positioning of 37 cosmid clones onto chromosomes (Brown et al. 1995). In addition, 21 cDNA genetic markers and 8 cosmid clones containing RFLP markers have been mapped to the chromosomes from this cell line (Brown et al. 2001). The map was distance-based, meaning that positions of the markers were determined by direct measurements of their locations on the chromosomes from the p terminus (FLpter). The availability of this physical map together with linkage mapping data (Severson et al. 1993, 2002) has placed

31% of the original *Aedes aegypti* assembly to the chromosomes without the order and orientation (Nene et al. 2007).

Recently, an alternative physical mapping approach for *Aedes aegypti* was introduced (Sharakhova et al. 2011b). Instead of cell lines, which usually accumulate chromosomal rearrangements (Brown et al. 1995), the new method used chromosomes from IDs of fourth instar larvae. The positions of the DNA probes have been determined using idiograms—schematic representations of the chromosomal banding patterns. The idiograms have been constructed for the chromosomes at early metaphase. The three chromosomes of the mosquito have been divided into 23 regions and 94 subdivisions. One hundred BAC clones carrying major genetic markers have been placed and ordered on chromosomes using FISH (Timoshevskiy et al. 2013). These BAC clones are carrying previously mapped major genetic markers determined by a PCR approach. All BAC clones have been ordered within each band by multicolor FISH. A linear regression analysis demonstrated a good correlation between positions of the markers on the physical and linkage maps. The genomic locations of the BAC clones have been linked to the genetic locations of QTL related to pathogen transmission (Timoshevskiy et al. 2013). In addition to 100 genetic markers and 183 Mbp of genomic sequences, this study also anchored to the exact chromosome positions of a marker for sex determination and 12 QTL associated with the transmission of dengue virus 2 (DEN2) (Gomez-Machorro et al. 2004); filarial nematode *Brugia malayi* (Severson et al. 1994); and the avian malaria parasite *Plasmodium gallinaceum* (Severson et al. 1995; Zhong et al. 2006). This study has developed the first integrated linkage, chromosome, and genome map—iMap—for the yellow fever mosquito. Interestingly, the mapping demonstrated that 12 QTL corresponding to the multiple pathogens, including DENV QTL, form only 5 major chromosomal clusters (Figure 4.22). The discovery of the localization of multiple "unrelated" QTL in a few major chromosome clusters suggests a possibility that the transmission of different pathogens is controlled by the same genomic loci.

A molecular landmark–guided mapping approach has placed additional 368 BAC clones from the largest genomic supercontigs to the *Aedes aegypti* chromosomes (Timoshevskiy et al. 2014). Two-color hybridization of BAC clones has been performed in the presence of three landmark probes with known locations in each of the three chromosomes. BAC clones that produced a strong unique signal in telomeric regions on the chromosomes have been used as landmarks. Together with previously generated data, this work has assigned 294 genomic supercontigs to chromosome

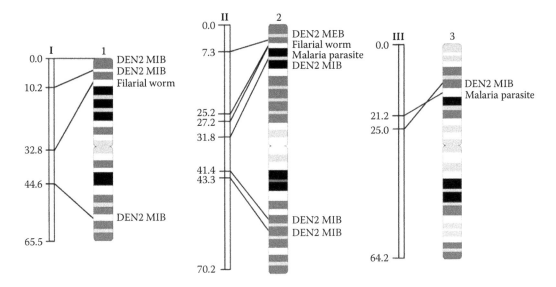

FIGURE 4.22 Localization of quantitative trait locus (QTL) to various pathogens on linkage and chromosome maps of *Aedes aegypti*. The linkage positions of QTL in cM on linkage groups I, II, and III are indicated on the left side. QTL to dengue virus 2 (DEN2), midgut infection barrier (MIB), DEN2 midgut escape barrier (MEB), malaria parasite, and filarial worm are shown on the right sides of chromosomes 1, 2, and 3. (From Timoshevskiy, V.A. et al., *PLoS Negl. Trop. Dis.* 7(2), e2052, 2013).

bands based on FISH results. The 619 Mb of *Aedes aegypti* genome has been mapped at the approximately 15 Mb resolution. The genomic supercontigs were assigned to the chromosome bands without ordering and orientation. This study has demonstrated that physical mapping can orient genomic supercontigs and identify potential mistakes in genome assembly. About 8% or 24 genomic supercontigs have been considered misassembled based on FISH results. In these cases, two or more BAC clones from the same genomic supercontig hybridized to the very different places on chromosomes.

The low-resolution physical map has guided further analysis of the distribution of genes, tandem repeats, and TEs along the chromosomes. The study has found that the q arm of homomorphic sex chromosome 1 and the areas around centromeres have the lowest gene content and the highest density of tandem repeats. In contrast to tandem repeats, TEs have been more abundant in autosomes and in euchromatic areas of the chromosomes. A comparative genomic analysis with *Anopheles gambiae* has demonstrated that, in addition to previously detected centromeric fusion between genomic material of chromosome X and a part of the 2R arm of *Anopheles gambiae* (Nene et al. 2007), several pericentric inversions may have reshuffled the genetic materials between chromosome arms 1p and 1q. Similar patterns of pericentric inversions and whole-arm translocations have also been found in autosomes. The study has shown that the gene order within chromosome arms of *Aedes aegypti* was poorly conserved due to multiple paracentric inversions. The homomorphic sex chromosome 1 had the highest rate of the genome rearrangements. These data suggest that tandem repeats rather than TEs played a major role in the rapid evolution of the homomorphic sex chromosome 1 in *Aedes* (Timoshevskiy et al. 2014). This map will facilitate the identification of genomic determinants of traits responsible for susceptibility or refractoriness of *Aedes aegypti* to diverse pathogens and will also contribute to a more complete understanding of the genome organization and function in the mosquitoes.

4.3.3   Advances in Physical Mapping and Genome Analysis of *Culex quinquefasciatus*

The lack of a high-quality chromosome-based genome assembly for *Culex quinquefasciatus* remains a significant impediment to further progress in *Culex* biology and comparative genomics of mosquitoes. The JHB strain of *Culex quinquefasciatus* has been selected for the genome sequencing, because it could yield relatively high-quality polytene chromosome spreads suitable for in situ hybridization studies (McAbee et al. 2007). However, a routine use of polytene chromosome preparations for genome mapping is challenging due to a low yield of mappable chromosome spreads. A recent study has found the correspondence between genetic map and mitotic chromosomes of *Culex quinquefasciatus* (A. N. Naumenko, V. A. Timoshevskiy, B. S. deBruyn, D. W. Severson, I. V. Sharakhov, and M. V. Sharakhova, unpublished data). This has been achieved by a direct placement of 12 BAC clones carrying genetic markers to the chromosomes using FISH. Markers from linkage groups II and III hybridize to the largest (9.2 μm) and intermediate (8.4 μm) chromosomes, respectively, which contradicts with the previous nomenclature. According to the genetic nomenclature for *Culex quinquefasciatus*, chromosomes 1, 2, and 3 have been renumbered as the smallest, largest, and intermediate chromosomes, respectively. This study sequenced 576 BAC clones from *Culex quinquefasciatus*; these BAC clones have been matched with 195 of the largest supercontigs in the genome assembly. Physical mapping of 195 BAC clones from the largest genomic supercontigs would place 23.4% of the *Culex quinquefasciatus* genome to chromosomes. Cytogenetic procedures required for obtaining high-quality chromosomal spreads and the FISH method have been optimized. The percentage of chromosomal slides suitable for further analyses is 80%. On the basis of chromosome banding patterns, idiograms for mitotic chromosomes from IDs at mid-metaphase have been constructed. Three chromosomes have been subdivided into 2 subdivisions and 112 bands. In addition to 12 BAC clones, carrying major genetic markers from previous linkage mapping, 10 BAC clones from the largest genomic supercontigs, 3 PCR-amplified fragments, 2 plasmids, and an 18S rDNA probe have been successfully hybridized to chromosomes. As a result, the cytogenetic map was integrated with the previous linkage map, and approximately 10 Mb of *Culex quinquefasciatus* genome has been placed to precise chromosomal positions. Thus, a cytogenetic tool for physically mapping the genome is now available for this species. Further mapping effort is required for the development of a detailed genome map for *Culex quinquefasciatus*.

4.4 CONCLUSION

We are now witnessing an explosion of genome sequencing projects for mosquito vectors. A set of reference genome assemblies for 16 species of *Anopheles* mosquitoes has been completed (https://olive.broadinstitute.org/comparisons/anopheles.3). These species include *Anopheles albimanus, Anopheles arabiensis, Anopheles atroparvus, Anopheles christyi, Anopheles culicifacies, Anopheles dirus, Anopheles epiroticus, Anopheles farauti, Anopheles funestus, Anopheles maculatus, Anopheles melas, Anopheles merus, Anopheles minimus, Anopheles quadriannulatus, Anopheles sinensis,* and *Anopheles stephensi.* With genome sequences now available, researchers have the unique opportunity to perform comparative analysis for inferring evolutionary changes relevant to vector ability. However, success of these comparative genomic analyses will be limited, and inferences about evolution of the vectorial capacity traits will be less informative if researchers deal with numerous sequencing scaffolds rather than with chromosome-based genome assemblies. The presence of readable polytene chromosomes in *Anopheles* species provides an uncommon opportunity for creating highly finished reference genome assemblies.

Finding the exact genomic positions of the QTL responsible for arbovirus transmission is impossible without the development of complete physical maps for *Aedes aegypti* and *Culex quinquefasciatus.* Knowledge about the chromosomal position of the QTL is also important for understanding the function of the specific genes associated with this particular QTL. A combination of next-generation sequencing and high-resolution cytogenetic mapping is needed to improve the quality of the *Aedes aegypti* and *Culex quinquefasciatus* genome assemblies. Both mitotic and polytene chromosomes are a great resource that can be used for creating genome physical maps. The iMap approach can potentially create a detailed chromosome-based physical map and will significantly enhance the quality of the existing genome assembly for culicine mosquitoes.

The availability of detailed physical maps for mosquito genomes can greatly enhance the genome assemblies, can help to identify potential haplotype polymorphisms within the genome, and can differentiate polymorphic scaffolds from regions of segmental duplication. A correct genome assembly is crucial for whole-genome association studies and for correct genome annotation. An improved genome assembly enables genome comparisons between different mosquito species as a guide for reconstructing their biological history and for studying mechanisms of genome evolution. Comparative genomics of mosquitoes and other insects is needed for our understanding of both the genetic differences among species and the genetic basis of vector competence and will promote development of novel disease control strategies. In addition, taxonomic and population studies by whole-genome resequencing of wild mosquitoes heavily rely on chromosome-based reference assemblies.

ACKNOWLEDGMENTS

This work was supported by the National Institutes of Health grants 1R21AI094289 to Igor V. Sharakhov and 1R21AI88035 and 1R21AI101345 to Maria V. Sharakhova. We thank Sergei Demin and Tatyana Karamysheva for their help with chromosome preparation and FISH on *Anopheles*, David Severson for providing us *Aedes* and *Culex* genomic DNA BAC clones, and Melissa Wade for editing the text. The Mopti and Sua2La strains of *Anopheles gambiae* were obtained from the Malaria Research and Reference Reagent Resource Center.

REFERENCES

Antolin, M. F., C. F. Bosio, J. Cotton, W. Sweeney, M. R. Strand, and W. C. Black IV. 1996. Intensive linkage mapping in a wasp (Bracon hebetor) and a mosquito (Aedes aegypti) with single-strand conformation polymorphism analysis of random amplified polymorphic DNA markers. *Genetics* 143 (4):1727–38.

Arensburger, P., K. Megy, R. M. Waterhouse, J. Abrudan, P. Amedeo, B. Antelo, L. Bartholomay et al. 2010. Sequencing of Culex quinquefasciatus establishes a platform for mosquito comparative genomics. *Science* 330 (6000):86–8.

Bailey, J. A., R. Baertsch, W. J. Kent, D. Haussler, and E. E. Eichler. 2004. Hotspots of mammalian chromosomal evolution. *Genome Biol* 5 (4):R23.

Barrett, A. D. and S. Higgs. 2007. Yellow fever: A disease that has yet to be conquered. *Annu Rev Entomol* 52:209–29.

Bartholomay, L. C., R. M. Waterhouse, G. F. Mayhew, C. L. Campbell, K. Michel, Z. Zou, J. L. Ramirez et al. 2010. Pathogenomics of Culex quinquefasciatus and meta-analysis of infection responses to diverse pathogens. *Science* 330 (6000):88–90.

Besansky, N. J. and J. R. Powell. 1992. Reassociation kinetics of Anopheles gambiae (Diptera: Culicidae) DNA. *J Med Entomol* 29 (1):125–8.

Bosio, C. F., R. E. Fulton, M. L. Salasek, B. J. Beaty, and W. C. Black IV. 2000. Quantitative trait loci that control vector competence for dengue-2 virus in the mosquito Aedes aegypti. *Genetics* 156 (2):687–98.

Bridi, L. C., M. V. Sharakhova, I. V. Sharakhov, J. Cordeiro, G. M. Azevedo Jr., W. P. Tadei, and M. S. Rafael. 2013. Chromosomal localization of actin genes in the malaria mosquito Anopheles darlingi. *Med Vet Entomol* 27 (1):118–21.

Brown, S. E., J. Menninger, M. Difillipantonio, B. J. Beaty, D. C. Ward, and D. L. Knudson. 1995. Toward a physical map of Aedes aegypti. *Insect Mol Biol* 4 (3):161–7.

Brown, S. E., D. W. Severson, L. A. Smith, and D. L. Knudson. 2001. Integration of the Aedes aegypti mosquito genetic linkage and physical maps. *Genetics* 157 (3):1299–305.

Bryan, J. H., M. Di Deco, V. Petrarca, and M. Coluzzi. 1982. Inversion polymorphism and incipient speciation in Anopheles gambiae s.str. in The Gambia, West Africa. *Genetica* 59:167–76.

Cai, W., Y. Jin, J. Girton, J. Johansen, and K. M. Johansen. 2010. Preparation of Drosophila polytene chromosome squashes for antibody labeling. *J Vis Exp* (36), e1748

Campos, J., C. F. Andrade, and S. M. Recco-Pimentel. 2003a. Malpighian tubule polytene chromosomes of Culex quinquefasciatus (Diptera, Culicinae). *Mem Inst Oswaldo Cruz* 98 (3):383–6.

Campos J., C. F. Andrade, and S. M. Recco-Pimentel. 2003b. A technique for preparing polytene chromosomes from Aedes aegypti (Diptera, Culicinae). *Mem Inst Oswaldo Cruz* 98 (3):387–90.

Chambers, E. W., J. K. Meece, J. A. McGowan, D. D. Lovin, R. R. Hemme, D. D. Chadee, K. McAbee, S. E. Brown, D. L. Knudson, and D. W. Severson. 2007. Microsatellite isolation and linkage group identification in the yellow fever mosquito Aedes aegypti. *J Hered* 98 (3):202–10.

Charlesworth, B., J. A. Coyne, and N. H. Barton. 1987. The relative rates of evolution of sex chromosomes and autosomes. *Am Nat* 130:113–46.

Cheng, C., B. J. White, C. Kamdem, K. Mockaitis, C. Costantini, M. W. Hahn, and N. J. Besansky. 2012. Ecological genomics of Anopheles gambiae along a latitudinal cline: A population-resequencing approach. *Genetics* 190 (4):1417–32.

Clements, A. N. 1992. *The Biology of Mosquitoes: Development, Nutrition and Reproduction.* London, United Kingdom: Chapman & Hall.

Coetzee, M., R. H. Hunt, R. Wilkerson, A. della Torre, M. B. Coulibaly, and N. J. Besansky. 2013. Anopheles coluzzii and Anopheles amharicus, new members of the Anopheles gambiae complex. *Zootaxa* 3619 (3):246–74.

Coker, W. Z. 1958. The inheritance of DDT resistance in Aedes aegypti. *Ann Trop Med Parasitol* 52:443–55.

Collins, F. H., M. A. Mendez, M. O. Rasmussen, P. C. Mehaffey, N. J. Besansky, and V. Finnerty. 1987. A ribosomal RNA gene probe differentiates member species of the Anopheles gambiae complex. *Am J Trop Med Hyg* 37 (1):37–41.

Coluzzi, M. 1968. Chromosomi politenici cellule nutrici ovariche nel complesso gambiae del genere Anopheles. *Parassitologia* 10 (2–3):179–83.

Coluzzi, M., G. Cancrini, and M. Di Deco. 1970. The polytene chromosomes of Anopheles superpictus and relationships with Anophels stephensi. *Parassitologia* 12:101–12.

Coluzzi, M. and S. Montalenti. 1966. Osservazioni comparative sul cromosoma X nelle specie A e B del comlesso Anopheles gambiae. *Citologia* 40:671–8.

Coluzzi, M., V. Petrarca, and M. A. Dideco. 1985. Chromosomal inversion intergradation and incipient speciation in Anopheles gambiae. *Bollettino Di Zoologia* 52 (1–2):45–63.

Coluzzi, M. and A. Sabatini. 1967. Cytotaxanomic observations on species A and B of the Anopheles gambiae complex. *Parassitologia* 9:73–88.

Coluzzi, M. and A. Sabatini. 1968. Cytogenetic observations on species C, merus and melas of the Anopheles gambiae complex. *Parassitilogia* 10 (2–3):156–64.

Coluzzi, M. and A. Sabatini. 1969. Cytogenetic observations on the salt water species, Anopheles merus and Anopheles melas, of the gambiae complex. *Parassitilogia* 11 (3):177–87.

Coluzzi, M., A. Sabatini, A. della Torre, M. A. Di Deco, and V. Petrarca. 2002. A polytene chromosome analysis of the Anopheles gambiae species complex. *Science* 298 (5597):1415–8.

Coluzzi, M., A. Sabatini, V. Petrarca, and M. A. Di Deco. 1979. Chromosomal differentiation and adaptation to human environments in the Anopheles gambiae complex. *Trans R Soc Trop Med Hyg* 73 (5):483–97.

Costantini, C., N. Sagnon, E. Ilboudo-Sanogo, M. Coluzzi, and D. Boccolini. 1999. Chromosomal and bionomic heterogeneities suggest incipient speciation in Anopheles funestus from Burkina Faso. *Parassitologia* 41 (4):595–611.

Coulibaly, M. B., N. F. Lobo, M. C. Fitzpatrick, M. Kern, O. Grushko, D. V. Thaner, S. F. Traore, F. H. Collins, and N. J. Besansky. 2007. Segmental duplication implicated in the genesis of inversion 2Rj of Anopheles gambiae. *PLoS ONE* 2 (9):e849.

Craig, G. B. Jr. and W. A. Hickey. 1967. *Genetics of Aedes aegypti*. Edited by J. W. Wright and R. Pal, *Genetics of insect vectors of disease*. Amsterdam, London, New York: Elsevier.

Daniel, A. 1985. The size of prometaphase chromosome segments. Tables using percentages of haploid autosome length (750 band stage) *Clin Genet* 28:216–24.

della Torre, A., Z. Tu, and V. Petrarca. 2005. On the distribution and genetic differentiation of Anopheles gambiae ss molecular forms. *Insect Biochem Mol Biol* 35 (7):755–69.

Dennhofer, L. 1968. Die speicheldrusenchromosomen der stechmucke Culex Pipiens. *Chromosoma* 25:365–76.

Drosopoulou, E., I. Nakou, J. Sichova, S. Kubickova, F. Marec, and P. Mavragani-Tsipidou. 2012. Sex chromosomes and associated rDNA form a heterochromatic network in the polytene nuclei of Bactrocera oleae (Diptera: Tephritidae). *Genetica* 140 (4–6):169–80.

Favia, G., A. della Torre, M. Bagayoko, A. Lanfrancotti, N. F. Sagnon, Y. T. Touré, and M. Coluzzi. 2003. Molecular identification of sympatric chromosomal forms of Anopheles gambiae and further evidence of their reproductive isolation. *Insect Mol Biol* 6 (4):377–83.

Frizzi, G. and M. Holstein. 1956. Etude cytogenetique d'Anopheles gambiae. *B World Health Organ* 15 (3–5):425–35.

Frumkin, D., A. Wasserstrom, S. Itzkovitz, A. Harmelin, G. Rechavi, and E. Shapiro. 2008. Amplification of multiple genomic loci from single cells isolated by laser micro-dissection of tissues. *BMC Biotechnol* 8:17.

Garimberti, E. and S. Tosi. 2010. Fluorescence in situ hybridization (FISH), basic principles and methodology. In *Fluorescence in situ hybridization (FISH)*. Edited by J. M. Bridger and E. V. Volpi. New York: Springer Science+Business Media.

Gentile, G., M. Slotman, V. Ketmaier, J. R. Powell, and A. Caccone. 2001. Attempts to molecularly distinguish cryptic taxa in Anopheles gambiae s.s. *Insect Mol Biol* 10 (1):25–32.

George, P., M. V. Sharakhova, and I. V. Sharakhov. 2010. High-resolution cyto-genetic map for the African malaria vector Anopheles gambiae. *Insect Mol Biol* 19 (5):675–82.

George, P., M. V. Sharakhova, and I. V. Sharakhov. 2012. High-throughput physi-cal mapping of chromosomes using automated in situ hybridization. *J Vis Exp* (64), e4007 10.3791/4007, DOI: 10.3791/4007.

George, P., A. Sharma, and I. V. Sharakhov. 2014. 2D and 3D chromosome paint-ing in malaria mosquitoes. *J Vis Exp* (83), e51173, doi:10.3791/51173.

Gomez-Machorro, C., K. E. Bennett, M. del Lourdes Munoz, and W. C. Black IV. 2004. Quantitative trait loci affecting dengue midgut infection barriers in an advanced intercross line of Aedes aegypti. *Insect Mol Biol* 13 (6):637–48.

Gould, E. A. and T. Solomon. 2008. Pathogenic flaviviruses. *Lancet* 371 (9611):500–9.

Gray, E. M., K. A. Rocca, C. Costantini, and N. J. Besansky. 2009. Inversion 2La is associated with enhanced desiccation resistance in Anopheles gambiae. *Malar J* 8:215.

Gubler, D. J. 2012. The economic burden of dengue. *Am J Trop Med Hyg* 86 (5):743–4.

Habtewold, T., M. Povelones, A. M. Blagborough, and G. K. Christophides. 2008. Transmission blocking immunity in the malaria non-vector mosquito Anopheles quadriannulatus species A. *PLoS Pathog* 4 (5):e1000070.

Halasa, Y. A., D. S. Shepard, and W. Zeng. 2012. Economic cost of dengue in puerto rico. *Am J Trop Med Hyg* 86 (5):745–52.

Hickner, P. V., B. Debruyn, D. D. Lovin, A. Mori, S. K. Behura, R. Pinger, and D. W. Severson. 2010. Genome-based microsatellite development in the Culex pipiens complex and comparative microsatellite frequency with Aedes aegypti and Anopheles gambiae. *PLoS ONE* 5 (9). doi:10.1371/journal.pone.0013062.

Hickner, P. V., A. Mori, D. D. Chadee, and D. W. Severson. 2013. Composite linkage map and enhanced genome map for Culex pipiens complex mos-quitoes. *J Hered* 104 (5):649–55.

Hockner, M., M. Erdel, A. Spreiz, G. Utermann, and D. Kotzot. 2009. Whole genome amplification from microdissected chromosomes. *Cytogenet Genome Res* 125 (2):98–102.

Holt, R. A., G. M. Subramanian, A. Halpern, G. G. Sutton, R. Charlab, D. R. Nusskern, P. Wincker et al. 2002. The genome sequence of the malaria mosquito Anopheles gambiae. *Science* 298 (5591):129–49.

Kaiser, P. E. and J. A. Seawright. 1987. The ovarian nurse cell polytene chromo-somes of Anopheles quadrimaculatus, species-A. *J Am Mosquito Contr* 3 (2):222–30.

Kamali, M., M. V. Sharakhova, E. Baricheva, D. Karagodin, Z. Tu, and I. V. Sharakhov. 2011. An integrated chromosome map of microsatel-lite markers and inversion breakpoints for an Asian malaria mosquito, Anopheles stephensi. *J Hered* 102 (6):719–26.

Kamali, M., A. Xia, Z. Tu, and I. V. Sharakhov. 2012. A new chromo-somal phylogeny supports the repeated origin of vectorial capacity in malaria mosquitoes of the Anopheles gambiae complex. *PLoS Pathog* 8 (10):e1002960.

Kirkpatrick, M. and N. Barton. 2006. Chromosome inversions, local adaptation and speciation. *Genetics* 173 (1):419–34.

Kitzmiller, J. B., R. D. Kreutzer, and M. G. Rabbani. 1974. The salivary gland chromosomes of Anopheles walkeri Theobald. *Mosq News* 34:22–8.

Knight, K. L. 1978. *Supplement to a Catalog of the Mosquitoes of the World.* College Park, MD: Entomological Society of America.

Knight, K. L. and A. Stone. 1977. *A Catalog of the Mosquitoes of the World (Diptera: Culicidae), 2nd ed.* College Park, MD: Entomological Society of America.

Krimbas, C. B. and J. R. Powell. 1992. *Drosophila Inversion Polymorphism.* Edited by C. B. Krimbas and J. R. Powell. Boca Raton, FL: CRC Press.

Krzywinski, J., O. G. Grushko, and N. J. Besansky. 2006. Analysis of the complete mitochondrial DNA from Anopheles funestus: An improved dipteran mitochondrial genome annotation and a temporal dimension of mosquito evolution. *Mol Phylogenet Evol* 39 (2):417–23.

Krzywinski, J., R. C. Wilkerson, and N. J. Besansky. 2001. Toward understanding Anophelinae (Diptera, Culicidae) phylogeny: Insights from nuclear single-copy genes and the weight of evidence. *Syst Biol* 50 (4):540–56.

Kumar, A. and K. S. Rai. 1990. Chromosomal localization and copy number of 18S+28S ribosomal RNA genes in evolutionary diverse mosquitoes (Diptera, Culicidae). *Hereditas* 113:277–89.

Labbe, P., C. Berticat, A. Berthomieu, S. Unal, C. Bernard, M. Weill, and T. Lenormand. 2007. Forty years of erratic insecticide resistance evolution in the mosquito Culex pipiens. *PLoS Genet* 3 (11):e205.

Labbe, P., N. Sidos, M. Raymond, and T. Lenormand. 2009. Resistance gene replacement in the mosquito Culex pipiens: Fitness estimation from long-term cline series. *Genetics* 182 (1):303–12.

Lawniczak, M. K., S. J. Emrich, A. K. Holloway, A. P. Regier, M. Olson, B. White, S. Redmond et al. 2010. Widespread divergence between incipient Anopheles gambiae species revealed by whole genome sequences. *Science* 330 (6003):512–4.

Levan, A., K. Fredga, and A. A. Sandberg. 1964. Nomenclature for centromeric possions on chromosomes. *Hereditas* 52:201–20.

Lobachev, K. S., A. Rattray, and V. Narayanan. 2007. Hairpin- and cruciform-mediated chromosome breakage: Causes and consequences in eukaryotic cells. *Front Biosci* 12:4208–20.

Mahmood, F. and R. K. Sakai. 1984. Inversion polymorphisms in natural populations of Anopheles stephensi. *Can J Genet Cytol* 26 (5):538–46.

Mahmood, F. and R. K. Sakai. 1985. An ovarian chromosome map of Anopheles stephensi. *Cytobios* 43 (171):79–86.

McAbee, R. D., J. A. Christiansen, and A. J. Cornel. 2007. A detailed larval salivary gland polytene chromosome photomap for Culex quinquefasciatus (Diptera: Culicidae) from Johannesburg, South Africa. *J Med Entomol* 44 (2):229–37.

McClelland, G. A. H. 1962. Sex-linkage in Aedes aegypti. *Trans Roy Soc Trop Med Hyg* 56:4.

McDonald, P. T. and K. S. Rai. 1970. Correlation of linkage groups with chromosomes in the mosquito Aedes aegypti. *Genetics* 66:475–85.

Morens, D. M. and A. S. Fauci. 2008. Dengue and hemorrhagic fever: A potential threat to public health in the United States. *JAMA* 299 (2):214–6.

Mori, A., J. Romero-Severson, and D. W. Severson. 2007. Genetic basis for reproductive diapause is correlated with life history traits within the Culex pipiens complex. *Insect Mol Biol* 16 (5):515–24.

Mori, A., D. W. Severson, and B. M. Christensen. 1999. Comparative linkage maps for the mosquitoes (Culex pipiens and Aedes aegypti) based on common RFLP loci. *J Hered* 90 (1):160–4.

Motara, M. A., S. Pathak, K. L. Satya-Prakash, and T. C. Hsu. 1985. Agrentophilic structures of spermatogenesis in the yellow fever mosquito. *J Hered* 76:295–300.

Motara, M. A. and K. S. Rai. 1977. Chromosmal differentiation in two species of Aedes and their hybrids revealed by Giemsa C-banding. *Chromosoma* 64:125–32.

Motara, M. A. and K. S. Rai. 1978. Giemsa C-banding patterns in Aedes (Stegomia) mosquitoes. *Chromosoma* 70:51–8.

Munstermann, L. E. 1990. Linkage Map for Yellow Fever Mosquito, Aedes aegypti. Edited by S. O'Brein, *Locus maps of complex geneomes.* New York: Cold Spring Harbor Laboratories, Cold Spring Harbor.

Munstermann, L. E. and G. B. Craig Jr. 1979. Genetics of Aedes aegypti. Updating the linkage map. *J Hered* 70:291–6.

Nadeau, J. H. and B. A. Taylor. 1984. Lengths of chromosomal segments-conserved since divergence of man and mouse. *Proc Natl Acad Sci USA* 81 (3):814–8.

Neafsey, D. E., G. K. Christophides, F. H. Collins, S. J. Emrich, M. C. Fontaine, W. Gelbart, M. W. Hahn et al. 2013. The evolution of the Anopheles 16 genomes project. *G3* (Bethesda) 3 (7):1191–4.

Nene, V., J. R. Wortman, D. Lawson, B. Haas, C. Kodira, Z. J. Tu, B. Loftus et al. 2007. Genome sequence of Aedes aegypti, a major arbovirus vector. *Science* 316 (5832):1718–23.

Newton, M. E., D. I. Southern, and R. J. Wood. 1974. X and Y chromosmes of Aedes aegypti (L.) distiguished by Giemsa C-banding. *Chromosoma* 49:41–9.

Novikov, D. V., I. Kireev, and A. S. Belmont. 2007. High-pressure treatment of polytene chromosomes improves structural resolution. *Nat Methods* 4 (6):483–5.

Peery, A., M. V. Sharakhova, C. Antonio-Nkondjio, C. Ndo, M. Weill, F. Simard, and I. V. Sharakhov. 2011. Improving the population genetics toolbox for the study of the African malaria vector Anopheles nili: Microsatellite mapping to chromosomes. *Parasit Vectors* 4:202.

Perez-Requejo, J. L., J. Aznar, M. T. Santos, and J. Valles. 1985. Early platelet-collagen interactions in whole blood and their modifications by aspirin and dipyridamole evaluated by a new method (BASIC wave). *Thromb Haemost* 54 (4):799–803.

Powell, J. R., V. Petrarca, A. della Torre, A. Caccone, and M. Coluzzi. 1999. Population structure, speciation, and introgression in the Anopheles gambiae complex. *Parassitologia* 41 (1–3):101–13.

Raghunathan, A., H. R. Ferguson Jr., C. J. Bornarth, W. Song, M. Driscoll, and R. S. Lasken. 2005. Genomic DNA amplification from a single bacterium. *Appl Environ Microbiol* 71 (6):3342–7.

Rai, K. S. 1963. A comparative study of mosquito karyotypes. *Ann Entomol Soc Amer* 56:160–70.

Rao, P. N. and K. S. Rai. 1990. Genome evolution in the mosquitoes and other closely related members of superfamily Culicoidea. *Hereditas* 113 (2):139–44.

Reddy, B. P., P. Labbe, and V. Corbel. 2012. Culex genome is not just another genome for comparative genomics. *Parasit Vectors* 5:63.

Rocca, K. A., E. M. Gray, C. Costantini, and N. J. Besansky. 2009. 2La chromosomal inversion enhances thermal tolerance of Anopheles gambiae larvae. *Malar J* 8:147.

Santos, M. J., J. Garrido, C. Oliver, A. R. Robbins, and F. Leighton. 1985. Characterization of peroxisomes in Chinese hamster ovary cells in culture. *Exp Cell Res* 161 (1):189–98.

Scott, J. A., W. G. Brogdon, and F. H. Collins. 1993. Identification of single specimens of the Anopheles gambiae complex by the polymerase chain reaction. *Am J Trop Med Hyg* 49 (4):520–9.

Scriven, P. N., T. L. Kirby, and C. M. Ogilvie. 2011. FISH for pre-implantation genetic diagnosis. *J Vis Exp* (48), e2570.

Severson, D. W. 2008. Mosquito. In *Genome Mapping and Genomics in Animals*. Edited by C. Kole and W. Hunter. Berlin Heidelberg, Germany: Springer-Verlag.

Severson, D. W., J. K. Meece, D. D. Lovin, G. Saha, and I. Morlais. 2002. Linkage map organization of expressed sequence tags and sequence tagged sites in the mosquito, Aedes aegypti. *Insect Mol Biol* 11 (4):371–8.

Severson, D. W., A. Mori, Y. Zhang, and B. M. Christensen. 1993. Linkage map for Aedes aegypti using restriction fragment length polymorphisms. *J Hered* 84 (4):241–7.

Severson, D. W., A. Mori, Y. Zhang, and B. M. Christensen. 1994. Chromosomal mapping of two loci affecting filarial worm susceptibility in Aedes aegypti. *Insect Mol Biol* 3 (2):67–72.

Severson, D. W., V. Thathy, A. Mori, Y. Zhang, and B. M. Christensen. 1995. Restriction fragment length polymorphism mapping of quantitative trait loci for malaria parasite susceptibility in the mosquito Aedes aegypti. *Genetics* 139 (4):1711–7.

Sharakhov, I. V., A. C. Serazin, O. G. Grushko, A. Dana, N. Lobo, M. E. Hillenmeyer, R. Westerman et al. 2002. Inversions and gene order shuffling in Anopheles gambiae and A. funestus. *Science* 298 (5591):182–5.

Sharakhov, I. V. and M. V. Sharakhova. 2008. Cytogenetic and physical mapping of mosquito genomes. In *Chromosome Mapping Research Developments*. Edited by J. F. Verrity and L. E. Abbington. New York: Nova Science Publishers.

Sharakhova, M. V., C. Antonio-Nkondjio, A. Xia, C. Ndo, P. Awono-Ambene, F. Simard, and I. V. Sharakhov. 2011a. Cytogenetic map for Anopheles nili: Application for population genetics and comparative physical mapping. *Infect Genet Evol* 11 (4):746–54.

Sharakhova, M. V., C. Antonio-Nkondjio, A. Xia, C. Ndo, P. Awono-Ambene, F. Simard, and I. V. Sharakhov. 2013. Polymorphic chromosomal inversions in Anopheles moucheti, a major malaria vector in Central Africa. *Med Vet Entomol*. doi: 10.1111/mve.12037.

Sharakhova, M. V., P. George, I. V. Brusentsova, S. C. Leman, J. A. Bailey, C. D. Smith, and I. V. Sharakhov. 2010a. Genome mapping and characterization of the Anopheles gambiae heterochromatin. *BMC Genomics* 11:459.

Sharakhova, M. V., M. P. Hammond, N. F. Lobo, J. Krzywinski, M. F. Unger, M. E. Hillenmeyer, R. V. Bruggner, E. Birney, and F. H. Collins. 2007. Update of the Anopheles gambiae PEST genome assembly. *Genome Biol* 8 (1):R5.

Sharakhova, M. V., V. A. Timoshevskiy, F. Yang, SIu Demin, D. W. Severson, and I. V. Sharakhov. 2011b. Imaginal discs—A new source of chromosomes for genome mapping of the yellow fever mosquito Aedes aegypti. *PLoS Negl Trop Dis* 5 (10):e1335.

Sharakhova, M. V., A. Xia, S. C. Leman, and I. V. Sharakhov. 2011c. Arm-specific dynamics of chromosome evolution in malaria mosquitoes. *BMC Evol Biol* 11 (1):91.

Sharakhova, M. V., A. Xia, Z. Tu, Y. S. Shouche, M. F. Unger, and I. V. Sharakhov. 2010b. A physical map for an Asian malaria mosquito, Anopheles stephensi. *Am J Trop Med Hyg* 83 (5):1023–7.

Sharma, G. P., O. P. Mittal, S. Chaudhry, and V. Pal. 1978. A preliminary map of the salivary gland chromosomes of Aedes (stegomyia) aegypti (Culicadae, Diptera). *Cytobios* 22 (87–88):169–78.

Sun, T., J. Jia, D. Zhong, and Y. Wang. 2006. Determination of 17 kinds of banned organochlorine pesticides in water by activated carbon fiber-solid phase microextraction coupled with GC-MS. *Anal Sci* 22 (2):293–8.

Takken, W., W. Eling, J. Hooghof, T. Dekker, R. Hunt, and M. Coetzee. 1999. Susceptibility of Anopheles quadriannulatus Theobald (Diptera: Culicidae) to Plasmodium falciparum. *Trans R Soc Trop Med Hyg* 93 (6):578–80.

Takken, W. and B. G. J. Knols. 2009. Malaria vector control: Current and future strategies. *Trends Parasitol* 25 (3):101–4.

Tesler, G. 2002. GRIMM: Genome rearrangements web server. *Bioinformatics* 18 (3):492–3.

Tewfik, H. R. and A. R. Barr. 1976. Paracentric inversion in Culex pipiens. *J Med Entomol* 13 (2):147–50.

Timoshevskiy, V. A, N. A. Kinney, B. S. deBruyn, C. Mao, Z. Tu, D. W. Severson, I. V. Sharakhov, and M. V. Sharakhova. 2014. Genomic composition and evolution of Aedes aegypti chromosomes revealed by the analysis of physically mapped supercontigs. *BMC Biol* 12:27. doi: 10.1186/1741-7007-12-27.

Timoshevskiy, V. A., D. W. Severson, B. S. Debruyn, W. C. Black, I. V. Sharakhov, and M. V. Sharakhova. 2013. An integrated linkage, chromosome, and genome map for the yellow fever mosquito Aedes aegypti. *PLoS Negl Trop Dis* 7 (2):e2052.

Timoshevskiy, V. A., A. Sharma, I. V. Sharakhov, and M. V. Sharakhova. 2012. Fluorescent in situ hybridization on mitotic chromosomes of mosquitoes. *J Vis Exp* (67), e4215 10.3791/4215.

Tolle, M. A. 2009. Mosquito-borne diseases. *Curr Probl Pediatr Adolesc Health Care* 39 (4):97–140.

Toure, Y. T., V. Petrarca, S. F. Traore, A. Coulibaly, H. M. Maiga, O. Sankare, M. Sow, M. A. DiDeco, and M. Coluzzi. 1998. The distribution and inversion polymorphism of chromosomally recognized taxa of the *Anopheles gambiae* complex in Mali, West Africa. *Parassitologia* 40:477–511.

Treff, N. R., J. Su, X. Tao, L. E. Northrop, and R. T. Scott Jr. 2011. Single-cell whole-genome amplification technique impacts the accuracy of SNP microarray-based genotyping and copy number analyses. *Mol Hum Reprod* 17 (6):335–43.

Trifonov, V. A., N. N. Vorobyeva, and W. Rens. 2009. FISH with and without COT1 DNA. In *Fluorescence* In Situ *Hybridization (FISH)*. Edited by T. Leiehr. Berlin, Germany: Springer-Verlag.

von Grotthuss, M., M. Ashburner, and J. M. Ranz. 2010. Fragile regions and not functional constraints predominate in shaping gene organization in the genus Drosophila. *Genome Res* 20:1084–96.

Wallace, A. J. and M. E. Newton. 1987. Heterochromatin diversity and cyclic responses to selective silver staining in Aedes aegypti (L.). *Chromosoma* 95:89–93.

WHO. 2009. Dengue. *Guidelines for Diagnosis, Treatment, Prevention and Control*. Geneva, Switzerland: World Health Organization.

World Malaria Report, WHO Global Malaria Programme. 2010.

Xia, A., M. Sharakhova, S. Leman, Z. Tu, J. Bailey, C. Smith, and I. V. Sharakhov. 2010. Genome landscape and evolutionary plasticity of chromosomes in malaria mosquitoes. *PLoS ONE* 5 (5):e10592.

Xia, A., M. V. Sharakhova, and I. V. Sharakhov. 2008. Reconstructing ancestral autosomal arrangements in the Anopheles gambiae complex. *J Comput Biol* 15 (8):965–80.

Zambetaki, A., N. Pasteur, and P. Mavragani-Tsipidou. 1998. Cytogenetic analysis of the Malpighian tubule polytene chromosomes of Culex pipiens (Diptera: Culicidae). *Genome* 41:751–5.

Zhimulev, I. F. 1996. Morphology and structure of polytene chromosomes. In *Advances in genetics*, Vol. 34, pp. 1–490. Edited by J. C. Hall. San Diego, CA: Academic Press.

Zhong, D., D. M. Menge, E. A. Temu, H. Chen, and G. Yan. 2006. Amplified fragment length polymorphism mapping of quantitative trait loci for malaria parasite susceptibility in the yellow fever mosquito Aedes aegypti. *Genetics* 173 (3):1337–45.

5

Beetles (Coleoptera)

Diogo C. Cabral-de-Mello

CONTENTS

LIST OF ABBREVIATIONS

C_0t, concentration of DNA and time
CMA_3, chromomycin A_3
DA, distamycin A
DAPI, 4,6 -diamidino-2-phenylindole
dATP, deoxyadenosine triphosphate
dCTP, deoxycytidine triphosphate
DDT, dithiothreitol
dGTP, deoxyguanosine triphosphate
DNA, deoxyribonucleic acid
dNTPs, deoxynucleotide triphosphates
dTTP, deoxythymidine triphosphate
dUTP, 2 -deoxyuridine 5 –triphosphate
EDTA, ethylenediaminetetraacetic acid
FISH, fluorescence in situ hybridization
Mb, megabases
NOR, Nucleolus Organizer Region
PCR, polymerase chain reaction
pg, picogram
rDNA, ribosomal DNA
rpm, revolutions per minute
SDS, sodium dodecyl sulfate
SSC, saline sodium citrate
TAE, Tris-acetate
TE, Tris-EDTA
WGA, whole genome amplification

5.1 INTRODUCTION

5.1.1 Taxonomy

The Coleoptera order (beetles) comprises the most species-rich group among the insect class, corresponding to approximately 40% of this class and 30% of all animals, encompassing more than 350,000 described species distributed worldwide (Lawrence 1982; Lawrence and Britton 1994; Arnett and Thomas 2001; Arnett et al. 2002). Although a large number of species have been cataloged, the number of beetle species described so far could be less than the real diversity of the group, and according to Terry Erwin (1982), the total number of beetles on the planet could be estimated at 12,000,000. In contrast, more recent revisions proposed the occurrence of 850,000–4,000,000 living species (Hammond 1995; Stork 1999; Nielsen and Mound 1999).

The evolutionary history of this group dates from approximately 285 million years ago (Crowson 1981; Grimaldi and Engel 2005). According to some authors, the evolutionary success and adaptive radiation of beetles were favored by the presence of their hardened forewings, named elytra, which in most species cover completely the abdominal region, and a pair of membranous wings. This structure is the most distinctive feature among beetles, and its occurrence likely favored the protection of beetles against predation, niche exploration, and other environmental stresses, without the loss of flight ability (Hammond 1979; Crowson 1981; Lawrence and Britton 1994; Costa 1999). Moreover, for some extant families, the great diversity could be explained by the association with angiosperms during periods of intense radiation of this group, in addition to the coradiation with mammals and speciation associated with climatic changes (Crowson 1981; Erwin 1985; Farrell 1998; Davis et al. 2002).

The order is subdivided into four suborders, Archostemata, Myxophaga, Adephaga, and Polyphaga, with most of the described representatives belonging to Polyphaga. The phylogenetic relationships among the Coleoptera subfamilies are controversial, but the monophyly of the order is accepted by some authors when considering at least the extant lineages. Among the families, Curculionidae, Staphylinidae, Chrysomelidae, Scarabaeidae and Cerambycidae belonging to Polyphaga, and Carabidae from Adephaga, are considered megadiverse and each presents at least 20,000 described species (Crowson 1955; Lawrence 1982; Lawrence and Newton 1995; Beutel and Haas 2000; Bouchard et al. 2009).

5.1.2 Importance of the Species

Beetles are important elements in natural environments and in cultivated and agricultural areas. They act as pollinators of flowering plants and play important roles in organic matter recycling. Cantharophily (pollination by beetles) occurs, for example, almost exclusively by beetles in more than 180 species of angiosperms representing 34 families (Bernhardt 2000). This association with flowering plants is considered primitive, and in some species the development of specialized mouthparts for nectar and pollen collecting are observed (Barth 1985; Bernhardt 2000).

Another beneficial effect of beetles can be observed in dung beetles, represented mainly by Scarabaeidae species, which play an important role in nutrient recycling through the consumption and burial of dung for reproduction and larval alimentation. This process protects agricultural areas by preventing the development or establishment of flies and other pests. Examples of introduced species that present this activity and are useful for fly control are *Digitonthophagus gazella* and *Euoniticellus intermedius* (Fincher 1981; Anderson et al. 1984). Another beneficial effect of beetles is the biological control of invasive plants by herbivory, observed in several Curculionidae and Chrysomelidae representatives.

The negative effects of beetles include the consumption of crop species and stored products, including dried fruits, cereals, and tobacco, in both the larval and adult stages. Moreover, some species are forestry pests. *Anthonomus grandis grandis* (cotton boll weevil, Curculionidae), *Tribolium castaneum* (red flour beetle, Tenebrionidae), *Dendroctonus ponderosae* (bark beetle, Curculionidae), *Xanthogaleruca luteola* (elm leaf beetle, Chrysomelidae) are a few examples of beetles considered pests that cause economic concerns.

In addition, the beetles also have importance as human food, being the most commonly consumed insect (31%) according to the FAO Forestry. Moreover, they have been used in forensic entomology and in religion and mythology, for example, the scarab beetle for the ancient Egyptians (Bouchard et al. 2009; van Huis et al. 2013).

5.1.3 Karyotype

Although highly diverse in the number of species, and from the chromosomal point of view, only approximately 1% of Coleoptera species have so far been studied using cytogenetic techniques. In addition, these studies have concentrated on the description of the general features of beetle karyotypes, including diploid number, chromosomal morphology, and identification of sex chromosomes The book published by Smith and Virkki (1978) is the main reference concerning the chromosomal structure under conventional cytogenetic analysis in beetles belonging to the four Coleoptera suborders. The main karyotypic structure observed in Coleoptera is the occurrence of $2n = 20$ with the presence, in most species, of an Xy sex chromosomal system associated during meiosis, such as a parachute-like structure named Xy_p. In this system, the X chromosome is a large bi-armed element representing the "canopy" associated to a punctate y chromosome that represents the "load."

The karyotype composed of $2n = 20$, Xy_p and bi-armed chromosomes, is considered modal and is an ancestral characteristic in beetles, at least for the Polyphaga, being observed in almost all families. In most groups, the presence of bi-armed chromosomes is common (Smith and Virkki 1978).

Banding techniques permitted a better understanding of the chromosomal organization in coleopterans. C-banding has revealed the occurrence of heterochromatin mainly in the pericentromeric areas of the chromosomes. In some other cases, the presence of large paracentromeric blocks and diphasic chromosomes (chromosomes with one euchromatic and one heterochromatic arm) was reported, for example, in Tenebrionidae and Scarabaeidae representatives, respectively. Moreover, additional heterochromatic blocks in telomeric and interstitial regions have been described (Juan and Petitpierre 1989; Pons 2004; Rozék et al. 2004; Cabral-de-Mello et al. 2010a). Silver nitrate ($AgNO_3$) staining revealed mostly autosomal nucleolus organizer regions (NORs) (Schneider et al. 2007), and CMA_3 (chromomycin A_3)/DAPI (4'6-diamidino-2-phenylindole) fluorochrome staining indicated variability for heterochromatin repeats compositions, with respect to the A+T and G+C compositions.

More recently, the advent of cytogenetic mapping through fluorescent in situ hybridization (FISH) has permitted a deeper knowledge of karyotype architecture and evolution in beetles, with regard to repetitive DNA, such as multigene families, satellite DNAs, transposable elements, and telomere repeats (Frydrychová and Marec 2002; Martínez-Navarro et al. 2004; Mravinac et al. 2004; Palomeque et al. 2005; Cabral-de-Mello et al. 2011b; Mravinac et al. 2011; Oliveira et al. 2012).

5.1.4 Genome Size

The genome size has been estimated in approximately 240 beetle species belonging to 24 families, but the data are highly biased toward Chrysomelidae (64 species), Tenebrionidae (64 species), and Coccinellidae (30 species), with the other families represented by less than 20 species and in most cases by fewer than 10 representatives. These genome sizes were estimated using distinct methods, such as Feulgen image analysis densitometry, flow cytometry, and biochemical analysis, using sperms, hemocytes, brain, or whole body, resulting, in general, in similar results (Animal Genome Size Database, www.genomesize.com, accessed on 13 July, 2013; Hanrahan and Johnston 2011). Beetle genome-size variation ranges from 0.16 pg in *Tribolium audax* (Tenebrionidae) and *Oryzaephilus surinamensis* (Silvanidae) to 3.69 pg in *Chrysolina carnifex* (Chrysomelidae), and, for the most part it is smaller than 1 pg. The species *Aramigus tessellatus* was not considered; it exhibits a genome of 5.02 pg, and is 3*n*. For comparison, the smallest genome of beetles is almost 20 times smaller than the human genome.

5.1.5   Genome Sequencing Project

Currently, only the complete coleopteran genomes of the pest of stored grain products species *Tribolium castaneum* (Richards et al. 2008) and the pest of pine forests *Dendroctonus ponderosae* (Keeling et al. 2013) were sequenced. These genomes are approximately 204 Mb in size and are rich in A+T content. One-third of the *T. castaneum* genome is composed of repetitive DNAs (the type of sequence primarily mapped in beetle chromosomes), whereas in *D. ponderosae* approximately 17% (male) and 23% (female) are repetitive elements (Richards et al. 2008; Keeling et al. 2013).

Considering the massive genome sequencing by modern methods and the small genome size of beetles compared to, for example, other insects such as grasshoppers, and even human, in the next few years, several additional sequenced genomes will likely be available. According to the i5K initiative, which has plans to sequence 5000 insect and related arthropod genomes, there are 69 species of beetle currently nominated as part of the initiative (accessed on 27 July, 2013, at www.arthropodgenomes.org/wiki/i5k; Evans et al. 2013).

5.2   PROTOCOLS

The protocols presented in this chapter are simple to follow and a beginning researcher in Coleoptera chromosomes could conduct the experiments. They were adapted for the purpose of obtaining better results using chromosomes of beetles as a source. Presented are some adaptations developed in the lab and also some useful commercially available kits for chromosome studies in Coleoptera. These protocols and kits are also useful for chromosomal studies of other insect orders.

Although the main focus of this chapter is chromosome mapping, some protocols for classical chromosomal banding are also presented. The integration of the results that can be obtained using these assays with the results of chromosome mapping is important for answering a myriad of questions about chromosome/genome organization and evolution.

5.2.1 Materials and Supplies

The necessary material and other supplies are presented in the text at the end of each protocol.

5.2.2 Equipment

The laboratory for the development of the protocols presented in this chapter requires a structure that combined equipment for classical cytogenetics and molecular biology experiments. Among the essential equipment are, for example, a stereomicroscope, microscope, hot plate, refrigerator, freezers (−20°C and −80°C), laboratory oven, water bath, analytical balance, nitrogen container, autoclave, thermocycler, refrigerated centrifuge, electrophoresis apparatus, bright field and epifluorescence microscope coupled to a digital camera and specific filters (e.g., for DAPI, rhodamine, Alexa Fluor® 488, Cy3, Cy5, and fluorescein isothiocyanate), spectrophotometer, hybridization oven, transilluminator, laminar flow cabinet, incubation shaker, inverted microscope coupled to a microdissector, micropipette puller, vortex mixers, pipettes, minispin, and centrifuge.

5.2.3 Species Culture

Because of the high diversity of beetles, for example, habitat, alimentation and behavior, and the distinct requirements along the life cycle, it is not easy to maintain distinct colonies of some species in the laboratory for chromosomal comparative analysis; moreover, it could be time consuming. However, some species can be successfully maintained in the laboratory, including *Tribolium castaneum* (Tenebrionidae), *Tenebrio molitor* (Tenebrionidae), *Callosobruchus maculatus* (Chrysomelidae), *Zophobas morio* (Tenebrionidae), *Anthonomus grandis* (Curculionidae), and *Digitonthophagus gazella* (Scarabaeidae), which has encouraged the use of some of these species as models for genetic studies, as in the case of *Tribolium castaneum*.

On the other hand, most chromosomal studies in Coleoptera have been performed using beetles captured in the field and used directly for karyotype analysis. These beetles are manually collected under the bark of trees, logs and stones, on leaves and flowers, in animal excrements, or by using specific apparatus, such as insect nets, pitfall traps, and light traps, among others. These beetles should be maintained alive until manipulation in the laboratory to obtain the tissues for chromosomal studies, or they should be stored in 100% ethanol to preserve the DNA for molecular cytogenetic analyses.

5.2.4 Source of Chromosomes

In adult beetles, a simple way to obtain chromosomes is from testicular follicles in males. However, it is also possible to use other tissues as chromosome sources, such as embryos, ovarioles, mid guts, and cerebral ganglion from prepupal larvae. The advantage of testis is that it provides good meiotic and mitotic metaphases, in addition to the occurrence of initial meiotic cells, which in some cases are important to understand the chromosomal behavior, mainly for the sex elements. This tissue can be directly dissected from the insect and fixed in Carnoy's solution (3:1, 100% ethanol or methanol:acetic acid); however, in some cases, colchicine pretreatment is required to increase the number of premeiotic spermatogonial metaphases.

1. Anesthetize and pin the animal in a petri dish containing solid paraffin and insect saline solution (Figure 5.1a).
2. Open the animal elytra and cut the membranous wing. Use an adequate scissors to cut the abdominal membrane, and visualize the internal organs (Figures 5.1b through d).
3. Dissect the follicular testes, place them in modified Carnoy's solution (3:1, 100% ethanol or methanol:acetic acid) and store in 1.5- to 2-mL microtubes at −20°C until use (Figure 5.1d through f).

Alternatively perform the colchicine treatment as follows:

4. After follicular testis dissection, place them in 2 mL of 0.05% colchicine solution for 90–120 minutes.
5. Add an equal amount of distilled water and wait for 15 minutes for hypotonic treatment.
6. Place the testes in Carnoy's solution (3:1, 100% ethanol or methanol:acetic acid) and store at −20°C until use.

    Materials/Reagents
        Glacial acetic acid
        Colchicine
        100% Ethanol
        Insect saline solution
        100% Methanol

    Instruments/Equipment
        Microtube
        Petri dish with paraffin
        Pin
        Scissors
        Stereomicroscope
        Tweezers

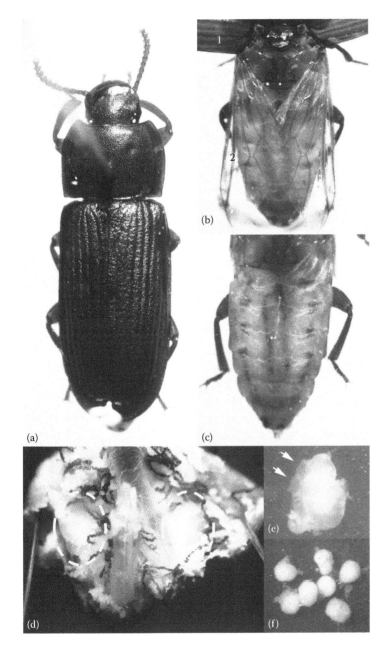

FIGURE 5.1   **(See color insert.)** Dissection of tissue useful for chromosomal analysis. (a) *Tenebrio molitor* exemplar pinned on a Petri dish containing insect saline solution; (b) dorsal view of the abdomen showing the opened elytra (1) and the membranous wing pair (2); (c) dorsal view of the abdomen without the membranous wings; (d) opened abdomen indicating the testis position (dashed yellow circle); (e) testis before fixation, the arrows indicate two follicular testis; (f) individualized testicular follicles after fixation in modified Carnoy's solution.

5.2.5   Chromosome Preparation, Staining, and Banding

5.2.5.1   Chromosome Preparation

The slides obtained using the following methods can be used for conventional, differential staining and in FISH experiments.

1. Place one testis follicle on a clean slide, add a drop of 50% acetic acid, and disintegrate the tissue with a flat iron macerator. Depending on the size of the follicle, it can be used for more than one slide.

2. Add another drop of 50% acetic acid and spread the solution on the slide.

3. Place the slide on a hot plate at 40°C–45°C until the solution evaporates completely. Tilt the slide to spread the solution with cells over the slide surface more efficiently. Avoid temperatures higher than 45°C because of the possibility of DNA/chromosome degradation, which could decrease the quality of results after FISH.

4. For the squashing technique, alternatively, after the tissue maceration (step 1), the material could be covered with a 24 × 24 or 24 × 32 mm glass coverslip and pressed using filter paper and the eraser side of a pencil or tweezers. To obtain chromosomes in a similar focal plane, the final pressure could be applied with the thumb.

5. Immerse the slide in liquid nitrogen for 30–60 seconds, and then remove the coverslip using a razor blade.

6. Check the quality of the spreads and analyze the slides using a 10 × objective of a light microscope to select the cells.

   Materials/Reagents
      Glacial acetic acid

   Instruments/Equipment
      Flat iron macerator
      Glass coverslip
      Hot plate
      Liquid nitrogen container
      Microscope
      Razor blade
      Slide
      Stereomicroscope
      Tweezers

5.2.5.2   Conventional Staining

1. Place one drop of 2% lacto–acetic orcein on a clean slide and add one or a portion of one follicular testis.

2. Using the flat iron macerator, macerate the material on the slide and wait at least 10 minutes for good material staining.

3. Cover the material with an adequate coverslip (24 × 24 or 24 × 32 mm) and press using filter paper and either the eraser side of a pencil or tweezers to squash the material and eliminate bubbles. To obtain chromosomes in a similar focal plane, the final pressure could be applied with the thumb.
4. Check the spread quality and analyze the cells using a light microscope.

Note: Alternatively the slides obtained as described in Section 5.2.5.1 could be stained with 5% Giemsa in phosphate buffer for 3 minutes for conventional analysis.

  Materials/Reagents
    Lacto–acetic orcein or Giemsa
    Phosphate buffer

  Instruments/Equipment
    Flat iron macerator
    Glass coverslip
    Microscope
    Slide
    Stereomicroscope
    Tweezers

5.2.5.3 Banding Techniques

5.2.5.3.1 *C-Banding*

The protocol for constitutive heterochromatin (C-positive blocks) detection is based on the description of Sumner (1972).

1. Place the slide in a Coplin jar containing 0.1 N HCl at room temperature for 30 minutes.
2. Wash the preparation with distilled water.
3. Place the slide in 5% barium hydroxide solution in a Coplin jar at 60°C in a water bath for 10 seconds to 3 minutes. Note: This time is variable for distinct species, and it is also dependent on the slide age. Old slides require longer incubation times in 5% barium hydroxide. Distinct times of incubation in 5% barium hydroxide solution should be tested to obtain good results.
4. Immediately place the slide in 0.1 N HCl at room temperature for 1 minute under agitation.
5. Wash the slide with distilled water.
6. Incubate the slide in 2 × SSC (saline sodium citrate) at 60°C in a water bath for 45 minutes.
7. Wash the slide with distilled water.
8. Stain the chromosome spread with 5% Giemsa diluted in phosphate buffer for 3 minutes.

9. Check the quality of the spread and C-bands, and analyze the slides using a 10 × objective of a light microscope. Note: Alternatively, after C-banding treatment, the chromosome spreading could be stained with DAPI (0.2 mg/mL) or propidium iodide (0.5 µg/mL) directly mixed in VECTASHIELD mounting medium (Cat. no. H-1000, Vector Laboratories, UK), proportion 0.5:15 µL (v:v), used for slide mounting. This type of staining facilitates the observation of C-positive blocks.

Materials/Reagents
  Barium hydroxide Ba(OH)2
  DAPI or propidium iodide
  Giemsa
  Hydrochloric acid (HCl)
  Phosphate buffer
  SSC
  VECTASHIELD mounting medium

Instruments/Equipment
  Coplin jar
  Microscope
  Water bath
  Micropipettes

5.2.5.3.2   *Triple Fluorescent Staining with CMA₃/DA/DAPI*

This technique is used to detect G+C– and A+T–rich chromatin, which stain with CMA_3 or DAPI, respectively. The protocol is based on the work of Schweizer (1980, 1981), and the whole procedure should be conducted under dark conditions. The time for each step presented is variable depending on the material.

1. Place 80 µL CMA_3 (0.5 mg/mL) on the slide, cover with a glass or parafilm coverslip of an adequate size, avoiding bubbles, and incubate at room temperature inside a dark box for 40 minutes.
2. Wash the slide with distilled water, eliminating the coverslip and the excess CMA_3 solution, and air-dry.
3. Place 80 µL distamycin A (DA) (0.1 mg/mL) on the slide, cover with a glass or parafilm coverslip, and incubate in a dark box for 30 minutes.
4. Wash the slide with distilled water, eliminating the coverslip and the excess DA solution, and air-dry the slide.
5. Pour 80 µL DAPI (0.5 mg/mL) on the slide, cover with a glass or parafilm coverslip, and incubate in a dark box for 30 minutes.

6. Wash the slide with distilled water, eliminating the cover-slip and the excess DAPI solution, and air-dry the slide.
7. Mount the slide with VECTASHIELD mounting medium using a glass coverslip and analyze under an epifluorescence microscope using the appropriate filters.

Materials/Reagents
    4′, 6-diamidino-2-phenylindole (DAPI)
    Chromomycin A_3 (CMA_3)
    Distamycin A (DA)
    VECTASHIELD mounting medium

Instruments/Equipment
    Dark box
    Epifluorescence microscope
    Glass and parafilm coverslips
    Micropipettes

5.2.5.3.3 *Silver Nitrate Staining*

Silver impregnation in beetles detects the nucleolar material attached to the chromosomes responsible for major rDNA cluster expression. In some cases, the specific determination of the chromosome responsible for nucleolus biogenesis is difficult to make because of the similarity of the autosomes and the not-well-spread initial meiosis in some beetles. Moreover, the nucleolus persists attached to the chromosome responsible for its biogenesis for only a short time. To facilitate the interpretation of the data, the analysis of mitotic cells is also useful. In addition, in some groups of beetles, this technique has revealed the heterochromatin distribution, and in the case of the occurrence of the Xy_p sex system, the lumen of this bivalent is also heavily stained by $AgNO_3$ because of the presence of argyrophilic substance, which according to Virkki et al. (1990) might have a function in association/disjunction during meiosis. The most commonly used protocol is based on the description from Rufas et al. (1982) as follows:

1. Select preparations rich in cells at initial meiosis.
2. Place the slides in 2 × SSC at 60°C in a water bath for 10 minutes.
3. Wash the slide with distilled water and air-dry.
4. Pour one drop of silver nitrate solution on the slide and cover using a glass coverslip. (Silver nitrate solution: Dissolve 0.5 g silver nitrate [$AgNO_3$] in 1 mL formic acid solution [add one drop of formic acid, pH 3–3.5, to 100 mL distilled water]).
5. Incubate the slide in a humid chamber at 70°C–80°C inside a laboratory oven. The time for good staining is variable, and it is necessary to check the color of the material during incubation. A dark brown color provides good results.

6. Wash the slide with distilled water and air-dry.
7. Check the quality of the spread and analyze the slides using a 10× objective using light microscope.

   Materials/Reagents
   > SSC
   > Formic acid
   > Silver nitrate (AgNO$_3$)

   Instruments/Equipment
   > Glass coverslip
   > Humid chamber
   > Laboratory oven
   > Microscope

5.2.6   Tissue Extraction

DNA samples can be obtained using commercially available kits or through a manual extraction methodology. The protocol presented is based on an extraction using phenol:chloroform:isoamyl alcohol (25:24:1). A good source for DNA is the muscle tissue, for example, from the legs or the pronotum region of fresh animals or animals stored in 100% ethanol at −20°C. In general, the protocol presented provides a good quality and quantity of DNA that is adequate for probe-based assays, such as polymerase chain reaction (PCR), C_0t-1 DNA isolation, and enzymatic digestion.

1. For each sample, prepare 500 μL solution in a 1.5-mL microtube as follows:

5 M sodium chloride (NaCl)	10 μL
1 M Tris-HCl, pH 8.00	5 μL
0.5 M ethylenediaminetetraacetic acid (EDTA), pH 8.00	25 μL
10% Sodium dodecyl sulfate (SDS)	25 μL
10 mg/mL proteinase K	10 μL
Distilled H$_2$O	425 μL

2. Drain the ethanol from the tissue and macerate the material inside the microtube containing the solution described in step 1 using scissors or adequate pestles for cell homogenization.
3. Incubate the microtube in a water bath at 45°C for 90–120 minutes or until the tissue is destroyed. Homogenize periodically during this period.
4. Add 500 μL phenol:chloroform:isoamyl alcohol (25:24:1) and homogenize with circular rotation for 15 minutes.

5. Centrifuge at 15,000 rpm for 15 minutes at 4°C.
6. Transfer the upper aqueous phase to a clean 1.5-mL microtube.
7. Add 0.2 volumes of 1 M NaCl and 2 volumes of 100% cold ethanol and homogenize by inversion to precipitate the DNA.
8. Centrifuge at 15,000 rpm for 15 minutes at 4°C.
9. Discard the supernatant and add 375 μL cold 70% ethanol, without agitation.
10. Centrifuge at 15,000 rpm for 15 minutes at 4°C.
11. Discard the supernatant and incubate at 37°C or at room temperature to dry the pellet.
12. Rehydrate the DNA in 100 μL ultrapure water for 1 hour.
13. Load 5 μL sample on a 0.8% agarose gel to check the quality of the DNA and measure the concentration using a spectrophotometer.
14. If necessary, proceed with the RNase treatment using 1 μL RNase A (10 mg/mL) for each 100 μL eluted DNA and incubate for 1 hour at 37°C.

Materials/Reagents
    $1 \times$ TAE (Tris-acetate-EDTA) buffer
    Agarose
    DNA gel stain
    Ethanol
    Ethylenediaminetetraacetic acid (EDTA)
    Ladder marker
    Phenol:chloroform:isoamyl alcohol (25:24:1)
    Proteinase K
    RNase
    Sodium dodecyl sulfate (SDS)
    Sodium chloride (NaCl)
    Tris-HCl

Instruments/Equipment
    Centrifuge
    Electrophoresis apparatus
    Microtube
    Scissors or pestles for cell homogenization
    Spectrophotometer
    Transilluminator
    Water bath
    Micropipettes

5.2.7 Types of Probes

Repetitive DNA comprises the type of sequence most often used as probes for physical chromosome mapping in coleopterans. Among these sequences, there are the multigene families, such as 45S rDNA (and the distinct fragments of this gene, such as 18S and 28S rDNA), 5S rDNA, and H3 histone genes. Besides satellite DNA and, to a lesser extent, transposable elements, the C_0t-1 DNA fraction, which comprises a pool of highly and moderately repetitive DNA, have been also mapped.

The methods for obtaining a probe are based mainly on three molecular methods: (1) PCR, (2) enzymatic restriction, and (3) DNA reassociation kinetics, and in addition the chromosome microdissection.

5.2.7.1 Multigene Families

PCR is a useful, low cost, and rapid method to obtain distinct fragments for multigene families, which are useful for chromosome mapping. The literature describes distinct universal primers that can be used to obtain partial sequences of multigene families in beetles. Some of the primers are presented in Table 5.1.

The fragments of the multigene families can be amplified by following reaction and thermal cycle conditions presented, and the pattern of amplification and the size of the fragments can then be confirmed on a 1% agarose gel.

TABLE 5.1

Examples of Primers to Amplify Fragments of Multigene Families through PCR Previously Used in Beetle

Target Sequence	Primer Sequences	Estimated Fragment Size	Reference
18S rDNA	F 5′ CCCCGTAATCGGAATGAGTA 3′ R 5′ GAGGTTTCCCGTGTTGAGTC 3′	822 bp	Cabral-de-Mello et al. 2010b
	F 5′ GTAGTCATATGC′ITGTCTC 3′ R 5′ GGCTGCTGGCACCAGAC′ITGC 3′	564 bp	White et al. 1990
5S rDNA	F 5′ AACGACCATACCACGCTGAA 3′ R 5′ AAGCGGTCCCCCATCTAAGT 3′	92 bp	Cabral-de-Mello et al. 2010b
H3 histone	F 5′ GGCNMGNACNAARCARAC 3′ R 5′ TGDATRTCYTTNGGCATDAT 3′	376 bp	Cabral-de-Mello et al. 2010b

Note: Other specific primers could be designated using the sequences of beetles or sequences from other insects deposited in the GenBank, including for other multigene families.

General reaction (for final volume of 25 µL):

10× *Taq* buffer	2.5 µL
Taq polymerase (5 U/µL)	0.25 µL
MgCl$_2$ (50 mM)	0.25 µL
dNTP set (8 mM)	0.5–0.8 µL
Forward primer (10µM)	1 µL
Reverse primer (10µM)	1 µL
Genomic DNA (100–200 ng/µL)	1 µL
DNase-free water	up to 25 µL

Note: In some cases, the volume of reagents in the solution can be adjusted to obtain good results.

Thermal cycle

Step	Temperature	Time
Initial denaturation	95°C	5 minutes
30 cycles	95°C	1 minute
	(45°C–60°C)[a]	30 seconds
	72°C	1 minute
Final extension	72°C	5 minutes
Hold	4°C	

[a]Distinct temperatures using the indicated range should be tested. The primer manufacturer provides the theoretical optimal temperature, but it should be tested in practice.

In addition, fragments obtained from other species can be successfully used as probes for chromosomal mapping, for example, the 28S rDNA from *Apis mellifera* (Bione et al. 2005) and 45S rDNA from *Arabidopsis thaliana* (Cabral-de-Mello et al. 2010a) and *Drosophila melanogaster* (Galián et al. 1995).

If it is isolate a conserved sequence, such as the rDNA and histone genes presented here, that could be used as probes in distinct species, the cloning of the fragments will permit long-term storage at −20°C. The main advantage of cloned DNA is the amount of DNA that can be obtained for labeling.

Materials/Reagents
    1 × TAE buffer
    Agarose
    DNA gel stain
    Ladder marker
    PCR supplies
    Primers

Instruments/Equipment
    Electrophoresis apparatus

Microtube
Thermocycler
Transilluminator
Micropipettes

5.2.7.2 C_0t-1 DNA Isolation

The isolation of highly and moderately repetitive DNA fractions is based on the reassociation kinetics of genomic DNA. Briefly, this methodology follows five main steps: (1) DNA extraction, (2) DNA fragmentation, (3) DNA denaturation, (4) DNA reannealing, and (5) treatment with a nuclease single-stranded DNA enzyme. After these procedures, the DNA is extracted from the resultant solution. The protocol for this assay in beetles is based on the work of Zwick et al. (1997) with modifications performed by Cabral-de-Mello et al. (2010b).

1. Dilute the genomic DNA to 100–500 ng/μL in 0.3 M NaCl using a 1.5-mL microtube. It is important to use nondegraded genomic DNA.
2. Fragment the DNA through autoclaving at 1.4 atm/120°C or by using DNase I. The time used for DNA fragmentation is variable, and several distinct times should be tested to determine the optimal conditions.
3. Apply 3 μL autoclaved or digested DNA in 1% agarose gel and check the size of the fragments. It is recommended that DNA fragments ranging from 100 to 1000 bp be used.
4. Denature at least three samples (tubes 1, 2, and 3) of 50 μL fragmented DNA using a thermocycler or water bath at 95°C for 10 minutes.
5. Place the tubes on ice for 10 seconds and add S1 nuclease enzyme to tube 1 and incubate at 37°C for 8 minutes. After 10 seconds on ice, immediately transfer tubes 2 and 3 to a water bath/thermocycler at 65°C to renature the DNA.
6. After 1 minute, add S1 nuclease enzyme to tube 2, and after 5 minutes, add S1 nuclease enzyme to tube 3 and incubate at 37°C for 8 minutes. Note: Multiple times should be tested to obtain a high quantity of the repetitive DNA fraction. (Use 1 U S1 nuclease enzyme for each 1 μg DNA and 5.5 μL of 10 × S1 nuclease buffer.)
7. Add an equal volume of phenol:chloroform:isoamyl alcohol (25:24:1) and rotate the tubes.
8. Centrifuge for 5 minutes at 15,000 rpm and transfer the upper aqueous phase to a clean 1.5-mL microtube.

9. Add 2.5 volumes cold 100% ethanol to precipitate the DNA and place the tube in a −70°C freezer for 30 minutes or alternatively place the tube in a −20°C freezer for 2 hours.
10. Centrifuge for 15 minutes at 15,000 rpm at 4°C.
11. Dry the pellet at room temperature and add 30–50 μL ultrapure DNase-free water.
12. Check the size of fragments in 1% agarose gel. The fragments should range from 50 to 500 bp (Figure 5.2a).
13. Quantify the DNA using a spectrophotometer.
14. Label the necessary DNA for FISH through nick translation (Section 5.2.8).

Materials/Reagents
  1 × TAE buffer
  Agarose
  DNA gel stain
  DNase I
  Ethanol
  Ladder marker
  Phenol:chloroform:isoamyl alcohol (25:24:1)
  S1 nuclease (with buffer)
  Sodium chloride (NaCl)

Instruments/Equipment
  Autoclave
  Electrophoresis apparatus
  Microtube
  Spectrophotometer
  Thermocycler
  Transilluminator
  Water bath
  Micropipettes

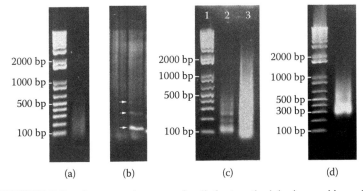

(a)   (b)   (c)   (d)

FIGURE 5.2   Agarose gel patterns for distinct methodologies used in probe obtaining. (a) C_0t-1 DNA; (b) genomic DNA digested with restriction enzyme presenting specific bands (arrows); (c) patterns of amplification using the whole genome amplification GenomePlex kits, ladder (lane 1), WGA4 amplification (lane 2), and WGA3 amplification (lane 3); (d) amplification of telomere motif.

5.2.7.3 Satellite DNAs

Satellite DNAs in beetles were classically isolated in Tenebrionidae and less commonly in Chrysomelidae representatives through enzymatic digestion with restriction endonucleases (RE), which is the most useful technique for isolating unknown repetitive sequences (in general, satellites) to characterize the heterochromatin. The genomic DNA should not be degraded, and the sample should be treated with RNase. For a simple enzymatic digestion, follow the recommendations below using, for example, enzymes from Fermentas, Waltham, MA, or following the recommendations of the supplier.

1. In a microtube pour

Genomic DNA	X µL (total amount 6 µg)
10× enzyme buffer	10 µL
Restriction enzyme	X µL (for each 1 µg genomic DNA use 5 U enzyme)
DNase-free ultrapure water	X µL (up to final volume of 100 µL)

2. Vortex the tube and centrifuge, then incubate at 37°C for 18 hours.
3. Precipitate the DNA by adding 200 µL cold 100% ethanol and 2 µL of 5 M NaCl.
4. Centrifuge at 14,000 rpm at 4°C for 10 minutes.
5. Drain the liquid phase and dry the pellet at 37°C for 40 minutes.
6. Rehydrate the DNA by adding 10 µL DNase-free ultrapure water.
7. Check the pattern of digestion, loading the 10 µL digested DNA in a 2% agarose gel.

If repetitive DNA with specific site for the applied RE is present in the genome of the species studied, it will be possible to observe a band in the agarose gel (Figure 5.2b). This band should be excised and purified using a cleanup DNA kit, such as the Zymoclean™ gel DNA recovery kit (Cat. no. D4001, The Epigenetics Company, Irvine, CA), and used for further assays, such as cloning, Southern blotting, sequencing, and FISH experiments, for sequence characterization.

5.2.7.4 Chromosome Probes

For Coleoptera, there is no published data using the chromosome microdissection/painting approach, although recently, it was successfully obtained a paint probe for the B chromosome of the species *Dichotomius sericeus* (Amorim IC, Cabral-de-Mello DC, Moura RC, unpublished data). The use of a GenomePlex® single cell whole genome amplification kit, WGA4 (Cat. no. 071MG105, Sigma-Aldrich, St Louis, MO), followed by a GenomePlex WGA reamplification kit, WGA3 (Cat. no. 089K6081, Sigma-Aldrich), provided good results for chromosome painting.

The greatest challenge for chromosome microdissection in Coleoptera is the recognition of the target chromosome, because of the high similarity among the chromosomes, which have similar morphology and size, although specific chromosomes can be easily recognizable, for example, the sex elements and the B chromosomes. The chromosomes for microdissection could be obtained from meiotic cells, preferentially from metaphase I, in which the chromosomes are well spread. The testis follicles could be fixed in Carnoy's solution (3:1 ethanol or methanol:acetic acid) and stored in a −20°C freezer until use.

Before microdissection, the chromosome spread should be obtained under a coverslip using a hot plate at 45°C.

1. Disaggregate one testis follicle in 50 µL of 50% acetic acid in a 0.6-mL microtube by pipetting.
2. Using a pipette, place the material on a 24 × 60 mm coverslip and aspirate the solution over a hot plate at 45°C. Repeat this process until the solution is evaporated, avoiding the border of the coverslips, which could complicate the microdissection process.
3. Stain the chromosomes using a 5% Giemsa solution and perform the microdissection of the target chromosome. The chromosome microdissection is performed using an inverted microscope and a glass needle coupled to a manipulator.

After microdissection, insert the chromosomes in a 0.2-mL PCR tube containing 9 µL DNase-free ultrapure water and follow the protocol indicated in the whole genome amplification kit (GenomePlex) WGA4. Note: The quantity of microdissected chromosomes could be variable for each type of chromosome.

Library construction and DNA amplification

1. In the tube containing the microdissected chromosomes with 9 µL DNase-free ultrapure water, add 1 µL freshly prepared working lysis and fragmentation buffer solution, and vortex thoroughly.

Working lysis and fragmentation buffer solution: Mix 32 µL of 10 × single-cell lysis and fragmentation buffer and 2 µL proteinase K solution.

2. Incubate at 50°C for 1 hour, then heat at 99°C for exactly 4 minutes and transfer the microtube to ice.

3. Add 2 μL of 1 × single-cell library preparation buffer and 1 μL library stabilization solution, then mix and incubate at 95°C for 2 minutes. Centrifuge and cool on ice.

4. Add 1 μL library preparation enzyme, vortex, centrifuge, and then incubate in a thermocycler following the reaction: 16°C for 20 minutes, 24°C for 20 minutes, 37°C for 20 minutes, and then 75°C for 5 minutes. Store at 4°C. Note: After this step, the sample can be stored at −20°C for 3 days.

5. Proceed with the amplification of the generated fragments associated to the adaptors. Add to the tube containing 14 μL reaction reagents as follows:

   7.5 μL of 10 × amplification master mix
   48.5 μL DNase-free ultrapure water
   5 μL WGA DNA polymerase.

6. Mix and centrifuge and then incubate in a thermocycler, following the reaction: 95°C for 3 minutes (initial denaturation), followed by 25 cycles (94°C for 30 seconds/denaturation step followed by 65°C for 5 minutes/annealing and extension step). Hold at 4°C.

7. Check the amplification in 1% agarose gel (load 4–8 μL sample) and store at −20°C until use for cleaning step and the reamplification reaction with WGA3. The size of the fragments should range from 100 to 1000 bp (Figure 5.2c).

8. The amplified fragments should be cleaned using a specific kit, as for example Zymoclean gel DNA recovery kit.

9. Quantify the DNA and proceed with the reamplification reaction.

DNA reamplification:

1. In a 0.2-mL microtube, pour:
   10 μL (at a concentration of 1 ng/μL) WGA4 amplified and purified chromosomal DNA

   49.5 μL DNase-free water
   7.5 μL amplification master mix
   3.0 μL of 10 mM dNTP mix
   5.0 μL WGA DNA polymerase

2. Vortex and conduct the following thermocycle reaction: 95°C for 3 minutes (initial denaturation), followed by 14 cycles (94°C for 15 seconds/denaturation step followed by 65°C for 5 minutes/annealing and extension step). Hold at 4°C.

3. Check the amplification pattern in 1% agarose gel (load 4–8 μL sample) and store at −20°C until labeling for the FISH experiments. The size of the fragments should concentrate in the range from 100 to 1000 bp (Figure 5.2c).

Materials/Reagents
  1 × TAE buffer
  Agarose
  DNA clean up kit
  DNA gel stain
  Giemsa
  Glass microneedle
  Ladder marker
  WGA3 kit
  WGA4 kit
Instruments/Equipment
  Electrophoresis apparatus
  Glass coverslip
  Glass needle
  Inverted microscope coupled to manual microdissector
  Micropipette puller
  Microtube
  Thermocycler
  Transilluminator
  Micropipettes

5.2.7.5 Telomere Repeat

In insects, the ancestral telomeric motif is TTAGG, but this repeat was lost in several insect orders (Sahara et al. 1999). In Coleoptera, the TTAGG motif is present, but it was also lost in some distinct groups, not presenting a phylogenetic relationship (Frydrychová and Marec 2002), being replaced, for example, by TCAGG (Mravinac et al. 2011). In species where this motif (TTAGG) is present, the probe for the FISH can be obtained through the reaction below, using the self-annealing primers F $(TTAGG)_5$ and R $(CCTAA)_5$ (Ijdo et al. 1991). Alternatively, telomeric probes could also be obtained as synthetic oligonucleotides labeled at 5' or 3'ends.

1. In a microtube add

Taq DNA polymerase buffer (10×)	5 μL
$MgCl_2$ (50 mM)	0.5 μL
F primer (10 mM)	2 μL
R primer (10 mM)	2 μL
dATP (2 mM)	1 μL
dCTP (2 mM)	1 μL
dGTP (2 mM)	1 μL
dTTP (2 mM)	0.7 μL
Labeled dUTP (1 mM)	0.6 μL
Taq DNA polymerase (5 U/μL)	0.4 μL
Sterile ultrapure water	up to 50 μL

2. Incubate in a thermocycler as follows:

Step	Temperature	Time
Initial denaturation	95°C	5 minutes
10 cycles	95°C	1 minute
	55°C	30 seconds
	72°C	1 minute
35 cycles	95°C	1 minute
	60°C	30 seconds
	72°C	1 minute and 30 seconds
Final extension	72°C	5 minutes
Hold	4°C	

3. Check the amplification product on a 1% agarose gel (load 2 μL sample). Fragments between 100 and 1000 bp that appear as a smear give good FISH results (Figure 5.2d).

Materials/Reagents
   1 × TAE buffer
   Agarose
   DNA gel stain
   Ladder marker
   Labeled nucleotide
   Nucleotide set (dATP, dGTP, dCTP, dTTP)
   PCR supplies
   Primer set: F $(TTAGG)_5$ and R $(CCTAA)_5$

Instruments/Equipment
   Electrophoresis apparatus
   Microtube
   Thermocycler
   Transilluminator
   Micropipettes

5.2.8   Labeling Probes

The probes currently used in chromosomal studies in Coleoptera are labeled by two main strategies: nick-translation and PCR.

5.2.8.1   Indirect Labeling through Nick Translation

For nick translation, a rapid and efficient strategy uses available commercial kits, such as DIG nick translation mix (Cat. no. 11 745 816 910, Roche, Basilea, Switzerland) or BioNick labeling system (Cat. no. 18247-015, Invitrogen, Carlsbad, CA). Other commercial kits with direct and indirect labeling are also useful. This strategy of labeling is convenient for labeling DNA fragments larger than approximately 600 bp, total genomic DNA, C_0t-1 DNA, and microdissected chromosomes.

Using the DIG nick translation mix (Digoxigenin-11-dUTP)

1. In a 0.2-mL microtube add

Template DNA	X µL (total amount 1 µg)
DIG nick translation mix	4 µL
Distilled water	up to 20 µL

2. Vortex the tube and centrifuge briefly.
3. Incubate at 15°C for 90 minutes.
4. Add 1 µL of 0.5 M EDTA (pH 8.0) and heat the solution at 65°C for 10 minutes to stop the reaction.
5. Proceed with the ethanol precipitation to remove unincorporated nucleotides by adding 1/10 volume of 3 M sodium acetate and 2 × volumes of cold 100% ethanol. Mix by inversion.
6. Incubate the tube at −70°C for 15 minutes or alternatively at −20°C for 2 hours.
7. Centrifuge at 14,000 rpm for 10 minutes, remove the supernatant and dry the pellet at room temperature or at 37°C.
8. Add 50 µL ultrapure autoclaved water.
9. Store the probes at −20°C until use.

Using the BioNick labeling system (Biotin-14-dATP)

1. In a 0.2-mL microtube add

10× dNTP mix	5 µL
10× enzyme mix	5 µL
Template DNA	X µL (1 µg)
Distilled water	up to 45 µL

2. Vortex the tube and centrifuge briefly.
3. Incubate at 16°C for 1 hour.
4. Add 5 µL stop buffer, vortex and centrifuge.
5. Perform the ethanol precipitation by adding 1/10 volume of 3 M sodium acetate and 2 × volumes of cold 100% ethanol, and mix by inverting.
6. Incubate the tube at −70°C for 15 minutes or alternatively at −20°C for 2 hours.
7. Centrifuge at 14,000 rpm for 10 minutes, remove the supernatant and dry the pellet at room temperature or at 37°C.
8. Add 50 µL ultrapure autoclaved water.
9. Store the probes at −20°C until use.

Alternatively, the nick translation for probe labeling can be performed using available enzyme mixtures (DNA polymerase I and DNase I) with distinct labeling times (from 30 to 90 minutes), producing probes with adequate size (200–500 bp) for FISH experiments.

1. In a 0.2-mL microtube, add

Unlabeled ACG nucleotide mixture (0.2 mM)	5 μL
Labeled dUTP (1 mM)	1 μL
DTT (dithiothreitol, 10 mM)	1 μL
10× nick translation buffer	5 μL
Template DNA	X μL (1 μg)
DNA polymerase I/DNase I (0.4 U/μL)	5 μL
dTTP (0.05 mM)	1 μL
Ultrapure water	up to 45 μL

2. Vortex the tube and centrifuge briefly.
3. Incubate at 15°C for 30–90 minutes.
4. Add 5 μL EDTA to stop the reaction.
5. Perform the ethanol precipitation by adding 1/10 volume of 3 M sodium acetate and 2 × volumes of cold 100% ethanol and mix by inverting.
6. Incubate the tube at −70°C for 15 minutes or alternatively at −20°C for 2 hours.
7. Centrifuge at 14,000 rpm for 10 minutes, remove the supernatant and dry the pellet at room temperature or at 37°C.
8. Add 50 μL ultrapure autoclaved water.
9. Store the probes at −20°C until use.

5.2.8.2 Polymerase Chain Reaction Labeling

The PCR approach is recommended for labeling DNA fragments smaller than approximately 600 bp. If the fragment is larger than approximately 600 bp, it is recommended to use DNase after the labeling process to obtain probes of an appropriate size for easy penetration into the nucleus. For this type of label, the same thermocycler conditions for obtaining PCR fragments are used, and the relation between the labeled and nonlabeled nucleotide should be approximately 30%–70%. See the following example:

In addition to the labeling of microdissected chromosomes through nick translation, these chromosomes can also be labeled efficiently through a thermocycler. For this propose, use the same protocol as the WGA3 kit (see Section 5.2.7.4) changing only 3 μL of 10 mM dNTP

Taq DNA polymerase buffer (10×)	2.5 μL
MgCl$_2$ (50 mM)	0.25 μL
F primer (10 mM)	1 μL
R primer (10 mM)	1 μL
dATP (2 mM)	0.5 μL
dCTP (2 mM)	0.5 μL
dGTP (2 mM)	0.5 μL
dTTP (2 mM)	0.35 μL
Labeled dUTP (1 mM)	0.3 μL
Taq DNA polymerase (5 U/μL)	0.1 μL
Genomic DNA	2 μL (concentration 50–100 ng/μL)
Sterile ultrapure water	up to 25 μL

mix to 3 μL mix containing a labeled nucleotide, such as dUTP, following the concentration indicated by the WGA3 kit.

Materials/Reagents
  1 × TAE buffer
  Agarose
  DTT
  DNA gel stain
  Labeled nucleotide
  Ladder marker
  Magnesium chloride (MgCl$_2$)
  Nick translation kit
  Nucleotide set (dATP, dGTP, dCTP, dTTP)
  PCR supplies
  Primers
  Tris-HCl
  WGA3 kit
Instruments/Equipment
  Electrophoresis apparatus
  Microtube
  Thermocycler
  Transilluminator
  Micropipettes

5.2.9 Hybridization and Detection

Since Pinkel et al. (1988) performed the FISH, several protocols and adaptations have been published, optimizing the quality of the results in distinct groups. The protocol presented here for FISH, which is routinely used and gives good results for beetle chromosomes, is divided into three main stages: (1) slide pretreatment (day one), (2) DNA denaturation/hybridization (day 1), and (3) washing/probe detection (day 2). Freshly or properly stored (at −20°C) slides can be used for FISH experiments.

5.2.9.1 Slide Pretreatment

1. Dehydrate the slide in an ethanol series (70%, 85%, and 100%) for 5 minutes each at room temperature and air-dry at 37°C.
2. Incubate the slide in 100 μg/mL RNase solution in 2 × SSC using a parafilm coverslip for 1 hour at 37°C in a humid chamber.
3. Wash the slide three times in 2 × SSC for 5 minutes each at room temperature.
4. Optional step: Incubate the slide in a 10 μg/mL pepsin solution in 0.1 N HCl using a parafilm coverslip for 20 minutes at 37°C in a humid chamber. Note: This step is important for material with a high quantity of cytoplasm.
5. Wash the slide three times in 2 × SSC for 5 minutes each at room temperature.
6. Place the slide in a Coplin jar with 3.7% formaldehyde solution diluted in wash-blocking buffer for 10 minutes.
7. Wash three times in 2 × SSC for 5 minutes each at room temperature.
8. Dehydrate slides in an ethanol series (70%, 85%, and 100%) for 5 minutes each and air-dry at 37°C.

5.2.9.2 Hybridization

1. Prepare the hybridization probe mixture: In a 0.2-mL microtube, add

   At least 100 ng labeled DNA
   Formamide (final concentration 50%)
   SSC (final concentration 2 × SSC)
   Dextran sulfate (final concentration 10%)
   Example of hybridization mix
   6 μL labeled DNA
   15 μL formamide 100%
   6 μL dextran sulfate 50%
   3 μL of 20 × SSC

2. Mix and denature the hybridization probe mixture at 95°C for 10 minutes and immediately place the tube on ice or in a −20°C freezer for 5 minutes.

3. Place the hybridization probe mixture on a slide and cover using a glass coverslip, preventing bubbles. The quantity of the hybridization mixture will depend on the coverslip size that is used.

4. Incubate the slides with the hybridization mixture at 75°C using a metal plate in a water bath, in a thermocycler or over a hot plate for 5 minutes. Note: In general, 5 minutes of incubation has given good results for FISH, but this time could be distinct for distinct species. Overdenaturation of chromosomal DNA gives a high background.

5. Incubate the slides overnight (18 hours) in a humid chamber at 37°C.

5.2.9.3   Posthybridization Washes

1. Remove the coverslip and incubate the slides in a Coplin jar with $2 \times$ SSC for 5 minutes at room temperature.

2. Wash the slides twice in $2 \times$ SSC at 42°C for 5 minutes each.

3. Wash the slides twice in $0.1 \times$ SSC at 42°C for 5 minutes each.

4. Wash the slides once in $2 \times$ SSC at 42°C for 5 minutes.

5. Place the slides in $2 \times$ SSC at room temperature for 10 minutes.

6. Transfer the slides to a Coplin jar containing wash-blocking buffer.

5.2.9.4   Probe Detection

Note: If the probes were directly labeled with fluorochrome, steps 1 through 3 are not required, go to step 4.

1. Dilute the streptavidin Alexa Fluor 488 conjugate (Cat. no. S32354, Life Technologies, Carlsbad, CA) for detection probes labeled with biotin or anti-digoxigenin-rhodamine (Cat. no. 11 207 750 910, Roche) for detection probes labeled with digoxigenin, in wash-blocking buffer as follows:

   1:100 µL of streptavidin Alexa Fluor 488 conjugate (initial concentration 2 mg/mL):wash-blocking buffer.

   0.5:100 µL of anti-digoxigenin-rhodamine (initial concentration 200 µg/mL):wash-blocking buffer.

Note: For two-color FISH experiments, dilute 0.5 µL anti-digoxigenin-rhodamine and 1 µL streptavidin Alexa Fluor 488 conjugate in 100 µL wash-blocking buffer solution.

2. Add the solution to the slide, cover with a parafilm coverslip and incubate at 37°C for 1 hour.

3. Wash the slide three times in wash-blocking buffer at 45°C for 5 minutes each.
4. Mount the slide using 0.5 μL DAPI (0.2 mg/mL) mixed in 15 μL VECTASHIELD antifade solution using a glass coverslip of an adequate size.
5. Store the slides in the dark at 4°C until the analysis.
6. Analyze the chromosome preparations under an epifluorescence microscope coupled to an adequate filter set.

Other available detection systems can be used in this step of the assay. The combination shown above provides reliable signals and low background interference in the results.

Materials/Reagents
Streptavidin Alexa Fluor 488 conjugate
Anti-digoxigenin-rhodamine
Bovine serum albumin (BSA)
DAPI
Dextran sulfate
Ethanol
Formaldehyde
Formamide
Hydrochloric acid (HCl)
Pepsin
RNase
SSC
Skimmed milk
Triton-X
VECTASHIELD antifade solution

Instruments/Equipment
Coplin jar
Epifluorescence microscope
Glass and parafilm coverslips
Humid chamber
Laboratory oven
Microtubes
Thermocycler
Water bath
Micropipettes

5.2.10 Solutions

0.05% Colchicine: Dissolve 0.05 g colchicine in 100 mL distilled water.
0.5 EDTA, pH 8.0: Dissolve 186.1 g EDTA in 800 mL distilled water, agitate and adjust to pH 8.0 with NaOH. The pH adjustment will contribute to the dissolution of the salt.

10% SDS: Dissolve 10 g SDS in distilled water to a final volume of 100 mL.

$10 \times$ Nick translation buffer: 0.5 M Tris-HCl pH 7.8, 50 mM $MgCl_2$, 5 mg/mL BSA.

1 M DTT: Dissolve 30.9 g DTT in 20 mL of 0.01 M sodium acetate (pH 5.2).

1 M Tris-HCl, pH 8.0: Dissolve 121.1 g Tris-base in 800 mL distilled water. Adjust the pH by adding NaOH/HCl to pH 8.0. Bring to a final volume of 1000 mL.

2% Lacto–acetic orcein: Dissolve 2 g orcein in 25 mL distilled water and add 25 mL glacial acetic acid. Then, add 25 mL of 85% lactic acid under constant stirring in 25 mL distilled water. Filter the solution before use to remove orcein particles.

$20 \times$ SSC, pH 7.0: Dissolve 175.3 g NaCl, 88.2 g $Na_3C_6H_5O_7$. H_2O in 800 mL distilled water and adjust to pH 7.0 using NaOH/HCl. Add distilled water to a final volume of 1000 mL.

3 M sodium acetate, pH 5.2: Dissolve 40.81 g $CH_3CO_2Na.3H_2O$ in 60 mL distilled water and adjust to pH 5.2. Bring to a final volume of 100 mL.

5% Barium hydroxide: Dissolve 5 g barium hydroxide in distilled water to a final volume of 100 mL.

50% Dextran sulfate: Dissolve 0.5 g dextran sulfate in 1 mL ultrapure autoclaved water. Heat at 60°C until dissolved completely.

$50 \times$ TAE buffer: Dissolve 242 g Tris base in 700 mL distilled water and add 57.1 mL glacial acetic acid and 100 mL of 0.5 M EDTA (pH 8.0). Complete to a final volume of 1000 mL.

5 M NaCl: Dissolve 29.22 g NaCl in distilled water to a final volume of 100 mL.

CMA_3 (0.5 mg/mL): Dissolve 5 mg chromomycin A_3 in 10 mL solution of 1:1 McIlvaine buffer pH 7.0 and distilled water. Add 10 μL of 5 M $MgCl_2$.

DA (0.1 mg/mL): Dissolve 1 mg DA in 10 mL McIlvaine buffer (pH 7.0).

DAPI (0.2 mg/mL): Dissolve 0.2 g DAPI in 1000 mL ultrapure autoclaved water.

DAPI (0.5 mg/mL): Dissolve 5 mg DAPI in 10 mL McIlvaine buffer (pH 7.0).

Formic acid solution: In 100 mL distilled water, add one drop of formic acid (pH 3.0–3.5).

Insect saline solution: Dissolve 9 g NaCl, 0.42 KCl, 0.33 $CaCl_2.2H_2O$, 0.2 g $NaHCO_3$ in distilled water up to a final volume of 1000 mL.

Carnoy's solution: Mix three parts of 100% ethanol or methanol with one part of glacial acetic acid (v:v).

0.1 N HCl: Add 9.9 mL hydrochloric acid to 990.1 mL distilled water.

Pepsin (1 mg/mL): Dissolve 0.01 g pepsin in 10 mL distilled water.

Phosphate buffer, pH 6.8: Solution A (dissolve 9.079 g KH_2PO_4 to a final volume of 1000 mL); solution B (dissolve 11.876 g $Na_2HPO_4.2H_2O$ to a final volume of 1000 mL). For use, mix 509 mL solution A with 491 mL solution B.

Propidium iodide (50 µg/mL): Dissolve 0.5 mg propidium iodide in 10 mL ultrapure autoclaved water.

Proteinase K solution (10 mg/mL): Dissolve 10 mg proteinase K in 1 mL ultrapure autoclaved water.

Wash-blocking buffer: $0.4 \times SSC$, 0.1% Triton X and 1% BSA or skimmed milk.

5.2.11   Representative Results

Figures 5.3 through 5.6 show examples of the results obtained through classical cytogenetic techniques. The modal and ancient diploid number in Coleoptera, at least for the Polyphaga representatives, is $2n = 20$, with the occurrence of the Xy sex system in males (meioformulae 9II+Xy_p), as in *Zophobas morio* (Figure 5.3a). The sex bivalent is associated during meiosis I as a parachute structure (Xy_p). The X chromosome is bi-armed and the y is a punctiform element (Figure 5.3a, inset). In general, the occurrence of bi-armed chromosomes is prevalent in most families, as observed in the karyotype of *Zophobas morio* stained with 5% Giemsa presented in Figure 5.3b. Figure 5.4 shows the patterns of C-positive blocks occurrence in *Dichotomius sericeus* (Figure 5.4a) and *Zophobas morio* (Figure 4b). *Dichotomius sericeus* presents heterochromatin restricted to the centromeric region, that is, pericentromeric blocks

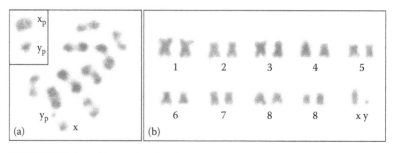

FIGURE 5.3   Conventional staining of (a) metaphase I and (b) karyotype of *Zophobas morio* using 5% Giemsa, showing the modal karyotype in beetles. Note: the karyotype constituted $2n = 20$, Xy_p and the bi-armed chromosomes and punctate y. The inset in (a) highlights the Xy_p sex system.

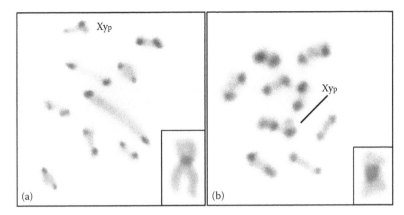

FIGURE 5.4 C-banded chromosomes in metaphase I of (a) *Dichotomius sericeus* and (b) *Zophobas morio*, showing small and large C-positive blocks, respectively, in the pericentromeric/paracentromeric regions. The insets are chromosome 1 of each species at metaphase II.

(Figure 5.4a), the most common pattern in Coleoptera. In Z. morio heterochromatin occupies a greater extension of the centromeric area, forming large paracentromeric blocks (Figure 5.4b). Note the clear difference in size of the C-positive blocks when metaphase I is analyzed and the difference of the blocks extension analyzing metaphase II chromosomes (Figure 5.4, insets).

Figure 5.5 shows distinct patterns staining by CMA_3 and DAPI fluorochrome dyes. In *Dichotomius laevicollis* (Figure 5.5a), the CMA_3^+ blocks (G+C rich) are restrict to one bivalent. In *Zophobas*

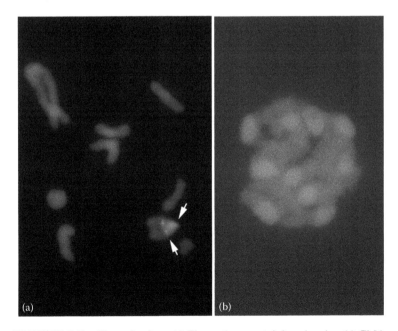

FIGURE 5.5 **(See color insert.)** Fluorochrome staining showing (a) CMA_3 positive blocks (G+C rich) in diacinesis of *Dichotomius laevicollis* (arrows) and (b) DAPI positive blocks (A+T rich) in initial meiosis I of *Zophobas morio*.

morio the C-blocks are A+T rich (DAPI⁺), as observed in initial meiosis (Figure 5.5b). Note in Figure 5.5b the occurrence of 10 positive blocks, 9 corresponding to autosomes and 1 to the sex bivalent. In Figure 5.6 it is possible to note distinct results that could be obtained through silver nitrate staining, such as analysis of the NORs (Figure 5.6a), the heterochromatin distribution, and the association of the sex bivalent (Figure 5.6b). The NOR (Figure 5.6a, asterisk) is easily recognized as associated with the sex chromosomes (Figure 5.6a, arrow) in *Euphoria* spp. in initial meiosis. The location of the major ribosomal genes is corroborated by the observation of the FISH results using the 18S rDNA probe (Figure 5.6a, insets). In *Dichotomius semisquamosus*, the staining of the heterochromatin/kinetochore and the lumen of the sex bivalent Xy_p could be observed (Figure 5.6b).

Figures 5.7 and 5.8 show the organization of distinct classes of repetitive DNA mapped in Coleoptera, and Figure 5.9 shows a chromosome painting example obtained through FISH. The probes were labeled indirectly with digoxigenin-11-dUTP or biotin-14-dATP through nick translation or PCR and detected with anti-digoxigenin-rhodamine or streptavidin Alexa Fluor 488 conjugate, respectively. The chromosomes were counterstained with DAPI. The cytogenetic mapping of three distinct multigene families, rDNAs and H3 histone genes, in distinct species of beetles belonging to the subfamily Scarabaeinae using probes obtained through PCR assay can be seen in Figure 5.7. The FISH experiments were performed through the use of only one probe and by combination of two and three probes in the

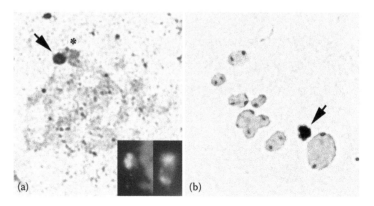

FIGURE 5.6 (**See color insert.**) Silver nitrate staining in an initial meiotic cell (Zygotene) and metaphase I of (a) *Euphoria* spp. and (b) *Dichotomius semisquamosus*, respectively. In (a) the arrow points to the sex bivalent and the asterisk to the nucleolar material associated with these chromosomes, and in (b) the arrow shows the sex bivalent (Xy_p) impregnated by the silver nitrate, a common pattern in Coleoptera. In (a) the insets show the position of the 18S rDNA clusters in the sex bivalent at the initial cell (left) and metaphase I (right). Note in (b) the staining of the pericentromeric heterochromatin/kinetochore.

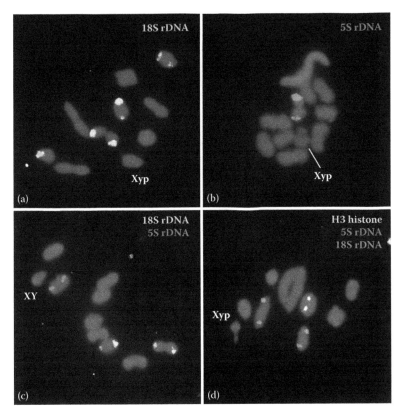

FIGURE 5.7 **(See color insert.)** FISH mapping of distinct multigene families in metaphase I obtained from four species of Coleoptera. (a) *Ontherus sulcator*, (b) *Digitonthophagus gazella*, (c) *Coprophanaeus dardanus*, and (d) *Dichotomius geminatus*. Each probe used is directly indicated in the cells. The sex chromosomes are also indicated.

same metaphase. In the case of the location of the three probes, two rounds of FISH were performed on the same slide and the signals were pseudocolored (Figure 5.7d).

The mapping of the highly and moderately repetitive DNA pool (C_0t-1 DNA) is presented in Figure 5.8. FISH mapping in the initial meiosis of *Dichotomius sericeus* (Figures 5.8a through c) shows the exclusive location of the C_0t-1 DNA in the heterochromatic blocks (arrows) and in the sex bivalent. In *Dichotomius bos*, although the C_0t-1 DNA fraction is associated with pericentromeric heterochromatin, the chromosome pairs 1–3 lack signals (Figure 5.8d). In *Coprophanaeus cyanescens*, a species with a high amount of heterochromatin (diphasic chromosomes), there are C_0t-1 DNA signals along the whole heterochromatic extension (Figure 5.8e). A cross species C_0t-1 DNA hybridization probe from *Dichotomius geminatus* hybridized in chromosomes of *Dichotomius bos*, shows the signals restricted to the terminal regions of the chromosomes (Figure 5.8f).

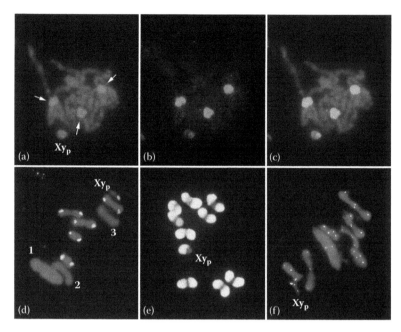

FIGURE 5.8 (**See color insert.**) Chromosomal mapping through FISH of the C_0t-1 DNA fraction in three species of beetles. Initial meiosis from *Dichotomius sericeus* (a) DAPI, (b) C_0t-1 DNA fraction, and (c) merge. Metaphases I from (d) *Dichotomius bos* and (e) *Coprophanaeus cyanescens*. (f) FISH using C_0t-1 DNA fraction obtained from *Dichotomius geminatus* genome in chromosomes in metaphase I of *Dichotomius bos*. Note: in (d, e) the C_0t-1 DNA signals are mainly in the heterochromatic blocks and in (f) the signals restrict to terminal regions of chromosomes. The arrows in (a) indicate the heterochromatic regions.

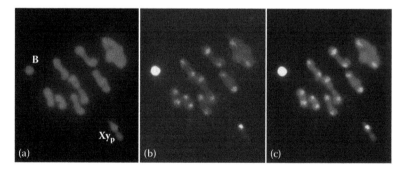

FIGURE 5.9 (**See color insert.**) FISH mapping in metaphase I using the B chromosome microdissected from *Dichotomius sericeus*: (a) DAPI, (b) B probe, (c) merge. The B and the sex bivalent are indicated.

Figure 5.9 shows the chromosome painting of a B chromosome in metaphase I in *Dichotomius sericeus*. The B chromosome was microdissected through manual microdissection and amplified using the GenomePlex, WGA4 and WGA3 kits.

5.2.12 Troubleshooting

Technique	Problem	Problem Solution
Chromosome preparations	Dense cytoplasm	Proceed with the change of the Carnoy's modified solution
	Degraded chromosomes	Take care with the temperature in the hot plate used for obtaining slide
Chromosome staining	Dark staining with Giemsa	Decrease the time of staining or dilute Giemsa solution
	Orcein precipitate	Filter the lacto–acetic orcein solution
	Some silver nitrate precipitate	Decrease the time of slide incubation
PCR	No or low amplification	Check the quality of DNA and possible contamination. Check the primers and temperature for annealing. Increase the amount of initial DNA template. Check the quality and concentration of reagents
Chromosome painting	Low signal intensity	Increase the amount of probe on the slide
	Spread signal in heterochromatin	Use the C_0t-1 DNA fraction to blocking repetitive DNAs
CMA_3/DA/DAPI fluorochrome staining	No evident fluorescence signals	Increase the time of DA counterstaining
Chromosome microdissection/ amplification	Low amount of DNA amplified	The number of microdissected chromosomes should be increased
FISH	No or faint fluorescence signal	Check the efficiency of probe labeling. Increase the time of incubation in pepsin. Increase the amount of probe on the slide
	Background	Increase the stringency in posthybridization washing. Reduce the amount of probe in the slide. Check the size of the probe in agarose gel and use probes smaller than 600 bp. Decrease the time of incubation for chromosome denaturation
	Degraded or overdenatured chromosomes	Decrease the time of pepsin incubation. Decrease the time of incubation for chromosome denaturation

5.3 DISCUSSION

5.3.1 Integration of Cytogenetic, Linkage, and Physical Maps and Genome Sequences

The knowledge concerning coleopteran genomes is currently scarce, and the results obtained from the distinct methodologies are in general fragmented, with some data available for karyotype structure (diploid number, chromosomal morphology, and sex chromosomes) and less information available for differential chromosome banding, physical maps through FISH, genomic analysis, and linkage maps. The integration of these types of maps is scarce, and only in a few cases distinct information were integrated, such as in studies of satellite DNA in Tenebrionidae. The scarcity of the distinct information integration or mapping to chromosomes using the FISH approach could be difficult for medium-sized or small chromosomes (0.5–6.5 µm long) in beetles if compared, for example, to orthopteran and to the difficulty of good spread obtaining in some groups.

Since the extensive revision of chromosomal data in Coleoptera performed by Smith and Virkki (1978), other karyotypes were described in the most diverse families comprising more than 3000 karyotyped species (Petitpierre 1996). However, the physical mapping of distinct DNA classes was not extensively performed, and reports of mapping of multigene families, satellite DNAs, transposable elements, and telomere repeats have been published. These studies, in general, did not involve a large number of species, preventing robust evolutionary perspective analysis. Although, for some specific groups, the scenario is divergent, for example, the understanding of the evolutionary dynamics of satellite DNA in Tenebrionidae karyotypes and for the major rDNA multigene family for some families (De la Rúa et al. 1996; Sánchez-Gea et al. 2000; Galián et al. 2002; Martínez-Navarro et al. 2004; Mravinac et al. 2004; Pons 2004; Bruvo-Madarić et al. 2007; Cabral-de-Mello et al. 2011b).

The DNA classes most mapped in beetle chromosomes are the multigene families, mainly the major rDNA in representatives from mostly a few families, such as Scarabaeidae (Cabral-de-Mello et al. 2011b), Carabidae, and Cicindelidae (De la Rúa et al. 1996; Sánchez-Gea et al. 2000; Martínez-Navarro et al. 2004). For other multigene families, such as 5S rDNA and H3 histone genes, the chromosomal mapping is restricted to Scarabaeidae representatives (Cabral-de-Mello et al. 2011b). The mapping of telomeric repeats and transposable elements are scarcer, but revealed interesting patterns of diversification in the telomeres, indicating the loss of the basic telomere insect motif TTAGG in some groups not presenting a phylogenetic relationship (Frydrychová and Marec 2002), being replaced, for example, by TCAGG (Mravinac et al. 2011). The transposable elements shed light on the understanding of the B chromosome and heterochromatin evolution (Oliveira et al. 2012).

The studies of linkage mapping and sequenced genomes are restricted to a few species based on random amplified DNA fingerprinting, amplified fragment length polymorphism, bacterial artificial chromosomes, expressed sequence tags, and random amplified polymorphic DNA (Hawthorne 2001; Schlipalius et al. 2002; Yezerski et al. 2003; Lorenzen et al. 2005). This scarcity of linkage maps in the group and the low availability of BAC libraries complicates the integration of the genomic information to chromosomes. In addition, the genomic information from sequenced genomes is available only for two species, *Tribolium castaneum* and *Dendroctonus ponderosae* (Richards et al. 2008; Keeling et al. 2013). One example of the difficulty for the integration of the distinct types of maps is that until the finalization of this chapter, the unique species in which the knowledge of distinct maps could be integrated was *Tribolium castaneum* (Tenebrionidae). For this species, there is information for the karyotype and banding, satellite DNA organization, the genome sequence, a rich linkage map, and a BAC library. Moreover, the genomic information for this species is available in an online database (BeetleBase, http://beetlebase.org/).

5.3.2   Chromosome and Genome Organization and Evolution

The karyotypic structure in Coleoptera is extremely variable if the order as a whole is considered. The diploid number $2n = 20$ Xy sex system (associated as a parachute, Xy_p) and the occurrence of bi-armed chromosomes is the most frequent karyotype, being modal and considered primitive, at least to Polyphaga. From this ancestral karyotype, some chromosomal rearrangements, such as inversions, fusions between autosomes and between autosomes and the X chromosomes, fissions, and loss of y, are known to possibly have occurred, generating great chromosomal diversity. These rearrangements are responsible for the diploid number variation from $2n = 4$ observed in *Chalcolepidius zonatus*, Elateridae, to $2n = 69$ in *Dixus capito*, Carabidae, the origin of the B (supernumerary) chromosomes, and in some cases, are also responsible for the intriguing variability regarding the sex chromosomes (Smith and Virkki 1978; Serrano 1981; Ferreira et al. 1984; Petitpierre 1996; Galián et al. 2002; Cabral-de-Mello et al. 2008; Dutrillaux and Dutrillaux 2009).

Because of the homogeneity of Coleoptera karyotypes, it is difficult to elucidate the specific chromosomal rearrangements responsible for the variability observed in some groups, even when using the mapping of DNA sequences. However, in specific groups, that is, families/subfamilies, the possible drifts in chromosomal changes have been established, for example, in Scarabaeidae (Yadav and Pillai 1979), Scarabaeinae (Cabral-de-Mello et al. 2008); Cicindelidae (Galián et al. 2002); Chrysomelidae (Petitpierre et al. 1988), Cassidinae (de Julio et al. 2010), Chrysomelinae (Petitpierre 2011); Elateridae (Schneider et al. 2007); Buprestidae (Karagyan et al. 2004); and Tenebrionidae (Juan and Petitpierre 1991).

Concerning the multigene families studied in Coleoptera (rDNAs and H3 histone gene), as observed in some other groups among plants and animals, the chromosome maps for major rDNA have revealed huge diversity in number and position, from 1 pair of chromosomes (some species) to 15 sites (in *Coprophanaeus ensifer*), with cases of remarkable intrageneric variability, in addition to intraspecific polymorphisms (Oliveira et al. 2010; Cabral-de-Mello et al. 2011b). Although, the occurrence of two autosomal sites (one bivalent) is common. In Scarabaeidae representatives, the dispersion of major rDNA is associated with the heterochromatin spreading rather than with macro-chromosomal rearrangements, indicating that in distinct lineages of this group, these genomic components could be constrained by similar evolutionary forces (Cabral-de-Mello et al. 2011b). On the contrary, another rDNA cluster, 5S rDNA, is highly conserved, occurring frequently on one autosomal pair and associated to the H3 histone cluster, suggesting that they could undergo similar mechanisms of evolution (Cabral-de-Mello et al. 2011b). Linked arrays of multigene families were also established from the molecular point of view in *Anthonomus* (Curculionidae) species for histone quintets and major rDNA (Roehrdanz et al. 2010). Other families that have been studied for major rDNA distribution are Carabidae and Cicindelidae, revealing as a common placement of the occurrence of autosomal clusters (De la Rúa et al. 1996; Sánchez-Gea et al. 2000; Martínez-Navarro et al. 2004). Examples of interesting chromosomal organization for major rDNA were related in the genus *Zabrus*, Carabidae (Sánchez-Gea et al. 2000) with occurrence of some polymorphism, and some cases of the translocation to sex chromosomes were reported in *Cicindela*, Cicindelidae (Galián et al. 2007).

Heterochromatic segments are also highly variable in Coleoptera considering the amount of C-positive blocks and their location on chromosomes, but they are poorly known concerning the specific constitution, for example, of satellite DNAs. Studies concerning this genomic content are concentrated in Tenebrionidae and, in the lesser extent, in Chrysomelidae representatives, in which distinct types of satellite DNAs were isolated, studied from the molecular point of view, and mapped on chromosomes comparatively (Lorite et al. 2001, 2002, 2013 Mravinac et al. 2004; Pons 2004; Palomeque et al. 2005; Bruvo-Madarić et al. 2007). For example, in Tenebrionidae, a group with species presenting large blocks of heterochromatin and a large amount of satellite DNA, the genera *Tribolium, Tenebrio, Palorus,* and *Pimelia* were studied in detail. This group presents A+T enrichment with the occurrence of species-specific sequences and also conserved satellite DNA described in congeneric and also in unrelated species (Meštrović et al. 1998; Mravinac et al. 2002, 2004, 2005; Pons 2004). Among these conserved satellite families, those described in the genus *Pimelia* share a common ancestral unit that occupies large blocks of the pericentromeric heterochromatin in all chromosomes (Pons et al. 2004). Another well-studied satellite DNA is the TCAST, a satellite isolated from *Tribolium castaneum,* which makes up 35% of its genome encompassing the pericentromeric area of all chromosomes, its major component. This satellite presents two types of related subfamilies (Tcast1a and Tcast1b) organized as higher order repeats, revealing differences in the mutational profiles among populations (laboratory strains from distinct geographic locations) (Feliciello et al. 2011). According to Feliciello et al. (2011), this change in the mutational profile could be the first step in the origin of a population-specific profile for Tcast. In addition to the occurrence of Tcast in heterochromatic blocks, by using computational analysis in the assembled genome of *Tribolium castaneum,* arrays for this sequence were found in euchromatin, suggesting a possible role in the gene regulation (Brajković et al. 2012).

Another approach to understand the repetitive DNA/heterochromatin variability was applied in *Dichotomius* (Scarabaeidae) representatives through the use of C_0t-1 DNA fraction (associated to heterochromatin), revealing high turnover of heterochromatin in the congeneric species (Cabral-de-Mello et al. 2011a). Finally, analysis of the repetitive DNA fraction using the two sequenced Coleoptera genomes revealed low similarity of novel repeats observed in the genome of *Dendroctonus ponderosae* with repeats of the *Tribolium castaneum* genome (only 0.15%), indicating little commonality for these repeats among beetles (Keeling et al. 2013).

5.3.3   Concluding Remarks

The aspects presented in this chapter are only examples of the chromosomal organization and genomic evolution of beetles obtained from cytogenetic mapping, showing that, in addition to the variability of the karyotypic structure, the group is also highly variable at the microgenomic point of view (i.e., for organization of distinct repetitive DNAs). The actual knowledge of genome organization on the chromosomal level in this group is, as presented, mainly focused on the organization of karyotypes and, to a lesser extent, for repetitive DNA using simple and reliable molecular technologies. We are far from a solid understanding of the chromosomal organization in beetles, and many more studies using mapping of multigene families, satellite DNA, transposable elements, microsatellites, and entire chromosomes should be addressed in distinct families.

The promise of an increased number of sequenced genomes through massive sequencing assays will provide more reliable information about the genomes of Coleoptera. The challenge is anchoring this information to the chromosomes and integrating the genome and karyotype information to understand the evolution of the group by means of distinct DNA classes. The completely sequenced genomes and the construction of a BAC library would permit comparison of the related karyotypes through mapping of orthologous genes or scaffolds by means of FISH or by computational comparisons. In addition, sequenced genomes will permit the selection of specific sequences to perform comparative analysis, such as tandem repeats and transposable elements, in addition to the characterization of gene cluster organizations.

Another aspect that could help in the understanding of the chromosomal evolution in beetles is obtaining robust phylogenies from distinct groups, such as families, which in some cases is complicated by the large number of species. This approach could increase the resolution of comparative cytogenetics, helping in the elucidation of the possible drifts of chromosomal/genomic changes. Undoubtedly, the integration of chromosomal and genome studies, and the increase of linkage map availability, with comprehensive phylogenies will provide information to test hypotheses related to the intriguing chromosomal diversity, concerning, for example, diploid number, sex chromosomes, and heterochromatin diversification in beetles.

ACKNOWLEDGMENTS

This study was supported by Fundação de Amparo a Pesquisa do Estado de São Paulo (FAPESP 2011/19481-3), Conselho Nacional de Desenvolvimento Científico e Tecnológico (CNPq 475308/2011-5), and Pró Reitoria de Pesquisa da UNESP (PROPE-UNESP). Also thanks to Dr. Sárah Gomes de Oliveira for Figure 5.8e.

REFERENCES

Anderson, J. R., Merritt, R. W., and Loomis, E. C. 1984. The insect-free cattle dropping and its relationship to increased dung-fouling of rangeland pastures. *J Econ Entomol* 77:133–141.

Arnett, R. H. Jr. and Thomas, M. C. 2001. *American beetles. Archostemata, Myxophaga, Adephaga, Polyphaga: Staphyliniformia*, Vol. 1. New York: CRC Press.

Arnett, R. H. Jr., Thomas, M. C., Skelley, P. E., and Frank, J. H. 2002. *American beetles. Polyphaga: Scarabaeoidea through Curculionoidea*, Vol. 2. Boca Raton, FL: CRC Press.

Barth, F. B. 1985. *Insects and Flowers: The Biology of a Partnership*. Princeton, NJ: Princeton University Press.

Bernhardt, P. 2000. Convergent evolution and adaptative radiation of beetle-pollinated angiosperms. *Plant Syst Evol* 222:293–320.

Beutel, R. G. and Haas, F. 2000. Phylogenetic relationships of the suborders of Coleoptera (Insecta). *Cladistics* 16:103–141.

Bione, E., Camparoto, M. L., and Simões, Z. L. P. 2005. A study of the constitutive heterochromatin and nucleolus organizer regions of *Isocopris inhiata* and *Diabroctis mimas* (Coleoptera: Scarabaeidae, Scarabaeinae) using C-banding, AgNO$_3$ staining and FISH techniques. *Genet Mol Biol* 28:111–116.

Bouchard, P., Grebennikov, V. V, Smith, A. B. T., and Douglas, H. 2009. Biodiversity of Coleoptera. In *Insect Biodiversity: Science and Society*, Foottit, R. G. and Adler, P. H. (eds.). Oxford, United Kingdom: Wiley-Blackwell.

Brajković, J., Feliciello, I., Bruvo-Madarić, B., and Ugarković, D. 2012. Satellite DNA-like elements associated with genes within euchromatin of the beetle *Tribolium castaneum*. *G3* 2:931–941.

Bruvo-Madarić, B., Plohl, M., and Ugarković, D. 2007. Wide distribution of related satellite DNA families within the genus *Pimelia* (Tenebrionidae). *Genetica* 130:35–42.

Cabral-de-Mello, D. C., Moura, R. C., Carvalho, R., and Souza, M. J. 2010a. Cytogenetic analysis of two related *Deltochilum* (Coleoptera, Scarabaeidae) species: Diploid number reduction, extensive heterochromatin addition and differentiation. *Micron* 41:112–117.

Cabral-de-Mello, D. C., Moura, R. C., and Martins, C. 2010b. Chromosomal mapping of repetitive DNAs in the beetle *Dichotomius geminatus* provides the first evidence for an association of 5S rRNA and histone H3 genes in insects, and repetitive DNA similarity between the B chromosome and A complement. *Heredity* 104:393–400.

Cabral-de-Mello, D. C., Moura, R. C., Melo, A. S., and Martins, C. 2011a. Evolutionary dynamics of heterochromatin in the genome of *Dichotomius* beetles based on chromosomal analysis. *Genetica* 139:315–325.

Cabral-de-Mello, D. C., Oliveira, S. G., Moura, R. C., and Martins, C. 2011b. Chromosomal organization of 18S and 5S rRNAs and H3 histone genes in Scarabaeinae coleopterans: Insights on the evolutionary dynamics of multigene families and heterochromatin. *BMC Genet* 12:88.

Cabral-de-Mello, D. C., Oliveira, S. G., Ramos, I. C., and Moura, R. C. 2008. Karyotype differentiation patterns in species of the subfamily Scarabaeinae (Scarabaeidae, Coleoptera). *Micron* 39:1243–1250.

Costa, C. 1999. Coleoptera. In *em Biodiversidade do Estado de São Paulo, Brasil. Invertebrados Terrestres*, Vol. 5. Joly C. A. and Bicudo C. E. M. (eds.). São Paulo, Brazil: Editora FAPESP.

Crowson, R. A. 1955. *The Natural Classification of the Families of Coleoptera*. London, United Kingdom: Nathaniel Lloyd.

Crowson, R. A. 1981. *The Biology of Coleoptera*. London, United Kingdom: Academic Press.

Davis, A. L. V., Scholtz, C. H., and Philips, T. K. 2002. Historical biogeography of Scarabaeine dung beetles. *J Biogeogr* 29:1217–1256.

de Julio, M., Fernandes, F. R., Costa, C., Almeida, M. C., and Cella, D. M. 2010. Mechanisms of karyotype differentiation in Cassidinae sensu lato (Coleoptera, Polyphaga, Chrysomelidae) based on seven species of the Brazilian fauna and an overview of the cytogenetic data. *Micron* 41:26–38.

De la Rúa, P., Serrano, J., Hewitt, G. M., and Galián, J. 1996. Physical mapping of rDNA genes in the ground beetle *Carabus* and related genera (Carabidae: Coleoptera). *J Zool Syst Evol Res* 34:95–101.

Dutrillaux, A. M. and Dutrillaux, B. 2009. Sex chromosome rearrangements in Polyphaga beetles. *Sex Dev* 3:43–54.

Erwin, T. L. 1982. Tropical forests: Their richness in Coleoptera and other arthropod species. *Coleopt Bull* 36:74–75.

Erwin, T. L. 1985. The taxon pulse: A general pattern of lineage radiation and extinction among carabid beeltes. In *Taxonomy, Phylogeny and Biogeography of Beetles and Ants*, Ball, G. E. (ed.). Dordrecht, The Netherlands: W. Junk.

Evans, J. D., Brown, S. J., Hackett, K. J. et al. 2013. The i5K initiative: Advancing arthropod genomics for knowledge, human health, agriculture, and the environment. *J Hered* 104:595–600.

Farrell, B. D. 1998. "Inordinate fondness" explained: Why are there so many beetles? *Science* 281:555–559.

Feliciello, I., Chinali, G., and Ugarković, D. 2011. Structure and population dynamics of the major satellite DNA in the red flour beetle *Tribolium castaneum*. *Genetica* 139:999–1008.

Ferreira, A., Cella, D. M., Tardivo, J. R., and Virkki, N. 1984. Two pairs of chromosomes: A new low record for Coleoptera. *Rev Bras Genet* 7:231–239.

Fincher, G. T. 1981. The potential value of dung beetles in pasture ecosystems. *J Georgia Entomol Soc* 16:316–333.

Frydrychová, R. and Marec, F. 2002. Repeated losses of TTAGG telomere repeats in evolution of beetles (Coleoptera). *Genetica* 115:179–187.

Galián, J., Hogan, J. E., and Vogler, A. P. 2002. The origin of multiple sex chromosomes in tiger beetles. *Mol Biol Evol* 19:1792–1796.

Galián, J., Proença, S. J. R., and Vogler, A. P. 2007. Evolutionary dynamics of autosomal-heterosomal rearrangements in a multiple-X chromosome system of tiger beetles (Cicindelidae). *BMC Evol Biol* 7:158.

Galián, J., Serrano, J., De la Rúa, P., Petitpierre, E., and Juan, C. 1995. Localization and activity of rDNA genes in tiger beetles (Coleoptera: Cicindelinae). *Heredity* 74:524–530.

Grimaldi, D. and Engel, M. S. 2005. *Evolution of the Insects.* New York: Cambridge University Press.

Hammond, P. M. 1979. Wing-folding mechanisms of beetles, with special reference to investigations of adephagan phylogeny (Coleoptera). In *Carabid Beetles: Their Evolution, Natural History, and Classification*, Erwin, T. L., Ball, G. E., Whitehead, D. R., and Halpern, A. L. (eds.). The Hague, The Netherlands: W. Junk.

Hammond, P. M. 1995. Described and estimated species numbers: An objective assessment of current knowledge. In *Microbial Diversity and Ecosystem Function*, Allspp, D., Hawkesworth, D. L., and Colwell, R. R. (eds.). Wallingford, United Kingdom: CAB International.

Hanrahan, S. J. and Johnston, J. S. 2011. New genome size estimates of 134 species of arthropods. *Chromosome Res* 19:809–823.

Hawthorne, D. J. 2001. AFLP-based genetic linkage map of the Colorado potato beetle *Leptinotarsa decemlineata*: Sex chromosomes and a pyrethroid-resistance candidate gene. *Genetics* 158:695–700.

Ijdo, J. W., Wells, R. A., Baldini, A., and Reeders, S. T. 1991. Improved telomere detection using a telomere repeat probe (TTAGGG)n generated by PCR. *Nucleic Acids Res* 19:4780.

Juan, C. and Petitpierre, E. 1989. C-banding and DNA content in seven species of Tenebrionidae (Coleoptera). *Genome* 32:834–839.

Juan, C. and Petitpierre, E. 1991. Chromosome number and sex-determining systems in Tenebrionidae (Coleoptera). In *Advances in Coleopterology*, Zunino, M. H., Belles, X., and Blas, M. (eds.). Barcelona, Spain: AEC.

Karagyan, G., Kuznetsova, V. G., and Lachowska, D. 2004. New cytogenetic data on Armenian buprestid (Coleoptera, Buprestidae) with a discussion of karyotype variation within the family. *Folia Biol* 52:151–158.

Keeling, C. I., Yuen, M. M. S., Liao, N. Y. et al. 2013. Draft genome of the mountain pine beetle, *Dendroctonus ponderosae* Hopkins, a major forest pest. *Genome Biol* 14:R27.

Lawrence, J. F. 1982. Coleoptera. In *Synopsis and Classification of Living Organisms*, Vol. 2, Parker, S. P. (ed.). New York: McGraw-Hill.

Lawrence, J. F. and Britton, E. B. 1994. *Australian Beetles*. Carlton, Australia: Melbourne University Press.

Lawrence, J. F. and Newton, A. F. Jr. 1995. Families and subfamilies of Coleoptera (with selected genera, notes, references and data on family-group names). In *Biology, Phylogeny, and Classification of Coleoptera: Papers Celebrating the 80th Birthday of Roy A. Crowson*, Pakaluk, J. and Ślipiński, S. A. (eds.). Warszawa, Poland: Muzeum i Instytut Zoologii PAN.

Lorenzen, M. D., Doyungan, Z., Savard, J. et al. 2005. Genetic linkage maps of the red flour beetle, *Tribolium castaneum*, based on bacterial artificial chromosomes and expressed sequence tags. *Genetics* 170:741–747.

Lorite, P., Carrillo, J. A., Garneria, I., Petitpierre, E., and Palomeque, T. 2002. Satellite DNA in the elm leaf beetle, *Xanthogaleruca luteola* (Coleoptera, Chrysomelidae): Characterization, interpopulation analysis, and chromosome location. *Cytogenet Genome Res* 98:302–307.

Lorite, P., Palomeque, T., Garnería, I., and Petitpierre, E. 2001. Characterization and chromosome location of satellite DNA in the leaf beetle *Chrysolina americana* (Coleoptera, Chrysomelidae). *Genetica* 110:143–150.

Lorite, P., Torres, M. I., and Palomeque, T. 2013. Characterization of two unrelated satellite DNA families in the Colorado potato beetle *Leptinotarsa decemlineata* (Coleoptera, Chrysomelidae). *Bull Entomol Res* 103:538–546.

Martínez-Navarro, E. M., Serrano, J., and Galián, J. 2004. Chromosome evolution in ground beetles: Localization of the rDNA loci in the tribe Harpalini (Coleoptera, Carabidae). *J Zool Syst Evol Res* 42:38–43.

Meštrović, N., Plohl, M., Mravinac, B., and Ugarković, D. 1998. Evolution of satellite DNAs from the genus *Palorus*—experimental evidence for the "library" hypothesis. *Mol Biol Evol* 15:1062–1068.

Mravinac, B., Meštrović, N., Čavrak, V. V., and Plohl, M. 2011. TCAGG, an alternative telomeric sequence in insects. *Chromosoma* 120:367–376.

Mravinac, B., Plohl, M., Meštrović, N., and Ugarković, D. 2002. Sequence of PRAT satellite DNA "frozen" in some coleopteran species. *J Mol Evol* 54:774–783.

Mravinac, B., Plohl, M., and Ugarković, D. 2004. Conserved patterns in the evolution of *Tribolium* satellite DNAs. *Gene* 332:169–177.

Mravinac, B., Plohl, M., and Ugarković, D. 2005. Preservation and high sequence conservation of satellite DNAs suggest functional constraints. *J Mol Evol* 61:542–550.

Nielsen, E. S. and Mound, L. A. 1999. Global diversity of insects: The problems of estimating numbers. In *Nature and Human Society: The Quest for a Sustainable World*, Raven, P. H. and Williams, T. (eds.). Washington, DC: National Academy Press.

Oliveira, S. G., Moura, R. C., and Martins, C. 2012. B chromosome in the beetle *Coprophanaeus cyanescens* (Scarabaeidae): Emphasis in the organization of repetitive DNA sequences. *BMC Genet* 13:96.

Oliveira, S. G., Moura, R. C., Silva, A. E. B., and Souza, M. J. 2010. Cytogenetic analysis of two *Coprophanaeus* species (Scarabaeidae) revealing wide constitutive heterochromatin variability and the largest number of 45S rDNA sites among Coleoptera. *Micron* 41:960–965.

Palomeque, T., Muñoz-López, M., Carrillo, J. A., and Lorite, P. 2005. Characterization and evolutionary dynamics of a complex family of satellite DNA in the leaf beetle *Chrysolina carnifex* (Coleoptera, Chrysomelidae). *Chromosome Res* 13:795–807.

Petitpierre, E. 1996. Molecular cytogenetics and taxonomy of insects, with particular reference to the Coleoptera. *Int J Insect Morphol Embryol* 25:115–133.

Petitpierre, E. 2011. Cytogenetics, cytotaxonomy and chromosomal evolution of Chrysomelinae revisited (Coleoptera, Chrysomelidae). *ZooKeys* 157:67–79.

Petitpierre, E., Segarra, C., Yadav, C. S., Virkki, N. 1988. Chromosome numbers and meioformulae of Chrysomelidae. In *Biology of Chrysomelidae*, Jolivet, P., Petitpierre, E., and Hsiao, T. H. (eds.). Dordrecht, The Netherlands: Kluwer Academic Publishers.

Pinkel, D., Landegent, J., Collins, C. et al. 1988. Fluorescence in situ hybridization with human chromosome specific libraries: Detection of trisomy 21 and translocation of chromosome 4. *Proc Natl Acad Sci USA* 85:9138–9142.

Pons, J. 2004. Evolution of diploid chromosome number, sex-determining systems, and heterochromatin in Western Mediterranean and Canarian species of the genus *Pimelia* (Coleoptera: Tenebrionidae). *J Zool Syst Evol Res* 42:81–85.

Pons, J., Bruvo, B., Petitpierre, E., Plohl, M., Ugarković, D., and Juan, C. 2004. Complex structural features of satellite DNA sequences in the genus *Pimelia* (Coleoptera: Tenebrionidae): Random differential amplification from a common 'satellite DNA library'. *Heredity* 92:418–427.

Richards, S., Gibbs, R. A., Weinstock, G. M. et al. 2008. The genome of the model beetle and pest *Tribolium castaneum*. *Nature* 452:949–955.

Roehrdanz, R., Heilmann, L., Senechal, P., Sears, S., and Evenson, P. 2010. Histone and ribosomal RNA repetitive gene clusters of the boll weevil are linked in a tandem array. *Insect Mol Biol* 19:463–471.

Rozék, M., Lachowska, D., Petitpierre, E., and Holecová, M. 2004. C-bands on chromosomes of 32 beetles species (Coleoptera: Elateridae, Cantharidae, Oedemeridae, Cerambycidae, Anthicidae, Chrysomelidae, Attelabidae, and Curculionidae). *Hereditas* 140:161–170.

Rufas, J. S., Iturra, P., de Souza, W., Esponda, P. 1982. Simple silver staining procedures for the localization of nucleolus and nucleolar organizer under light and electron microscopy. *Arch Biol* 93:267–274.

Sahara, K., Marec, F., and Traut, W. 1999. TTAGG telomeric repeats in chromosomes of some insects and other arthropods. *Chromosome Res* 7:449–460.

Sánchez-Gea, J. F., Serrano, J., and Galián, J. 2000. Variability in rDNA loci in Iberian species of the genus *Zabrus* (Coleoptera: Carabidae) detected by fluorescence in situ hybridization. *Genome* 43:22–28.

Schlipalius, D. I., Cheng, Q., Reilly, P. E., Collins, P. J., and Ebert, P. R. 2002. Genetic linkage analysis of the lesser grain borer *Rhyzopertha dominica* identifies two loci that confer high-level resistance to the fumigant phosphine. *Genetics* 161:773–782.

Schneider, M. C., Rosa, S. P., Almeida, M. C., Costa, C., and Cella, D. M. 2007. Chromosomal similarities and differences among four Neotropical Elateridae (Conoderini and Pyrophorini) and other related species, with comments on the NOR patterns in Coleoptera. *J Zool Syst Evol Res* 45:308–316.

Schweizer, D. 1980. Simultaneous fluorescent staining of R bands and specific heterochromatic regions (DA-DAPI bands) in human chromosomes. *Cytogenet Cell Genet* 27:190–193.

Schweizer, D. 1981. Counterstain-enhanced chromosome banding. *Hum Genet* 57:1–14.

Serrano, J. 1981. Chromosome numbers and karyotypic evolution of Caraboidea. *Genetica* 55:51–60.

Smith, S. G. and Virkki, N. 1978. *Animal Cytogenetics. Insecta 5. Coleoptera*, Vol. 3. Berlin, Germany: Gebrüder Borntraeger.

Stork, N. E. 1999. Estimating the number of species on Earth. In *The Other 99%: The Conservation and Biodiversity of Invertebrates*, Ponder, W. and Lunney, D. (eds.), pp. 1–7. Sydney, Australia: Royal Zoological Society of New South Wales.

Sumner, A. T. 1972. A simple technique for demonstrating centromeric heterochromatin. *Exp Cell Res* 75:304–306.

van Huis, A., Van Itterbeeck, J., Klunder, H. et al. 2013. Edible insects: Future prospects for food and feed security. *FAO Forestry paper* 171.

Virkki, N., Mazzella, C., and Denton, A. 1990. Staining of substances adjacent to the sex bivalent in certain weevils of Puerto Rico. *J Agric Univ P R* 74:405–418.

White, T. J., Bruns, T., Lee, S., and Taylor, J. 1990. Amplification and direct sequencing of fungal ribosomal RNA genes for phylogenetics. In *PCR Protocols: A Guide to Methods and Applications*, Innis, M. A., Gelfand, D. H., Sninsky, J. J., and White, T. J. (eds.). New York: Academic Press.

Yadav, J. S. and Pillai, R. K. 1979. Evolution of karyotypes and phylogenetic relationships in Scarabaeidae (Coleoptera). *Zool Anz* 202:105–118.

Yezerski, A., Stevens, L., and Ametrano, J. 2003. A genetic linkage map for *Tribolium confusum* based on random amplified polymorphic DNAs and recombinant inbred lines. *Insect Mol Biol* 12:517–526.

Zwick, M. S., Hanson, R. E., McKnight, T. D. et al. 1997. A rapid procedure for the isolation of C_0t-1 DNA from plants. *Genome* 40:138–142.

6

Silk Moths (Lepidoptera)

Atsuo Yoshido, Ken Sahara, and Yuji Yasukochi

CONTENTS

LIST OF ABBREVIATIONS

AFLP, amplified fragment length polymorphism
BAC, bacterial artificial chromosome
BSA, bovine serum albumin
CGH, comparative genomic hybridization
DABCO, 1,4-Diazabicyclo[2.2.2]octane
DAPI, 4',6-diamidino-2-phenylindole
DOP-PCR, degenerate oligonucleotide primed-polymerase
   chain reaction
EDTA, 2,2',2'',2'''-(Ethane-1,2-diyldinitrilo)tetraacetic acid
EST, expressed sequence tag
FISH, fluorescence in situ hybridization
GISH, genomic in situ hybridization
HDR, high-density replica
NGS, next-generation sequencing
PCR, polymerase chain reaction
RAPD, random amplified fragment polymorphism
RFLP, restriction fragment length polymorphism
SNP, single nucleotide polymorphism
SSC, saline sodium citrate
SSR, simple sequence repeats
STS, sequence-tagged site

6.1   INTRODUCTION

6.1.1   Taxonomy and Importance of the Species

Lepidoptera, moths and butterflies, is the second species-richest
insect order, which comprise more than 130,000 species (Beccaloni
et al. 2003). The holokinetic chromosome nature of Lepidoptera
as well as Trichoptera (cf., caddis flies: sister clade of Lepidoptera)
is shared with Hemiptera (aphids and bugs) within Insecta (see

Section 6.1.2). However, sex chromosome systems of Lepidoptera and Trichoptera are female heterogamety (WZ/ZZ, female/male), unlike male heterogametic system (XX/XY, female/male) of all others.

The silkworm, *Bombyx mori*, is one of the most well-known insects by its commercial value of "silk" production from 5000 years ago. *B. mori* has been believed to be domesticated from the very close relative species, *B. mandarina*. Domestication made *B. mori* no longer surviving in the field by itself. Hence, intermediate type from *B. mandarina* to *B. mori* is thought to be extinct during the domestication process. Instead, more than 3000 economically and scientifically important strains have been maintained in the world by feeding mulberry leaves and/or artificial diet (Yamamoto 2000).

B. mori is, meanwhile, the representative of Lepidoptera in the scientific communities. *B. mori* contributed to rediscovery of Mendel's law together with the first empirical research of heterosis (Toyama 1906), the first characterization of sex pheromone, bombykol (Butenandt et al. 1961), the identification of diapause hormone (Imai et al. 1991), and the first development of lepidopteran transgenesis method (Tamura et al. 2000). In addition, the reference genome sequence is now available (The International Silkworm Genome Consortium 2008) (see Section 6.1.3). National BioResource Project in Japan can serve 456 mutant strains kept in Kyushu University to worldwide researchers through http://www.shigen.nig.ac.jp/silkwormbase/index.jsp.

6.1.2 Karyotype and Sex Chromosome System

Lepidoptera are organisms harboring holokinetic chromosomes, and most species have diploid chromosome numbers close to 62 (Robinson 1971). Besides the typical haploid chromosome number ($n = 31$), lepidopteran species possess numerical karyotype variations from $n = 5$ to $n = 223$ (de Lesse 1970; De Prins and Saitoh 2003; Brown et al. 2004), probably caused by multiple chromosomal rearrangements. In addition, the karyotype variations within closely related species have been reported for some species (Lukhtanov et al. 2005; Yoshido et al. 2013).

The lepidopteran holokinetic chromosomes are generally small, uniform in shape and size in mitotic metaphase, and lacking primary constrictions (Figure 6.1a). For a long time, the absence of morphological landmarks and the lack of a banding technique prevented the identification of individual chromosomes. In contrast to mitotic chromosomes, pachytene prophase stage during meiosis is available for the identification of a few chromosomes with the help of specific chromomere patterns by conventional staining (Traut 1976). The use of pachytene mapping allowed the identification of a bivalent carrying the nucleolus organizer region (NOR) (Figure 6.1b) and sex

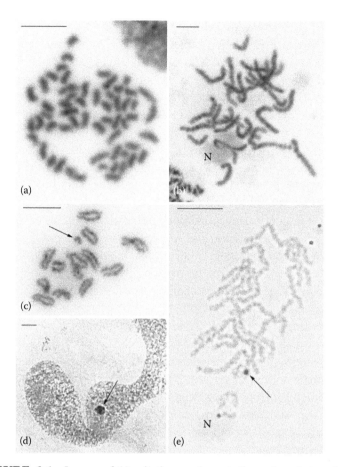

FIGURE 6.1 Images of (a) mitotic complement, (b, c, e) pachytene biva-
lents complements, and (d) a polyploid nucleus, from lepidopteran females.
Bar = 10 μm and N, nucleolus. *Bombyx mori*, (a) fluorescent images (black and
white inverted) of DAPI-stained mitotic metaphase and (b) meiotic prophase
(pachytene stage) complements showing $2n = 56$ and $n = 28$, respectively.
Samia cynthia ricini, (c) fluorescent images (black and white inverted) of
DAPI-stained late pachytene complement showing $2n = 27$, arrow represents
Z chromosome univalent. *S. cynthia pryeri*, (d) phase contrast microscopic
images of Giemsa-stained polyploid nucleus, arrow represents W chromatin
body and (e) meiotic pachytene complement showing $2n = 28$, arrow represents
W chromosome.

chromosome bivalent (Traut and Marec 1997) (Figure 6.1e, arrow)
or sex chromosome univalent (Figure 6.1c arrow) only under the
best chromosome condition in some species.

Lepidopteran species only share the sex chromosome system of
female heterogamety with caddis flies (Trichoptera) among insects
(Traut et al. 2007). The majority of lepidopteran species possess a
WZ♀/ZZ♂ sex chromosome system. The W chromosome occurs
in the majority of Lepidoptera, specifically in the clades Ditrysia
and Tischeriina (Traut and Marec 1996, 1997; Lukhtanov 2000;
Marec et al. 2010). The presence or absence of W chromosomes
can be confirmed by W-chromatin bodies from highly polyploid

nuclei, for example, those of Malpighian tubules in several species (Traut and Marec 1996) (Figure 6.1d). The complete absence of meiotic recombination in females has resulted in the accelerated molecular divergence and heterochromatin-rich W chromosomes. Besides the common WZ/ZZ system, variants without the W chromosome (Z0♀/ZZ♂) (Figure 6.1c) and multiple sex chromosomes, such as $W_1W_2Z♀/ZZ♂$ and $WZ_1Z_2♀/Z_1Z_1Z_2Z_2♂$, have been found (Traut et al. 2007; Sahara et al. 2012). In some species, sex chromosome differentiation can be studied by using a pachytene mapping technique (Traut and Marec 1997). In chromosome preparations of pachytene oocytes, a simple cytogenetic method makes it possible to identify sex chromosomes bivalent (or trivalent) by heterochromatinization of the W chromosomes (Figure 6.1e), but fails if the W heterochromatin is not seen as in *B. mori* (Figure 6.1b). This situation makes it difficult to distinguish simple sex chromosome and complex multiple sex chromosomes in most species. Thus, little is known about the origin and evolution of lepidopteran sex chromosomes.

Advanced molecular cytogenetic methods have been applied to resolve cytogenetic problems in Lepidoptera. Comparative genome hybridization (CGH) and genomic in situ hybridization (GISH) using genomic DNA as probes, can visualize all the lepidopteran W chromosomes so far investigated (Traut et al. 1999; Sahara et al. 2003a,b; Yoshido et al. 2006). Fluorescence in situ hybridization with bacterial artificial chromosome (BAC-FISH) or fosmid probes (fosmid-FISH) approach opened the gate to complete resolution of the cytogenetic problems in Lepidoptera. BAC-FISH mapping enabled to identify all the chromosomes and establish the complete karyotype, and assign them to linkage groups in the silkworm, *B. mori* (Yoshido et al. 2005a). This method as well as fosmid-FISH is also applicable to other lepidopteran species and currently available for gene-based comparative cytogenetic mapping (Sahara et al. 2007; Yasukochi et al. 2009; Yoshido et al. 2011b; Sahara et al. 2013). In this chapter, we present molecular cytogenetic protocols of chromosome identification and gene mapping in Lepidoptera.

6.1.3 Genome Sequencing Project and Genetic Map

B. mori is appropriate for genetic analysis, as artificial selection eliminated the capability of escape and flight from the wild ancestor, *B. mandarina*, and rearing technique is established during the history of sericulture. In addition, genetic diversity of *B. mori* races is quite low compared with *B. mandarina* (Xia et al. 2009) because of the bottleneck effect of domestication and inbreeding over thousands of generations. Furthermore, genetic recombination does not occur in lepidopteran oogenesis, which greatly facilitates interpretation of the results obtained by linkage analysis as genotyping reveals crossing-over events on paternal chromosomes exactly (Yasukochi 1998).

It was in 1906 that K. Toyama first made clear that traits of larval spots and yellow cocoon were inherited in *B. mori* in a Mendelian fashion. Since then, many spontaneous and artificially induced phenotypic mutants of *B. mori* have been isolated, stored (Banno et al. 2010), and used to construct linkage maps (Banno et al. 2005). However, the gene order and genetic distances between loci are not necessarily precise as the results of multiple, independent three-point crosses are integrated to construct the phenotypic map.

Polymerase chain reaction (PCR)–based methods for mapping greatly improved the accuracy and resolution of linkage maps as progeny of single pair mating was sufficient for map construction of Lepidoptera. Several linkage maps are published by use of restriction fragment length polymorphism (RFLP) (Shi et al. 1995), random amplified fragment polymorphism (RAPD) (Promboon et al. 1995; Yasukochi 1998), amplified fragment length polymorphisms (AFLP) (Tan et al. 2001), simple sequence repeats (SSR) (Miao et al. 2005), and single-nucleotide polymorphism (SNP)–detection assay (Yamamoto et al. 2006). We constructed an integrated map anchoring genes, BAC contigs, and cytogenetic markers (Yasukochi et al. 2006).

Construction of linkage maps for other Lepidoptera faced difficulties in establishment of two nearly inbred lines with high frequency of fixed polymorphisms that critically determine the efficiency of linkage analysis. In addition, lack of genome information made it difficult to map genes and expressed sequence tags (ESTs). Consequently, few maps were constructed mainly based on anonymous markers such as SSR and AFLP for butterflies (Wang and Porter 2004; Jiggins et al. 2005; Van't Hof et al. 2008; Winter and Porter 2010) and moths (Dopman et al. 2004; Van't Hof et al. 2013).

Current spread of next-generation sequencing (NGS) is greatly reducing cost, time, and labor required for genetic marker discovery and genotyping (Davey et al. 2011). Therefore, it becomes feasible to map enormous genes and ESTs of genetically less characterized species as far as substantial financial resources are available. Detailed gene-based linkage maps were published solely (Beldade et al. 2009; Baxter et al. 2011) or as a part of genome sequencing project (The *Heliconius* Genome Consortium 2012).

Genome sequencing of *B. mori* also preceded that of other Lepidoptera. In 2004, two independent projects published draft sequences (Xia et al. 2004; Mita et al. 2004). Then, these sequences were merged and assigned onto 28 *B. mori* chromosomes (The International Silkworm Genome Consortium 2008).

Recently, detailed genome sequences are available for three lepidopteran species. First, a draft sequence of the migratory monarch butterfly, *Danaus plexippus*, was published (Zhan et al. 2011). Then, the genome of a mimetic butterfly, *Heliconius melpomene*, was sequenced and ordered onto 21 *Heliconius melpomene* chromosomes (The *Heliconius* Genome Consortium 2012). Finally, the genome of the diamondback moth, *Plutella xylostella*, was deeply sequenced mainly based on fosmids and assigned to chromosomes (You et al. 2013) using a previously constructed linkage map (Baxter et al. 2011).

6.2 PROTOCOLS

6.2.1 *Bombyx mori* Culture

B. mori is monophagous and waits feeding of mulberry leaves without escape (Figure 6.2a). Favorable condition for silkworm rearing is at room temperature (25°C) and 60%–80% humidity. Alternatively, artificial diet (Figure 6.2b) is also applicable to the feeding (Fukuda et al. 1962). Commercially artificial diet such as Silkmate 2S (Life Tech Department, Nosan Corporation, Yokohama, Japan) is available, however please note that the ability of feeding artificial diets is different among strains (http://www.shigen.nig.ac.jp/silkwormbase/feeding_synthetic.jsp).

6.2.2 Sex Discrimination of *Bombyx mori* Larvae and Pupae by Outer Appearances

Larval sex can be distinguished from Herold's gland for males (Figure 6.3a) and Ishiwata's fore and hind glands for females (Figure 6.3b). To discriminate male and female larvae easily, sex-limited strains were developed, whose W chromosome is fused with autosomal fragments responsible for larval

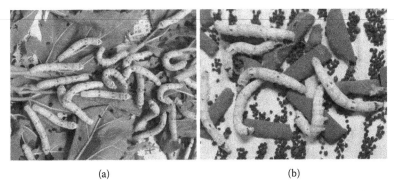

(a)   (b)

FIGURE 6.2 (**See color insert.**) *Bombyx mori* rearing by (a) mulberry leaves and (b) artificial diets.

FIGURE 6.3 **(See color insert.)** Sex discrimination in the larval and pupal stages. (a) A male larva can be discriminated by a Herald's gland (black arrow) appearing in ventral tale part; (b) a female shows Ishiwata's fore (F) and hinder (H) glands in the similar part of the early stage of last instar larva; (c) a female and a male of a sex-limited strain. Females of the strain have a second chromosome fragment carrying normal marking $(+^p)$ locus onto the W chromosome. Both females and males have second chromosome pairs with plain (p) loci. In the pupal stage, one can easily discriminate (d) a male from (e) a female, by the different morphology pointed out by white arrows.

morphological phenotypes. For instance, females of sex-limited normal pattern $(+^p)$ strain have W chromosome that is fused with a translocated autosomal fragment carrying $+^p$ locus of chromosome 2. The locus of the autosome in the strain is a recessive gene, plane (p). Hence, one can discriminate the sex by larval phenotypes: normal pattern and plane as females and males, respectively (Figure 6.3c). Sex-limited strains are also developed for egg and cocoon color phenotypes. Sex discrimination becomes easier in pupal stage (Figure 6.3d and e).

6.2.3 Tissue and Good Stage for Chromosome Preparation

We usually use ovaries and testes for mitotic metaphase (Figure 6.1a) and pachytene (a stage of prophase in meiosis) (Figure 6.1b) preparations. The good stage for *B. mori* chromosome preparation is not so strict, but is dependent on strains. Here we describe a case of the standard strain, p50, alternatively called Daizo or Dazao, that was used to determine the reference genome sequence (http://sgp.dna.affrc.go.jp/KAIKObase/). For chromosome preparation from male, testes are usually dissected from 12 to 36 hours of the fifth instar larvae. Whereas adequate stage for the ovary is between 24 and 72 hours of the fifth instar larvae.

If one needs more mitotic chromosome in a preparation, one can dissect wing discs from wandering to early spinning stages. We have less experience in this preparation procedure in *B. mori*, but this organ may be good to obtain more mitotic chromosome in Lepidoptera.

6.2.4 Chromosome Preparation

In a Ringer solution (150 mM NaCl, 5.6 mM KCl, 2.4 mM NaHCO$_3$, 2.2 mM CaCl$_2$·2H$_2$O) (Glaser 1917), we usually dissect larval bodies from the ventral side after fixing the head and tail parts with marking or insect pins. By means of dissection scissors cut ventral skin from the tail part and pin the skin to open the body (Figure 6.4a). Ovaries (Figure 6.4b) and testes (Figure 6.4c) are observed in the dorsal side (under the midgut) of fifth abdominal segment; third abdominal legs (Figure 6.4a, arrows) are the marker of the segment. Testis is larger than ovary if larvae are at the exactly same stage. Testis looks elliptical in shape (Figure 6.4c). Gonads are connected with trachea and a duct, so we cut them and transfer the gonads to a hole filled with Ringer solution of three-hole glass equipment (Figure 6.5a) (if no such material, one may use small petri dish).

For chromosome preparation from female, we peal out ovarioles from an ovary and then carry out a hypotonic treatment (100 mM solution mixed with 75 mL KCl and 15 mL NaCl) for 10 minutes in the center hole. Then the ovarioles are transferred into the last hole to fix with Carnoy's solution (ethanol:chloroform:acetic acid, 6:3:1) for 10 minutes. Less than one-fourth from the tip of an ovariole seems to be good for the chromosome preparation. We normally use two ovarioles for one preparation. Pick them by attaching the tip of insect pin stacked into a wooden chopstick (Figure 6.5b). A target part of an ovariole is shown in Figure 6.6a. They are then transferred into a drop of 60% acetic acid put on a glass slide heated at 45°C–55°C. We use a hotplate (Hp-4530 hotplate, ASONE,

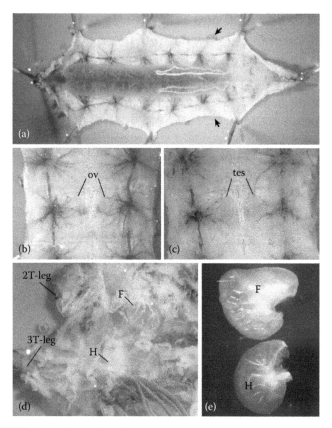

FIGURE 6.4 **(See color insert.)** *Bombyx mori* gonads and wing discs. (a) A fifth instar larva cut open from ventral side, arrows indicate third abdominal legs where is a marker of the gonad segment; (b) positions of a pair of ovary and (c) testis; (d) cut-opened thorax (left part from ventral view); and (e) fore- and hind-wing discs. 2T-leg, second thoracic leg; 3T-leg, third thoracic leg; F, fore-wing disc; H, hinder wing disc; ov, ovary; tes, testis.

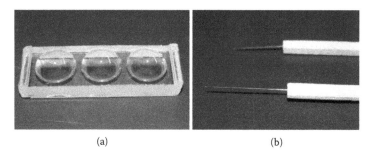

(a)   (b)

FIGURE 6.5 **(See color insert.)** (a) A 3-hole glass and (b) dissecting insect-pin stacked to wooden chopstick.

Osaka, Japan) for this purpose. We macerate cells in the drop and spread them on the glass slide and air-dry the preparation. Original methods for the preparation were written by Traut (1976).

We usually treat testes with another hypotonic solution (75 mM KCl) for 10–15 minutes in the center hole. Then the testes are

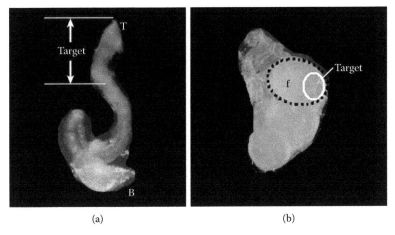

(a) (b)

FIGURE 6.6 (**See color insert.**) Target cells for chromosome preparation. (a) An ovariole and a part harboring cells with adequate stages (target) for preparation; (b) a testis with 3 follicles. A representative follicle surrounded with broken oval. A part in the oval shows the target.

transferred to the last hole to fix with Carnoy's solution for 10 minutes. A target part of cells is shown in Figure 6.6b. One should peal the skin of testis to avoid dirty material except for spermatocyte and spermatogonia. We use less than half mass of them in a single follicle for one chromosome preparation. After transferring the cells into a drop of 60% acetic acid on a glass slide by means of the equipment shown in Figure 6.5b, the same procedure as described earlier should be done (Sahara et al. 1999).

Fore- and hind-wing discs (Figure 6.4d and e) can be dissected from second and third segments of the thorax (Figure 6.4d). The hypotonic treatment, fixation, and cell preparation procedure are the same as that of spermatocytes and spermatogonia. One-fifth to one-tenth of the whole disc cells for one preparation seems to be enough.

6.2.5 Types of Probes

6.2.5.1 Bacterial Artificial Chromosomes and Fosmids

BAC (Shizuya et al. 1992) and fosmid (Kim et al. 1992) are ideal probes for cytogenetic mapping, as sufficient amount of 40–150 kb genomic fragments cloned into the vectors can be easily prepared by a commonly used alkaline lysis method without the risk of off-target DNA contamination. Compared with fosmids, BACs generate stronger and more specific signals in general, presumably because maximum insert size of BAC is far larger than that of fosmid, which enables BACs to stain longer chromosomal region and alleviate negative position effects of chromatin structure. However, only a few laboratories have the skill to construct high-quality BAC libraries of insects. Fosmids can be used as good probes if experimental conditions are adequately tuned (Yoshido et al. 2011b; Sahara et al. 2013), and it is worth to consider fosmid libraries instead of BAC.

It is a critical step to isolate BAC or fosmid clones that contain target DNA sequences. Two methods are widely used for this purpose. One is colony hybridization using high-density replica (HDR) filters. Typically, 16 colonies derived from the same well of 8 microplates are dotted in the 4 × 4 array on a nylon membrane placed on a solid medium by a gridding robot. The colonies are cultured, lysed, and fixed on the membrane. HDR filters are provided from genomic resource centers at reasonable price, however, it costs high to prepare HDR filters from custom-made libraries newly. In addition, several filters are needed to screen relatively small-insert libraries like fosmids. Thus, we recommend an alternative method, PCR-based screening, that is a high throughput and can be performed using common apparatus in laboratories of molecular genetics (Yasukochi et al. 2011). Here, we describe the method briefly, as it is not the main topic of this chapter and a detailed protocol was described earlier (Yasukochi 2002).

Compared with other methods, PCR is very robust to the dilution of target DNA and the disturbance by nontarget DNA as far as primers does not contain sequence motives involved in repetitive sequences. PCR-based screening uses this feature by pooling liquid cultures of multiple clones to save labor required for DNA preparation. In case of *de novo* library construction, it is preferable to perform picking up of clones and pooling of cultures, simultaneously, because frozen stocked cells might be damaged by thawing and freezing. DNA preparation can be performed afterward if pooled cultures are centrifuged and collected bacterial cells are frozen.

There are two strategies for screening clones stocked in multiple microplates. One is single-step "three-dimensional" (3-D) screen (Figure 6.7a) and the other is two-step "two-dimensional" (2-D) screen (Figure 6.7b). Two dimensions are row and column and the third dimension is plate. In 3-D screen, DNA pools are prepared from the mixture of cultures from wells in the same row, column, or plate, independently, whereas in 2-D screen, DNA pools are prepared from the mixture of cultures from wells in the same row or column of the same plate, and plate pools are prepared secondary by assemble of all the row or column pools.

In the 3-D screening process, PCR reactions are performed against row, column, and plate DNA pools, simultaneously (Figure 6.7a), whereas plate DNA pools are first screened and row and column DNA pools of the positive plate are then checked in

3-D screening

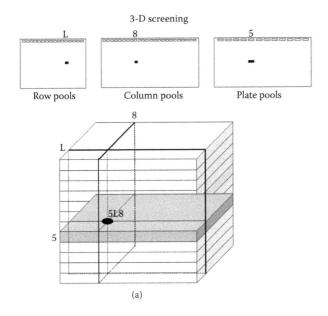

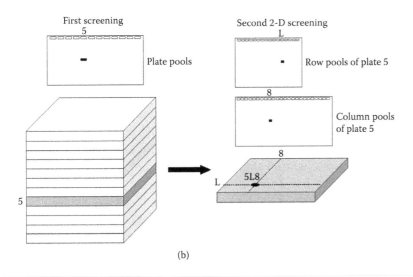

FIGURE 6.7 Schematic representation of two strategies in PCR-based screening of twelve 384-well microplates where a positive clone is located in L8 well of plate 5. (a) Three-dimensional screening; (b) two-step 2-D screening.

the 2-D screening process. The evident advantage of 3-D screen is reduced number of DNA preparations. In the case of screening twelve 384-well microplates (Figure 6.7), only 52 (16 rows, 24 columns, and 12 plates) preparations are needed. In contrast, 480 (16 rows and 24 columns for each plate) preparations are needed for 2-D screen.

However, additional confirmation process is needed if there are multiple positive clones in the screened unit. For example, when three positive clones are located on three different microplates, 27 candidate clones must be screened to identify 3 positive clones, which is not necessary in 2-D screen. Therefore, we recommend two-step 2-D screen for isolation of genome-wide cytogenetic markers.

The identified positive clones must be spread on solid media to develop into single colony and the isolated clones must be confirmed to contain the target sequence. It is not rare that bacterial cells in the same well are the mixture of multiple clones because of picking up overlapped clones or cross-contamination between neighboring wells. Use of nonisolated clones might lead to signals generated by contaminated probes that seriously confuse the conclusion of experiments.

6.2.5.2   Genomic DNA

Genomic DNA is used as probes for CGH and GISH. In lepidopteran species, GISH and CGH are the main methods for identification of W chromosomes and examination of molecular composition of the W chromosomes. Genomic DNA for both GISH and CGH can be obtained from eggs, larvae, pupae, and adults for any purpose by standard phenol–chloroform extraction (Blin and Stafford 1976) with slight modification as described.

1. Homogenize tissues from specimens in liquid nitrogen and add lysis buffer (20 mM Tris-HCl pH 8.0, 20 mM EDTA pH 8.0, 1% SDS) with approximately 200 µg/mL proteinase K, and then incubate the sample at 37°C for more than 3 hours up to overnight.
2. After incubation, add equal volume of Tris-EDTA (TE)–saturated phenol and invert several times.
3. Centrifuge the tube at 5000 rpm for 10 minutes and transfer aqueous (top) phase to new tube.
4. Add equal volume of phenol–chloroform and mix vigorously.
5. Centrifuge at 5000 rpm for 10 minutes and transfer aqueous (top) phase to new tube.
6. Add 0.1 volume of 3 M sodium acetate (pH 5.2) and 2.5-fold absolute ethanol, and then mix well by vortex.
7. Centrifuge at 5000 rpm for 10 minutes at 4°C and remove the supernatant.
8. After rinsing with 70% ethanol, centrifuge at 5000 rpm for 10 minutes at 4°C to remove the supernatant and air-dry the pellet.
9. Resuspend the pellet to TE buffer.

10. Add RNase A (final concentration ~100 µg/mL) and incubate the sample at 37°C for 2 hours.
11. After incubation, add proteinase K and incubate the sample at 37°C for 2 hours.
12. Repeat steps 2 through 9.

6.2.5.3 Polymerase Chain Reaction Products

PCR products are also available as probes for cytogenetic mapping in Lepidoptera. We successfully applied PCR products of repetitive sequences as well as single gene fragments in Lepidoptera as FISH probes (Yoshido et al. 2011a). Repetitive sequences and gene fragments are amplified by PCR using specific primers from genomic DNA or cDNA as templates and then labeled by PCR or nick translation system (see Section 6.2.6).

6.2.5.4 Telomeric Repeat

Lepidopteran telomere consists of (TTAGG)n, widely shared in many arthropods (Sahara et al. 1999), and retrotransposons (TRAS and SART) (Okazaki et al. 1995; Takahashi et al. 1997). FISH using the (TTAGG)n telomeric probes detects all chromosome ends of Lepidoptera, except for some special cases (Rego and Marec 2003). The method is also powerful tool for researching multiple sex chromosome systems (Yoshido et al. 2005b).

The (TTAGG)n telomeric probes are generated by nontemplate PCR (Ijdo et al. 1991; Sahara et al. 1999). There are two types of nontemplate PCR with different labeling methods (indirect and direct). The procedure of direct nontemplate PCR labeling for telomeric repeat is described in Section 6.2.6.2. Here we describe nontemplate PCR reaction for indirect labeling.

1. Set up reaction mixture to a final volume of 100 µL (Table 6.1).
2. An initial period of 90 seconds at 94°C is followed by 30 cycles of 45 seconds at 94°C, 30 seconds at 52°C, and

TABLE 6.1
Nontemplate PCR Reaction Mix for Telomeric Repeats

0.5 µL	100 µM TELO 1 (5′-TAGGTTAGGTTAGGTTAGGT-3′) primer
0.5 µL	100 µM TELO 2 (5′-CTAACCTAACCTAACCTAAC-3′) primer
10.0 µL	10× *Ex-Taq* buffer
10.0 µL	dNTP mix (2.5 mM each)
0.5 µL	*Ex-Taq* polymerase (5U/µL)

Nuclease-free water, fill up to 100 µL

1 minute at 72°C, and concluded by a final extension step of 10 minutes at 72°C.

3. Purify PCR products using PCR purification column (e.g., Wizard SV Gel and PCR Clean-Up System, Promega, Fitchburg, Wisconsin) or by ethanol precipitation and then proceed to indirect labeling (see Section 6.2.6.2).

6.2.5.5 W Chromosome

The sequences derived from lepidopteran W chromosomes can be obtained by either RAPD (Abe et al. 2010) or laser microdissection (Fuková et al. 2007). BAC clones selected by W-RAPD primers in *B. mori* is also a powerful tool to visualize and analyze the W chromosomes (Sahara et al. 2003b; Yoshido et al. 2007). Laser microdissection of the W chromatin is the universally applicable approach to analyze the W chromosomes in some lepidopteran species (Fuková et al. 2007; Vitková et al. 2007; Yoshido et al. 2013). In this section, we present the procedure of laser microdissection of the W chromosome.

W-chromatin bodies from highly polyploid nuclei of Malpighian tubules are usually used for laser microdissection of W chromosome (Figure 6.1d, arrow) (Fuková et al. 2007). The W-chromatin bodies consist of many W chromosomes that allow to obtain large number of sequences. If lepidopteran species of interest have clear and easy recognizable W chromosomes, it is also possible to dissect from interphase, mitotic and meiotic nuclei (Figure 6.1e, arrow). The procedure of preparation for laser microdissection is similar to that of chromosome preparation (see Section 6.2.4) but with some modifications.

1. Dissect out tissues from lepidopteran species of interest in a physiological solution.
2. Swell tissues for 10 minutes in a hypotonic solution (75 mM KCl).
3. Fix tissues for 15 minutes in methanol:acetic acid (3:1).
4. Spread tissues in 60% acetic acid on a glass slide coated with a polyethylene naphthalate membrane at 45°C–50°C.
5. Dehydrate the preparations through ethanol series (70%, 80%, and 98%) at room temperature, 30 seconds each, and air-dry.
6. Stain the preparations with 4% Giemsa in Sørensen's buffer (pH 6.8) for 10 minutes.
7. Cutout W chromosome (W chromatin) with a laser microbeam (1.5–1.7 µJ/pulse, 1 µm in diameter). Note that microdissections are performed with the help of a microlaser system, for example, a PALM MicroLaser System

(Carl Zeiss MicroImaging GmbH, Munich, Germany) as described by Kubickova et al. (2002).

8. Catapult by a single laser pulse (2 μJ/pulse) into a micro-tube cap containing 2 μL PCR oil.

9. DNA of microdissected samples, each containing some W chromosomes (usually 10–20), are amplified by PCR using a WGA4 GenomePlex Single Cell Whole Genome Amplification Kit (Sigma-Aldrich, St. Louis, Missouri). The PCR products are reamplified using WGA3 GenomePlex WGA Reamplification Kit (Sigma-Aldrich) and then labeled with fluorescent dUTP.

6.2.6 Labeling

There are two types of labeling methods: nick translation and PCR. In addition, users can make a choice either direct- (e.g., Cy3-dUTP or Cy5-dUTP) or indirect- (e.g., Biotin-dUTP or DIG-dUTP) fluorescent labeling. Choice of labeling methods depends on types of probes. In the following, we represent preferable combinations between types of probes and labeling methods used in FISH of lepidopteran species (Table 6.2).

6.2.6.1 Nick Translation

Nick translation systems are mainly used in labeling of large size DNA (genomic DNA or genomic clones of either BAC or fosmid libraries). Here, we describe the protocol using Nick translation kit (Cat. no. 32-801300, Abbott Molecular, Des Plaines, Illinois), with one modification of dNTP mix (see Table 6.3)

1. Set up reaction mixture to a final volume of 50 μL (Table 6.3). Note that our choices of fluorescent dyes are mainly Cy3-dUTP, Cy5-dUTP (GE Healthcare Life Sciences, Cleveland, Ohio), Green-dUTP, Orange-dUTP, and

TABLE 6.2

Preferable Combinations between Type of Probes and Labeling Methods

Type of Probes	Labeling Methods	Direct or Indirect Labeling
BAC or fosmid clones	Nick translation	Direct
Genomic DNA	Nick translation	Direct
Repetitive DNA[a]	Nick translation or PCR	Direct or indirect
Gene fragment	PCR	Direct or indirect

[a] Repetitive DNA contains telomeric repeat, rDNA, and W chromatin.

TABLE 6.3

Reaction Mix for Nick Translation

30.0 µL	DNA 1.0 µg (500 ng)
5.0 µL	10× Nick translation buffer
5.0 µL	10× dNTP mix for dUTP labeling: standard mix (0.25 mM d[ACG]TP, 0.17 mM dTTP, 0.08 mM fluorescent-dUTP) or high mix (0.25 mM d[ACG]TP, 0.09 mM dTTP, 0.16 mM fluorescent dUTP)
10.0 µL	Nick translation enzyme
50.0 µL	Total

Red-dUTP (Abbott Molecular). The difference between "standard" and "high" of 10× dNTP mix for dUTP labeling is volumes of fluorescent dyes. One can use "standard" for better cost performance. If FISH signal is weak or nothing, "high" of 10× dNTP mix for dUTP labeling could result in an improvement.

2. Mix thoroughly with pipette and/or vortex (then briefly centrifuge if needed). Incubate in water bath for 4–5 hours at 15°C. (The efficiency of labeling depends on purity of template DNA and/or incubation time.)

3. Stop the reaction by heating at 70°C for 10 minutes and then put on ice (or stored in –20°C).

6.2.6.2 Polymerase Chain Reaction

PCR labeling is mainly used in FISH by means of probes from PCR products of repetitive sequences or single-gene fragments. Here, we describe the PCR-labeling protocol using *Ex-Taq* polymerase and WGA3 kit.

1. Set up reaction mix to a final volume of 100 µL in case of usual PCR labeling with *Ex-Taq* polymerase (Table 6.4) or 18.75 µL in case of PCR labeling using WGA3 kit (Table 6.5).

   Note that for direct nontemplate PCR labeling of telomeric sequences, remove template from above reaction mix (see Table 6.1).

2. Mix thoroughly with pipetting.

3. Carry out PCR using below conditions.

   In case of usual PCR labeling using *Ex-Taq* polymerase, an initial period of 2 minutes at 94°C is followed by 40 cycles of 30 seconds at 94°C, 1 minute at 50°C–65°C (depends on primer), and 4 minutes at 72°C, and concluded by a final extension step of 10 minutes at 72°C. PCR condition for direct nontemplate PCR labeling of telomeric sequences is very similar to that described in

TABLE 6.4

Reaction Mix for Usual PCR Labeling

1.0 µL	Template (10–100 ng; PCR product or its clone)
0.5 µL	Primer forward (100 µM)
0.5 µL	Primer reverse (100 µM)
10.0 µL	10× *Ex-Taq* buffer
10.0 µL	dNTP mix for PCR labeling containing 1 mM d(AGC)TP and 0.3 mM dTTP
2.0 µL	1.0 mM fluorescent-dUTP (recommend Cy3-, Orange-, Green-dUTP)
0.5 µL	*Ex-Taq* polymerase

Nuclease-free water, fill up to 100 µL

TABLE 6.5

Reaction Mix for PCR Labeling Using WGA3 Kit

1.0 µL	Template (DNA obtained by laser microdissection, 15 ng)
1.875 µL	10× Amplification master mix
0.75 µL	dNTP mix containing 10 mM d(AGC)TP and 8.4 mM dTTP
0.75 µL	1.0 mM fluorescent-dUTP (recommend Cy3-, Orange-, Green-dUTP)
1.25 µL	WGA polymerase

Nuclease-free water, fill up to 18.75 µL

Section 6.2.5.4, but should change from 1 to 4 minutes at 72°C of cycle extension step.

In case of PCR labeling using WGA3 kit, an initial period of 3 minutes at 95°C is followed by 14 cycles of 15 seconds at 94°C, 5 minutes at 65°C.

4. Purify labeled PCR product using PCR purification column (e.g., Wizard SV Gel and PCR Clean-Up System, Promega).

6.2.7 Hybridization and Detection

6.2.7.1 Probe Purification and Denaturation

1. For one slide, one should generally prepare the following probe cocktail:
   a. Labeled DNA probes (generally 50–1000 ng).
   b. Unlabeled sonicated genomic DNA (generally 1–100 µg; the amounts depend on type of probes or species or chromosome condition).
   c. A 25 µg sonicated salmon sperm DNA (Sigma-Aldrich) as coprecipitating agent.
2. Purify the above probe cocktail by ethanol precipitation.
3. After air-dry, resuspend pellets in 10 µL hybridization solution (50% formamide, 10% dextran sulfate, 2× saline sodium citrate [SSC]).

4. The probe cocktail is denatured for 5 minutes at 90°C and chilled immediately on ice.

6.2.7.2   Slide Pretreatment and Denaturation

1. After removal from freezer (–30°C or –80°C), pass the chromosome preparations through graded ethanol series (70%, 80%, and 98%) and air-dry.
2. Denature in a buffer (70% formamide, 2× SSC), with coverslip, at 72°C for 3.5 minutes.
3. After denaturation, snap off the coverslip and place the slide immediately into 70% ethanol prechilled at –30°C.
4. Dehydrate through 80% and 98% ethanol series at room temperature, 30 seconds each, and air-dry.

6.2.7.3   Hybridization

1. Spin probes for 2 minutes at room temperature and put on ice to leave any precipitate behind.
2. Spot 10 μL hybridization mix on a slide (chromosome preparation) and cover immediately with a 24 × 32-mm coverslip. Gently press the coverslip with forceps to evenly distribute hybridization mix avoiding bubbles.
3. Seal edges of coverslip completely with rubber cement.
4. Incubate for 12–72 hours* at 37°C in a moist chamber.

6.2.7.4   Wash and Detection

The procedures after hybridization differ in labeling methods (direct or indirect labeling) of probes as follows:

Posthybridization wash for direct labeling: Detection step is not necessary for FISH using direct-labeled probes.
1. After removal of coverslips, wash the slides in 0.1× SSC containing 1% Triton X at 62°C for 5 minutes in Coplin jar with water bath.
2. Dip briefly in 2× SSC for a few minutes and then proceed to chromosome staining step (Section 6.2.7.5).
Posthybridization wash for indirect labeling: Detection step is inevitable for FISH using indirect-labeled probes.
1. After removing coverslips, wash slides in 50% formamide, 2× SSC at 46°C for 5 minutes in Coplin jar with shaking water bath. Repeat this step twice.

---

* Incubation times for hybridization depend on type of probes and/or labeling methods. It is generally recommended that the incubation is 3 days (over 2 nights) when one uses direct-labeled large DNA (genomic DNA and/or BAC, etc.) probes.

2. Wash slides in 2× SSC at 46°C for 5 minutes. Repeat this step once more.
3. Wash slides in 0.1× SSC at 62°C for 5 minutes. Repeat this step twice.
4. Wash in 4× SSC containing 0.1% Tween 20 at room temperature for 5–30 minutes.

Detection: Here we describe the procedure of biotin–streptavidin using biotin-labeling.

1. Incubate in blocking solution (2.5% BSA [Bovine serum albumin] in 4× SSC) at room temperature for 10–20 minutes.
2. Treat the slides with primary antibody* (fluorescent dye–conjugated streptavidin) and incubate for 30 minutes at 37°C.
3. Wash slides in 4× SSC containing 0.1% Tween 20 at 37°C for 5 minutes. Repeat this step once more.

After step (3) proceed to chromosome staining (Section 6.2.7.5). Optionally, the procedure of biotin–streptavidin can enhance hybridization signals by the following step:

4. Incubate in blocking solution (2.5% BSA in 4× SSC) at room temperature for 10–20 minutes.
5. Treat the slides with secondary antibody* (anti-streptavidin biotinylated) and incubate for 30 minutes at 37°C.
6. Wash slides in 4× SSC containing 0.1% Tween 20 at 37°C for 5 minutes. Repeat this step once more.
7. Repeat steps (1) through (3) once again.

6.2.7.5 Chromosome Staining

Mount in 20–50 μL antifade based on DABCO (1,4-diazabicyclo (2.2.2)-octane) containing 0.5 μg/mL DAPI (4′,6-diamidino-2-phenylindole) (Sigma-Aldrich, Tokyo, Japan) or VECTASHIELD mounting medium with DAPI (Vector Laboratories, Burlingame, California) and cover with 24 × 50-mm coverslip. Press gently with a Kimwipe to squeeze out excess mountant and seal coverslip with clear nail polish.

6.2.7.6 Reprobing

A reprobing technique of FISH is also available in lepidopteran chromosome preparations (Shibata et al. 2009). The following is the procedure:

1. After the observation of the first round FISH, peel nail polish and remove coverslips.
2. Rinse slides in 2 × SSC at room temperature for a few minutes.

---

* Antibody should be diluted with blocking solution.

3. Wash slides in probe-stripping solution (50% formamide, 0.1× SSC, 1% Triton X-100) at 70°C for 10 minutes.
4. Rinse slides in distilled water at room temperature for a few minutes.
5. Dehydrate through 70%, 80%, and 98% ethanol series at room temperature for 30 seconds each, and air-dry, and then proceed to second round FISH.

6.2.8   Visualization and Mapping

We observed the FISH preparations in a Leica DM6000B fluorescence microscope (Leica Microsystems, Tokyo, Japan). Digital images are captured with a DFC350FX B&W CCD camera (Leica Microsystems) by filter sets, A4 (UV for DAPI), L5 (for Green-dUTP dye), N3 (for Orange-dUTP dye), Y5 (for Cy5-dUTP dye), and an ordered filter composed of XF1207, XF2020, and XF3023 (Omega Optical, Brattleboro, Vermont) (for Red-dUTP). Adobe Photoshop is used to pseudocolor and superimpose the fluorescent images.

Here we describe gene-based comparative mapping between *B. mori* and *S. cynthia* (Figure 6.8) as an example of lepidopteran cytogenetic mapping. Geographic populations of *S. cynthia* show a unique polymorphism in chromosome number, resulting in variations in the sex chromosome systems (Yoshido et al. 2005b, 2013). By mapping eight anchor loci, we revealed the relationship between sex chromosomes of *S. cynthia* subspecies (*S. cynthia ricini* and *S. cynthia walkeri*) and that of *B. mori*. We also showed the gene order among species, based on *B. mori* genome information. Here is the procedure of gene-based comparative mapping.

1. Selection of target sequences for comparative mapping between *S. cynthia* and *B. mori*, from *B. mori* genome database (KAIKO.base, http://sgp.dna.affrc.go.jp/KAIKObase/).
2. Search and determine *S. cynthia* orthologs of *B. mori* genes from the *S. cynthia* EST database (SilkBase, http://silkbase.ab.a.u-tokyo.ac.jp/cgi-bin/index.cgi) or NCBI GenBank. If sequences of target genes have not yet been recorded in any public database, determine target sequences by DOP–PCR or constructing EST library. Then design sequence-tagged site (STS) primers to isolate clones containing target sequences from constructed genomic library.
3. Screening of clones containing target sequences by PCR from genomic library (in this case, use *S. cynthia* fosmid library) (see Section 6.2.5.1).
4. Isolation of clones containing target sequences.

(1) Selection of genes for comparative mapping between *B. mori* and *S. cynthia*.

(2) Using database, design of STS primer sets for *S. cynthia* orthologs of *B. mori* genes.

(3) By PCR using STS primers, screening of clones containing target sequences.

B. mori		*S. cynthia*	
Gene symbol	Chromosome position	Database	Fosmid clone
kettin	Chr 1: 6,505,696-6,533,895	AB543309	45A6
imp	Chr 1: 11,424,630-11,431,079	S06A01NCLL0014_I05	19B8
HP	Chr 8: 205,759-211,533	S13A01NGRL0004_C20	56J8
GRP2	Chr 8: 4,927,144-4,928,289	S06A01NCLL0014_M19	32B23
BR-C	Chr 8: 18,077,270-18,081,471	AB564751	56J22
P109	Chr 12: 4,989,327-5,006,556	I10A02NGRL0003_B23	21P14
JDCP	Chr 12: 7,728,258-7,729,559	S13A02NGRL0001_G21	44E23
XDH1	Chr 12: 8,493,509-8,511,254	I09A02NGRL0003_I02	14J3

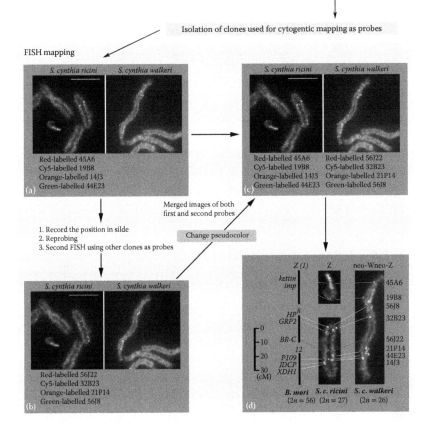

Isolation of clones used for cytogentic mapping as probes

FISH mapping

S. cynthia ricini · *S. cynthia walkeri*

Red-labelled 45A6
Cy5-labelled 19B8
Orange-labelled 14J3
Green-labelled 44E23
(a)

S. cynthia ricini · *S. cynthia walkeri*

Red-labelled 45A6 · Red-labelled 56J22
Cy5-labelled 19B8 · Cy5-labelled 32B23
Orange-labelled 14J3 · Orange-labelled 21P14
Green-labelled 44E23 · Green-labelled 56J8
(c)

Merged images of both first and second probes

1. Record the position in silde
2. Reprobing
3. Second FISH using other clones as probes

Change pseudocolor

S. cynthia ricini · *S. cynthia walkeri*

Red-labelled 56J22
Cy5-labelled 32B23
Orange-labelled 21P14
Green-labelled 56J8
(b)

Z (1) · Z · neo-Wneo-Z
kettin · 45A6
imp · 19B8
· 56J8
· 32B23
HP · 56J22
GRP2 · 21P14
—0 · 44E23
—10 · *BR-C* · 14J3
—20 · *P109*
—30 · *JDCP*
(cM) · *XDH1*

B. mori · *S. c. ricini* · *S. c. walkeri*
(2n = 56) · (2n = 27) · (2n = 26)
(d)

FIGURE 6.8 (See color insert.) Flowchart of comparative gene-based FISH mapping between *Bombyx mori* and *Samia cynthia* ssp.

5. Labeling of clones.
6. First round FISH using four probes labeled with different fluorescent dye (Figure 6.8a). The signals of four probes containing *S. cynthia* orthologs of *B. mori* Z chromosomal and autosomal genes were detected in an univalent and a bivalent of *S. cynthia ricini* female, respectively, although their signals appeared only on a bivalent of *S. cynthia walkeri* female (Figure 6.8a).

7. After recording the position of chromosomes captured, remove four probes by reprobing procedure (see Section 6.2.7.6).
8. Second round FISH using four other probes labeled with different fluorescent dye (Figure 6.8b). The new signals appear but signals from the first round FISH have not been observed in the second FISH image.
9. Merge FISH images from first and second round FISH, and use pseudocolor for discriminating eight probes (Figure 6.8c).
10. Summarize FISH data of gene-based mapping (Figure 6.8d).

As a result, FISH mapping reveals that the Z chromosome and chromosomes 8 and 12 of *B. mori* (Figure 6.8d, black bar) correspond to the Z chromosome and an autosome (we designed as chromosome 13) (Yoshido et al. 2011b) of *S. cynthia ricini*, respectively, and the homologues of *S. cynthia ricini* Z chromosome and chromosome 13 forms neo-sex chromosomes (the segment of autosomal origin) of *S. cynthia walkeri*. Furthermore, the results show that the gene order is well conserved between the respective segments of autosomal origin in neo-sex chromosomes of *S. cynthia walkeri* and the homologous autosomes of *S. cynthia ricini* (Figure 6.8d).

6.2.9   Representative Results

6.2.9.1   FISH Using Repetitive Sequences as Probes

We represent CGH results in *Manduca sexta* and *Samia cynthia pryeri* (Figure 6.9a through j). In *M. sexta* with WZ/ZZ sex chromosome system, we cannot identify the W chromosome by DAPI staining (Figure 6.9a, arrow). However CGH enable us to identify the W chromosome. Both probes strongly highlighted the W chromosome at similar intensities (Figure 6.9b through e). In *S. cynthia pryeri* with WZ/ZZ sex chromosome system, the W chromosome is recognized by DAPI staining (Figure 6.9f) and consists of a highly heterochromatic part and a less condensed euchromatin-like part (Yoshido et al. 2013). Both the female (green) and male (red) genomic probes hybridize to all chromosomes, resulting in yellowish coloration. Both probes highlighted a part of the NOR bivalent at similar intensities (Figure 6.9g). The highly heterochromatic part of the W chromosome was preferentially and strongly differentiated by the female genomic probe (Figure 6.9h through j).

The probes of (TTAGG)n telomeric repeats highlighted all chromosomal ends of 13 bivalents and Z univalent (Figure 6.9k, arrow)

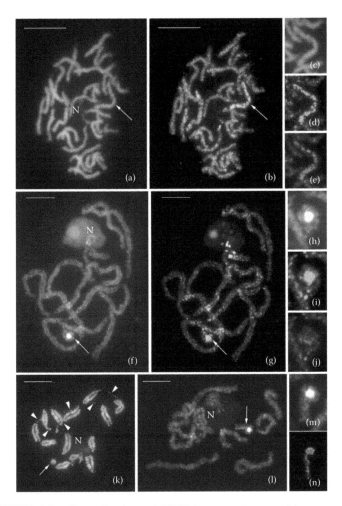

FIGURE 6.9 (See color insert.) FISH images using repetitive sequences as probes in some lepidopteran species, (a–e) *Manduca sexta*, (f–j and l–n) *Samia cynthia pryeri*, (k) *S. cynthia ricini*. Respective chromosomes were counterstained with DAPI (white). N, nucleolus; Bar = 10 μm. Comparative genomic hybridization (CGH) between respective sexes in female of (a–e) *M. sexta* and (f–j) *S. cynthia pryeri*. Female-derived genomic DNA probes were labeled with Green-dUTP (green), male-derived genomic DNA probes with Cy3-dUTP (red). (a and f) DAPI images. (b and g) Merged images of both probes. (c–e and h–j) A detail of the WZ bivalents. Arrow represents WZ bivalents. (k) FISH with the Cy3-labeled (TTAGG)n telomeric probe (red signals) in female of *S. cynthia ricini*. Arrow represents Z chromosome univalent. (l–n) FISH with W chromosome painting probes in female of *S. cynthia pryeri*. Pachytene complement (l). DAPI image (m) and W-probe (n) in a detail of the WZ bivalent. Arrow represents WZ bivalent.

in female *S. cynthia ricini*. It has been known that postpachytene bivalents are interconnected by telomeric DNA in female lepidopteran species (Rego and Marec 2003). Telomere-FISH in this postpachytene nucleus also shows some telomeric associations between nonhomologous chromosomes (Figure 6.9k, arrowhead).

The probes prepared by laser microdissection of the W chromosome from females of *S. cynthia pryeri* can be used in FISH to paint the whole W chromosome (Figure 6.9l). The W chromosome of *S. cynthia pryeri* is composed of a highly heterochromatic part, which is distinguished by strong DAPI staining, and a euchromatin-like part (Figure 6.9m and n) (Yoshido et al. 2013).

6.2.9.2 BAC- or Fosmid-FISH and Gene Mapping

In Lepidoptera, FISH using clones of genomic library (BAC or fosmid) as probes is powerful tool for chromosome identification, karyotyping, gene-based mapping, and comparative mapping among species. In this section, we present representative results of lepidopteran BAC- and fosmid-FISH and gene mapping.

It is important to use competitor DNA for lepidopteran BAC- and fosmid-FISH. Sonicated genomic DNA is adequate as the competitor for FISH in lepidopteran species. We represent FISH images using same probe (*B. mori* BAC clone no. 9D6C), which is selected by RAPD primer (Yoshido et al. 2005a) in pachytene complement of *B. mori* without (Figure 6.10a) and with competitor (Figure 6.10b). The probe without competitor detects many hybridization signals that have disturbed chromosome identification. In contrast, the probe with competitor detects a specific signal and identify individual chromosome (Figure 6.10b, arrow). The amount of competitor should be changed depending on the lepidopteran species, the physical condition, and the target chromosomes (meiosis or mitosis) of preparation and so forth. *B. mori* genome consists of many repetitive sequences (~40%) (The International Silkworm Genome Consortium 2008). Because clones of large genomic library (BAC or fosmid) contain some repetitive sequences, the probe without competitor may bring repetitive signals.

We present FISH results using *B. mori* W and Z chromosome-derived-BAC clones (respective W- and Z-BAC) as probes (Figure 6.10c and d). The W-BAC probe highlighted the whole W chromosome in both meiotic and mitotic stage (green signals in Figure 6.10c and d). In female pachytene complement, the Z-BAC probe indeed hybridizes to the Z chromosome, which identified independently as the pairing partner of the W (red signal in Figure 6.10c) and to a chromosome in mitotic complements (red signal in Figure 6.10d).

BAC- and/or fosmid-FISH enable us to identify all chromosomes in a nucleus (karyotyping). Here we describe karyotype of *S. cynthia walkeri* by using fosmid-FISH (Figure 6.11). For identifying all individual chromosomes, we used a total of 21 fosmid probes. The FISH identifies all 12 autosome and sex chromosome bivalents (neo-Wneo-Z) and, hence, definitively karyotype the *S. cynthia walkeri* female pachytene complement.

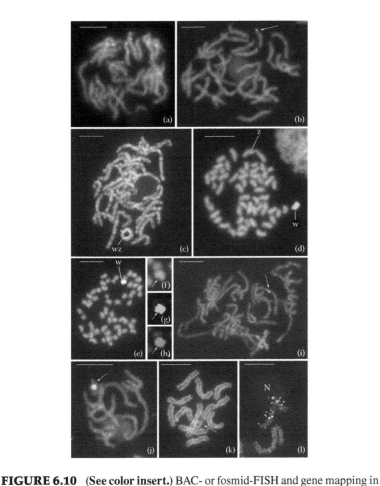

FIGURE 6.10 **(See color insert.)** BAC- or fosmid-FISH and gene mapping in lepidopteran species, (a–d) *B. mori*, (e–h) *B. mandarina*, (i) *Helicoverpa armigera*, (j–l) *S. cynthia* subspecies. Respective chromosomes were counterstained with DAPI (white). Bar = 10 μm. (a) FISH with *B. mori* chromosome 2–derived BAC (9D6C) probe without competitor and (b) with competitor, arrow represents *B. mori* chromosome 2; (c) FISH with *B. mori* W and Z chromosome-derived-BAC probes in *B. mori* females, pachytene complement and (d) mitotic metaphase complement, Green-labeled probe of the 19L6H clone (green signal) and Cy3-labeled probe of the 9A5H clone (red signal), respectively; (e) FISH with Green-labeled probe of *B. mori* 19L6H clone (green signal) and Cy3-labeled probe of *B. mandarina* female genomic DNA (red signal) in *B. mandarina* female mitotic metaphase complement; (f–h) a detail of the W chromosome, arrow represents the W chromosome: (i) FISH with Red-labeled probe of *Heliothis virescens* 55I09 (red signal) and Green-labeled probe of *Helicoverpa armigera* 26P10 (green signal) in *H. armigera* pachytene complement; FISH with Cy3-labeled *S. cynthia* ortholog of *B. mori* Z-chromosome-linked gene, (j) *BYB* in female pachytene chromosomes of *S. cynthia pryeri* and (k) *S. cynthia ricini*, arrows represent WZ bivalent (j) and Z univalent (k), respectively; (l) FISH with different fluorescence dye-labeled probes of four *S. cynthia* fosmid clones, 11P18 (Red-labeled probe, red signal), 60G11 (Orange-labeled probe, yellow signal), 15C21 (Cy5-labeled probe, purple signal), 32H9 (Green-labeled probe, green signal) and Green-labeled probe of *S. cynthia* ortholog of *RpL4* (cyan signal, pseudocolor) and Cy3-labeled probe of 18S rDNA (orange signal, pseudocolor) in *S. cynthia ricini* pachytene chromosome. N, nucleolus.

BAC- or fosmid-FISH is also applicable to cross-hybridization among closely related species. We show two representative results of cross-hybridization. The first is cross-hybridization between *Bombyx* genus (Yoshido et al. 2007). In female *B. mandarina* mitotic complement ($2n = 54$), the *B. mori* W-BAC probes (green) highlights the *B. mandarina* W chromosome that is also identified by GISH (red) using *B. mandarina* females genomic probes (Figure 6.10e). Genomic (green) and *B. mori* W-BAC (red) probes display a similar hybridization signals on the W chromosome of *B. mandarina* (Figure 6.10f through h). As another result, we present BAC-FISH mapping between a closely related species, *Heliothis virescens* and *Helicoverpa armigera* (Figure 6.10i) (Sahara et al. 2013). Two BAC clones, 55I09 of *Heliothis virescens* and 26P10 of *Helicoverpa armigera* contain respective ortholog of *Ultraspiracle* gene. FISH mapping using two BAC probes, 55I09 (red) and 26P10 (green) shows colocalization at the identical position of the bivalent in *Helicoverpa armigera* pachytene chromosome (Figure 6.10i, arrow).

We also present an alternative strategy of gene mapping without genomic library (BAC or fosmid). A PCR-generated probe made from *S. cynthia* ortholog of *B. mori* Z-chromosome-linked gene, *BYB*, hybridizes to the Z paired with the W chromosome (Figure 6.10j, arrow) in *S. cynthia pryeri* female and to the Z univalent in *S. cynthia ricini* female (Figure 6.10k, arrow). The gene mapping using single-gene fragment has the advantage of being able to apply in wide range species. The procedure can omit the construction of genomic libraries and screening of clones carrying target genes (Yoshido et al. 2011a). However this strategy is unsuitable for cytogenetical mapping with numbers of probes. Thus, we mainly combine direct gene mapping and BAC- and/or fosmid-FISH mapping. When there is no clone containing target sequence in genomic library, above combination is good method for cytogenetical mapping. We provide the representative result of the combined FISH mapping in *S. cynthia* (Figure 6.10l). In first round of FISH, the signals of four fosmid probes labeled by different fluorescent dye were detected in the same chromosome (Figure 6.10l, red, yellow, purple, and green). The *S. cynthia* ortholog of *RpL4* was recovered by direct PCR labeling, because we could not isolate a suitable clone containing the *RpL4* ortholog in our fosmid library. Similarly, 18S rDNA was recovered by direct PCR labeling. In second round of FISH after reprobing, PCR-generated probes made from *RpL4* ortholog (Figure 6.10l, cyan) and 18S rDNA (Figure 6.10l, orange) hybridize to the same chromosome with four fosmid probes. The combined FISH mapping clearly shows that four fosmids, *RpL4* ortholog, and 18S rDNA probes are located on the chromosome carrying NOR (Figure 6.10l, N), and reveals the gene order in *S. cynthia*.

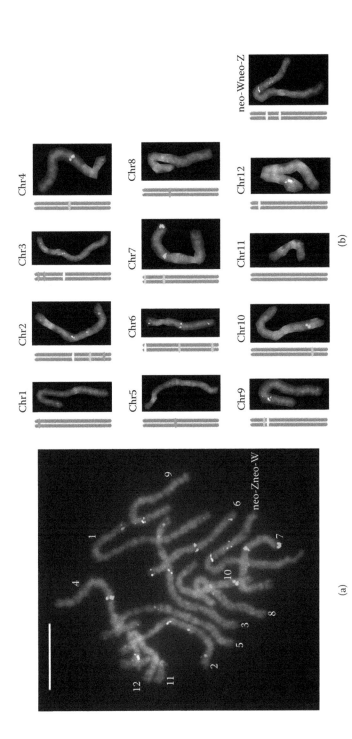

FIGURE 6.11 (See color insert.) Fosmid-FISH karyotype of *Samia cynthia walkeri*. (a) A female pachytene complement of 12 autosomal bivalents and a sex chromosome bivalent (neo-W and neo-Z), each identified by 1–3 pseudocolored hybridization signals of fosmid probes. Chromosomes were counterstained with DAPI (white); (b) 12 autosomal bivalents and neo-Wneo-Z bivalent of same pachytene complement as in (a) arranged according to their chromosome numbers. Bar = 10 μm.

6.2.10 Troubleshooting

We describe a few potential problem and possible solution that may help researchers troubleshoot during cytogenetic mapping in Lepidoptera.

Problem 1: No or few chromosome is observed in the preparation.
Solution 1: The adequate stages are fully strain and/or species dependent. They are also different between sexes. For example; in *Manduca sexta*, the good stage for chromosome preparation of testes is the last instar larvae but that of ovaries is the pupal stage. Hence, one should alter the dissecting stage for preparation.

Problem 2: Only clumped chromosomes are in the preparation.
Solution 2: One should prolong the hypotonic treatment and/or change a hypotonic solution to that of lower concentration.

Problem 3: Chromosomes are too spread to obtain complements.
Solution 3: One can avoid hypotonic treatment. None hypotonic preparation keeps more heterochromatinized segments in pachytene bivalents.

Problem 4: Detection of many hybridization signals on chromosomes but not in the background.
Solution 4: We recommend using more amount of competitor (sonicated genomic DNA). In many lepidopteran species, BAC- and/or fosmid-FISH without competitor detects many hybridization signals on chromosomes, because lepidopteran genome is composed of many scattered repetitive sequences, such that the clones of genomic library have possibility to carry them (Figure 6.10a). In this case more genomic competitor DNA is required for BAC- and/or fosmid-FISH. Amount of competitor may depend on species, chromosome condition, and clones used as probes. In *B. mori*, we have an experience with BAC-FISH using maximum 150 µg of competitor per preparation. Increasing the amount of competitor DNA is usually effective for solving the problem. When the clones contain too many repetitive sequences to suppress scattered signals, we have not yet recognized any effective solution.

Problem 5: Detection of many hybridization signals in the background.
Solution 5: The remnants of cytoplasm or nuclear protein in chromosome preparations may result the detection of high background signals. Contamination of bacteria also causes the high background signals. You should make clean chromosome preparation as much as possible

(see Section 6.2.4) and then store at under –30°C or –80°C. An additional high-stringency wash in posthybridization step may be effective; if one uses indirect-labeled probes, this would be more effective but it is ineffective in direct-labeling probes. However the better solution is to retry the FISH using another chromosome preparation.

Problem 6: Detection of scattered weak hybridization signals on chromosomes.

Solution 6: This problem occurs in FISH using the clones of genomic libraries (BAC and/or fosmid) or GISH/CGH using genomic DNA as probes. First it may be solved by increasing amount of probes and/or decreasing competitor DNAs. If there is no improvement by changed amount and/or purity of probes and competitor DNAs, you should distrust the labeling step by the nick translation system. Nick translation enzyme consists of DNase I and *Escherichia coli* polymerase I. If the activity rate of two enzymes differs from optimal value, negative effects appear on the efficiency of fluorescent-labeling and/or the final length of probes labeled. These may result in a reduction in intensity of hybridization signals. We recommend checking the probe length by agarose gel electrophoresis. If you recognize only very short DNA length, you should change the enzyme system.

Problem 7: No detection of hybridization signals.

Solution 7: First you should perform FISH using a positive control (optimal) probe. After you get exact signal by the probe, you have to check the origin, insert size, quantity, and purity of the labeled DNAs. If you do not find any problem, change chromosome preparations, reagents, and amount of competitor DNA.

6.3 DISCUSSION

6.3.1 Integration of Linkage Maps, Chromosomes, and Genomic Sequences

Needless to say, linkage analysis and genome sequencing cannot reveal chromosomal location by themselves. FISH analysis using BAC or fosmid clones harboring genetic markers or particular genomic sequences can clearly visualize chromosomal location of them (Figure 6.8), which identifies the coverage of genetic markers and genome sequence scaffolds to the whole chromosomes, as well as disorder of them derived from misinterpretation of the results. Small scale rearrangements within a chromosome are reported to occur frequently (d'Alençon et al. 2010; The *Heliconius* Genome

Consortium 2012), however, it is necessary to confirm that local rearrangements are not the artifacts because of technical errors in genome assembly of either species.

As described above, NGS-based-mapping methods make it easy to construct linkage maps of genetically uncharacterized species using heterozygous wild populations. However, the resolution of these maps is not so high because the number of analyzed progeny is relatively small compared with those using conventional mapping methods because of high cost. Thus, fine mapping by FISH analysis well completes this disadvantage especially for cold spots where crossing-over events rarely occur. Sequencing of BAC or fosmid clones used as probes is useful to design polymorphic markers for extended scale of linkage analysis.

In addition, polymorphisms genotyped by NGS-based-mapping methods are randomly selected in many cases, and there are not so many consensus markers among different experiments even if the same strains are used. We showed that BAC or fosmid probes containing well-conserved genes generate stable (Figure 6.8) and reproducible results against chromosomes of other species belonging to a different genus (Figure 6.10e through i) (Yasukochi et al. 2009; Sahara et al. 2013). Thus, this technique can be used to integrate the results of independent linkage analyses including related species.

6.3.2   Chromosome Organization and Evolution

Lepidopteran species have variations of karyotype (chromosome numbers) and sex chromosome systems. However classical cytogenetic methods could not reveal the karyotype and genome evolution in the species. Advanced cytogenetic mapping overcomes the cytogenetic problems.

CGH, GISH, and FISH with W-BAC or W-painting probes made from laser microdissection are powerful methods for not only identifying the W chromosomes but also supplying gross information of molecular composition in lepidopteran W chromosomes (Figures 6.9 and 6.10). CGH and GISH experiments suggested that the W chromosomes in some species are predominantly composed of accumulated repetitive sequences, also occurring but scattered in other autosomes and Z chromosomes (Figure 6.9a through e) (Traut et al. 1999; Sahara et al. 2003a; Yoshido et al. 2006). Furthermore, CGH and GISH also revealed that W chromosomes of some lepidopteran species consist of a large block of female-specific DNA (Figure 6.9f through j) (Vítková et al. 2007; Yoshido et al. 2013). FISH with W-painting probes showed that the W chromosomes share homologous sequences in only closely related species (Vítková et al. 2007; Yoshido et al.

2013). The divergence of W chromosomes in respective species suggested that the complete absence of meiotic recombination in lepidopteran females resulted in acceleration of specific molecular differentiation and degeneration of the W chromosomes in the respective species.

Lepidoptera have some variants of sex chromosome system besides WZ/ZZ. They are unique models for studying sex chromosome evolution in Lepidoptera. The combination of GISH and telomere-FISH is a simple, fast, and reliable method for resolution of multiple sex chromosome constitution in lepidopteran species (Yoshido et al. 2005b). This method can detect multiple sex chromosome constitution in all species examined. For example, this method has newly showed sex chromosome polymorphisms among closely related species, *S. cynthia* subspecies and *Orgyia* genus (Yoshido et al. 2005b). The derived sex chromosome constitutions in Lepidoptera are thought to come either from sex chromosome–autosome fusion or fission (Marec et al. 2010; Sahara et al. 2012). Advanced cytogenetic method has also proved an above hypothesis. Gene-based comparative FISH mapping has identified the gene order in autosomal segment of neo-sex chromosomes and showed that neo-sex chromosome evolutions occurred by autosome–sex chromosome fusions in a closely related species, *S. cynthia* subspecies (Figure 6.8) (Yoshido et al. 2011a,b) or tortricid moths (Nguyen et al. 2013). Hence, advanced cytogenetic techniques significantly contribute to find unique models for studying sex chromosome evolution and understanding its evolutional history in Lepidoptera.

The gene-based comparative FISH mapping also enables us to research chromosome (genome) rearrangements and the karyotype evolution among lepidopteran species. FISH mapping using respective BAC or fosmid clones containing orthologs of single *B. mori* genes has revealed that several Lepidoptera exhibit a synteny and conserved gene order. These do not depend on chromosome numbers in respective species, $n = 28$ with *M. sexta* (Sahara et al. 2007; Yasukochi et al. 2009), $n = 12–14$ with *S. cynthia* subspecies (Yoshido et al. 2011b), $n = 31$ with *Mamestra brassicae, Heliothis virescens,* and *Helicoverpa armigera* (Sahara et al. 2013). This is also shown in a study by linkage mapping in other lepidopteran species (Pringle et al. 2007; Beldade et al. 2009; Baxter et al. 2011; Van't Hof et al. 2013). The haploid chromosome number $n = 31$ is considered the basal number of Lepidoptera. Chromosome synteny with little rearrangement between species with $n = 31$ and others suggested that even though chromosome numbers had undergone fusion/fission events in lepidopteran evolutionary process, their gene orders were conserved in the corresponding chromosome parts.

Comparative mapping with the help of advanced cytogenetic techniques showed high degree of internal stability in the wide range of lepidopteran chromosomes (genomes). Lepidoptera have holokinetic chromosomes (Wolf 1996). Chromosomal rearrangements in Lepidoptera should be theoretically more flexible than those in organisms with common monocentric structure. Mutagenesis experiments in lepidopteran species support the idea that the risk for lethality by fragmentations of holokinetic chromosome is critically low (Fujiwara et al. 2000; Marec et al. 2001). Hence, it is more surprising that little inter- and/or intrachromosomal rearrangements by a gross view have been occurred in evolutionary history of Lepidoptera.

It is suggested the karyotype diversification contributes to speciation of lepidopteran species (Lukhtanov et al. 2005; Kandul et al. 2007; Yoshido et al. 2011a; Nguyen et al. 2013). Furthermore, sex chromosomes appear to play a disproportionally large role in speciation (Presgraves 2008; Qvarnström and Bailey 2009). Advanced cytogenetic methods provide us to analyze detailed chromosome and genome organization in Lepidoptera. Further researches should facilitate elucidating the role of lepidopteran chromosome evolution in their speciation.

REFERENCES

Abe, H., T. Fujii, and T. Shimada. 2010. Sex chromosomes and sex determination in *Bombyx mori*. In *Molecular Biology and Genetics of the Lepidoptera*, edited by M. R. Goldsmith and F. Marec, pp. 65–87. CRC Press, Boca Raton, FL.

Banno, Y., H. Fujii, Y. Kawaguchi et al. 2005. *A guide to the silkworm mutants: 2005 Gene name and gene symbol*. Institute of Genetic Resources, Kyushu University, Fukuoka, Japan.

Banno, Y., T. Shimada, Z. Kajiura, and H. Sezutsu. 2010. The silkworm—an attractive BioResource supplied by Japan. *Exp. Anim.* 59: 139–146.

Baxter, S. W., J. W. Davey, J. S. Johnston et al. 2011. Linkage mapping and comparative genomics using next-generation RAD sequencing of a non-model organism. *PLoS One* 6: e19315.

Beccaloni, G., M. Scoble, I. Kitching et al. (eds.). 2003. The Global Lepidoptera Names Index (LepIndex). World Wide Web electronic publication. http://www.nhm.ac.uk/entomology/lepindex, accessed June 20, 2013.

Beldade, P., S. V. Saenko, N. Pul, and A. D. Long. 2009. A gene-based linkage map for *Bicyclus anynana* butterflies allows for a comprehensive analysis of synteny with the lepidopteran reference genome. *PLoS Genet.* 5: e1000366.

Blin, N. and D. W. Stafford. 1976. A general method for isolation of high molecular weight DNA from eukaryotes. *Nucleic Acids Res.* 3: 2303–2308.

Brown, K. S. J., B. Von Schoultz, and E. Suomalainen. 2004. Chromosome evolution in Neotropical Danainae and Ithomiinae (Lepidoptera). *Hereditas* 141: 216–236.

Butenandt, A., R. Beckmann, and E. Hecker. 1961. Uber den sexuallockstoff des seidenspinners.1. der biologische test und die isolierung des reinen sexuallockstoffes [On the sex-attractant of silk-moths. I. The biological test and the isolation of the pure sex-attractant bombykol]. *Hoppe-Seylers Zeitschrift fur Physiologische Chemie.* 324: 71–83.

d'Alençon, E., H. Sezutsu, F. Legeai, et al. 2010. Extensive synteny conservation of holocentric chromosomes in Lepidoptera despite high rates of local genome rearrangements. *Proc. Natl. Acad. Sci. U S A* 107: 7680–7685.

Davey, J. W., P. A. Hohenlohe, P. D. Etter et al. 2011. Genome-wide genetic marker discovery and genotyping using next-generation sequencing. *Nature Rev. Genet.* 12: 499–510.

de Lesse, H. 1970. Les nombres de chromosomes dans le groupe de Lysandra argester et leur incidence sur la taxonomie [Chromosome numbers in the *Lysandra argester* group and their impact on taxonomy]. *Bull. Soc. Entomol. Fr.* 75: 64–68.

De Prins, J. and K. Saitoh. 2003. Karyology and sex determination. In *Lepidoptera, Moths and Butterflies. 2. Morphology, Physiology, and Development*, edited by N. P. Kristensen, pp. 449–468. Walter de Gruyter, Berlin, Germany, and New York.

Dopman, E. B., S. M. Bogdanowicz, and R. G. Harrison. 2004. Genetic mapping of sexual isolation between E and Z pheromone strains of the European corn borer (*Ostrinia nubilalis*). *Genetics* 167: 301–309.

Fujiwara, H., Y. Nakazato, S. Okazaki, and O. Ninaki. 2000. Stability and telomere structure of chromosomal fragments in two different mosaic strains of the silkworm, *Bombyx mori. Zool. Sci.* 17: 743–750.

Fuková, I., W. Traut, M. Vítková et al. 2007. Probing the W chromosome of the codling moth, *Cydia pomonella*, with sequences from microdissected sex chromatin. *Chromosoma* 116: 135–145.

Fukuda, T., M. Suto, T. Kameyama, and S. Kawasugi. 1962. Synthetic diet for silkworm raising. *Nature* 196: 53–54.

Glaser, R. W. 1917. "Ringer" solutions and some notes on the physiological basis of their ionic composition. *Comp. Biochem. Physiol.* 2: 241–289.

The *Heliconius* Genome Consortium. 2012. Butterfly genome reveals promiscuous exchange of mimicry adaptations among species. *Nature* 487: 94–98.

Ijdo, J. W., R. A. Wells, A. Baldini, and S. T. Reeders. 1991. Improved telomere detection using a telomere repeat probe (TTAGGG)$_n$ generated by PCR. *Nucleic Acids Res.* 19: 4780.

Imai, K., T. Konno, Y. Nakazawa et al. 1991. Isolation and structure of diapause hormone of the silkworm, *Bombyx mori. Proc. Jap. Acad. B* 67: 98–101.

The International Silkworm Genome Consortium. 2008. The genome of a lepidopteran model insect, the silkworm *Bombyx mori. Insect Biochem. Mol. Biol.* 38: 1036–1045.

Jiggins, C. D., J. Mavarez, M. Beltrán et al. 2005. A genetic linkage map of the mimetic butterfly, *Heliconius melpomene. Genetics* 171: 557–570.

Kandul, N. P., V. A. Lukhtanov, and N. E. Pierce. 2007. Karyotypic diversity and speciation in Agrodiaetus butterflies. *Evolution* 61: 546–559.

Kim, U-J., H. Shizuya, P. J. de-Jong, B. Birren, and M. I. Simon. 1992. Stable propagation of cosmid sized human DNA inserts in an F factor based vector. *Nucleic Acids Res.* 20: 1083–1085.

Kubickova, S., H. Cernohorska, P. Musilova, and J. Rubes. 2002. The use of laser microdissection for the preparation of chromosome-specific painting probes in farm animals. *Chromosome Res.* 10: 571–577.

Lukhtanov, V. A. 2000. Sex chromatin and sex chromosome systems in nonditrysian Lepidoptera (Insecta). *J. Zool. Syst. Evol. Res.* 38: 73–79.

Lukhtanov, V. A., N. P. Kandul, J. B. Plotkin et al. 2005. Reinforcement of pre-zygotic isolation and karyotype evolution in *Agrodiaetus* butterflies. *Nature* 436: 385–389.

Marec, F., K. Sahara, and W. Traut. 2010. Rise and fall of the W chromosome in Lepidoptera. In *Molecular Biology and Genetics of the Lepidoptera*, edited by M. R. Goldsmith and F. Marec, pp. 49–63. CRC Press, Boca Raton, FL.

Marec, F., A. Tothová, K. Sahara, and W. Traut. 2001. Meiotic pairing of sex chromosome fragments and its relation to atypical transmission of a sex-linked marker in *Ephestia kuehniella* (Insecta: Lepidoptera). *Heredity* 87: 659–671.

Miao, X. X., S. J. Xub, M. H. Li et al. 2005. Simple sequence repeat-based consensus linkage map of *Bombyx mori*. *Proc. Natl. Acad. Sci. U S A* 102: 16303–16308.

Mita, K., M. Kasahara, S. Sasaki et al. 2004. The genome sequence of silkworm, *Bombyx mori*. *DNA Res.* 11: 27–35.

Nguyen, P., M. Sýkorová, J. Šíchová et al. 2013. Neo-sex chromosomes and adaptive potential in tortricid pests. *Proc. Natl. Acad. Sci. U S A* 110: 6931–6936.

Okazaki, S., H. Ishikawa, and H. Fujiwara. 1995. Structural analysis of TRAS1, a novel family of telomeric repeat-associated retrotransposons in the silkworm, *Bombyx mori*. *Mol. Cell. Biol.* 15: 4545–4552.

Presgraves, D. C. 2008. Sex chromosomes and speciation in *Drosophila*. *Trends Genet.* 24: 336–343.

Pringle, E. G., S. W. Baxter, C. L. Webster et al. 2007. Synteny and chromosome evolution in the Lepidoptera: Evidence from mapping in *Heliconius melpomene*. *Genetics* 177: 417–426.

Promboon, A., T. Shimada, H. Fujiwara, and M. Kobayashi. 1995. Linkage map of random amplified polymorphic DNAs (RAPDs) in the silkworm, *Bombyx mori*. *Genet. Res.* 66: 1–7.

Qvarnström, A. and R. I. Bailey. 2009. Speciation through evolution of sex-linked genes. *Heredity* 102: 4–15.

Rego, A. and F. Marec. 2003. Telomeric and interstitial telomeric sequences in holokinetic chromosomes of Lepidoptera: Telomeric DNA mediates association between postpachytene bivalents in achiasmatic meiosis of females. *Chromosome Res.* 11: 681–684.

Robinson, R. 1971. *Lepidoptera Genetics*. Pergamon, Oxford, United Kingdom.

Sahara, K., F. Marec, U. Eickhoff, and W. Traut. 2003a. Moth sex chromatin probed by comparative genomic hybridization (CGH). *Genome* 46: 339–342.

Sahara, K., F. Marec, and W. Traut. 1999. TTAGG telomeric repeats in chromosomes of some insects and other arthropods. *Chromosome Res.* 7: 449–460.

Sahara, K., A. Yoshido, N. Kawamura et al. 2003b. W-derived BAC probes as a new tool for identification of the W chromosome and its aberrations in *Bombyx mori*. *Chromosoma* 112: 48–55.

Sahara, K., A. Yoshido, F. Marec et al. 2007. Conserved synteny of genes between chromosome 15 of *Bombyx mori* and a chromosome of *Manduca sexta* shown by five-color BAC-FISH. *Genome* 50: 1061–1065.

Sahara, K., A. Yoshido, F. Shibata et al. 2013. FISH identification of *Helicoverpa armigera* and *Mamestra brassicae* chromosomes by BAC and fosmid probes. *Insect Biochem. Mol. Biol.* 43: 644–653.

Sahara, K., A. Yoshido, and W. Traut. 2012. Sex chromosome evolution in moths and butterflies. *Chromosome Res.* 20: 83–94.

Shi, J., D. G. Heckel, and M. R. Goldsmith. 1995. A genetic linkage map for the domesticated silkworm, *Bombyx mori*, based on restriction fragment length polymorphisms. *Genet. Res.* 66: 109–126.

Shibata, F., K. Sahara, Y. Naito, and Y. Yasukochi. 2009. Reprobing of multicolor FISH preparations of lepidopteran chromosomes. *Zool. Sci.* 26: 187–190.

Shizuya, H., B. Birren, U. J. Kim et al. 1992. Cloning and stable maintenance of 300-kilobase-pair fragments of human DNA in *Escherichia coli* using an F-factor-based vector. *Proc. Natl. Acad. Sci. U S A* 89: 8794–8797.

Takahashi, H., S. Okazaki, and H. Fujiwara. 1997. A new family of site-specific retrotransposons, SART1, is inserted into telomeric repeats of the silkworm, *Bombyx mori*. *Nucleic Acids Res.* 25: 1578–1584.

Tamura, T., C. Thibert, C. Royer et al. 2000. Germline transformation of the silkworm *Bombyx mori* L. using a *piggyBac* transposon-derived vector. *Nature Biotech.* 18: 81–84.

Tan, Y., C. Wan, Y. Zhu et al. 2001. An amplified fragment length polymorphisms map of the silkworm. *Genetics* 157: 1277–1284.

Toyama, K. 1906. Studies on the hybridology of insects. I. On some silkworm crosses, with special reference to Mendel's law of heredity. *Bulletin of the College of Agriculture, Tokyo Imperial University* 7: 259–393.

Traut, W. 1976. Pachytene mapping in the female silkworm, *Bombyx mori* L. (Lepidoptera). *Chromosoma* 58: 275–284.

Traut, W. and F. Marec. 1996. Sex chromatin in Lepidoptera. *Q. Rev. Biol.* 71: 239–256.

Traut, W. and F. Marec. 1997. Sex chromosome differentiation in some species of Lepidoptera (Insecta). *Chromosome Res.* 5: 283–291.

Traut, W., K. Sahara, and F. Marec. 2007. Sex chromosomes and sex determination in Lepidoptera. *Sex Dev.* 1: 332–346.

Traut, W., K. Sahara, T. D. Otto, and F. Marec. 1999. Molecular differentiation of sex chromosomes probed by comparative genomic hybridization. *Chromosoma* 108: 173–180.

Van't Hof, A. E., F. Marec, I. J. Saccheri, P. M. Brakefield, and B. J. Zwaan. 2008. Cytogenetic characterization and AFLP-based genetic linkage mapping for the butterfly *Bicyclus anynana*, covering all 28 karyotyped chromosomes. *PLoS One* 3: e3882.

Van't Hof, A. E., P. Nguyen, M. Dalíková et al. 2013. Linkage map of the peppered moth, *Biston betularia* (Lepidoptera, Geometridae): a model of industrial melanism. *Heredity* 110: 283–295.

Vítková, M., I. Fuková, S. Kubíčková, and F. Marec. 2007. Molecular divergence of the W chromosomes in pyralid moths (Lepidoptera). *Chromosome Res.* 15: 917–930.

Wang, B. and A. H. Porter. 2004. An AFLP-based interspecific linkage map of sympatric, hybridizing Colias butterflies. *Genetics* 168: 215–225.

Winter, C. B. and A. H. Porter. 2010. AFLP linkage map of hybridizing swallowtail butterflies, *Papilio glaucus* and *Papilio canadensis*. *J. Hered.* 101: 83–90.

Wolf, K. W. 1996. The structure of condensed chromosomes in mitosis and meiosis of insects. *Int. J. Insect Morphol. Embryol.* 25: 37–62.

Xia, Q., Y. Guo, Z. Zhang et al. 2009. Complete resequencing of 40 genomes reveals domestication events and genes in silkworm (*Bombyx*). *Science* 326: 433–436.

Xia, Q., Z. Zhou, C. Lu et al. 2004. A draft sequence for the genome of the domesticated silkworm (*Bombyx mori*). *Science* 306: 1937–1940.

Yamamoto, K., J. Narukawa, K. Kadono-Okuda et al. 2006. Construction of a single nucleotide polymorphism linkage map for the silkworm, *Bombyx mori*, based on bacterial artificial chromosome end sequences. *Genetics* 173: 151–161.

Yamamoto, T. 2000. Silkworm strains. In *Strain Maintenance and Databank for Life Science*, edited by N. Nakatsuji, pp. 45–49. Kyoritsu Shuppan, Tokyo, Japan.

Yasukochi, Y. 1998. A dense genetic map of the silkworm, *Bombyx mori*, covering all chromosomes based on 1018 molecular markers. *Genetics* 150: 1513–1525.

Yasukochi, Y. 2002. PCR-based screening for bacterial artificial chromosome libraries. *Methods Mol. Biol.* 192: 401–410.

Yasukochi, Y., L. A. Ashakumary, K. Baba, A. Yoshido, and K. Sahara. 2006. A second-generation integrated map of the silkworm reveals synteny and conserved gene order between lepidopteran insects. *Genetics* 173: 1319–1328.

Yasukochi, Y., M. Tanaka-Okuyama, M. Kamimura et al. 2011. Isolation of BAC clones containing conserved genes from libraries of three distantly related moths: A useful resource for comparative genomics of Lepidoptera. *J. Biomed. Biotechnol.* 2011: 165894.

Yasukochi, Y., M. Tanaka-Okuyama, F. Shibata et al. 2009. Extensive conserved synteny of genes between the karyotypes of *Manduca sexta* and *Bombyx mori* revealed by BAC-FISH mapping. *PLoS One* 4: e7465.

Yoshido, A., H. Bando, Y. Yasukochi, and K. Sahara. 2005a. The *Bombyx mori* karyotype and the assignment of linkage groups. *Genetics* 170: 675–685.

Yoshido, A., F. Marec, and K. Sahara. 2005b. Resolution of sex chromosome constitution by genomic in situ hybridization and fluorescence in situ hybridization with (TTAGG)$_n$ telomeric probe in some species of Lepidoptera. *Chromosoma* 114: 193–202.

Yoshido, A., K. Sahara, F. Marec, and Y. Matsuda. 2011a. Step-by-step evolution of neo-sex chromosomes in geographical populations of wild silkmoths, *Samia cynthia* ssp. *Heredity* 106: 614–624.

Yoshido, A., J. Šíchová, S. Kubíčková, F. Marec, and K. Sahara. 2013. Rapid turnover of the W chromosome in geographical populations of wild silkmoths, *Samia cynthia* ssp. *Chromosome Res.* 21: 149–164.

Yoshido, A., Y. Yamada, and K. Sahara. 2006. The W chromosome detection in several lepidopteran species by genomic in situ hybridization (GISH). *J. Insect Biotech. Sericol.* 75: 147–151.

Yoshido, A., Y. Yasukochi, F. Marec, H. Abe, and K. Sahara. 2007. FISH analysis of the W chromosome in *Bombyx mandarina* and several other species of Lepidoptera by means of *B. mori* W-BAC probes. *J. Insect Biotech. Sericol.* 76: 1–7.

Yoshido, A., Y. Yasukochi, and K. Sahara. 2011b. *Samia cynthia* versus *Bombyx mori*: Comparative gene mapping between a species with a low-number karyotype and the model species of Lepidoptera. *Insect Biochem Mol Biol* 41: 370–377.

You, M., Z. Yue, W. He et al. 2013. A heterozygous moth genome provides insights into herbivory and detoxification. *Nature Genet.* 45: 220–225.

Zhan, S., C. Merlin, J. L. Boore, and S. M. Reppert. 2011. The monarch butterfly genome yields insights into long-distance migration. *Cell* 147: 1171–1185.

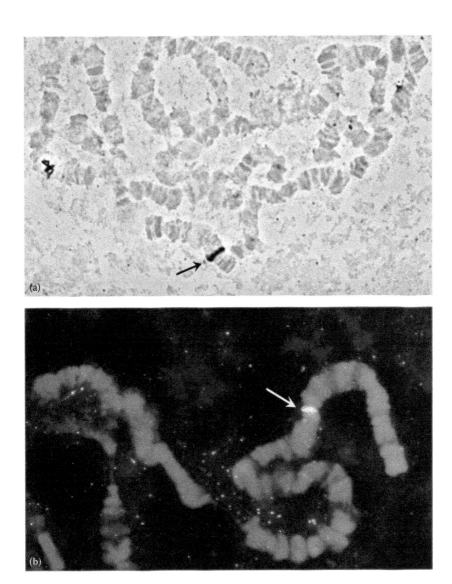

FIGURE 1.4 (a) In situ hybridization on the polytene chromosomes of *Bactrocera oleae* using a homologous probe (*ovo* gene, cDNA clone). (b) Fluorescence in situ hybridization on the polytene chromosomes of a transgenic strain of *Ceratitis capitata* using as marker the DsRed. Arrows indicate hybridization signals.

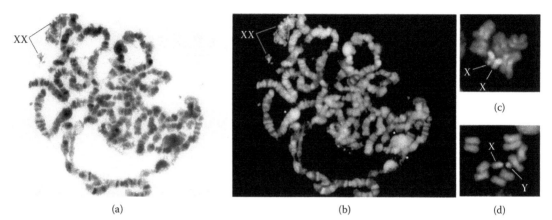

FIGURE 1.5 Chromosome painting of sex chromosomes in *Bactrocera oleae*. Chromosomes were counterstained with DAPI (blue); hybridization signals of the X- or Y-chromosome-derived probes are red. (a) Phase contrast image of female polytene nucleus before hybridization. Lines indicate the granular network corresponding to the X chromosomes. (b) The same polytene nucleus as in (a) after fluorescence in situ hybridization with the X-painting probe. (c) Female metaphase showing blocks of strong hybridization signals of the X-painting probe on the X chromosomes. (d) Male metaphase showing strong hybridization signals of the Y-painting probe covering the entire Y chromosome. (From Drosopoulou, E. et al., *Genetica.*, 140, 169–180, 2012.)

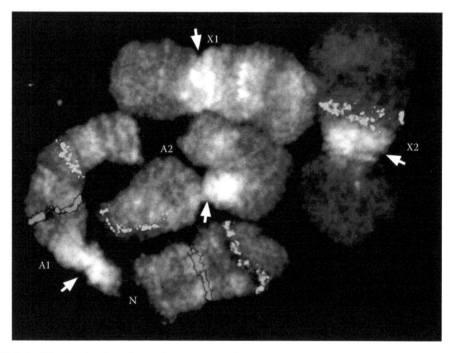

FIGURE 2.1 The Hessian fly salivary gland polytene chromosomes (8 S chromosomes). Shown is an example of in situ hybridization of Hessian fly polytene chromosomes A1, A2, X1, and X2. As in most dipteran genomes, the Hessian fly diploid chromosome number ($2n = 8$) is low and the homologs are often paired in diploid polytene nuclei, as they are here. Four biotin-labeled BAC clones (green) and two digoxigenin-labeled BAC clones (red) are visible on the chromosomes. The position of the nucleolus (N) on chromosome A1 is indicated. Centromeric heterochromatin is visible as brighter staining DNA near constrictions (arrows) that correspond to the chromosome centromeres.

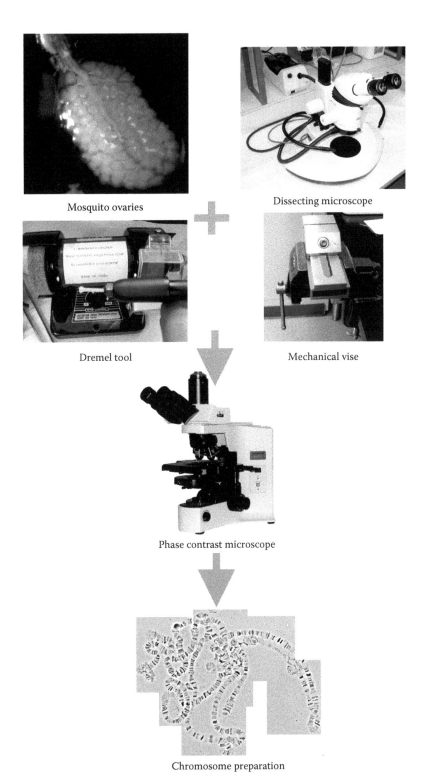

Mosquito ovaries

Dissecting microscope

Dremel tool

Mechanical vise

Phase contrast microscope

Chromosome preparation

FIGURE 4.1 Schematic representation of high-pressure chromosome preparation. Mosquito ovaries are shown at the correct stage of development. (From George, P. et al., *J. Vis. Exp.*, (64), e4007 10.3791/4007, 2012.)

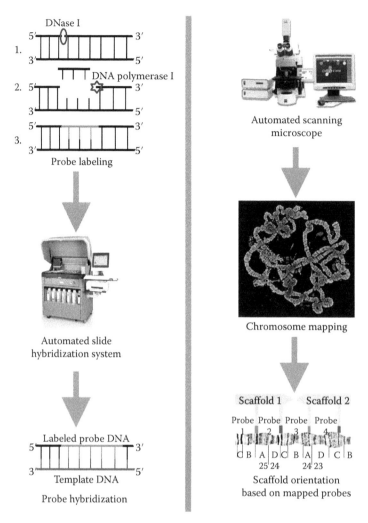

FIGURE 4.2 A scheme representing automated fluorescent in situ hybridization, slide scanning, and chromosome mapping of genomic scaffolds. (From George, P. et al., *J. Vis. Exp.*, (64), e4007 10.3791/4007, 2012.)

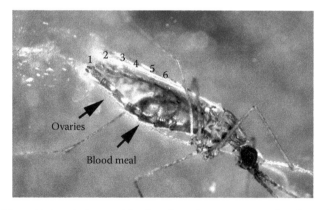

FIGURE 4.3 A half-gravid *Anopheles gambiae* female at the correct stage for dissection. Arrows show the light area occupied by the developing ovaries and the dark area with a blood meal. The numbers indicate dorsal abdominal segments.

FIGURE 4.4 Stages of ovarian development in *Anopheles*. (a) Ovaries at the Christophers' II stage. (b) Ovaries at the Christophers' III stage (the correct stage for chromosome preparations). (c) Ovaries at the Christophers' IV stage.

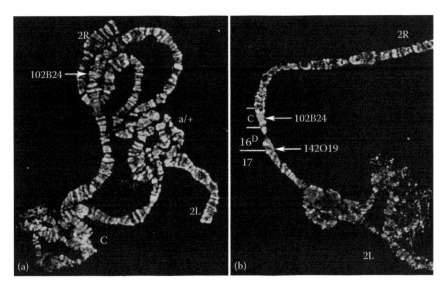

FIGURE 4.7 Fluorescent in situ hybridization (FISH) of bacterial artificial chromosome (BAC) clones to polytene chromosomes of *Anopheles gambiae*. (a) Hybridization of 102B24 (red signal) with the 2R arm. (b) Dual-color FISH of 102B24 (red signal) and 142O19 (blue signal) to subdivisions 16C and 16D of the stretched 2R arm, respectively. Arrows indicate signals of hybridization of BAC clones labeled with Cy3 (red) and Cy5 (blue). (c) The centromeric region. a/+ shows the heterozygote 2La inversion. Chromosomes were counterstained with the fluorophore YOYO-1. (From George, P. et al., *J. Vis. Exp.*, (64), e4007 10.3791/4007, 2012.)

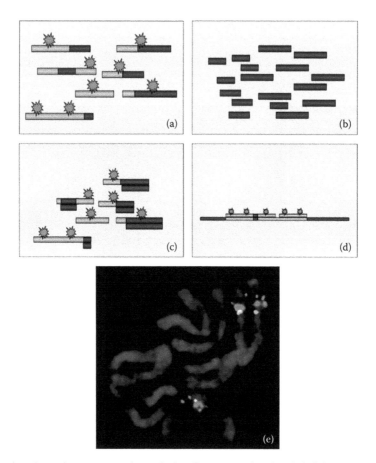

FIGURE 4.8 A schematic representation of the fluorescent in situ hybridization (FISH) procedure. (a) Preparation of fluorescently labeled DNA probe. (b) Preparation of unlabeled repetitive DNA fraction. (c) Blocking unspecific hybridization of the probe with unlabeled repetitive DNA fraction. (d) Hybridization of fluorescently labeled DNA probe with chromosomes. (e) Visualization of FISH signals on mitotic chromosomes. (From Timoshevskiy, V.A. et al., *J. Vis. Exp.,* (67), e4215 10.3791/4215, 2012.)

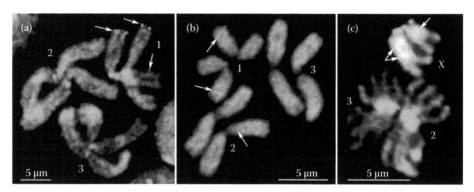

FIGURE 4.12 Examples of fluorescent in situ hybridization (FISH) result with mitotic chromosomes. (a) FISH of bacterial artificial chromosome (BAC) clones with chromosomes of *Aedes aegypti.* (b) FISH of BAC clones with chromosomes of *Culex quinquefasciatus.* (c) FISH of intergenic spacer rDNA with chromosomes of *Anopheles gambiae.* 1, 2, and 3 are numbers of chromosomes; X—female sex chromosome in *Anopheles gambiae.* (From Timoshevskiy, V.A. et al., *J. Vis. Exp.,* (67), e4215 10.3791/4215, 2012.)

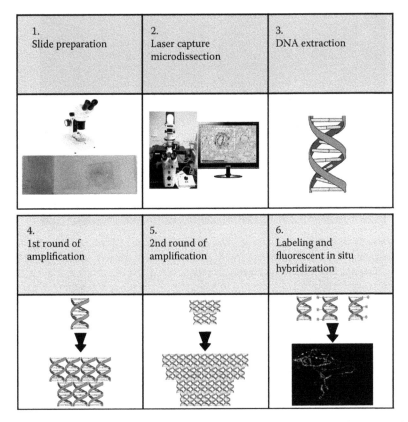

1. Slide preparation	2. Laser capture microdissection	3. DNA extraction
4. 1st round of amplification	5. 2nd round of amplification	6. Labeling and fluorescent in situ hybridization

FIGURE 4.13 Schematic representation of the experimental procedures toward the preparation of chromosome paints. (From George, P. et al., *J. Vis. Exp.,* (83), e51173, 2014.)

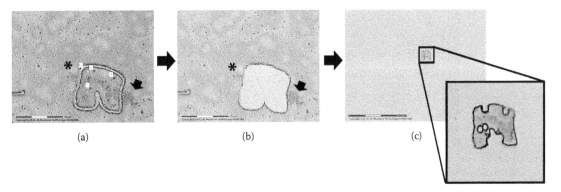

(a)   (b)   (c)

FIGURE 4.14 The major steps in chromosome microdissection. (a) Laser-assisted cutting of the chromosomal region of interest through the membrane. (b) The membrane with a hole after the catapulting is performed. (c) The view of the catapulted piece of the membrane with a chromosomal segment in it attached to the adhesive cap. The arrow indicates the heterochromatin of the X chromosome that remained on the slide. The asterisk shows a piece of another chromosome that remained on the slide. (From George, P. et al., *J. Vis. Exp.,* (83), e51173, 2014.)

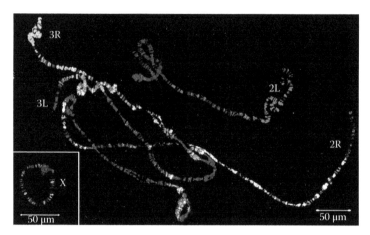

FIGURE 4.16 Painting of polytene chromosomes from ovarian nurse cells of *Anopheles gambiae* using four probes generated from microdissected material. The X chromosome is labeled in orange (Cy3) by nick translation of the REPLI-g material. The 2R arm is labeled in yellow (Cy5); the 2L arm is in red (Cy3); the 3R arm is labeled in green (fluorescein); the 3L arm is labeled in a mixture of red (Cy3) and yellow (Cy5). Autosomes are labeled with the WGA3 amplification kit. Chromatin is stained in blue (DAPI). Chromosome names are placed near telomeric regions. (From George, P. et al., *J. Vis. Exp.*, (83), e51173, 2014.)

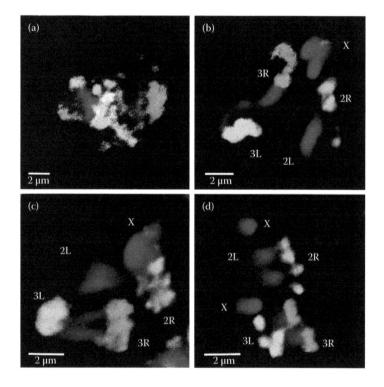

FIGURE 4.17 Painting of nonpolytene chromosomes from larval imaginal disks (IDs) of *Anopheles gambiae*. (a) Interphase nucleus. (b) Prophase chromosomes. (c) Prometaphase chromosomes. (d) Metaphase chromosomes. Three probes were generated from microdissected material labeled by WGA3. The 2R arm is labeled in green (fluorescein); the 2L arm is unlabeled; the 3R arm is in pink, a mixture of red (Cy3) and orange (Cy5); the 3L arm is labeled in orange (Cy5). The X chromosome has a red label corresponding to the 18S rDNA probe. Chromatin is stained in blue (DAPI). Brightly stained regions of chromosomes correspond to the heterochromatin. (From George, P. et al., *J. Vis. Exp.*, (83), e51173, 2014.)

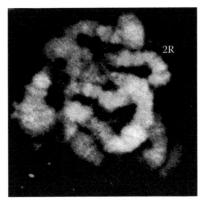

FIGURE 4.18 Whole-mount three-dimensional fluorescent in situ hybridization performed on *Anopheles gambiae* ovarian nurse cells. The probe is labeled in Cy3 (depicted in red) and was made from a microdissected 2R chromosome arm. Chromatin is stained with DAPI and is depicted by cyan pseudo-coloring. (From George, P. et al., *J. Vis. Exp.*, (83), e51173, 2014.)

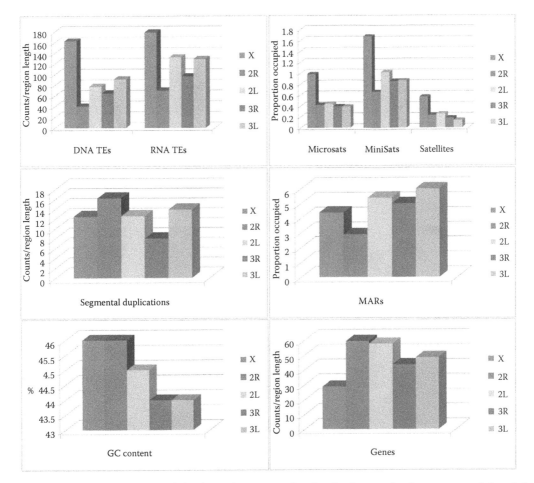

FIGURE 4.21 Median values of density and coverage of molecular features in chromosomes of *Anopheles gambiae*. Counts per 1 Mb are given for DNA transposable elements (DNA TEs), RNA TEs, regions involved in segmental duplications (SDs), and genes. Percentage of region length occupied per 1 Mb is indicated for microsatellites, minisatellites, satellites, and matrix-associated regions (MARs). (From Xia, A. et al., *PLoS ONE*, 5(5), e10592, 2010.)

FIGURE 5.1 Dissection of tissue useful for chromosomal analysis. (a) *Tenebrio molitor* exemplar pinned on a Petri dish containing insect saline solution; (b) dorsal view of the abdomen showing the opened elytra (1) and the membranous wing pair (2); (c) dorsal view of the abdomen without the membranous wings; (d) opened abdomen indicating the testis position (dashed yellow circle); (e) testis before fixation, the arrows indicate two follicular testis; (f) individualized testicular follicles after fixation in modified Carnoy's solution.

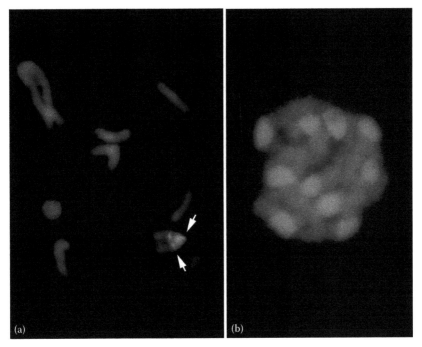

FIGURE 5.5 Fluorochrome staining showing (a) CMA$_3$ positive blocks (G+C rich) in diacinesis of *Dichotomius laevicollis* (arrows) and (b) DAPI positive blocks (A+T rich) in initial meiosis I of *Zophobas morio*.

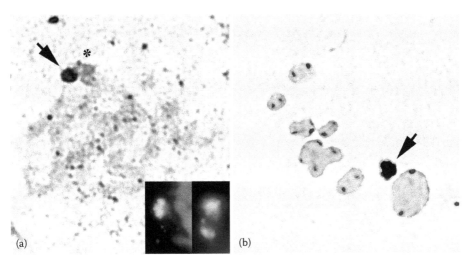

FIGURE 5.6 Silver nitrate staining in an initial meiotic cell (Zygotene) and metaphase I of (a) *Euphoria* spp. and (b) *Dichotomius semisquamosus*, respectively. In (a) the arrow points to the sex bivalent and the asterisk to the nucleolar material associated with these chromosomes, and in (b) the arrow shows the sex bivalent (Xy$_p$) impregnated by the silver nitrate, a common pattern in Coleoptera. In (a) the inserts show the position of the 18S rDNA clusters in the sex bivalent at the initial cell (left) and metaphase I (right). Note in (b) the staining of the pericentromeric heterochromatin/kinetochore.

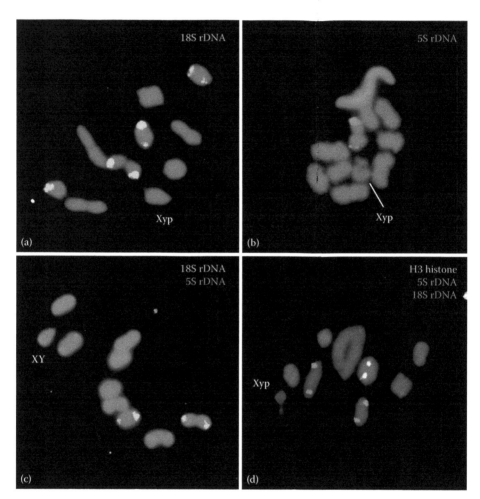

FIGURE 5.7 FISH mapping of distinct multigene families in metaphase I obtained from four species of Coleoptera. (a) *Ontherus sulcator*, (b) *Digitonthophagus gazella*, (c) *Coprophanaeus dardanus*, and (d) *Dichotomius geminatus*. Each probe used is directly indicated in the cells. The sex chromosomes are also indicated.

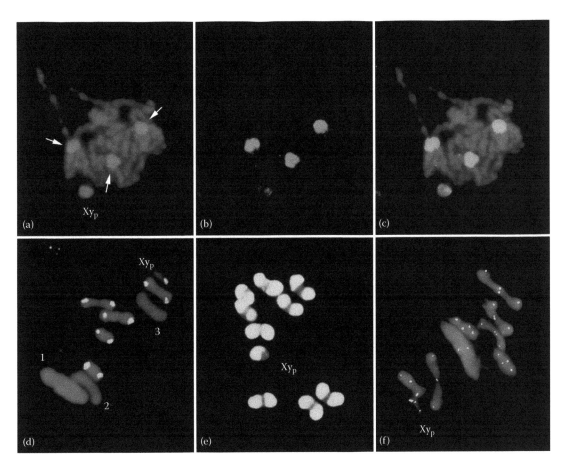

FIGURE 5.8 Chromosomal mapping through FISH of the C_0t-1 DNA fraction in three species of beetles. Initial meiosis from *Dichotomius sericeus* (a) DAPI, (b) C_0t-1 DNA fraction, and (c) merge. Metaphases I from (d) *Dichotomius bos* and (e) *Coprophanaeus cyanescens*. (f) FISH using C_0t-1 DNA fraction obtained from *Dichotomius geminatus* genome in chromosomes in metaphase I of *Dichotomius bos*. Note: in (d, e) the C_0t-1 DNA signals are mainly in the heterochromatic blocks and in (f) the signals restrict to terminal regions of chromosomes. The arrows in (a) indicate the heterochromatic regions.

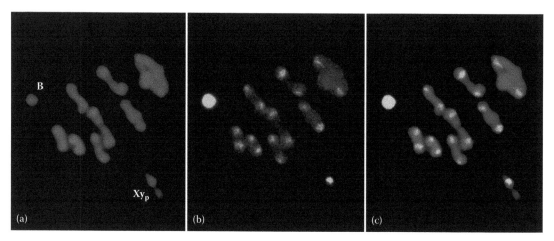

FIGURE 5.9 FISH mapping in metaphase I using the B chromosome microdissected from *Dichotomius sericeus*: (a) DAPI, (b) B probe, (c) merge. The B and the sex bivalent are indicated.

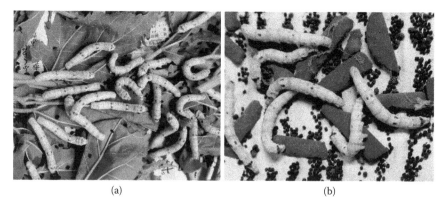

FIGURE 6.2 *Bombyx mori* rearing by (a) mulberry leaves and (b) artificial diets.

FIGURE 6.3 Sex discrimination in the larval and pupal stages. (a) A male larva can be discriminated by a Herald's gland (black arrow) appearing in ventral tale part; (b) a female shows Ishiwata's fore (F) and hinder (H) glands in the similar part of the early stage of last instar larva; (c) a female and a male of a sex-limited strain. Females of the strain have a second chromosome fragment carrying normal marking (+^{p}) locus onto the W chromosome. Both females and males have second chromosome pairs with plain (*p*) loci. In the pupal stage, one can easily discriminate (d) a male from (e) a female, by the different morphology pointed out by white arrows.

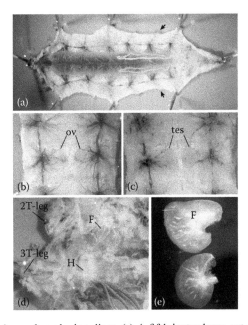

FIGURE 6.4 *Bombyx mori* gonads and wing discs. (a) A fifth instar larva cut open from ventral side, arrows indicate third abdominal legs where is a marker of the gonad segment; (b) positions of a pair of ovary and (c) testis; (d) cut-opened thorax (left part from ventral view); and (e) fore- and hind-wing discs. 2T-leg, second thoracic leg; 3T-leg, third thoracic leg; F, fore-wing disc; H, hinder wing disc; ov, ovary; tes, testis.

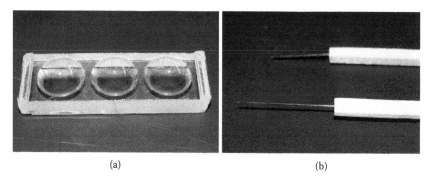

(a)   (b)

FIGURE 6.5   (a) A 3-hole glass and (b) dissecting insect-pin stacked to wooden chopstick.

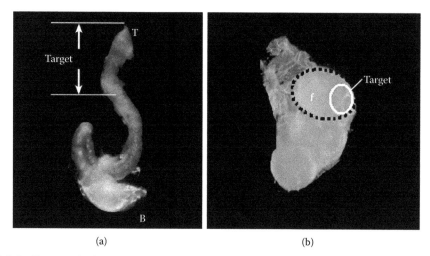

(a)   (b)

FIGURE 6.6   Target cells for chromosome preparation. (a) An ovariole and a part harboring cells with adequate stages (target) for preparation; (b) a testis with 3 follicles. A representative follicle surrounded with broken oval. A part in the oval shows the target.

(1) Selection of genes for comparative mapping between *B. mori* and *S. cynthia*.

(2) Using database, design of STS primer sets for *S. cynthia* orthologs of *B. mori* genes.

(3) By PCR using STS primers, screening of clones containing target sequences.

B. mori		S. cynthia	
Gene symbol	Chromosome position	Database	Fosmid clone
Kettin	Chr 1: 6,505,696-6,533,895	AB543309	45A6
imp	Chr 1: 11,424,630-11,431,079	S06A01NCLL0014_I05	19B8
HP	Chr 8: 205,759-211,533	S13A01NGRL0004_C20	56J8
GRP2	Chr 8: 4,927,144-4,928,289	S06A01NCLL0014_M19	32B23
BR-C	Chr 8: 18,077,270-18,081,471	AB564751	56J22
P109	Chr 12: 4,989,327-5,006,556	I10A02NGRL0003_B23	21P14
JDCP	Chr 12: 7,728,258-7,729,559	S13A02NGRL0001_G21	44E23
XDH1	Chr 12: 8,493,509-8,511,254	I09A02NGRL0003_I02	14J3

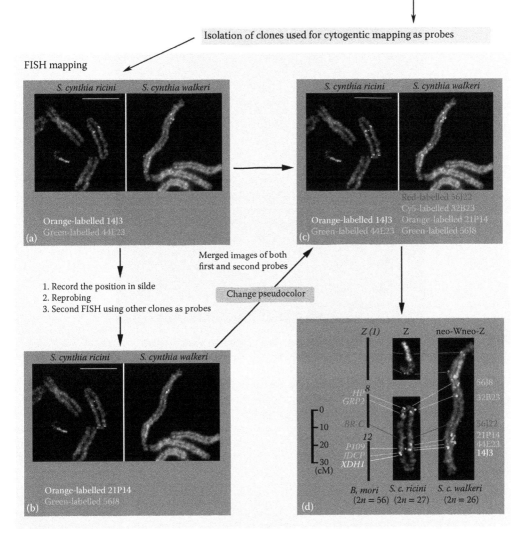

FIGURE 6.8 Flowchart of comparative gene-based FISH mapping between *Bombyx mori* and *Samia cynthia* ssp.

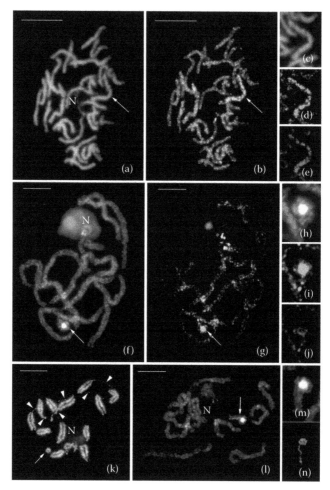

FIGURE 6.9 FISH images using repetitive sequences as probes in some lepidopteran species, (a–e) *Manduca sexta*, (f–j and l–n) *Samia cynthia pryeri*, (k) *S. cynthia ricini*. Respective chromosomes were counterstained with DAPI (white). N, nucleolus; Bar = 10 μm. Comparative genomic hybridization (CGH) between respective sexes in female of (a–e) *M. sexta* and (f–j) *S. cynthia pryeri*. Female-derived genomic DNA probes were labeled with Green-dUTP (green), male-derived genomic DNA probes with Cy3-dUTP (red). (a and f) DAPI images. (b and g) Merged images of both probes. (c–e and h–j) A detail of the WZ bivalents. Arrow represents WZ bivalents. (k) FISH with the Cy3-labeled (TTAGG)n telomeric probe (red signals) in female of *S. cynthia ricini*. Arrow represents Z chromosome univalent. (l-n) FISH with W chromosome painting probes in female of *S. cynthia pryeri*. Pachytene complement (l). DAPI image (m) and W-probe (n) in a detail of the WZ bivalent. Arrow represents WZ bivalent.

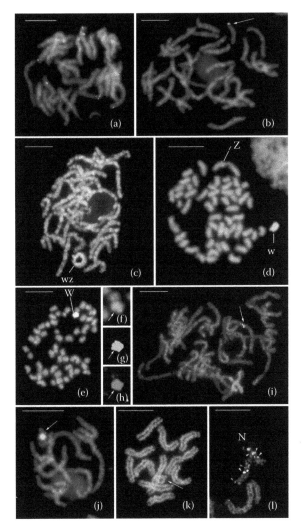

FIGURE 6.10 BAC- or fosmid-FISH and gene mapping in lepidopteran species, (a–d) *B. mori*, (e–h) *B. mandarina*, (i) *Helicoverpa armigera*, (j–l) *S. cynthia* subspecies. Respective chromosomes were counterstained with DAPI (white). Bar = 10 μm. (a) FISH with *B. mori* chromosome 2–derived BAC (9D6C) probe without competitor and (b) with competitor, arrow represents *B. mori* chromosome 2; (c) FISH with *B. mori* W and Z chromosome-derived-BAC probes in *B. mori* females, pachytene complement and (d) mitotic metaphase complement, Green-labeled probe of the 19L6H clone (green signal) and Cy3-labeled probe of the 9A5H clone (red signal), respectively; (e) FISH with Green-labeled probe of *B. mori* 19L6H clone (green signal) and Cy3-labeled probe of *B. mandarina* female genomic DNA (red signal) in *B. mandarina* female mitotic metaphase complement; (f–h) a detail of the W chromosome, arrow represents the W chromosome: (i) FISH with Red-labeled probe of *Heliothis virescens* 55I09 (red signal) and Green-labeled probe of *Helicoverpa armigera* 26P10 (green signal) in *H. armigera* pachytene complement; FISH with Cy3-labeled *S. cynthia* ortholog of *B. mori* Z-chromosome-linked gene, (j) *BYB* in female pachytene chromosomes of *S. cynthia pryeri* and (k) *S. cynthia ricini*, arrows represent WZ bivalent (j) and Z univalent (k), respectively; (l) FISH with different fluorescence dye-labeled probes of four *S. cynthia* fosmid clones, 11P18 (Red-labeled probe, red signal), 60G11 (Orange-labeled probe, yellow signal), 15C21 (Cy5-labeled probe, purple signal), 32H9 (Green-labeled probe, green signal) and Green-labeled probe of *S. cynthia* ortholog of *RpL4* (cyan signal, pseudocolor) and Cy3-labeled probe of 18S rDNA (orange signal, pseudocolor) in *S. cynthia ricini* pachytene chromosome. N, nucleolus.

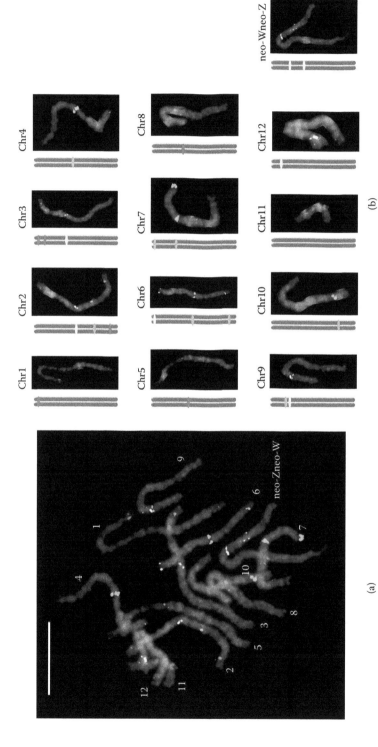

FIGURE 6.11 Fosmid-FISH karyotype of *Samia cynthia walkeri*. (a) A female pachytene complement of 12 autosomal bivalents and a sex chromosome bivalent (neo-W and neo-Z), each identified by 1–3 pseudocolored hybridization signals of fosmid probes. Chromosomes were counterstained with DAPI (white); (b) 12 autosomal bivalents and neo-Wneo-Z bivalent of same pachytene complement as in (a) arranged according to their chromosome numbers. Bar =10 μm.

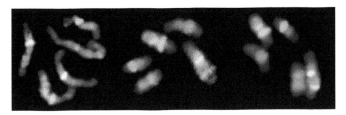

FIGURE 7.2 ReFISH with chromosome-specific probes of three different metaphase plates (chr. 1, yellow; chr. 2, purple; chr. 3, red; chr. 4, light blue; chr. 5, green). A combination of images recorded after the first and second hybridization allow discrimination of all five *Nasonia vitripennis* chromosomes by different color tags. Note, the consistent banding pattern of the probes, for example, double bands for chromosome1 (yellow) or single centromeric band for chromosome 2 (purple).

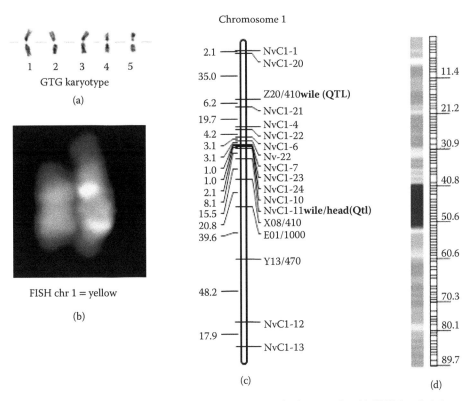

FIGURE 7.3 Connecting cytology, linkage mapping, and quantitative genetics. (a) GTG-banded chromosomes of a *Nasonia vitripennis* male. Chromosomes are numbered and ordered according to size. (From Gokhman, V.E. and M. Westendorff, *Beitr. Ent.,* 50, 193–198, 2000.) (b) Multicolor fluorescence in situ hybridization with chromosome-specific DOP-PCR products as probes (chr. 1, yellow). (c) Chromsomally anchored linkage maps based on a mapping population of *N. longicornis* × *N. vitripennis* hybrid F2-males. This linkage groups for chromosome 1 is predominantly based on microsatellite markers (chromosome-specific [NV-C1×] and unspecific randomly amplified polymorphic DNA markers [letter+number]). Quantitative trait loci for wing size are indicated toward the right of the markers with the largest effect. (For details see Ruetten et al. 2009). (d) Latest linkage map of chromosome 1 for *Nasonia* based on 19,708 loci, the color codes indicate recombination (blue, low recombination and red, high recombination). (For details see Desjardin, C.A. et al., *G3,* 5(2), 439–455, 2013.)

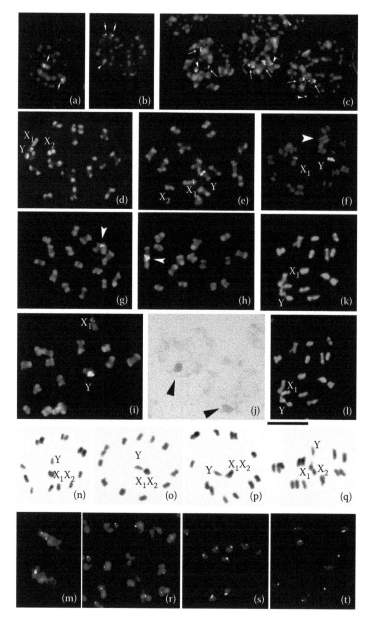

FIGURE 8.1 Different stages of male meiosis in *Cimex lectularius* after fluorescent in situ hybridization with an 18S ribosomal DNA probe (a–i, m, r–t), $AgNO_3$ (j), 4′,6-diamidino-2-phenylindole (k), chromomycin A_3 binding to guanine–cytosine-base pairs of DNA (l), and Schiff–Giemsa (n–q). (a–c)—spermatogonial metaphases: arrows point to X_1 and Y with signals, arrowheads to additional signals in some plates; (d–f)—meiotic prometaphase I: X_1 and Y with signals lie together (d) or separately (e, f); (g–i)—metaphase I: X_1 and Y with signals lie together or separately. Autosomal bivalents are condensed and consist of parallel-aligned chromosomes; (j)—diffuse stage: nucleolar proteins are localized on the sex chromatin body; (k–l)—metaphase I: X_1 and Y with CMA_3-positive signals (l); (m)—anaphase I: there are signals in both daughter cells; (n–q)—metaphase II: radial plates with sex chromosomes placed inside the ring formed by autosomes; (r–t)—consecutive stages of sperm formation: every sperm with a signal. Bar = 10 μm. (From Grozeva, S. et al., *Comp Cytogen* 4(2), 151–160, 2010.)

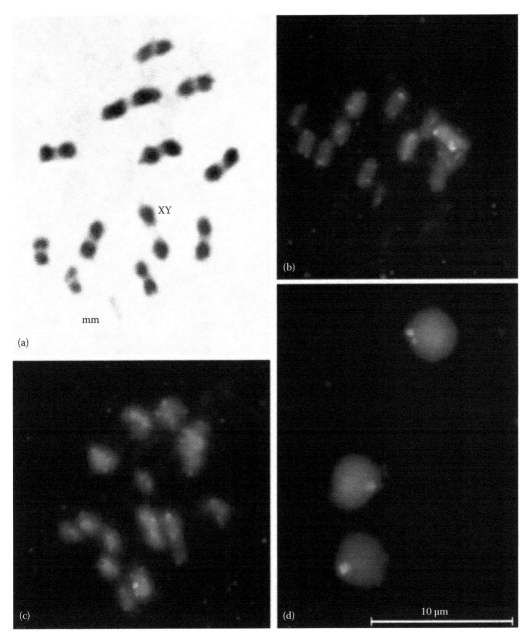

FIGURE 8.11 (a–d): Meiotic chromosomes of *Lethocerus patruelis* after standard staining (a) and fluorescence in situ hybridization (b–d). (a)—metaphase I showing $n = 11AA + mm + XY$; (b–d)—representative fluorescence in situ hybridization images of metaphase I, (b, c) hybridized with 18S rDNA and TTAGG probes and spermatids, (d) hybridized with TTAGG probe. Ribosomal clusters (green) in X and Y chromosomes (b, c), and TTAGG repeats (red) at the ends of chromosomes (b, c) and clustered at the periphery of spermatid nuclei (d). (From Kuznetsova, V. G. et al., *Comp Cytogen*, 6(4), 341–346, 2012.)

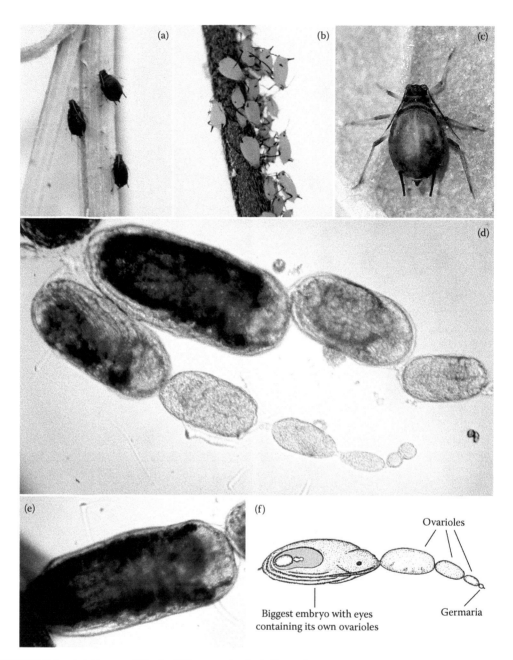

FIGURE 9.1 Aphids may differ in different traits, including color, as evident comparing *Uroleucon grossum* (a), *Aphis nerii* (b), and *Toxoptera aurantii* (c). Dissecting parthenogenetic females, it is possible to easily isolate embryos of different developmental stages and sizes that represent a useful source for mitotic cells, with the exception of the biggest embryos that are generally discarded (marked with arrowheads in (d, e)).

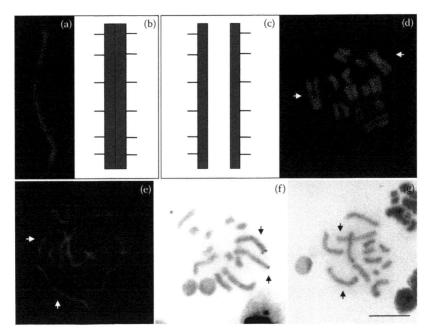

FIGURE 9.2 In aphid holocentric chromosomes, chromatids adhere to one another without any prominent structure detectable between them, as evident after staining with propidium iodide (a) and in the schematic representation (b). During mitotic anaphase (c and d), chromatids move apart in parallel and do not form the classical V-shaped figures typical of monocentric chromosomes. Aphid chromosomes may greatly differ in their condensation level, as evident comparing panels (d) and (e) and they can be stained with conventional staining, such as silver + Giemsa stainings (f) and Giemsa staining alone (g) to quickly verify the quality of the chromosome slides. Arrows indicated X chromosomes. Bar corresponds to 10 μm.

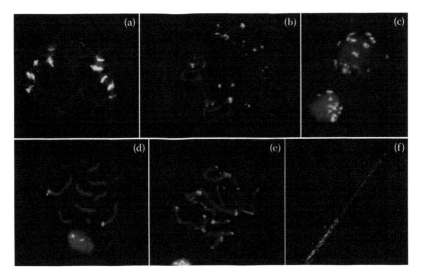

FIGURE 9.3 Fluorescent in situ hybridization has been frequently used to localize satellite DNAs, such as Hind200 (a) and the subtelomeric DNA repeat of the peach potato aphid *M. persicae* not only in chromosomes (b), but also in the interphase nuclei (c). At the same time, fluorescent in situ hybridization gave a good chromosomal mapping of different gene families, such as the major rDNA array (d), and other repeated sequences, including the telomeric (TTAGG)$_n$ sequence (e) in *M. persicae*. A fine mapping has been also obtained at higher resolution by fiber fluorescent in situ hybridization mapping of specific DNA sequences, such as the green FITC-labeled subtelomeric DNA repeat in the red propidium-stained fiber in *M. persicae* (f).

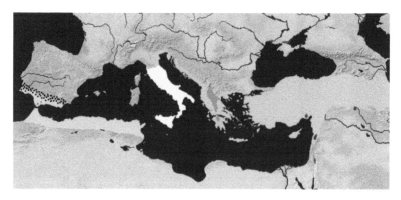

FIGURE 10.1 Map showing distribution of *Philaenus* species in the Mediterranean region: light blue—*P. maghresignus*; red—*P. tarifa*; yellow—*P. italosignus*; pink—*P. loukasi*; brown—*P. arslani*; orange—*P. signatus*; black dots—*P. tesselatus*. *P. spumarius* is sympatrically distributed with other species.

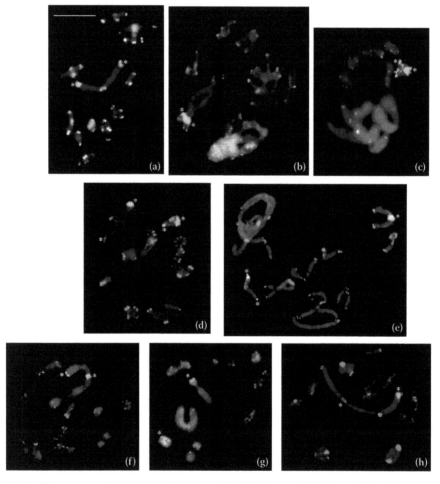

FIGURE 10.6 Fluorescence in situ hybridization with 18S ribosomal DNA and TTAGG telomere repeats as probes in *Philaenus* species: *P. tesselatus* (a), *P. arslani* (b), *P. signatus* (c), *P. spumarius* (d), *P. tarifa* (e), *P. italosignus* (f), *P. loukasi* (g), and *P. maghresignus* (h). *, 18S rDNA arrays in chromosomes. (From Maryańska-Nadachowska, A. et al., *Eur J Entomol* 110 (3), 411–8, 2013.)

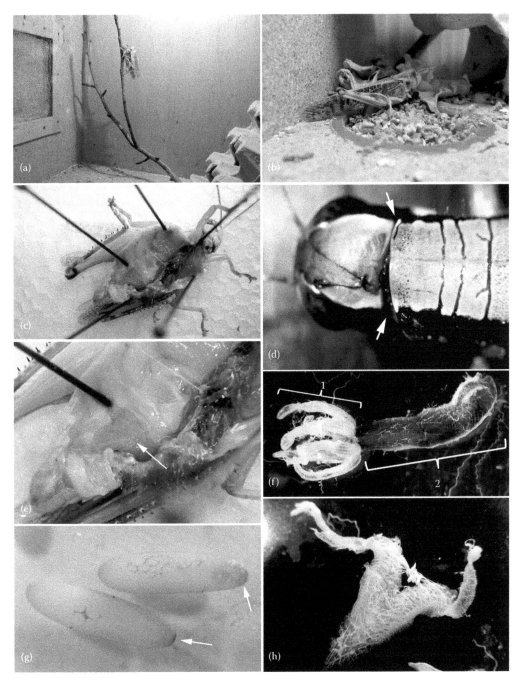

FIGURE 11.1 Grasshopper culture and dissection of appropriate tissues for chromosome analysis. (a) Mating pair of *Locusta migratoria*; (b) laying female of *L. migratoria*; (c) dissected male of *Eyprepocnemis plorans* showing a yellowish mass (indicated by an arrow in e) corresponding to the testes; (d) dorsal view of the head and pronotum of the grasshopper *Parascopas sanguineus* showing where to cut (arrows) for a rapid dissection of gastric caeca (1 in f) and gizzard (2 in f); (g) eggs showing the micropyle end (arrows); (h) ovaries.

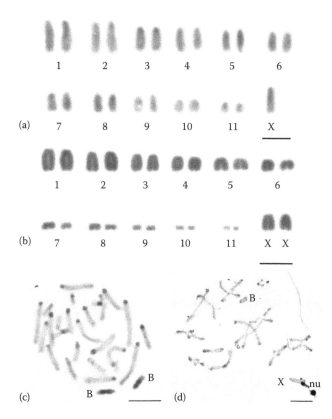

FIGURE 11.3 Typical grasshopper karyotypes (a and b) and C-banded (c) and silver stained (d) chromosomes. (a) Male $2n = 23$, X0 karyotype obtained from an embryo cell of *Eyprepocnemis plorans*; (b) female $2n = 24$, XX karyotype obtained from a gastric caecum cell of *Adimantos ornatissinus*; (c) C-banded embryo mitotic metaphase cell from *E. plorans*; (d) silver-stained diplotene cell from an *E. plorans* male. nu = nucleolus, Bar = 5 μm.

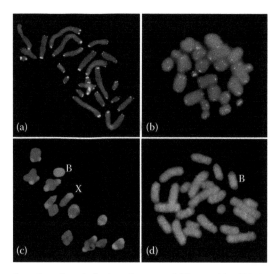

FIGURE 11.5 FISH mapping of tandem (a, b, c) and scattered (d) repetitive DNA sequences in L. migratoria (a), Adimantos arnatissimus (b) and E. plorans (c, d). (a, b) Telomeric DNA, (c) a 180-bp satDNA, (d) Gypsy transposable element. (a, d) Embryonic mitotic cells, (c) first meiotic metaphase cell, (b) mitotic cell obtained from gastric caeca. Note the presence of B chromosomes in c and d.

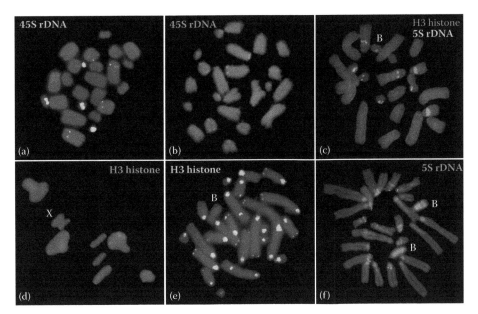

FIGURE 11.6 FISH using DNA probes for distinct multigene families in five grasshopper species, (a) *Adimantos arnatissimus*, (b and c) *Locusta migratoria*, (d) *Chorthippus jacobsi*, (e) *Abracris flavolineata*, and (f) *Eyprepocnemis plorans*. (a and e) Mitotic cells obtained from gastric caeca; (b, c, f) embryo mitotic cells, (d) meiotic metaphase I cell. The probes used are indicated in the cells. Note the presence of B chromosomes in c, e, f.

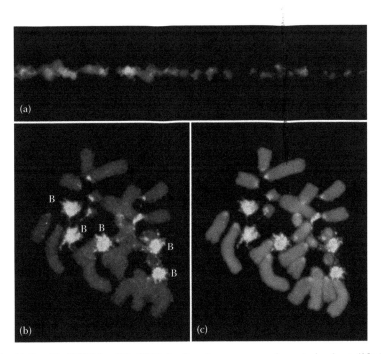

FIGURE 11.7 (a) Double FISH for 45S rDNA (red) and a sequence-characterized amplified region (SCAR) marker specific to B chromosomes in *E. plorans* was performed on chromatin fiber. Note the alternating arrangement of the two sequences; chromosome painting on an embryonic mitotic cell of *Locusta migratoria* using a B-chromosome DNA probe obtained through microdissection, (b) paint probe signal, (c) DAPI pattern in gray scale merged with paint probe signal.

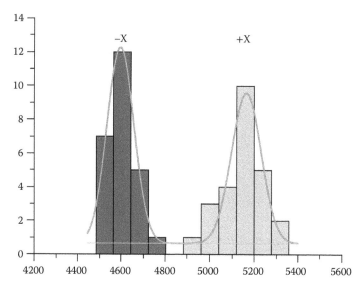

FIGURE 11.8 Histogram of IOD values obtained from 50 spermatids of an *E. plorans* male. A bimodal distribution shows that one peak belongs to spermatids without (left) or with (right) the X chromosome. The latter peak corresponds to the C-value of the species, and the difference between the two peaks is equivalent to the DNA content of the X chromosome.

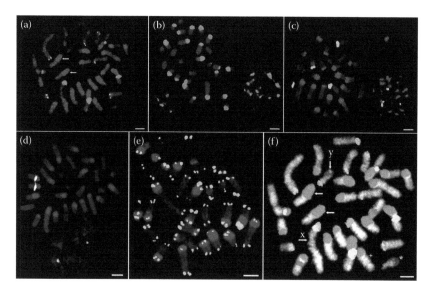

FIGURE 12.1 Fluorescence in situ hybridization of probes containing tandemly repetitive DNA to mitotic *Ixodes scapularis* chromosomes and cell nuclei (see panels b through d) prepared from cell line ISE18. Chromatin was stained with DAPI (a through e, blue; f, gray): (a) nucleolar regions (NORs, ribosomal DNA) (red); *I. scapularis* repeat-1 (ISR-1, 90 bp tandem repeat) (green); (b) ISR-2a (95 bp tandem repeat) (red); (c) ISR-2b (96 bp tandem repeat) (green); (d) ISR-3 (385 bp tandem repeat) (yellow); (e) (TTAGG)$_n$ telomere-localizing tandem repeat (green); and (f) preliminary *I. scapularis* karyotype based on the relative localization pattern of probes hybridizing to the NORs (yellow), ISR-1 (light blue), ISR-2a (red), ISR-2b (green), and ISR-3 (purple). The putative X and Y chromosomes are marked accordingly and shown with arrows. The asterisk indicates a chromosome fragment. The two arrows in the center of the panel show a chromosome pair consistently identified based on a strong signal for ISR-2a. The plus symbol shows an extra chromosome in this spread, commonly observed in ISE18. Scale bars = 5 μm. (With kind permission from Springer Science+Business Media: *Chromosome Res.*, Genome organization of major tandem repeats in the hard tick, *Ixodes scapularis*, 18, 2010, 357–70, Meyer, J. M., Kurtti, T. J., Van Zee, J. P., and Hill, C. A., Figures 2 through 4.)

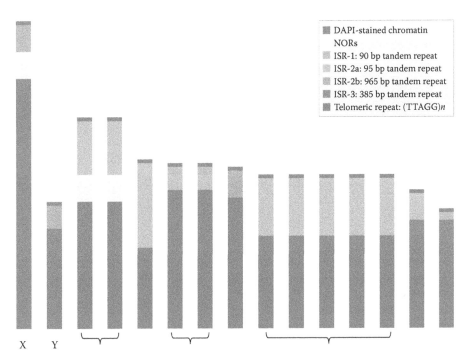

FIGURE 12.2 Ideogram showing the relative arrangement of tandemly repetitive DNA based on hybridization of *Ixodes scapularis* ISE18 cell line chromosomes. The X and Y sex-determining chromosomes are labeled, and groups of chromosomes sharing similar hybridization patterns are shown with brackets. The individual chromosomes within these groups could not be readily distinguished from one another based on their relative sizes or distribution of the tandemly repetitive DNA markers examined. Chromosomes are drawn to scale based on the representative example provided by Meyer et al. (2010). The considerable amount of variability observed in the relative sizes of ISE18 chromosomes among different chromosome spreads did not permit generation of a true karyotype where chromosomes are assigned numbers based on size and FISH marker distribution.

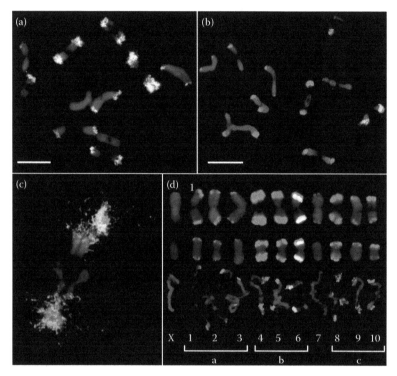

FIGURE 12.3 Fluorescence in situ hybridization of probes containing tandemly repetitive DNA to meiotic chromosomes of *Rhipicephalus (Boophilus) microplus*. Chromatin was stained with DAPI (blue); (a) colocalization of two probes (red + green = yellow signal) containing the *R. microplus* repeat-1 (RMR-1) (149 bp); (b) localization of a probe containing RMR-2 (178–216 bp) (red) and a probe for the 28S rDNA (green); (c) cohybridization of RMR-1 (green) and RMR-2 (red) to a bivalent; and (d) preliminary *R. microplus* karyotype based on the relative localization of probes hybridizing to the 28S rDNA (green) and RMR-1 (red). Chromosomes were separated into three groups (a through c) and ordered according to relative descending length and RMR-1 DNA quantity (red). Bivalent 6 also shows hybridization to an rDNA probe (green), and the X chromosome is identified. Scale bars = 10 μm. (With kind permission from Springer Science+Business Media: *Chromosome Res.*, The position of repetitive DNA sequence in the southern cattle tick genome permits chromosome identification, 17, 2009, 77–89, Hill, C.A. et al., Figures 4 and 5.)

7

Parasitoid Wasps (Hymenoptera)

Jürgen Gadau, Karsten Rütten, and Michaela Neusser

CONTENTS

LIST OF ABBREVIATIONS

BSA, bovine serum albumin
dH_2O, double distilled H_2O
dNTP, deoxyribonucleotide triphosphate
DOP-PCR, degenerate oligonucleotide primed PCR
FISH, fluorescence in situ hybridization
M-FISH, multiplex FISH
PBS, phosphate buffered saline
PCR, polymerase chain reaction
QTL, quantitative trait loci
RT, room temperature
SDS, sodium dodecylsulfate
sl-CSD, single-locus complementary sex determination
SSC, saline sodium citrate

7.1 INTRODUCTION

Hymenoptera is one of the four hyperdiverse holometabolic insect orders and its members show a wide range of life history features. However, the two most prominent and, maybe, successful features of the order are parasitism (e.g., Ichneumonidae, Braconidae, and Chalcidoidea) and social behavior (ants, bees, and wasps). This chapter focuses on one of the genetically and cytologically best-studied hymenopteran taxon the parasitoid genus *Nasonia* (Chalcidoidea), and discusses cytological research associated with social insects in general and ants in particular. The combination of linkage mapping, cytological analysis, and genome sequencing contributed significantly to the advance of honey bee and *Nasonia* genetics/genomics (Beye et al. 1994; Gadau et al. 1999; Rütten et al. 2004; The Honeybee Genome Sequencing Consortium 2006; Werren et al. 2010; Desjardins et al. 2013).

Here, we present protocols for chromosome preparation and staining optimized for *Nasonia*. However, these techniques should also work in most hymenopterans, such as ants and bumble bees (Gadau Juergen unpublished results; see Imai [1966] and Crozier [1968] for additional protocols). In addition, we report on chromosome microdissection and single chromosome amplification in *Nasonia vitripennis*. These chromosome-specific DNA templates were then used to produce chromosome-specific probes for fluorescence in situ hybridization (FISH) studies and to develop chromosome-specific microsatellite markers for genotyping and chromosomal anchoring of linkage maps. A combination of linkage mapping using the chromosome-specific microsatellites, FISH, and genome sequencing allowed us ultimately to connect cytology, linkage mapping, quantitative trait loci (QTL) studies, and genomic data (Rütten et al. 2004; Gadau et al. 2008; Werren et al. 2010; Loehlin and Werren 2012; Desjardins et al. 2013; Niehuis et al. 2013).

7.1.1 Taxonomy of *Nasonia vitripennis*

N. vitripennis is a parasitoid wasp that belongs to one of the largest families within the superfamily Chalcidoidea (Pteromalidae, 587 genera in 31 subfamilies). Currently, four *Nasonia* species have been described (*N. vitripennis, N. giraulti, N. longicornis,* and *N. oneida*). The sister taxon of *Nasonia* is *Trichomalopsis* with 54 described species. The age of the genus *Nasonia* has been estimated to approximately 4 million years (Werren et al. 2010).

7.1.2 Importance of *Nasonia vitripennis*

N. vitripennis has been introduced to genetic studies by P. Whiting in the 60s (Whiting 1967). *Nasonia* species are 2–3 mm small gregarious parasitoid wasps that parasitize pupae of several dipteran species (e.g., *Sarcophaga bullata* or *Caliphora* spp.). The genus *Nasonia* consists of the cosmopolitan species *N. vitripennis* and three North American species *N. longicornis, N. giraulti,* and *N. oneida* (Darling and Werren 1990; Raychoudhury et al. 2010). *N. vitripennis* is sympatric with the other three species. All species, with the exception of *N. oneida* and *N. giraulti*, are reproductively isolated by cytoplasmic *Wolbachia* infections, which cause nucleocytoplasmic incompatibilities (Breeuwer and Werren 1990; Raychoudhury et al. 2010). Once cured from *Wolbachia, Nasonia* species are able to interbreed despite various degrees of postzygotic isolation mechanisms (Niehuis et al. 2008; Raychoudhury et al. 2010; Gibson et al. 2013). Because of the ease of positional cloning, availability of haploid males, and the ability to generate hybrids, *Nasonia* has become a model organism in speciation genetics and the genetics of adaptive traits (Loehlin and Werren 2012; Niehuis et al. 2013), and the second model for hymenopteran genomics. In addition, two species *N. giraulti* and *N. vitripennis* are commercially available as fly control agents and can be ordered online (https://www.spalding-labs.com/products/fly_control_products/fly_control_for_horses/p/what_are_fly_predators.aspx).

7.1.3 Karyotype and Genome Size

The long history of cytogenetics in *Nasonia* species (previously known as *Mormoniella*) started with Gershenzon's (1946) initial description of *Nasonia's* five chromosomes. Later, cytogenetics helped to resolve the puzzling observation of the production of all male families in particular crosses. Using cytogenetic techniques Nur et al. (1988) discovered a parasitic B chromosome that destroyed the paternal chromosomes accompanying the B chromosome in the zygote. Reed (1993) described the cytogenetic characteristics of the B chromosome in detail, but we still do not know how this chromosome manages to specifically target the paternal chromosomes it traveled with for destruction but leaves the maternal chromosomes already present in the oocyte intact. Gokhman and Westendorf (2000) presented a first key for the identification of the five metacentric/submetacentric chromosomes of the three species of the genus *Nasonia* known by then. For their key, they used a very precise measurement of the chromosomal arms after Giemsa staining and/or C-banding. The first triploid females in

Hymenoptera were also described by Whiting (1960) based on karyotyping and a line that regularly produces triploid females in *N. vitripennis* has been retained for decades (Beukeboom Laboratory). This triploid female line (no triploid males were ever reported) has recently be used to clarify the sex-determination system in *N. vitripennis*, which is different from the better known complementary sex-determination system described for honey bees and *Bracon* spp. (Verhulst et al. 2010).

7.1.4 Genome Sequencing Project

The genome of *N. vitripennis* has been sequenced and published (Werren et al. 2010). Additional genomes from the congeners, *N. giraulti* and *N. longicornis*, have been sequenced but at a much lower coverage (1×). Therefore, these two genomes were only assembled using the *N. vitripennis* sequence as scaffold and have no independent assembly and annotation. The genome and additional genomic information and tools can be viewed, blasted, and/or downloaded at NasoniaBase (http://hymenopteragenome.org/nasonia/).

7.2 PROTOCOLS

7.2.1 Materials and Supplies

Tables 7.1 through 7.5 summarize a generic list of materials and supplies used to conduct the experiments presented in this chapter. More specific materials and methods used for particular experiments are listed throughout the text when mentioned for the first time.

7.2.2 Equipment

Specialized equipment needed besides a good dissecting and high resolution inverted microscope are polymerase chain reaction (PCR) machines, incubation chambers, and hybridization ovens. For the microdissection, it is essential to have a dedicated clean room to avoid spurious contamination and a vibration free table, because otherwise it is impossible to maneuver the needle close enough to the chromosome without bending or breaking the needle.

7.2.3 *Nasonia* Culture and Rearing

N. vitripennis strains (*Wolbachia* infected or uninfected) can be requested from any of the major *Nasonia* laboratories either in the United States or in Europe. *N. vitripennis* can also be ordered in the United States from Ward's Science (https://www.wardsci.com/store/catalog/product.jsp?catalog_number=876753) or Spalding (called fly

TABLE 7.1

Chemicals Utilized in Fluorescence In Situ Hybridization Experiments to Visualize Individual Chromosomes in *Nasonia vitripennis*

Chemicals	Company	URL
Biotin (bio)	Molecular Probes	www.invitrogen.com
Cy3-dUTP	GE Healthcare	www.gelifesciences.com
Cy5-dUTP	GE Healthcare	www.gelifesciences.com
Dextran sulfate	Sigma-Aldrich	www.sigmaaldrich.com
Digoxigenin-dUTP (DIG)	Molecular Probes	www.invitrogen.com
Dinitrophenyl aminohexanoid acid-dUTP (dnp)	Molecular Probes	www.invitrogen.com
Disodium hydrogen phosphate dihydrate	Merck	www.merck.com
Ethanol 100% (p. A)	Merck	www.merck.com
FITC-dUTP	Molecular Probes	www.invitrogen.com
Formamide	Merck	www.merck.com
Hydrochloric acid	Merck	www.merck.com
Magnesium chloride	Merck	www.merck.com
$NaHCO_3$	Sigma-Aldrich	www.sigmaaldrich.com
Potassium dihydrogen phosphate	Merck	www.merck.com
Salmon sperm DNA	Invitrogen	www.invitrogen.com
Sodium chloride	Merck	www.merck.com
Sodium citrate dihydrate	Merck	www.merck.com
Tamra-dUTP	Genaxxon bioscience	www.genaxxon.de
Texas Red	Molecular Probes	www.invitrogen.com
Tris-HCl	Sigma-Aldrich	www.sigmaaldrich.com
Tween 20	Merck	www.merck.com
VECTASHIELD antifade medium	Vector Laboratories	www.vectorlabs.com
W1 (Polyoxyethylene ether W1)	Sigma-Aldrich	www.sigmaaldrich.com

TABLE 7.2

List of Enzymes, Buffers, and Kits Used for DNA Amplifications

Enzymes, Buffers, and Kits	Company	URL
GeneAmp PCR buffer 10×	Applied Biosystems	www.appliedbiosystems.com
$MgCl_2$ solution (25 mM) PCR	PerkinElmer	www.perkinelmer.com
PCR buffer D 5×	Invitrogen	www.invitrogen.com
PCR buffer I (10×) w/o $MgCl_2$	PerkinElmer	www.perkinelmer.com
Pepsin	Sigma-Aldrich	www.sigmaaldrich.com
Taq polymerase	GE Healthcare	www.gelifesciences.com

Note: Enzymes, buffers, and kits should be stored according to manufacturer's instructions.

TABLE 7.3

Antibodies Used in Fluorescence In Situ Hybridization Experiments

Antibodies	Company	URL	Dilution
Streptavidin-Cy3	Dianova	www.dianova.com	1:500 in 4× SSCT/1% BSA
Streptavidin-Cy5	Dianova	www.dianova.com	1:100 in 4× SSCT/1% BSA
Avidin-Alexa 488	Molecular Probes	www.invitrogen.com	1:200 in 4× SSCT/1% BSA
Goat-α-DNP	Molecular Probes	www.invitrogen.com	1:200 in 4× SSCT/1% BSA
Mouse-α-Digoxigenin-Cy3	Dianova	www.dianova.com	1:100 in 4× SSCT/1% BSA
Mouse-α-Digoxigenin-Cy5	Dianova	www.dianova.com	1:100 in 4× SSCT/1% BSA

TABLE 7.4

Solutions Used in Fluorescence In Situ Hybridization Experiments

Solutions	Components	Protocol
(4x saline-sodium citrate buffer, 0,2% Tween)	0.2% Tween 20 in 4× SSC	2 mL Tween 20 in 1000 mL 4× SSC, store at room temperature
ACG-Mix for label DOP-PCR	2 mM dATP, dCTP, and dGTP	10 μL dATP, dCTP, dGTP (100 mM) each + 470 μL ddH$_2$O (autoclaved), store at −20°C
dTTP for label DOP-PCR	1 mM dTTP	10 μL dTTP + 990 μL ddH$_2$O (autoclaved), store at −20°C
EDTA (0.5 M)	EDTA (0.5 M)	Dissolve 186.12 g EDTA in 700 mL ddH$_2$O, adjust pH to 8.0 with NaOH, add ddH$_2$O to 1000 mL, store at RT
HCl (0.1 M)		50 mL HCl (1 M) + 450 mL ddH$_2$O, store at RT
Pepsinization solution	0.005% Pepsin in 0.01 M HCl	50 μL Pepsin (10%) + 10 mL 0.1 M HCl, add ddH$_2$O (37°C warm) to 100 mL, store at −20°C
Saline-sodium citrate (SSC)-buffer (pH 7.0)	150 mM NaCl, 15 mM Na-citrate	20× SSC: 175.3 g NaCl + 88.2 g Na-citrate, add ddH$_2$O to 1000 mL, adjust pH to 7.0 with NaOH, dilute to 4×, 2×, or 0.1× SSC with ddH$_2$O, store at RT

TABLE 7.5

Chemicals and Materials Used for Chromosome Preparation, Chromosomal Microdissection, and Microsatellite Fishing and Cloning

Chemicals	Company	URL
Dynabeads M 270 Streptavidin	Dynal	www.invitrogen.com
dNTPs, DIG-11-dUTP, Digoxigenin-FAB-Fragments	Roche	www.roche.com
Dichloromethylsilane	Merck	www.merck.com
Chloroform	Amresco	www.amrescoinc.com
Colchicine	Serva	www.serva.de
Sodium citrate	Merck	www.merck.com
Borosilicate pipettes	Hilgenberg	www.hilgenberg-gmbh.de
CDP-Star	Roche	www.roche.com
Antibodies		
Anti-Dig-Fluorescence-Fab Fragments	Roche	www.roche.com
Enzymes, buffers, and kits		
TOPO TA PCR Cloning kit	Invitrogen	www.invitrogen.com
DOP-PCR Master Kit	Roche	www.roche.com
Trypsin	Invitrogen	www.invitrogen.com

predators see Section 7.1.2). *Nasonia* cultures need very little space and can be kept at room temperature (RT, 18°C–25°C). *N. vitripennis* can be reared on a range of different hosts (dipteran from the genus *Sarcophaga* or *Calliphora*). A detailed manual *How to Keep and Rear* Nasonia has just been published (Werren and Loehlin 2009).

7.2.4   Chromosome Preparation and Single Chromosome Dissection in *Nasonia vitripennis*

To anchor the *N. vitripennis* linkage groups (Gadau et al. 1999; Niehuis et al. 2008) to specific chromosomes and eventually to genome sequences (Werren et al. 2010), we developed microsatellite markers and FISH probes from clones derived from single microdissected chromosomes (Figure 7.1). The microdissected genetic material of each of the five individual chromosomes (based on a single male haploid chromosome preparation) was amplified by a degenerate oligonucleotide primer PCR (DOP-PCR). In a subsequent second DOP-PCR three different templates were generated (Figure 7.1). From this template, we generated first a probe used in our dot blot experiment to verify the quality and purity of the template. Then the template was used to isolate microsatellite-containing sequences and develop primers to amplify chromosome-specific microsatellites for linkage mapping and finally, the same template was also used to generate probes for FISH experiments.

7.2.4.1   Preparation of Coverslips

All chromosomes of *N. vitripennis* were prepared on 60 × 24-mm coverslips for later microdissection. Smaller coverslips (24 × 24 mm), which were also used during the procedure, were siliconized to avoid chromosomes sticking to them. Siliconization was achieved by

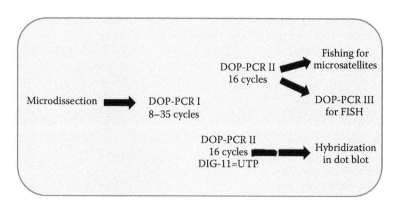

FIGURE 7.1   Flowchart showing the steps from microdissection to the isolation of chromosome-specific probes/markers.

incubating the coverslips in 70% ethanol (EtOH) for 30 minutes and in 1 M HCl overnight. After washing with distilled water (dH_2O) the coverslips have to be incubated in a suspension of 2.5 mL dichlormethylsilan and 47.5 mL chloroform for 1 minute, finally again washed with double distilled water (ddH_2O) and air-dried.

7.2.4.2 Preparation of Chromosomes

1. Dissect cerebral ganglia of male pupae (red eye stage) and incubate in 1% sodium citrate with 0.005% colchicine for 1 hour.
2. Transfer the probe on cleaned (briefly washed in 70% EtOH) 60 × 24-mm coverslip and remove as much as possible of the solution that came with the tissue.
3. Add several drops of 50% acetic acid (enough to cover all tissue) and incubate for 1 minute.
4. Macerate thoroughly with two dissection needles.
5. Cover with a 24 × 24-mm siliconized coverslip.
6. Squeeze as hard as possible with a tissue paper on top to spread the chromosomes.
7. Incubate for 15 minutes in Carnoy's fixative.
8. Air-dry at 37°C for 30 minutes.

The described procedure generated metaphase plates on the non-siliconized coverslip. For microdissection we chose an isolated metaphase plate where all chromosomes were spread well without touching each other.

7.2.4.3 Chromosome Staining

A GTG (G-bands by trypsin using Giemsa)-banding technique was used to obtain characteristic structures on different chromosomes helping to distinguish individual chromosomes according to Gokhman and Westendorff (2000). This technique uses trypsin to degenerate specific protein associations on the chromosomes depending on the binding affinity of the proteins and the DNA. A Giemsa staining after the trypsin digestion stains AT-rich regions darker and produces a specific banding pattern on the condensed chromosome.

1. Warm up the coverslips with the metaphase plates at 37°C.
2. Incubate in 0.05% trypsin in phosphate buffered saline (PBS) for 5 minutes at 37°C.
3. Incubate in 70% EtOH.
4. Incubate in PBS at 4°C.
5. Stain in 5% filtered Giemsa in PBS for 3 minutes (times differs between the species and the optimal staining time to produce the best staining needs to be tested out for each species).
6. Wash with dH_2O and air-dry.

7.2.4.4 Preparation of the Needles

Needles for microdissection were pulled out of 1 mm Borosilicate pipettes using a puller (H. Sauer, Reutlingen, Germany) in two stages, generating a tip diameter of less than 0.5 µm. The needles were sterilized under ultraviolet light.

7.2.4.5 Microdissection

The microdissection was performed using an inverted microscope with an object table that was able to fulfill complete rotation and a micromanipulator that allowed to record the needle's position and had a motor that automatically repositioned the needle after removing the needle to secure a dissected chromosome. Chromosomes of an ideal metaphase plate for microdissection are sufficiently widespread. The chromosomes on the periphery should be collected first to avoid touching other chromosomes when the target chromosome is removed. The focal chromosome should be placed in longitudinal direction of the needle. With 100× magnification the needle can then be placed close to the focal chromosome. Using 1000× magnification the tip of the needle should be positioned very close and slightly above the focal chromosome without touching the coverslip. This position is to be recorded and the needle tip should afterward be put out of focus. The focus should be on the chromosome. Shortly, before the dissection starts a drop of dH$_2$O is placed on top of the metaphase plate. For the actual dissection the needle will be automatically driven back to the initially recorded position. Now the tip of the needle can be brought onto the top of the coverslip. The chromosome but not the needle should now be in focus. By moving the needle tip a little bit further down it should now come into focus and touch the chromosome. Using the micromanipulator the needle tip is now moved slowly forward into the chromosome until the whole chromosome has been scratched of the slide and collected into the needle. The micromanipulator is then moved upward to remove the collected chromosome from the coverslip. Refocusing on the tip of the needle the chromosome material should be visible at the tip of the needle.

This position should be recorded. To control whether the chromosome adheres securely to the needle, it can be slowly swung back and forth. Note there is always a high risk of losing the chromosome when breaking the surface tension between air and water. Finally, the tip of the needle with the attached chromosome is broken off in 5 µL collecting buffer (10 µL 40 mM Tris-HCl, pH 7.5, 20 mM MgCl$_2$, 50 mM NaCl) at the bottom of a 500-µL microcentrifuge tube.

Troubleshooting: The closed reaction tube and gloves should be treated with antistatic spray before collecting the chromosome, because static forces might induce the chromosome to jump off the needle tip when lowering the needle into the tube.

As a negative control, the procedure of microdissection is repeated with a needle tip that was placed on the slide next to the site of the chromosome dissection but outside of the actual metaphase plate. Both the positive and the negative sample are then tested in a dot blot experiment (see Section 7.2.6).

7.2.5   Probe Preparation

7.2.5.1   Amplification of Microdissected DNA by Degenerated Oligonucleotide Priming (DOP)-PCR (I)

After breaking off the needle tip the probe should be centrifuged shortly and incubated overnight at 37°C in a master mix (MM) for proteinase K treatment (2.5 µL DOP primer [25 pmol], 10 µL dNTP [2 mM], 1.3 µL proteinase K 2.5 µg [purified for PCR], dH$_2$O adjust to 26.2 µL). Proteinase K is inactivated after the incubation by a heat treatment (96°C for 10 minutes). The single chromosome sample is now ready for the first round of DOP-PCR (I) amplification. This amplification is conducted in two steps using different polymerases, at different temperatures and at different cycle numbers. In the first round, 0.2 µL of a 1:8 diluted T7-sequenase is added at each of eight cycles (first cycle of initial denaturation: 95°C 3 minutes; eight cycles of denaturation: 95°C 1 minute, annealing: 30°C 2 minutes, extension: 30°C 2 minutes; cooling down: 5°C 5 minutes). The first DOP-PCR generates the template for a second DOP-PCR (II) with regular polymerase (DOP-PCR (I) Master Mix: 22.5 µL [2.5 U *Taq* polymerase, 200 µM each dNTP, 10 mM Tris-HCl, 50 mM KCl, 1.5 mM MgCl$_2$—everything refers to final concentrations] + 1 µL DOP-PCR primer [2 µM—final concentration] + 21.5 µL dH$_2$O) and 35 cycles (first cycle of initial denaturation: 95°C 3 minutes; 35 cycles of denaturation: 95°C 1 minute, annealing: 56°C 1 minute, extension: 72°C 2 minutes; one final extension: 72°C 5 minutes).

To test the success of microdissection, DOP-PCR (I), and control for contaminations, 5 µL DOP-PCR products of the microdissected DNA and negative control are analyzed using a standard 1% agarose gel and ethidium bromide staining. Only in case that there is no PCR product visible in the negative control, the probes should be used for the second DOP-PCR (II).

Troubleshooting: A faint smear in the negative probes is acceptable depending on the follow-up experiments. In our case, slight contaminations with other DNA sources were not problematic because we cross-referenced the results of our main experiments FISH and linkage mapping.

7.2.5.2 Reamplification of Primary DOP-PCR (I)
Amplification Product

This second amplification is necessary to obtain templates for the isolation of microsatellite markers and a further control experiment, which we dubbed "dot blot experiment." The dot blot experiment quantifies the amount of DNA that is similar to human DNA (i.e., amount of contamination). The same procedure can be used to generate probes for FISH.

Because the dot blot experiment allows us to estimate the amount of human DNA contamination during previous experimental steps, the PCR generating the template for microsatellite fishing and the PCR generating the probe for the dot blot experiment should be performed in parallel. An MM of the PCR reagents missing only the template DNA and the dNTP mix is convenient and should be prepared well in advance of the experiments. For the dot blot probe digoxigenin-labeled dUTP is used instead of regular dUTP. The following reaction mixture and PCR protocol was used: 1 μL DOP-PCR (I) product; 2 μL PCR buffer 10× (1× [16 mM $(NH_4)_2SO_4$, 7 mM Tris-HCl pH 8.8, 0.01% Tween]); 2 μL dNTP mix (2 mM each nucleotide—dNTP mix in case of the probe for dot blot 2 μL [2 mM dGTP, 2 mM dCTP, 13 mM dTTP, 0.7 mM digoxigenin-11-dUTP]); 0.6 μL $MgCl_2$ (150 μM); DOP-PCR primer 4 μL (20 μM); 0.2 μL *Taq* polymerase (1 U); 10.2 μL dH_2O. The sequence of the primer is the same as used in DOP-PCR (I). We used the following PCR conditions for DOP-PCR (II) reactions: initial denaturation 95°C 5 minutes; 16 cycles (denaturation: 95°C 1 minute, annealing: 56°C 2 minutes, extension: 72°C 2 minutes); final extension: 72°C 10 minutes. On a 1.5% agarose gel the PCR product should be visible as a smear in the area between 100 and 500 bp. The DIG-labeled probe seems to have a slightly larger product because of its lower velocity in the gel. The negative control should not show any PCR product.

7.2.6 Hybridization and Detection
(Dot Blot, Microsatellites)

7.2.6.1 Control for Contaminations: Dot Blot Experiment

To control the degree of contamination dot blot experiments can be performed. Dots of single-stranded human and *Nasonia* DNA are placed on a membrane and the single-stranded digoxigenin-labeled PCR product (II) is hybridized. After several washing steps, the detection of hybridized DNA takes place using anti-Dig-Fluorescence Fab fragments that bind onto the digoxigenin labeling and CDP-Star incubation that starts the fluorescence reaction. When signals mostly indicate hybridized *Nasonia*-DNA at *Nasonia* dots, the parallel performed probe for fishing microsatellites can be used for further treatment.

7.2.6.2   Preparing the Membrane for the Dot Blot Experiment

We used a vacuum blotter to fix the single-stranded DNA on the membrane. Here, a 120×80-mm nylone membrane is incubated for 2 minutes in 2× saline sodium citrate (SSC). For denaturation of DNA, 7.2 μg human and *Nasonia* DNA is diluted with dH$_2$O to 984 μL. A volume of 120 μL 20× SSC and 96 μL 5 M NaOH are added and mixed. Denaturation is then performed at RT for 10 minutes.

100 μL single-stranded DNA are placed onto opposite positions on the membrane.

A vacuum was generated for 2 minutes and the DNA was fixed by baking the membrane and DNA at 80°C for 2 hours.

Troubleshooting: The membrane can be stored at 4°C now and for each microdissection one piece can be cut off.

7.2.6.3   Hybridization and Detection (Dot Blot)

All incubation steps for the hybridization are conveniently performed in shrink-wrapped plastic bags on a shaker. For the steps of washing, it is comfortable to use 50 mL Greiner-Röhrchen (Greiner Bio-One, Frickenhausen, Germany).

1. Incubate nylon membrane in 5 mL prehybridization buffer (2% blocking reagent, 50% formamide in 2× SSC) for 1 hour at 42°C.
2. Denature the DIG-labeled DOP-PCR (II) product by heating it in a thermocycler for 5 minutes at 95°C. Place on ice afterward.
3. Add 10 μL denatured DIG-labeled DOP-PCR (II) product to 5 mL hybridization buffer (same as the prehybridization buffer).
4. Discard the prehybridization buffer and replace it with the hybridization buffer.
5. Hybridize overnight at 42°C on a shaker.
6. Wash the membrane twice for 5 minutes with 10 mL low-stringency buffer (2× SSC and 0.1% sodium dodecyl sulfate [SDS]) at RT.
7. Wash the membrane twice for 15 minutes with 10 mL low-stringency buffer (0.1× SSC and 0.1% SDS) at 68°C.
8. Incubate the membrane in 10 mL detection buffer 1 (0.1 M maleic acid, 0.15 M NaCl, pH 7.5).

9. To avoid the binding of antibodies on free areas of the membrane, incubate now for 30 minutes in 10 mL detection buffer 2 (detection buffer 1 with 1% blocking reagent) on a shaker.

10. Discard the blocking buffer and incubate in detection buffer 3 with antibodies (10 mL detection buffer 2 and 1 μL antibodies (0.75 U/μL) for 30 minutes on a shaker.

11. Discard the antibody buffer and wash off excessive antibodies by incubating the membrane twice for 15 minutes in 30 mL washing buffer (detection buffer 1 with 0.3% Tween 20).

12. Incubate the membrane in 5 mL detection buffer 4 (0.1 M Tris-HCl, 0.1 M NaCl, pH 9.5) for 5 minutes.

13. Discard detection buffer 4 and incubate in detection buffer 5 (10 mL detection buffer 4 and 25 μM CDP-Star) for 5 minutes.

14. Let the membrane become touch dry by placing it on a filter paper and shrink-wrap it in a clear plastic bag.

15. Place the plastic bag in a film cassette with an x-ray film and exposure for 2–10 minutes.

16. Develop the film as described in the manufacturer's instruction.

7.2.6.4   Isolation of Microsatellites from Chromosome-Specific DOP-PCR Templates

Troubleshooting: The procedure described below does not result in an assembly of pure microsatellites. Different repetitive genetic elements are actually fished out of the sample. However, a high percentage (50%) of them contains microsatellites. The same procedure has also been used by us successfully in conventional searching for new microsatellites in *Nasonia* using genomic DNA (Pietsch et al. 2004).

The isolation procedure according to the standard protocol uses streptavidin-coated magnetic beads (Dynal®, Invitrogen, Frankfurt, Germany) that bind to biotin-labeled oligonucleotides with a repetitive structure. In our case, we used a $(CA)_{10}$-oligonucleotide. DNA fragments that hybridized with our microsatellite probe were captured using a magnetic particle separator (Dynal) and washed three times, discarding all nonhybridized PCR fragments. Afterward, the fragments bound to the beads were eluted and amplified in a second PCR and directly cloned into the TOPO® TA cloning vector (Life technologies™, Darmstadt, Germany).

7.2.6.5   Protocol for the Isolation of Repetitive DNA/Microsatellite Sequences Using Streptavidin-Coupled Dynabeads

1. Add 1 µg of biotin-labeled $(CA)_{10}$ oligonucleotide that is 3′-labeled with biotin to the DOP-PCR (II) product without DIG-dUTP and adjust to 100 µL with 6× SSC.
2. Denature the PCR product by incubating at 96°C for 10 minutes in a thermocycler.
3. Place on ice for 2 minutes.
4. For hybridization, incubate for 5 minutes at RT.
5. Add 50 µL of Dynabead solution (for preparation see manufacture's instruction) and mix for 15 minutes at RT.
6. Place sample in magnetic particle separator and remove the unbound rest by aspiration.
7. Wash with 100 µL 2× SSC for 3 minutes at RT and repeat step 6.
8. Wash with 100 µL 1× SSC for 3 minutes at RT and repeat step 6.
9. Wash with 100 µL 1× SSC for 5 minutes at 57°C and repeat step 6 again.
10. Add 50 µL dH_2O to the captured DNA fragments and incubate for 5 minutes at 90°C.
11. Place the sample quickly back into the magnetic particle concentrator and transfer the supernatant into a new reaction tube.

The isolated chromosome-specific single-stranded DNA fragments with repetitive elements inside can now be amplified via the following PCR reactions: initial denaturation 95°C 5 minutes; 20 cycles of denaturation: 95°C 1 minute; annealing: 56°C 1 minute; extension: 72°C 3 minutes; final extension: 72°C 20 minutes (15 µL isolated DNA fragments, 25 µL DOP-PCR Master Mix [2.5 U *Taq* polymerase in Brij, 200 µM each dNTP, 10 mM Tris-HCl, 50 mM KCl, 1.5 mM $MgCl_2$], 1 µL DOP-PCR primer [100 µM], 9 µL dH_2O). The excessive final extension time should result in an overlap of an Adenine nucleotide at the 3′ for subsequent direct cloning into TOPO TA cloning vector (Life Technologies).

7.2.7   Fluorescence In Situ Hybridization

Chromosome-specific probes obtained by reamplification (DOP-PCR [II]) of the primary DOP-PCR (I) amplification product (see Section 7.2.5.2) can then be used for further FISH experiments.

7.2.7.1   Labeling of DNA Probes for Fluorescence In Situ Hybridization

For convenient handling, especially in case of frequent use of amplification reactions, MM aliquots of a stock-labeling reaction mix containing all reagents except for *Taq* polymerase and template DNA can be prepared and stored at −20°C. Ready to use, DNA and *Taq* polymerase is added to the MM prior to DOP-PCR labeling. To set up the MM for label DOP-PCR of a single 50 μL amplification reaction mix together,

1. 5 μL 10× PCR buffer (50 mM KCl, 10 mM Tris, pH 8.3)
2. 1 μL DOP-PCR primer (100 μM) (2 μM final concentration)
3. 2.5 μL ACG-mix (each 2 mM) (100 μM final concentration)
4. 4 μl dTTP (1 mM) (80 μM final concentration)
5. 1 μL (2–3 μL for fluor-dUTPs) Bio (or DIG or DNP)-dUTP or fluor-dUTP (e.g., FITC-dUTP), (1 mM) (final concentrations 20–60 μM)
6. Adjust with ddH$_2$O to total volume of 48.5 μL

For a standard reaction for a single label DOP amplification reaction mix together on ice in a 0.6 mL PCR tube:

- 48.5 μL label MM
- 1 μL (usually corresponds to 30–200 ng DOP amplified DNA)
- 0.5 μL *Taq* polymerase (5 U/μL)

Perform label PCR in a thermocycler using the following protocol: 22–27 cycles (1 cycle of initial denaturation: 94°C 3 minutes; 20–25 cycles of denaturation: 94°C 1 minute; annealing: 56°C 1 minute; extension: 72°C 30 seconds; final extension: 72°C 5 minutes).

Troubleshooting: Check 1 μL of PCR product on a 1% agarose gel with appropriate size marker.

7.2.7.2   Probe Preparation, Precipitation, and Setup

The labeling scheme of the probe or probe set depends on the filter setting of the epifluorescence microscope available in the laboratory. A basic epifluorescence microscope is usually equipped with three band-pass filter sets by which the chromosomal counterstain 4′,6-diamidino-2-phenylindole (DAPI) and two FISH probes labeled with green and orange fluorochromes (e.g., FITC and Cy3), respectively, can be spectrally discriminated. More sophisticated microscopic setups presently include up to eight narrow band-pass filters, which can be used in Multiplex FISH

(M-FISH) experiments. To discriminate more FISH probes in a single experiment than fluorescence filters available, probes can be labeled in a combinatorial manner, for example, probe 1 in FITC, probe 2 in Cy3, and probe 3 using a mixture of FITC and Cy3. Alternatively, multiple probes can be visualized by sequential hybridization to the same specimen.

Troubleshooting: The amount of DNA used for hybridization depends on the complexity of the probe and the sequence divergence between target and probe DNA. For repetitive probes, 1–10 ng/µL DNA/hybridization mixture is recommended and 20–100 ng/µL for nonrepetitive probes. Because exact measurement of DNA probe concentration may be somewhat tedious, we suggest to use 2 µL labeled PCR product per 1 µL hybridization mixture in case of chromosome painting probes or locus-specific probes. If euchromatic regions are to be highlighted, it is further recommended to add unlabeled competitor DNA (e.g., C_0t-1 DNA or unlabeled genomic DNA sheared to a fragment size of 200–1000 bp) to the probe for suppression of nonspecific hybridization. Because for the enrichment of C_0t-1 DNA, a large amount of starting DNA is required, which may be difficult to obtain from some arthropods, sheared whole genomic DNA may be preferable. In this case, we recommend to incubate high-molecular weight genomic DNA for 30–60 minutes at 94°C in a water bath or thermo shaker to shear the DNA. The concentration of unlabeled competitor DNA necessary depends on the abundance of repetitive sequences in the probe as well as in the chromosomal target species and has to be determined empirically in each case. As a rule of thumb, the recommended concentration of the competitor DNA should be approximately 10- to 50-fold of the concentration of the probe DNA. The sample area covered by 18 × 18-mm coverslip requires 5–8 µL of hybridization mixture. In case of smaller or larger hybridization areas the required amount should be adjusted accordingly.

7.2.7.3 Preparation of the Hybridization Solution (Mix in 1.5 mL Tube)

1. Add all labeled DNA probes that will be hybridized together.
2. Add unlabeled competitor DNA.
3. Add 20 µg unlabeled salmon sperm DNA (10 mg/mL).
4. Mix probe DNA with 2.5× volume ice-cold 100% EtOH.
5. Incubate at least 30 minutes at −20°C.
6. Spin down at 13,000 rpm for 20 minutes.
7. Discard supernatant.
8. Dry the DNA pellet (using vacuum centrifuge if available).

TABLE 7.6

Reagents to Compose the Hybridization Buffer

Hybridization buffer	Formamide, 50% dextran sulfate, 20× SSC, 1 M NAPO$_4$ buffer, 10% SDS, 50× Denhardt's, H$_2$O bidest	For 10 mL stock solution: 5 mL formamide + 2 mL 50% dextran sulfate + 1 mL 20× SSC + 400 μL 1 M sodium phosphate buffer + 100 μL 10% SDS + 200 μL 50× Denhardt's + 1.3 mL H$_2$O
50% Dextran sulfate	Dextran sulfate, H$_2$O bidest	For 100 mL stock solution: 50 g dextran sulfate to 100 mL H$_2$O, resuspend at 60°C
1 M Sodium phosphate buffer	1 M Na$_2$HPO$_4$, 1 M NaH$_2$PO$_4$	577 μL 1 M Na$_2$HPO$_4$ + 423 μL 1 M NaH$_2$PO$_4$
10% SDS solution	SDS, H$_2$O	For 100 mL stock solution: 10 g SDS to 100 mL H$_2$O

9. Resuspend the pellet in 50% formamide/2× SSC/10% dextran sulfate as follows.
    a. Resolve the pellet in the appropriate amount of 100% formamide.
    b. Shake at 37°C until DNA pellet is resolved (this can take up to a few hours).
    c. Add equal volume of 4× SSC/20% dextran sulfate.
    d. Briefly mix and incubate at 37°C for 10 minutes.
10. Alternatively, resuspend the pellet in complete hybridization buffer (see Table 7.6). Incubate preferably 3–5 hours on a thermo shaker at 37°C. Hybridization probes can be stored at −20°C for up to several years.

7.2.7.4   Denaturation and Hybridization

For the hybridization of metaphase chromosome preparations fixed with methanol/acetic acid, it is recommended to prepare the target slide some hours prior to hybridization to allow thorough drying and aging of the chromatin on the slide, for example, by incubation at 60°C for 1 hour.

1. Prepare a Coplin jar containing 70% formamide/2× SSC (pH 7) in water bath at 72°C.
2. Preheat a second water bath to 37°C.
3. Prepare a third Coplin jar with ice-cold 70% EtOH.
4. Denature DNA probe mix in water bath at 72°C for 7 minutes.
5. Incubate DNA probe mix at 37°C in water bath for at least 30 minutes and up to 3 hours, if repetitive sequences have to be suppressed by competitor DNA.
6. Denature slide for 1 minute 30 seconds in Coplin jar containing 70% formamide/2× SSC (pH 7) at 72°C.
7. Quickly transfer slide to Coplin jar with ice-cold 70% EtOH, incubate for 3 minutes.
8. Incubate 3 minutes each in an ascending EtOH series with 90% and 100% EtOH.

9. Air-dry slide till EtOH is completely evaporated.
10. Pipette probe solution onto the target area of the slide.
11. Place a coverslip of appropriate size onto probe solution and seal with rubber cement.
12. Incubate overnight and up to 72 hours in a metal box floating in a 37°C water bath.
    Troubleshooting: Incubation of the slide in pepsin solution to digest cytoplasm is often necessary for efficient binding of the probe to the target DNA.

7.2.7.5 Washing and Detection

Troubleshooting: All further steps should preferably be performed under light protection, because fluorochrome-coupled probes are used, which bleach under extended light exposure.

1. Prepare a Coplin jar with 4× SSCT (saline-sodium citrate, 0.2% Tween) at RT.
2. Preheat water bath to 62°C with two Coplin jars containing 0.1× SSC.
3. Preheat a second water bath to 42°C with two Coplin jars containing 4× SSCT.
4. Preheat a third water bath to 37°C.
5. After hybridization, peel off rubber cement, remove coverslip carefully, and transfer the slide to Coplin jar with 4× SSCT at RT, incubate for 5 minutes.
6. Wash the slide two times in 0.1× SSC at 62°C, each for 7.5 minutes.
7. Rinse the slide in 4× SSCT at RT.
8. Place slides in a metal box, pipette 1 mL 3% (w/v) BSA blocking solution (3 g Bovine serum albumin in 100 mL 4× SSCT).
9. Incubate metal box for 20 minutes floating in a 37°C water bath.
10. Prepare antibody solution in 1% BSA/4× SSCT, when hapten-coupled probes are used: dilute the required antibodies or avidin conjugates to the appropriate working concentration in this solution.
11. Pipette 200 µL antibody solution on the slide, cover with a 24 × 60-mm coverslip.
12. Place slide in a metal box and incubate metal box for 45–60 minutes floating in a 37°C water bath.
13. Remove coverslip, briefly rinse slide in 4× SSCT at RT.
14. Incubate slides two times 7.5 minutes in a coplin jar with 4× SSCT, heated to 42°C.
15. For DNA counterstaining use either a prepared DAPI solution (0.2 µg/mL DAPI in 4× SSCT) for 10 minutes

at 37°C or a ready-to-use antifading solution containing DAPI (e.g., VECTASHIELD with DAPI).

16. In case of using the separate DAPI staining method, wash briefly in 2× SSC after DAPI staining and mount hybridized area in antifading solution (e.g., VECTASHIELD without DAPI).

17. Apply 24 × 60-mm coverslip and optionally seal coverslip edges with colorless nail polish to avoid skin contact with DAPI.

7.2.7.6 ReFISH (Sequential Hybridization of Multiple Probe Sets to the Same Slide)

Troubleshooting: When following this protocol, up to five consecutive hybridizations to the same slide could be performed, and previously hybridized slides stored at 4°C for 2 years could be rehybridized (Müller et al. 2002). Essential for complete removal of previously hybridized probe and preservation of chromosome morphology is the combination of postfixation and increase of slide denaturing time in each consecutive round of FISH. For each round of rehybridization, the slide denaturation time is increased by 30 seconds. Further, the use of hapten-coupled dUTPs should be avoided in case of a third round of hybridization; direct fluorochrome-labeled probes should be used instead. In all other rounds, both hapten-dUTPs and fluor-dUTPs may be used.

First hybridization (standard FISH procedure)

- Denature slide (1 minute 30 seconds in coplin jar containing 70% formamide/2× SSC [pH 7], at 72°C), denature probe for 7 minutes, 72°C.
- Mark hybridization area with diamond pen,optionally.
- Hybridize 24–72 hours (depending on the probe).
- Capture images + record metaphase position.

Postfixation (essential)

- Wash off coverslip (4× SSCT) at RT, approximately 15 minutes.
- Incubate in 4× SSCT at RT, 60 minutes until antifading solution is completely removed.
- Dehydrate in ascending EtOH series: 70%, 90%, 100% EtOH, 3 minutes each.
- Incubate 30 minutes in fixative (methanol/acetic acid 3:1) at RT.
- Air-dry.
- Incubate overnight at 37°C in a dry oven.

Second hybridization

- Rehybridize (same slide/probe denaturation temperature, but the slide denaturation time was increased by 30
  seconds
- Hybridize 24–72 hours (depending on the probe).
- Capture images + record metaphase position.

Figure 7.2 shows the hybridization of all five chromosomes
of *N. vitripennis* using chromosome-specific probes established
by microdissection according to the ReFISH protocol described
earlier (Rütten et al. 2004). For this FISH experiment, the five
probes were divided in two subsets that were sequentially hybridized. Subset 1 was composed of chromosome 1 (Tamra-dUTP), 2
(Biotin-dUTP), and 3 (Digoxigenin-dUTP). Biotin was detected
by one layer Avidin-Cy5, Digoxigenin by one layer of FITC-
conjugated sheep anti-Digoxigenin antibody. After hybridization and detection of the first probe subset, microscopic images
were acquired and the metaphase coordinates were recorded.
Subsequently, the metaphases were rehybridized with the second
probe subset (chromosome 4, biotin-dUTP and 5, Digoxigenin-
dUTP), followed again by posthybridization washings, detection,
and microscopy. The overlay of both images of each metaphase,
recorded after the first and second hybridization, allows to define
each chromosome by a unique color code.

7.2.8   Genetic Mapping and Representative Results

N. vitripennis is a small parasitic hymenopteran with a 50-year history
of genetic work, including linkage mapping with mutant and molecular markers (Whiting 1967; Gadau et al. 1999; Beukeboom et al.
2010; Desjardins et al. 2013). However, it took more than 50 years
to anchor linkage groups to specific chromosomes, because in
contrast to *Drosophila melanogaster*, *Nasonia* has five very similar

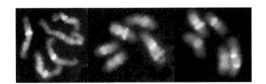

FIGURE 7.2   (**See color insert.**) ReFISH with chromosome-specific probes
of three different metaphase plates (chr. 1, yellow; chr. 2, purple; chr. 3, red;
chr. 4, light blue; chr. 5, green). A combination of images recorded after the first
and second hybridization allow discrimination of all five *Nasonia vitripennis*
chromosomes by different color tags. Note, the consistent banding pattern of the
probes, for example, double bands for chromosome1 (yellow) or single centromeric band for chromosome 2 (purple).

meta/submetacentric chromosomes (Gokhman and Westendorff 2000; Ruetten et al. 2004). Genetic mapping in *Nasonia* and other Hymenoptera is straightforward due to haplodiploidy in males even if phase is unknown (see Gadau et al. 1999 or Gadau et al. 2001 for a phase unknown mapping protocol). The linkage map of chromosome 1 in Figure 7.3 was generated based on 100 male offspring of an F1 female from a hybrid cross between *N. vitripennis* and *N. longicornis*. For genotyping we used site tagged sequences, microsatellites (both based on DNA templates from chromosome-specific microdissections), and randomly amplified polymorphic DNA markers (Rütten et al. 2004). On the basis of the microdissected markers and markers used in other linkage mapping projects (Gadau et al. 1999), we anchored previous linkage groups to the five chromosomes of *N. vitripennis*. The combination of linkage mapping and FISH using the same DNA

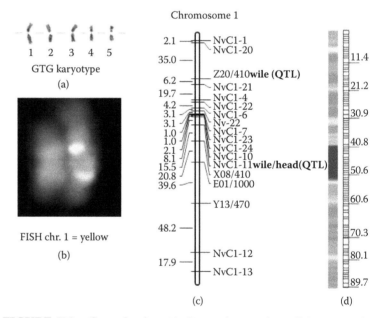

FIGURE 7.3 (**See color insert.**) Connecting cytology, linkage mapping, and quantitative genetics. (a) GTG-banded chromosomes of a *Nasonia vitripennis* male. Chromosomes are numbered and ordered according to size. (From Gokhman, V.E. and M. Westendorff, *Beitr. Ent.*, 50, 193–198, 2000.) (b) Multicolor fluorescence in situ hybridization with chromosome-specific DOP-PCR products as probes (chr. 1, yellow). (c) Chromsomally anchored linkage maps based on a mapping population of *N. longicornis* × *N. vitripennis* hybrid F2-males. This linkage groups for chromosome 1 is predominantly based on microsatellite markers (chromosome-specific [NV-C1×] and unspecific randomly amplified polymorphic DNA markers [letter+number]). Quantitative trait loci for wing size are indicated toward the right of the markers with the largest effect. (Modified after Ruetten et al. 2004). (d) Latest linkage map of chromosome 1 for *Nasonia* based on 19,708 loci, the color codes indicate recombination (blue, low recombination and red, high recombination). (For details see Desjardin, C.A. et al., *G3*, 5(2), 439–455, 2013.)

templates allowed us not only to verify chromosomal specificity of the microdissected markers but also to generate an ordered sequence of the genomic scaffolds. This detailed positional information made it then possible for the *Nasonia* community to very rapidly go from a phenotype or QTL to the genotype or gene (Loehelin et al. 2012; Niehuis et al. 2013). To test this approach in our initial publication, we asked whether QTL responsible for the reduction of male wing size in two different hybrid crosses (*N. vitripennis* × *N. longicornis* and *N. vitripennis* × *N.giraulti*) map to the same location (Rütten et al. 2004). One QTL with a major effect was found to map to the centromere region of chromosome 3 in both crosses, hence we hypothesized that the same gene(s) could be involved in the reduction of male wing size in *N. vitripennis* and *N. longicornis* (Figure 7.3). Note that, eventually the mutation responsible for one of the wing-size reduction QTL was discovered to be in a noncoding region (Gadau et al. 2002; Loehlin and Werren 2012).

7.3 DISCUSSION

7.3.1 Chromosome Organization and Evolution in Hymenoptera and *Nasonia vitripennis*

Hymenopteran cytogenetics was driven from its beginning by the peculiarities of the haplodiploid sex-determination system. Under haplodiploidy, haploid males are produced from unfertilized eggs (arrhenotoky) and diploid females are born from fertilized eggs (Cook 1993). This type of sex determination was first suggested by Dzierzon (1845) for honey bees, but it took a long time till the scientific community accepted Dzierzon's hypothesis (Siebold 1856; Nachtsheim 1916). Cytogenetics was also used successfully to confirm certain idiosyncrasies of the haplodiploid sex-determination system, for example, the occurrence of diploid males due to a "matched" mating in honey bees leading to a particular colony-level phenotype (shotgun brood). Hymenoptera such as *Habrobracon* spp. and *Apis melifera*, which have a single-locus complementary sex determination (sl-CSD) system, produce diploid males if the offspring inherited the same sex allele from both parents (Whiting 1933; Whiting 1961). These results in combination with inbreeding experiments were used to confirm sl-CSD in other species or reject this particular sex-determination mechanism for other Hymenoptera. Note, so far two additional sex-determination mechanisms have been confirmed in Hymenoptera, namely multilocus sex determination and maternal imprinting (Cook 1993; Verhulst 2010). The predictability of diploid male production in *Apis mellifera* was also crucial for the determination of the genetic (csd locus) and molecular basis of the sl-CSD (Whiting 1961; Beye 2003).

Crozier (1975) reviewed all relevant cytogenetic work in Hymenoptera up to this point in time. Imai, Crozier, and their students and collaborators dominated hymenopteran cytogenetics for the rest of the century and to some degree till now. They developed novel chromosome preparation and staining techniques, which could also be used in the field. Most of their work focused on karyotype evolution in ants (Imai 1966; Crozier 1968; Imai et al. 1988, 1994; Imai and Taylor 1989; Lorite and Palomeque 2010). Hence, the hymenopteran family Formicidae is probably one of the few insect families that is well studied in terms of cytogenetics and karyotypes. The last review listed karyotypes for 750 ants (Lorite and Palomeque 2010). Since then, hymenopteran cytogenetics has mostly been used to resolve phylogenetic relationships or systematic problems (Cardoso et al. 2012; Cristiano et al. 2013) with the exception of the work presented here that connected cytogenetics, QTL analysis, and genomics in the parasitoid wasp *N. vitripennis* (Gadau et al. 2008; Pietsch et al. 2004; Ruetten et al. 2008; Werren et al. 2010).

7.3.2 Integration of Cytogenetic, Linkage, and Genome Sequence

The integration of cytology (Gokhman and Westendorff 2000; Rütten et al. 2004) and genetic linkage maps using chromosomally anchored markers from microdissected and cloned DNA sequences allowed us to homologize the results of a diverse set of mapping projects that used mutant morphological, molecular, or life history markers (Gadau et al. [2008] gives an overview of linkage maps in *N. vitripennis*). Once molecular markers were available, we could use their sequence to find their location in scaffolds of the published genome sequence (Werren et al. 2010) and consecutively use the linkage information to orient and link genome scaffolds into superscaffolds or chromosomes (Desjardins 2013). This made a huge difference because once we knew the sequence of scaffolds and their position relative to QTL or mapped qualitative markers (e.g., eye color markers or *123* or developmental markers, such as *antennapedia*), we could identify candidate genes. For the latest update on the combination of linkage maps and genome sequence see Desjardins et al. (2013). This manuscript ordered and aligned 86% of the sequenced genome of *N. vitripennis* and 100% of the annotated genes. This newest mapping also revealed a small but significant subset of wrongly assembled scaffolds (see table in Desjardins [2013]).

In particular, parts of one large scaffold (>5Mb) mapped actually to chromosome 1, whereas other parts of the same scaffold mapped to chromosome 5. Banking on this result we could refine the genome assembly. This latest linkage map makes *Nasonia* arguably the animal with the highest density of markers on a linkage map.

7.3.3   Practical and Scientific Benefits of Genome Mapping

The ease of generating linkage maps, availability of a high number of markers, and haploid individuals make *Nasonia* an ideal organism to quickly go from a phenotype to the underlying gene (forward genetics). For example, we were able to map a small genomic region (less than 30 candidate genes) for the major gene(s) for a male sex pheromone difference between *N. vitripennis* and *N. giraulti* within 2 months. Using marker-assisted introgression we could further narrow the region down to seven genes within another 3 months and finally confirm the genes by using dsRNAi (Niehuis et al. 2013).

In addition, anchored linkage maps are the only method to determine recombination frequency with a high enough resolution to estimate the range and diversity of recombination frequencies within a genome and between different populations and crosses (Niehuis et al. 2010; Desjardins et al. 2013). Knowing about the actual recombination frequency in one particular part of the genome is also important if someone wants to use Genome Wide Association Studies or Selective Sweep approaches to narrow down or identify the genetic basis or architecture of a trait. If recombination frequency is low in the region of interest, many more individuals are needed to be genotyped to allow fine mapping or the genomic region that is impacted by a selective sweep is too large to come up with a useful set of candidate genes. For example, the heat map for chromosome 1 in Figure 7.3 identifies regions of low recombination (blue) in the middle of the linkage map for chromosome 1, which is also indicating position of the centromere (a similar pattern can be seen in the other chromosomes). The locus responsible for a significant reduction in wing size described by Loehlin and Werren (2012) was located in one of these recombinational cold-spots (low recombination relative to the rest of the genome), and it took the authors much longer to narrow down and eventually find the genomic region responsible for the observed phenotypic differences compared to our pheromone study (Niehuis et al. 2013).

ACKNOWLEDGMENT

This work was supported by funds from the DFG (SFB-554-TP Gadau) to JG.

REFERENCES

Beukeboom L. W., O. Niehui, B. Pannebakker, T. Koevoets, J. Gibson, D. Shuker, L. van de Zande, and J. Gadau. 2010. A comparison of recombination frequencies in intra- versus interspecific mapping populations of *Nasonia*. *Heredity* 104: 302–309.

Beye M., C. Epplen, and R. F. A. Moritz. 1994. Sex linkage in the honey bee *Apis mellifera* L. detected by multilocus DNA fingerprinting. *Naturwissenschaften* 81: 460–462.

Beye M., M. Hasselmann, M. K. Fondrk, R. E. Page Jr., and S. W. Omholt. 2003. The gene csd is the primary signal for sexual development in the honeybee and encodes an ST-type protein. *Cell* 114: 419–429.

Breeuwer J. A. J. and J. H. Werren. 1990. Microorganisms associated with chromosome destruction and reproductive isolation between two insect species. *Nature* 346: 558–560.

Cardoso D. C., M. P. Cristiano, L. A. C. Barros, D. M. Lopes, and S. D. G. Pompolo. 2012. First cytogenetic characterization of a species of the arboreal ant genus *Azteca* Forel, 1978 (Dolichoderinae, Formicidae). *Comparative Cytogenetics* 6(2): 107–114. doi: 10.3897/CompCytogen. v6i2.2397.

Cook J. M. 1993. Sex determination in the Hymenoptera: A review of models and evidence. *Heredity* 71: 421–435.

Cristiano M. P., D. C. Cardoso, and T. M. Fernandes-Salomaõ. 2013. Cytogenetic and molecular analyses reveal a divergence between Acromyrmex striatus (Roger, 1863) and other congeneric species: Taxonomic implications. *PLoS One* 8(3): e59784. doi:10.1371/journal.pone.0059784.

Crozier R. H. 1975. *Animal Cytogenetics. 3. Insecta (7) Hymenoptera*. P. 95. Gebrüder Bornträger, Berlin and Stuttgart, Germany.

Crozier R. H. 1968. An acetic acid dissociation, air-drying technique for insect chromosomes, with acetolactic orcein staining. *Stain Technology* 43: 171–173.

Darling D. C. and J. H. Werren. 1990. Biosystematics of *Nasonia* (Hymenoptera: Pteromalidae): Two new species reared from birds' nests in North America. *Annals of the Entomological Society of America* 83: 352–368.

Desjardins C. A., J. Gadau, J. A. Lopez, O. Niehuis, A. R. Avery, D. W. Loehlin, S. Richards, J. K. Colbourne, and J. H. Werren 2013. Fine scale mapping of the *Nasonia* genome to chromosomes using a high-density genotyping microarray. *G3* 5(2): 439–455. doi:10.1093/gbe/evt009.

Dzierzon J. 1845. Gutachten über die von Herrn Direktor Stöhr im ersten und zweiten Kapitel des General-Gutachtens aufgestellten Fragen. *Eichstädter Bienenzeitung* 1: 109–113; 119–121.

Gadau J., C. U. Gerloff, N. Krüger, H. Chan, P. Schmid-Hempel, A. Wille, and R. E. Page Jr. 2001. A linkage analysis of sex determination in *Bombus terrestris* (L.) (Hymenoptera: Apidea). *Journal of Heredity* 87: 234–242.

Gadau J., O. Niehuis, A. Peire, J. H. Werren, E. Baudry, and L. W. Beukeboom. 2008. Jewel wasp, *Nasonia* spp. In *Genome Mapping and Genomics in Animals*, Edited by Wayne Hunter and Chittaranjan Kole, pp. 27–41. Springer Verlag, Berlin and Heidelberg, Germany.

Gadau J., R. E. Page Jr., and J. H. Werren. 1999. Mapping hybrid incompatibility loci in Nasonia. *Genetics* 153: 1731–1741.

Gadau J., R. E. Page Jr., and J. H. Werren. 2002. The genetic basis of the interspecific differences in wing size in *Nasonia* (Hymenoptea; Pteromalidae)—Major QTL and epistasis. *Genetics* 161: 673–684.

Gershenzon S. M. 1946. The genetic structure of the natural populations of *Mormoniella vitripennis* Wlk. (Chalcididae: Hymenoptera). [in Russian] *Zh Obshchei Biol* 7: 165–173.

Gibson J. D., O. Niehuis, B. R. E. Peirson, E. I. Cash, and J. Gadau. 2013. Genetic and developmental basis of F2 hybrid breakdown in *Nasonia* parasitoid wasps. *Evolution* 67: 2124–2132.

Gokhman V. E. and M. Westendorff. 2000. The chromosomes of three species of the *Nasonia* complex (Hymenoptera, Pteromalidae). *Beiträge zur Entomologie* 50:193–198.

The Honeybee Genome Sequencing Consortium et al. 2006. Chromosomal anchoring of linkage groups and identification of wing size QTL using markers and FISH probes derived from microdissected chromosomes in *Nasonia* (Pteromalidae: Hymenoptera). *Nature* 443: 931–949.

Imai H. T. 1966. The chromosome observation techniques of ants and the chromosomes of Formicinae and Myrmicinae. *Acta Hymenopterologica* 2(3): 119–131.

Imai H. T. and R. W. Taylor. 1989. Chromosomal polymorphisms involving telomere fusion, centromeric inactivation and centromere shift in the ant *Myrmecia* (*pilosula*) $n = 1$. *Chromosoma* 98: 456–460.

Imai H. T., R. W. Taylor, M. W. J. Crossland, and R. H. Crozier. 1988. Modes of spontaneous chromosomal mutation and karyotype evolution in ants with reference to the minimum inter-action hypothesis. *Japanese Journal of Genetics* 63: 159–185.

Imai H. T., R. W Taylor, and R. H. Crozier. 1994. Experimental bases for the minimum interaction theory. I. Chromosome evolution in ants of the *Myrmecia pilosula* species complex (Hymenoptera: Formicidae: Myrmeciinae). *Japanese Journal of Genetics* 69: 137–182.

Loehlin D. W. and J. H. Werren. 2012. Evolution of shape by multiple regulatory changes to a growth gene. *Science* 335: 943–947.

Lorite P. and T. Palomeque. 2010. Karyotype evolution in ants (Hymenoptera: Formicidae), with a review of the known ant chromosome numbers. *Myrmecological News* 13: 89–102.

Müller S., Neusser M., and Wienberg J. 2002. Towards unlimited colors for fluorescence in-*situ* hybridization (FISH). *Chromosome Research* 10: 223–232.

Nachtsheim H. 1916. Zytologische Studien über die Geschlechtsbestimmung bei der Honigbiene (Apis mellifera). *Arch Zellforsch* 11: 169–241.

Niehuis O., J. Buellesbach, J. D. Gibson, D. Pothmann, C. Hanner, N. Mutti, A. K. Judson, J. Gadau, J. Ruther, and T. Schmitt. 2013. Behavioural and genetic analyses on *Nasonia* shed light on the evolution of sex pheromones. *Nature* 494: 345–348.

Niehuis O., J. D. Gibson, M. S. Rosenberg, B. A. Pannebakker, T. Koevoets, A. K. Judson, C. A. Desjardins et al. 2010. Recombination and its impact on the genome of the haplodiploid parasitoid wasp *Nasonia*. *PLoS One* 5(1): e8597. doi:10.1371/journal.pone.0008597.

Niehuis O., A. K. Judson, and J. Gadau. 2008. Cytonuclear genic incompatibilities cause increased mortality in male F2 hybrids of *Nasonia giraulti* and *N vitripennis*. *Genetics* 178: 413–426.

Nur, U., J. H. Werren, D. Eickbush, W. Burke, and T. Eickbush. 1988. A "selfish" B chromosome that enhances its transmission by eliminating the paternal chromosomes. *Science* 240: 512–514.

Pietsch C., K. B. Rütten, and J. Gadau. 2004. Eleven microsatellite markers in *Nasonia*, Ahmead 1904 (Hymenoptera; Pteromalidae). *Molecular Ecology Notes* 4: 43–45.

Raychoudhury R., C. A. Desjardins, J. Buellesbach, D. W. Loehlin, B. K. Grillenberger, L. Beukeboom, T. Schmitt, and J. H. Werren. 2010. Behavioral and genetic characteristics of a new species of *Nasonia*. *Heredity* 104: 278–288.

Reed K. M. 1993. Cytogenetic analysis of the paternal sex ratio chromosome of *Nasonia vitripennis*. *Genome* 36:157–161.

Rütten K. B., C. Pietsch, K. Olek, M. Neusser, L. W. Beukeboom, and J. Gadau. 2004. Chromsomal anchoring of linkage groups and identification of wing size QTL using markers and FISH probes derived from microdissected chromosomes in Nasonia (Pteromalidae: Hymenoptera). *Cytogenetic and Genome Research* 105: 126–133.

Siebold CTh. 1856. Die Drohneneier sind nicht befruchtet. *Bienenzeitung* 12: 181–184.

Verhulst E. C., L. W. Beukeboom, and L. van de Zande. 2010. Maternal control of haplodiploid sex determination in the wasp *Nasonia*. *Science* 328: 620–623.

Werren J. H. and D. W. Loehlin. 2009. The parasitoid wasp *Nasonia*: An emerging model system with haploid male genetics. *Cold Spring Harbor Protocols* 2009: pdb.emo134. doi:10.1101/pdb.emo134.

Werren J. H., S. Richards, C. A. Desjardins, O. Niehuis, J. Gadau, J. K. Colbourne, and The Nasonia Genome Working Group. 2010. Functional and evolutionary insights from the genomes of three parasitoid *Nasonia* species. *Science* 327: 343–348. doi:10.1126/science.1178028.

Whiting A. R. 1961. Genetics of Habrobracon. *Advances in Genetics* 10: 295–348.

Whiting A. R. 1967. The biology of the parasitic wasp *Mormoniella vitripennis* [=*Nasonia brevicornis*] (Walker). *The Quarterly Review of Biology* 42: 333–406.

Whiting P. W. 1933. Selective fertilization and sex determination in Hyemnoptera. *Science* 78 :537–538.

Whiting P. W. 1960. Polyploidy in Mormoniella. *Genetics* 45: 949–970.

8

Bedbugs (Hemiptera)

Snejana Grozeva, Boris A. Anokhin,
and Valentina G. Kuznetsova

CONTENTS

LIST OF ABBREVIATIONS

AgNO$_3$, silver nitrate
Ag-positive, argentum positive
AT, adenine-thymine
BSA, bovine serum albumin
C-banding, differential staining of the heterochromatin regions
CCD, (charge-coupled device) camera
CMA$_3$, chromomycin A$_3$ binding to GC-base pairs of DNA
DAPI, 4′,6-diamidino-2-phenylindole
dNTP, deoxyribose nucleoside triphosphate
dUTP, deoxyuridine nucleotide triphosphate
FISH, fluorescence in situ hybridization
FITC, fluorescein isothiocyanate
GC, guanine–cytosine
NORs, nucleolus organizing regions
PBS, phosphate-buffered saline
PCR, polymerase chain reaction
PFA, paraformaldehyde
rDNA, ribosomal DNA
RFBR, Russian Foundation for Basic Research
RNase A, ribonuclease A
rRNA, ribosomal RNA
RT, room temperature
SDS, sodium dodecyl sulfate
SSC, saline sodium citrate

Good night, sleep tight; don't let the bed bugs bite!

Nidhi Pandey (2005)
This common nighttime verse now has become a precautionary catch phrase around the globe.

Bai et al. (2011)

8.1 INTRODUCTION

8.1.1 Taxonomy

The bedbug genus *Cimex* Linnaeus, 1978, is a relatively small group of highly specialized hematophagous ectoparasites belonging to the worldwide spread family Cimicidae Latreille, 1802, (Hemiptera: Cimicomorpha) with 110 species hitherto described (Henry 2009). The genus includes 17 species distributed primarily across the Holarctic and associated with humans, bats, and birds (Schuh and Slater 1995; Simov et al. 2006). Most species feed primarily on bats and birds, but three species feed on humans: the tropical bedbug *Cimex hemipterus* Fabricius, 1803; *Leptocimex boueti* (Brumpt, 1910) found in the tropics of West Africa; and the common bedbug *Cimex lectularius* Linnaeus, 1758.

8.1.2 Origin and Distribution

Bedbugs have been known as human parasites for thousands of years. The association of the common bedbug and humans dates back to 3350 years ago or earlier, as evidenced by well-preserved bedbug remains recovered from the Workmen's Village at el-Amarna, Egypt (Panagiotakopulu and Buckland 1999). *C. lectularius* was associated with man and bats when all three lived together in caves somewhere in the Middle East (Sailer 1952). The hypothesis suggesting bats to be the original hosts is usually accepted as the most plausible (Sailer 1952; Ueshima 1966). When humans moved from caves into houses, the bugs went with them. Nowadays, *C. lectularius* is connected with man, chickens, and rarely, other domesticated animals. Due to its association with human beings, the distribution of *C. lectularius* is currently nearly cosmopolitan. Bedbugs can be found almost anywhere humans have established homes and cities, humans being the most common hosts nowadays (Ueshima 1966). *C. lectularius* was practically extirpated by a mass use of DDT (dichlorodiphenyltrichloroethane) in the 1940s and 1950s, but it has restarted new expansion in all developed countries of the Temperate Zone including North America, Europe, Australia, and Eastern Asia with an estimated 100%–500% annual increase in bedbug populations during the last decades. Resistance of *C. lectularius* to insecticides/pesticides is one factor thought to be involved in its sudden resurgence (Hwang et al. 2005; Romero et al. 2007; Bai et al. 2011).

8.1.3 Morphology

Adult bedbugs are 6 to 7 mm long, broadly oval, flat, brown to reddish-brown, with a three-segmented rostrum, four-segmented antennae, and vestigial wings. They are flightless; have mouthparts designed for piercing and sucking; dorsoventrally flattened bodies covered with short, golden-colored hairs; and give off a distinctive, musty, sweetish odor containing various aldehydes that are produced by glands located in the ventral metathorax (Weatherston and Percy 1978).

8.1.4   Life Cycle

Bedbugs exhibit incomplete or gradual metamorphosis, from egg, through five nymphal stages, to adult. The whole life cycle embraces 5 weeks at 75%–80% RH and 28°C–32°C. Each female may produce 200–500 eggs in her lifetime, which can last approximately 2 years. Each active instar may feed multiple times if hosts are readily available. Adults need at least one blood meal of adequate volume for nutrition and reproduction (Ueshima 1966; Krinsky 2002).

8.1.5   Reproductive Behavior

C. lectularius bugs display a unique reproductive behavior (shared with all the Cimicidae and some other true bugs) when males mate by extragenital insemination (internal insemination without the involvement of the female's genitalia). The insemination process is termed "traumatic" because it involves the male fertilizing the female through an integumental wound. When the male mounts the female, it uses the copulatory organ (paramere) to penetrate through the female's integument and inject sperm into the abdomen (Usinger 1966; Haynes et al. 2010). Females have evolved a specialized paragenital organ to accept the paramere through its body wall called an ectospermalage (Ueshima 1966; Stutt and Siva-Jothy 2001; Reinhardt and Siva-Jothy 2007).

8.1.6   Importance

The Heteroptera, or true bugs, include many species of economical and medical importance, the common bedbug being among them as one of the particularly relevant. Bedbug infestations are rapidly increasing worldwide. Bedbugs affect people of all social and economic levels, and infestations have been found in most every human-made structure, including hotels, apartments, hospitals, homeless shelters, single-family homes, nursing homes, office buildings, and schools (Potter et al. 2010). They inject anticoagulant as well as other pharmacologically active substances, and withdraw blood painlessly. Health consequences include nuisance biting and cutaneous and systemic reactions. Although evidence of disease transmission by bedbugs is lacking (Goddard and deShazo 2009), they are important public health pests as their bites can cause discomfort and anxiety (Reinhardt and Siva-Jothy 2007). The control of these important insect pests in urban environments costs billions of dollars annually and typically requires the use of large quantities of pesticides/insecticides (Bai et al. 2011).

Because of its high-impact status, *C. lectularius* is the subject of significant media attention (Wang et al. 2010). Norihiro Ueshima believed "scientifically, because bedbugs are easy to rear they are ideal subjects for laboratory research" (Ueshima 1966). *C. lectularius* does constitute a very interesting species including from a cytogenetic point of view.

8.1.7   Classical Cytogenetics

C. lectularius was the first bedbug species studied in respect to karyotype and male meiosis (Slack 1938, 1939a,b; Darlington 1939). Since then, many cytogenetic studies were focused on the genus *Cimex*, in particular on *C. lectularius* (see for review Ueshima 1966, 1979; Grozeva et al. 2010, 2011; Sadílek et al. 2013). As a result, a number of peculiar cytogenetic features inherent in this species were discovered and described.

8.1.7.1   Type of Chromosomes

As is typical in the Heteroptera, *C. lectularius* bugs display holokinetic chromosomes (Ueshima 1966; Grozeva et al. 2010; Sadílek et al. 2013). These chromosomes have, instead of localized centromere, a kinetochore plate spread along their whole or almost whole length and attach to spindle microtubules along their entire length during cell divisions (Ueshima 1966, 1979; Grozeva et al. 2010; Kuznetsova et al. 2011). As a result, during mitotic anaphase, the sister chromatids migrate in parallel to the spindle poles, in contrast to monocentric species in which pulling forces are exerted on a single chromosomal point and chromosome arms trail behind (Melters et al. 2012). Holokinetic chromosomes have the unusual property of fusing (complete or partial symploidy) or undergoing fragmentation (complete or partial agmatoploidy) with no abnormalities in mitotic and meiotic divisions because each chromosome fragment behaves as an intact chromosome (White 1973).

Holokinetic chromosomes occur in certain scattered groups of plants and animals, being particularly widespread in insects, including dragonflies (Odonata), earwigs (Dermaptera), barklice and booklice (Psocoptera), chewing lice (Mallophaga), sucking lice (Anoplura), leafhoppers, planthoppers, treehoppers, cicadas, aphids, psyllids, whiteflies, scale insects (Homoptera), true bugs (Heteroptera), butterflies (Lepidoptera), caddisflies (Trichoptera), and zorapterans or angel insects (Zoraptera). Thus, holokinetic chromosomes occur in every cohort of Pterygota, suggesting that they are likely to have evolved at least four times independently in insect evolution (Kuznetsova et al. 2011).

8.1.7.2 Standard Karyotype

Slack (1938) was first to study chromosome cytology of *C. lectularius*, one of the most popular insects all over the world. Shortly afterwards based on the observations of Darlington (1939) and then Ueshima (1966, 1967, 1979) who studied and discussed the unique aspects of male meiosis in this species, the standard complement of *C. lectularius* males was interpreted as $2n = 26 + X_1X_2Y$. In the recent studies on this species originating from 14 European countries, the United States, Mexico, Egypt, and Japan (Table 8.1), this karyotype formula was confirmed (Grozeva et al. 2010, 2011; Sadílek et al. 2013). The chromosomes gradually decrease in size from 5.3% to 1.7% and the sex chromosomes are medium sized, the X_1 being clearly larger than the X_2 while of similar size with the Y (Grozeva et al. 2010; Sadílek et al. 2013).

8.1.7.3 Polymorphism for X Chromosome Number

The X chromosomes were found to vary in number from 2 (X_1X_2Y, the standard) to 20 (X_1X_2Y + 18 extra Xs) in different populations of *C. lectularius* while occasionally between specimens of the same population and even between cells of a male or a female either between germinal cells or occasionally between its germinal and midgut cells (Ueshima 1967; Sadílek et al. 2013) (Table 8.1). Multiple (above two) X chromosomes have been described in both natural populations and laboratory stocks of *C. lectularius*. According to Darlington (1939), the average number of Xs is higher in wild populations than in laboratory cultures; however, a closer look at the presently available data on this species is called for (Grozeva et al. 2010).

Ueshima (1966) has investigated males and females in laboratory stocks originated from six populations of the United States (Berkeley, California; Columbus, Ohio), Mexico (Monterey; La Piedad), Japan (Nagasaki), France (Durtal), Egypt (Cairo), and Czech Republic (Moravia), respectively. He showed that the number of X chromosomes was variable between populations being, however, stable within every population (2X's in Berkeley, La Piedad, Nagasaki, and Durtal; 6X's in Cairo and Moravia) except for the Ohio population, in which males had either 7X's or 9X's (Table 8.1). Notice that the transmission of additional sex chromosomes throughout meiosis was, except in a very few cases, quite regular, and they seemed not to be important for sex determination.

More recently, Sadílek et al. (2013) studied 116 males and females of *C. lectularius* from 61 localities within 10 European countries (Austria, Switzerland, Czech Republic, France, Great Britain, Italy,

TABLE 8.1
Collection Sites and Sex Chromosomes in Different Samplings of *Cimex Lectularius*

Country[a]	Locality	♂	♀	Sex Chromosomes	References
		Specimens			*lectularius*
A	Melk	1	2	$X_7Y, X_{10}Y$	Sadílek et al. 2013
BG	Sofia	4		X_2Y	Grozeva et al. 2010
CH	?		1	X_2Y	Sadílek et al. 2013
CH	Fribourg—Rue de l'Hôpital	1	1	X_2Y	Sadílek et al. 2013
CH	Luzern	1		X_2Y	Sadílek et al. 2013
CZ	Bílá Lhota		1	X_2Y	Sadílek et al. 2013
CZ	Bohumín—Studentská	3		X_3Y, X_4Y	Sadílek et al. 2013
CZ	Bruntál	1		X_2Y	Sadílek et al. 2013
CZ	Česká Lípa—Svárovská	3	1	X_2Y, X_6Y, X_7Y	Sadílek et al. 2013
CZ	České Budějovice (1)—Puklicova		1	X_4Y	Sadílek et al. 2013
CZ	České Budějovice (3)—Okružní		1	X_4Y	Sadílek et al. 2013
CZ	České Budějovice (4)—Netolická	1	1	X_3Y, X_5Y, X_6Y, X_7Y	Sadílek et al. 2013
CZ	České Budějovice (5)—J. Bendy	1		X_2Y	Sadílek et al. 2013
CZ	České Budějovice (6)—M. Chlajna	2		X_2Y	Sadílek et al. 2013
CZ	Chomutov—Dřínovská	1	1	X_2Y	Sadílek et al. 2013
CZ	Chvalšiny		2	X_5Y	Sadílek et al. 2013
CZ	Humpolec	2	1	X_4Y	Sadílek et al. 2013
CZ	Janov	1	2	X_3Y, X_5Y, X_6Y, X_7Y	Sadílek et al. 2013
CZ	Jaroměřice nad Rokytnou	1		X_3Y	Sadílek et al. 2013
CZ	Jirkov—Na Borku	1	1	X_2Y	Sadílek et al. 2013
CZ	Liberec (1)—KrejČího	2	1	X_4Y, X_7Y	Sadílek et al. 2013
CZ	Liberec (2)—KrejČího	3		$X_8Y, X_{15}Y, X_{20}Y$	Sadílek et al. 2013
CZ	Moravia	?	?	X_6Y	Ueshima 1966[b]
CZ	MoraviČany	1		X_2Y	Sadílek et al. 2013
CZ	Plzeň (1)	2	1	X_2Y	Sadílek et al. 2013
CZ	Plzeň (2)—Na Vinicích		2	X_2Y	Sadílek et al. 2013
CZ	Plzeň (3)—Na Slovanech	1	1	X_2Y	Sadílek et al. 2013
CZ	Plzeň (4)—Na Slovanech	2	1	X_2Y	Sadílek et al. 2013
CZ	Plzeň (5)	1		X_3Y	Sadílek et al. 2013
CZ	Plzeň (6)—Na Slovanech	2		X_2Y	Sadílek et al. 2013
CZ	Plzeň (7)	2		X_2Y, X_5Y	Sadílek et al. 2013
CZ	Praha (1)	2		X_3Y	Sadílek et al. 2013
CZ	Praha (2)	1		X_4Y	Sadílek et al. 2013
CZ	Praha (3)		1	X_3Y	Sadílek et al. 2013
CZ	Praha (4)	3		X_3Y	Sadílek et al. 2013
CZ	Praha (5)—Křížíkova		1	X_3Y	Sadílek et al. 2013
CZ	Štědrákova Lhota	1	1	X_2Y, X_3Y	Sadílek et al. 2013
CZ	Strakonice—Bezděkovská	1		X_2Y	Sadílek et al. 2013
CZ	Stráž pod Ralskem	1		X_2Y	Sadílek et al. 2013
CZ	Šumperk	1	1	X_2Y	Sadílek et al. 2013
CZ	Týn nad Vltavou—Hlinecká	1		X_2Y	Sadílek et al. 2013

(Continued)

TABLE 8.1 (*Continued*)
Collection Sites and Sex Chromosomes in Different Samplings of *Cimex Lectularius*

Country[a]	Locality	Specimens ♂	♀	Sex Chromosomes	*lectularius* References
CZ	Žďár nad Sázavou	1		X_2Y	Sadílek et al. 2013
EG	Cairo	?	?	X_6Y	Ueshima 1966[b]
F	Aire/Adour		2	X_3Y	Sadílek et al. 2013
F	Durtal	?	?	X_2Y	Ueshima 1966[b]
GB	Brighton		1	X_2Y	Sadílek et al. 2013
I	Mestre	1	2	X_3Y	Sadílek et al. 2013
I	Venezia (1)	1		X_2Y	Sadílek et al. 2013
I	Venezia (2)	2	1	X_2Y, X_3Y	Sadílek et al. 2013
I	Venezia (3)	1		X_2Y, X_5Y	Sadílek et al. 2013
J	Nagasaki	?	?	X_2Y	Ueshima 1966[b]
M	Monterrey, La Piedad	?	?	X_2Y	Ueshima 1966[b]
N	Ottestad	1		X_2Y	Sadílek et al. 2013
PL	Białystok (1)	1		X_2Y	Sadílek et al. 2013
PL	Białystok (2)	1		X_2Y	Sadílek et al. 2013
PL	Gdansk (1)	1		X_3Y	Sadílek et al. 2013
PL	Gdansk (2)	2	1	X_3Y, X_4Y	Sadílek et al. 2013
PL	Świnoujscie		1	X_2Y	Sadílek et al. 2013
PL	Wroclaw—Grabiszynska		1	X_2Y	Sadílek et al. 2013
RU	St Petersburg	5		X_2Y	Grozeva et al. 2010
S	Borlänge (1)	2		X_6Y	Sadílek et al. 2013
S	Borlänge (2)	1	1	$X_6Y, X_{10}Y$	Sadílek et al. 2013
S	Stockholm—Vårber	1	2	X_4Y, X_5Y	Sadílek et al. 2013
SK	Banská Bystrica	2		X_2Y	Sadílek et al. 2013
SK	Hosťovce	2	1	X_2Y	Sadílek et al. 2013
SK	Krásnohorské Podhradie	2		X_2Y	Sadílek et al. 2013
SK	Trnava	5	1	$X_4Y, X_7Y, X_8Y, X_9Y, X_{10}Y, X_{13}Y$	Sadílek et al. 2013
UK	Edinbourgh, Glasgow, South London	35		$X_2\text{-}X_{16}Y$	Slack 1938
UK	London	11		$X_2Y, X_6Y, X_7Y, X_8Y, X_9Y, X_{10}Y$	Darlington 1939
UK	Cork	7		$X_2Y, X_6Y, X_7Y, X_{10}Y$	Darlington 1939
UK	Glasgow	15		$X_4Y, X_7Y, X_8Y, X_{10}Y, X_{11}Y, X_{12}Y, X_{14}Y$	Darlington 1939
UK	Lamberth	7		$X_7Y, X_9Y, X_{10}Y, X_{11}Y, X_{12}Y, X_{13}Y$	Darlington 1939
UK	Mitcham	11		$X_{10}Y, X_{13}Y, X_{14}Y$	Darlington 1939
UK	Sheffield	3		X_7Y	Darlington 1939
UK	Others	18		$X_2Y, X_6Y, X_7Y, X_8Y, X_9Y, X_{10}Y, X_{11}Y, X_{12}Y, X_{13}Y, X_{14}Y$	Darlington 1939
USA	Barkley, California	?	?	X_2Y	Ueshima 1966[b]
USA	Columbus, Ohio	?	?	X_7Y	Ueshima 1966[b]
USA	Pittsburgh, Pennsylvania	?	?	X_6Y	Ueshima 1966[b]

[a]A, Austria; BG, Bulgaria; CH, Switzerland; CZ, Czech Republic; EG, Egypt; F, France; GB, Great Britain; I, Italy; J, Japan; M, Mexico; N, Norway; PL, Poland; RU, Russia; S, Sweden; SK, Slovakia; UK, United Kingdom.
[b]All populations after Ueshima 1966 were presented at least 20 specimens each.

Norway, Poland, Sweden, and Slovakia). Among the specimens studied, 12 distinct chromosomal complements were identified. The standard complement, $2n = 26 + X_1X_2Y$, was found in 57.4% of the samples and in 44% of the specimens studied. The remaining specimens showed a great diversity of chromosome number due to the presence of extra X chromosomes varying in number from 1 to 18 with separate gaps only (Table 8.1). The number of autosomes was consistently 26. Within a locality, males and females usually had the same pattern whereas in some cases had not. Variability within a single specimen (the mosaicism) occurred rarely (4.3%).

The origin of multiple systems in the Heteroptera is usually ascribed to simple transverse fragmentations of the original X chromosome, the process that is facilitated by the holokinetic nature of the bugs' chromosomes (Schrader 1947; Ueshima 1966, 1979; Kuznetsova et al. 2011). The distinguishing feature of multiple sex chromosome systems formed by fragmentation is that as the number of X chromosomes increases, their size decreases (Ueshima 1966; Sadílek et al. 2013). However, this problem calls for further investigation using modern cytological techniques (Grozeva et al. 2010). For example, the application of C-banding showed a great variability in size and C-banding patterns of X chromosomes in triatomine bugs, which suggests that autosomal rearrangements may also be involved in the formation of the multiple sex mechanisms (Pérez et al. 2004; Panzera et al. 2010).

8.1.7.4 Male Meiosis

In *C. lectularius*, male meiosis is of a peculiar type with a reverse sequence of sex chromosomes' divisions in males (Ueshima 1979; Grozeva et al. 2010). The sex chromosomes, regardless of their number in a male, behave as univalents during the first round of meiosis and undergo equational separation at anaphase I. In X_1X_2Y males, X_1 and Y chromosomes at metaphase I tend to be located close to each other or even connected by ends (Figure 8.1d and e, Figure 8.1g and h). At metaphase II, the two X's and the Y appear associated end-to-end to form a pseudotrivalent, which is located inside the ring of autosomes (Figure 8.1n though q, and Figure 8.2a). During anaphase II, X's and Y chromosomes undergo reductional division and segregate to opposite poles (Figure 8.1q). This inverted meiosis, the so-called "the sex chromosome post-reduction," is known to occur in the great majority of Cimicomorpha (Kuznetsova et al. 2011) and the Heteroptera as a whole (Ueshima 1979; Papeschi and Bressa 2006). As in other heteropterans, the autosomes in *C. lectularius* show normal sequence of meiotic divisions (Figures 8.1n through q and 8.2) with homologous chromosomes segregating in the first round of meiosis, and sister chromatids separating in the second (Slack 1939b; Darlington 1939; Ueshima 1966, 1979; Grozeva et al. 2010).

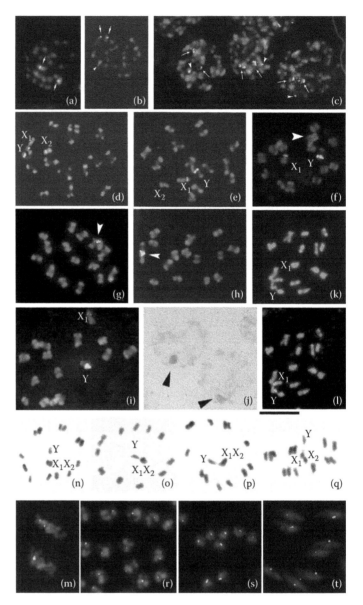

FIGURE 8.1 **(See color insert.)** Different stages of male meiosis in *Cimex lectularius* after fluorescence in situ hybridization with an 18S ribosomal DNA probe (a–i, m, r–t), $AgNO_3$ (j), 4′,6-diamidino-2-phenylindole (k), chromomycin A_3 binding to guanine–cytosine-base pairs of DNA (l), and Schiff–Giemsa (n–q). (a–c)—spermatogonial metaphases: arrows point to X_1 and Y with signals, arrowheads to additional signals in some plates; (d–f)—meiotic prometaphase I: X_1 and Y with signals lie together (d) or separately (e, f); (g–i)—metaphase I: X_1 and Y with signals lie together or separately. Autosomal bivalents are condensed and consist of parallel-aligned chromosomes; (j)—diffuse stage: nucleolar proteins are localized on the sex chromatin body; (k–l)—metaphase I: X_1 and Y with CMA_3-positive signals (l); (m)—anaphase I: there are signals in both daughter cells; (n–q)—metaphase II: radial plates with sex chromosomes placed inside the ring formed by autosomes; (r–t)—consecutive stages of sperm formation: every sperm with a signal. Bar = 10 μm. (From Grozeva, S. et al., *Comp Cytogen* 4(2), 151–160, 2010.)

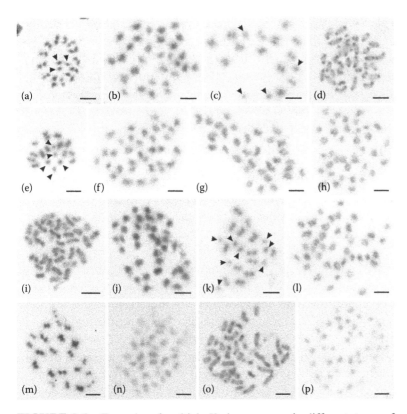

FIGURE 8.2 Examples of multiple X chromosomes in different stages of cell division in *C. lectularius* (Giemsa). (a)—metaphase II ♂, 2n = 29 (note the radial form of MII plate); (b)—mitotic metaphase ♀, 2n = 30; (c)—metaphase II ♂, 2n = 30; (d)—mitotic prometaphase ♀, 2n = 32; (e)—metaphase II ♂, 2n = 31; (f)—mitotic metaphase ♀, 2n = 34; (g)—mitotic metaphase ♂, 2n = 32; (h)—mitotic metaphase ♀, 2n = 36; (i)—mitotic prometaphase ♂, 2n = 33; (j)—mitotic metaphase ♀, 2n = 38; (k)—metaphase II ♂, 2n = 34; (l)—mitotic metaphase ♀, 2n = 40; (m)—metaphase I ♂, 2n = 35; (n)—mitotic metaphase ♂, 2n = 36; (o)—mitotic prometaphase ♂, 2n = 37; (p)—mitotic metaphase ♂, 2n = 40. Arrows indicate sex chromosomes. Bar = 5 μm. (From Sadílek, D. et al., *Comp Cytogen* 7(4), 253–269, 2013.)

According to Ueshima (1966, 1967), male meiosis in *C. lectularius* is chiasmate; in diakinesis, the homologues lay parallel and bivalents have one chiasma each. However, our observations (Grozeva et al. 2010) did not support this generalization suggesting the occurrence of achiasmate meiosis in *C. lectularius*. This meiotic pattern is probably characteristic of the family Cimicidae as a whole (Grozeva and Nokkala 2002; Poggio et al. 2009; Kuznetsova et al. 2011). Strong support for the achiasmate type of meiosis is known to come from the absence of diplotene and diakinesis stages (Nokkala and Nokkala 1986) and such is the case in all cimicid species studied so far, including *C. lectularius* (Grozeva et al. 2010). Notably, in cimicids, achiasmate meiosis is of the specific collochore type (Grozeva and Nokkala 2002; Grozeva et al. 2010).

In such meiosis, the homologous chromosomes are not physically aligned along their length during prophase; however, after synapsis, they appear physically associated in one-two sites by tenacious threads, the so-called collochores (Figure 8.1d through f, k and l), which hold homologous chromosomes together in the absence of chiasmata (Nokkala and Nokkala 1986; Kuznetsova and Grozeva 2010; Kuznetsova et al. 2011).

8.1.7.5 Localization of Chromosomal Nucleolus Organizing Regions

The nucleolus represents a subnuclear compartment of eukaryotic cells in which the synthesis of ribosomal RNA (rRNA) and formation of ribosomes take place (Busch and Smetana 1970). Silver nitrate ($AgNO_3$) is known to stain nucleolus organizing regions (NORs); however, this technique is able to reveal only active NORs (Hubbell 1985) being therefore inadequate to the study of NOR location into chromosomes. In the last few decades, the ability to identify separate chromosomes and specific regions in a chromosome has been markedly improved by the development of molecular cytogenetic techniques. In insects, fluorescence in situ hybridization (FISH) with ribosomal DNA (rDNA) probes is predominantly applied to directly detect the location of rRNA genes on a chromosome, regardless of their activity.

$AgNO_3$ staining being applied to *C. lectularius* males with $2n = 26 + X_1X_2Y$ originated from Bulgaria and Russia revealed argentum-positive sex chromosome body (Figure 8.1j) at diffuse stage of first meiosis (Grozeva et al. 2010). After CMA_3 (chromomycin A_3 binding to guanine–cytosine [GC]-base pairs of DNA) staining, signals were observed at the telomeric regions of X_1 and Y chromosomes (Figure 8.1l). Because NORs are largely GC-rich in the Heteroptera (Grozeva et al. 2004; Kuznetsova et al. 2007; Bressa et al. 2009; Bardella et al. 2010), CMA_3-positive regions in *C. lectularius* chromosomes were interpreted as the sites of NORs (Grozeva et al. 2010).

8.1.8 Molecular Cytogenetics

Since the first study in 1891 on *Pyrrhocoris apterus* (Pyrrhocoridae) (Henking 1891), all cytogenetic investigations in the Heteroptera have been predominantly restricted to chromosome counts of about 2,000 species (5% of approximately 40,000 described species) (Weirauch and Schuh 2011) and to gross karyotype and meiotic descriptions of some of them. In groups with holokinetic chromosomes, the main problem is to identify individual chromosomes and chromosomal regions in a karyotype. Differential cytogenetic techniques, such as C-banding, DNA-specific fluorochrome staining, $AgNO_3$ staining, make possible only a few markers to be revealed in true bugs' karyotypes (Papeschi and Bressa 2006; Kuznetsova et al. 2011).

In the last three decades, the major advance in bug cytogenetics has come with the application of molecular cytogenetic techniques such as FISH to detect DNA sequences in the chromosomes, southern-blot and dot-blot hybridization techniques to verify the presence or absence of a certain DNA sequence within a genomic DNA (gDNA) sample.

In two of our recent publications (Grozeva et al. 2010, 2011), FISH with 18S rDNA (Figure 8.1a through i and m through t) and insect $(TTAGG)_n$ telomere probes as well dot-blotting with a number of telomere repeat sequences characteristic for different groups of eukaryotes were applied for the first time to investigate genome composition in *C. lectularius*. Both studies were performed on males with standard karyotype of $2n = 26 + X_1X_2Y$ from Sofia (Bulgaria) and St. Petersburg (Russia).

8.2 PROTOCOLS

8.2.1 Fluorescence In Situ Hybridization Protocol

In situ hybridization experiments were performed following in general the protocols by Schwarzacher and Heslop-Harrison (2000) and Kuznetsova et al. (Chapter 10) with some modifications.

8.2.1.1 Material and Supplies

Use ultrapure water (ddH$_2$O) and analytical grade chemicals for preparing all stock solutions.

"Standard solutions" (such as ethanol, 20× saline sodium citrate [SSC], phosphate-buffered saline [PBS], formamide, formaldehyde) are needed. Remember that some of the chemicals are environmental toxins (as formamide and formaldehyde). Work in a chemical fume hood. Ensure that these substances are collected and treated as hazardous waste after use.

1. Formamide (Cat. No. F47671, Sigma-Aldrich, Germany)
2. Deionized formamide for hybridization buffer (Cat. No. F9037, Sigma-Aldrich)
3. PBS, pH 7.4 (Cat. No. 18912-014, phosphate buffered saline—Gibco, Invitrogen, UK)
4. 20× SSC (175.3 g NaCl, 88.2 g sodium citrate, water to 1000 mL), adjust to pH 7.0 autoclave before storage at room temperature (RT)
5. Deoxyribonucleic acid, single stranded from salmon testes (stock ~10 mg/mL) (Cat. No. D7656, Sigma-Aldrich, Germany)

6. Ribonuclease A (RNase A) from bovine pancreas (Cat. No. R4642, Sigma-Aldrich)

7. Pepsin from porcine gastric mucosa (Cat. No. P7000, Sigma-Aldrich, Germany)

8. 1N HCl

9. Rubber cement

10. Deoxyribose nucleoside triphosphate (dNTP): 100 mM stock solutions of deoxyadenosine triphosphate, deoxy-cytidine triphosphate, deoxyguanosine triphosphate, deoxythymidine triphosphate (dTTP) (Cat. No. R0181, Fermentas, EU)

11. Modified labeled nucleotide at 1 mM, such as digoxi-genin-11-deoxyuridine nucleotide triphosphate (dUTP) (Cat. No. 11 093 088 910, Roche Diagnostics GmbH, Mannheim, Germany), biotin-16-dUTP (Cat. No. 11 093 070, Roche Diagnostics GmbH, Mannheim, Germany), or biotin-11-dUTP (Cat. No. R0081, Fermentas)

12. Taq DNA polymerase with 10× Taq buffer (Cat. No. EP0402, Fermentas)

13. High-grade paraformaldehyde (37% PFA) (e.g., Merch or Fisher)

14. 1 mM stock $MgCl_2$

15. Bovine Serum Albumin (BSA) (Cat. No. 05473; Fluka, Sigma-Aldrich, St. Louis, MO)

16. Avidin–Fluorescein conjugate (Cat. No. A2662, Invitrogen)

17. Anti-Digoxigenin-Fluorescein, Fab fragments (Cat. No. 11207750910, Roche) or Anti-Digoxigenin-Rhodamine, Fab fragments (Cat. No. 11207750910, Roche)

18. 0.1 mM Rho-5-dUTP (Tamra-5-dUTP) (Cat. No. GC-013-013, Genecraft, Ares Bioscience GmbH, Köln, Germany)

19. ProLong Gold Antifade reagent with 4′,6-diamidino-2-phenylindole (DAPI) (Cat. No. P36931, Invitrogen, Molecular Probes)

Stock solutions to be prepared for FISH:

20. Hybridization buffer: dissolve 1 g dextran sulfate in 10 mL 50% deionized formamide/2× SSC/1% Tween 20, store in aliquots at −20°C

21. 0.01 N HCl for pepsin digestion (mix 0.5 mL 1 N HCl with 49.5 mL water)

22. 10% Pepsin stock (w/v): dissolve 1 g pepsin in 10 mL water at 37°C; store in aliquots at −20°C

23. Pepsin solution: add 50 μL of 10% Pepsin stock in 950 μL 0.01N HCl, make fresh

24. 1 M $MgCl_2$: dissolve 20.33 g of $MgCl_2$ in ddH_2O to a final volume of 100 mL

25. Postfix solution (1% PFA/1×PBS/0.05 M MgCl$_2$): dilute 25 mL of 4% PFA stock solution + 50 mL 2× PBS + 5 mL 1 M MgCl$_2$ with 20 mL water

26. RNase A solution (100 µg/mL) for slide pretreatment: dilute 3.3 µL RNase A solution (30 mg/mL) with 997 µL 2× SSC, make fresh

27. Washing buffers (diluted from stock 20× SSC): 2× SSC (dilute 100 mL 20× SSC with 900 mL water); 0.2× SSC (dilute 10 mL 20× SSC with 990 mL water); 4× SSC/0.02% Tween-20 (dilute 200 mL 20× SSC + 0.2 mL Tween-20 with 800 mL water)

28. Blocking buffer (1.5% BSA in 4× SSC/0.1% Tween 20): mix 0.15 g BSA + 10 µL Tween-20 + 2 mL 20× SSC + 8 mL water), store in aliquots at −20°C

29. 1 mg/mL Avidin–Fluorescein (or Avidin Rhodamine) conjugate stock (w/v): dissolve 1 mg Avidin–Fluorescein conjugate in 1 mL water or 1× PBS; store in aliquots at −20°C

30. 5 µg/mL Avidin–Fluorescein conjugate solution: dilute 5 µL Avidin–Fluorescein conjugate stock solution (1 mg/mL) with 995 µL blocking buffer, make fresh

31. 200 µg/mL Anti-Digoxigenin-Fluorescein or Anti-Digoxigenin-Rhodamine conjugate stock (w/v): dissolve 200 µg Anti-Digoxigenin-Fluorescein conjugate in 1 mL water; store in aliquots at −20°C

32. 5 µg/mL Anti-Digoxigenin-Fluorescein conjugate solution: dilute 25 µL Anti-Digoxigenin-Fluorescein conjugate with 975 µL blocking buffer, make fresh

8.2.1.2 Equipment

The standard molecular cytogenetic equipment and some specialized items are needed: charge-coupled device camera with image capture and hard- and software; fluorescence microscope, hot plate with digital temperature control for slide warming (see Section 8.2.1.6 for details).

1. Thermomixer compact (e.g., Eppendorf, Germany)
2. Polymerase chain reaction (PCR) cycler (e.g., Mastercycler personal; Eppendorf)
3. Centrifuge with maximum speed 14,000–16,000 rcf (e.g., Centrifuge 5415C, Eppendorf, Germany)
4. Agarose gel electrophoresis apparatus
5. UV transilluminator
6. Phase-contrast microscope (e.g., Leica DM6000 B microscope with a 100× objective, Leica DFC345 FX camera, and Leica Application Suite 3.7 software with an Image

Overlay module [Leica Microsystems, Wetzlar GmbH, Germany]. The filter sets we applied are A, L5, N21 [Leica Microsystems, Wetzlar GmbH, Germany])

8.2.1.3 Source of Chromosomes and Chromosome Spread Preparation

Extract testes from male adults and fix them in 3:1 ethanol/glacial acetic acid fixative solution:

1. Dissect gonads and squash in a drop of 45% acetic acid under coverslips (20 × 20 mm).
2. Remove cover slips by the dry ice method.
3. Dehydrate slides in fresh 3:1 fixative and air-dry.
4. Analyze preparations under a phase contrast microscope at 400.
5. Store slides in slide storage boxes at 4°C or −20°C for aging.

8.2.1.4 Probe Preparation (Labeling)

8.2.1.4.1 Labeling of Ribosomal 18S Ribosomal DNA Probe

For rDNA sequences detection on chromosomes by FISH, an 1.1 kb fragment of 18S rDNA was amplified by PCR using gDNA isolated by NucleoSpin Tissue Kit (Cat. No. 740952.50, Macherey-Nagel, Düren, Germany) from a specimen of a bug species *Pyrrhocoris apterus* (Heteroptera: Pyrrhocoridae).

Prepare 50 µL PCR reaction mixture containing the following:

1. 50–100 ng of gDNA
2. 5 µL of 1 mM modified labeled nucleotides, such as Fluorescein-12-, or Biotin-11-dUTP, or Digoxigenin-11-dUTP (0.1 mM final concentration [f.c.])
3. 5 µL of 1 mM dTTP/2 mM each of the other dNTPs (0.1 mM f.c. for dTTP, 0.2 mM f.c. for each of the other dNTPs)
4. 5 µL of 25 mM $MgCl_2$ (2.5 mM f.c.)
5. 5 µL of 10 pmol of each heterologous primer: 18S_R 5′-CGATACGCGAATGGCTCAAT-3′, 18S_F 5′-ACAAGGGGCACGGACGTAATCAAC-3′ (1 pmol f.c.)
6. 1 U Taq DNA polymerase in 1× buffer supplied by the polymerase manufacture
7. ddH_2O to a final volume of 50 µL

Perform PCR with thermal conditions—an initial denaturation period of 3 minutes at 94°C followed by 30–35 cycles of 30 seconds each at 94°C, annealing for 30 seconds at 50°C, 1.5 minutes extension step at 72°C, and final extension step of 5 minutes at 72°C.

8.2.1.4.2   Labeling of Telomeric (TTAGG)$_n$ Probe

To visualize clusters of telomeric repeats, the pentanucleotide (TTAGG)$_n$-specific telomeric DNA probe was amplified by PCR.

Prepare 50 µL PCR reaction mixture containing the following:

1. 0.1 mM (f.c.) Tamra-5-dUTP (Cat. No. GC-013-013, Genecraft)
2. 0.1 mM (f.c.) dTTP, 0.2 mM each of the other dNTPs
3. 2.5 mM (f.c.) MgCl$_2$
4. 1 pmol (f.c.) the two heterologous primers TTAGG_F 5′-TAACCTAACCTAACCTAACCTAA-3′ and TTAGG_R 5′-GGTTAGGTTAGGTTAGGTTAGG-3′)
5. 1U of Taq polymerase in 1× buffer supplied by polymerase manufacture.

Perform PCR with thermal conditions—an initial denaturation period of 3 minutes at 94°C followed by 30 cycles of 45 seconds each at 94°C, annealing for 30 seconds at 50°C, 50 seconds extension step at 72°C, and final extension step of 5 minutes at 72°C.

Using 2–5 µL of PCR product, check the quality of amplified DNA by agarose gel electrophoresis.

For 18S rDNA probe, a single band of DNA about 1100 bp in size should be visible.

For (TTAGG)$_n$ probe, a smear of DNA in size range between 100 and 1000 bp should be visible.

8.2.1.4.3   Postlabeling Treatment of the Probe

The PCR-labeled DNA should be prepared as follows:

1. Precipitated the PCR product containing labeled probe with 10 volumes of ethanol (96%–100%, −20°C) and 0.1 volumes of 3 M NaCl.
2. Precipitation can be done for 60 minutes at −80°C or overnight at −20°C.
3. Pellet the DNA by centrifugation at 14,000 rcf for 30 minutes at 4°C. After centrifugation discard the supernatant.
4. Untreated dNTP is then removed by washing the DNA pellet with 500–700 µL of ice-cold 70% ethanol: add ethanol and centrifuge at 14,000 rcf at 4°C for 15–20 minutes and discard the supernatant without removing the pellet.
5. Allow to air-dry at 37°C in a heat block.
6. Resuspend pellet in 20–30 µL of water. The concentration should be not less 200 ng/µL.
7. DNA can be kept at −20°C for long periods of time.

8.2.1.5 Hybridization and Detection Method

8.2.1.5.1 Slide Pretreatment Making of Hybridization Mixture

Pretreatment with RNase A

1. Apply 100–200 µL of RNase A solution (100 µg/mL) per slide and incubate for 0.5–1 hour at 37°C under coverslip in a humid chamber.
2. Rinse slides three times for 5 minutes in 2× SSC (pH 7.0–7.4) at RT.
3. Dehydrate slides through an alcohol series (make up fresh): 2 minutes in 70% ethanol, 2 minutes in 80% ethanol, and 2 minutes in 96%–100% ethanol.
4. Allow to air-dry.

Pretreatment with pepsin

5. Apply 100–200 µL of pepsin solution (5 mg/mL) and incubate for 5–12 minutes at 37°C under coverslip in a humid chamber.
6. Wash slides at RT sequentially for 5 minutes in 1× PBS, for 5 minutes in PBS× 1/0.05 M $MgCl_2$, for 10 minutes in 1% PFA in PBS× 1/0.05 M $MgCl_2$, for 5 minutes in 1× PBS, for 5 minutes in PBS× 1/0.05 M $MgCl_2$.
7. Dehydrate slides through 70%/80%/96% ethanol.
8. Air-dry slides.

To prepare hybridization mixture for one slide, add about 100 ng of labeled probe (not more than 0.5 µL, therefore the concentration should be not less than 200 ng/µL), 10 µg salmon-sperm DNA to 7 µL hybridization buffer (50% formamide, 2× SSC, 10% [w/v] dextran sulfate, 1% [w/v] Tween-20), and mix well for 1 hour at 37°C –45°C, spin down before dropping. The use of high temperatures should be avoided to prevent the probe denaturation.

8.2.1.5.2 Fluorescence In Situ Hybridization

After pretreatment, add 6.5–7 µL of hybridization mixture (about 100 ng of labeled probe, 50% formamide, 2× SSC, 10% [w/v] dextran sulfate, 1% [w/v] Tween-20, and 10 µg salmon-sperm DNA) directly on to each slide.

1. Put 20 × 20 mm coverslip (avoiding bubbles) on the hybridization mixture drop.
2. Mount with rubber cement.
3. Incubate slides on a warming plate for 5 minutes at 75°C.
4. Incubate slides for 42–44 hours at 37°C (overnight incubation at 42°C is also possible).

5. Take slides out of incubator.
6. Remove gently the rubber cement with forceps avoiding movements of the coverslips that can result in damage of chromosomes and nuclei.

8.2.1.5.3 *Slide Washing and Mounting*

To remove unbound or nonspecifically bound probe fragments:

1. Wash slides in prewarmed Coplin staining jar with 2× SSC for 3 minutes at 45°C.
2. Transfer slides in a Coplin staining jar with 50% formamide in 2× SSC for 10 minutes at 45°C.
3. Wash twice in 2× SSC (10 minutes each).
4. Wash twice in 0.2× SSC (5–10 minutes each) at 45°C.
5. Block in 1.5% (w/v) BSA/4× SSC/0.1% Tween-20 for 30 minutes at 37°C in a humid chamber.

Detection

1. Detect biotin labeled probe with 5 µg/mL Avidin–Fluorescein isothiocyanate in 1.5% BSA/4× SSC/0.1% Tween-20 (for digoxigenin-labeled probe, use 5 µg/mL solution of Anti-digoxigenin fluorophores, e.g., Anti-Digoxigenin-Fluorescein or Rhodamine) for 1 hour at 37°C.
2. Wash slides three times in 4× SSC/0.02% Tween-20 (10 minutes each) at 45°C.
3. Dehydrate slides through 70%/80%/96% ethanol at RT.
4. Add 15 µL DAPI/Antifade solution (e.g., ProLong Gold Antifade reagent with DAPI).
5. Cover slides with coverslips (24 × 32 mm) to examine the results under a fluorescence microscope.

8.2.1.6 Visualization and Mapping

Use fluorescence microscope equipped with appropriate filter set specific for the fluorochromes for visualization of fluorescence hybridization signals and a cool digital camera to document the results.

In our laboratory, the system in use for FISH analysis is a Leica DM6000 B microscope with a 100× objective, Leica DFC345 FX camera, and Leica Application Suite 3.7 software with an Image Overlay module (Leica Microsystems). The filter sets we applied are A, L5, N21 (Leica Microsystems).

8.2.2 Dot-Blot Protocol

1. Isolate gDNA using NucleoSpin Tissue Kit (Macherey-Nagel).
2. Label by PCR the probes by biotin (see above).

3. Add drop (about 20–100 ng) of denatured (5 minutes at 96°C) gDNA to Hybond N+ nylon membranes (Amersham, GE Healthcare, UK) and dry.

4. Carry out hybridization overnight in hybridization mixture containing about 100–200 ng of labeled probe, 50% formamide, 4× SSC, 0.5% (w/v) sodium dodecyl sulfate (SDS), and 10 µg salmon-sperm DNA at 40°C.

5. Wash membranes two times in 2× SSC/0.1% SDS (10 minutes each) at RT and two times in 0.2× SSC/0.1% SDS (10 minutes each) at RT.

6. Perform detection procedure using the Biotin Chromogenic Detection Kit protocol (Fermentas, Germany).

8.2.3   Troubleshooting

1. The quality of preparations is absolutely crucial for good hybridization results. A background should be avoided on the preparation. The preparation should be carefully squashed, single layer, and rich in chromosome spreads with well differentiated chromosomes. The preparations should be initially examined under a phase-contrast microscope to check the quality. The best slides to be aged for further FISH treatments should be stored at –20°C.

2. If no or weak fluorescence signal is observed, then, modify some steps. Notice, the water quality has a stronger effect on the signal-to-noise ratio than the purity of the chemicals. For most FISH solutions, it is necessary to autoclave ddH$_2$O for 20 minutes to destroy a DNase activity and ensure sterility.

   RNase/pepsin pretreatment is recommended. Usually, without pretreatment chromosomes show poor FISH signals, because cytoplasm makes the target DNA inaccessible to the probe. Normally, the pretreatment with RNase A is done at 37°C for between 30 minutes and 1 hour. If it lasts less than 30 minutes, the digestion is not complete whereas if it lasts longer than 1 hour, RNase A may begin to influence the chromosome morphology. Pepsin pretreatment is recommended for 5–12 minutes, the older material the longer time of pretreatment.

   Check the size of the labeled probe. In our laboratory, we use PCR product with labeled DNA fragments in size range between 100 and 1000 bp for (TTAGG)$_n$ probe and in size of 1100 bp for 18S rDNA probe. In every case, larger DNA fragments would result in bright fluorescence signals all over the slide. If the size of labeled DNA is larger, it should be recut using DNase I. When labeled DNA fragments are too small, they might not hybridize efficiently to chromosomal DNA; hence, FISH signals would not be visible.

3. If there is starry background fluorescence, then spin the detection solution 4 minutes at 5000 rcf and take only supernatant.

4. If the chromosomes and nuclei are overstained, then check the size of the labeled probe on an agarose gel with a suitable size marker. If the probe is too small, you have to relabel the probe following the same protocol.

8.2.4 Representative Results

8.2.4.1 Fluorescence In Situ Hybridization Mapping of Ribosomal RNA Genes

The majority of mitotic metaphase cells in *C. lectularius* males showed the presence of 18S rDNA clusters in subtelomeric regions of the chromosomes X_1 and Y (Figure 8.1a, f, and g); however, hybridization signals were over two in some prometaphase cells (Figure 8.1c). FISH signals were used as markers for the precise identification of the X_1 and Y chromosomes both in mitotic cells (Figure 8.1a through c) and at different stages of meiosis (Figure 8.1d through i). Notice that such identification is embarrassed when routine staining is applied. FISH signals were coincident with CMA_3-positive sites indicating thus rDNA arrays to be GC-rich (Figure 8.1l). A size polymorphism for the 18S rDNA clusters was occasionally inherent even within the same male (Figure 8.1f through i). As expected, FISH/rDNA signals were observed in every spermatid, confirming thus the location of NORs on one of the X chromosomes (X_1) and on the Y chromosome, which underwent a reductional division at anaphase II.

8.2.4.2 Fluorescence In Situ Hybridization Mapping with TTAGG and Dot-Blotting with Different Telomeric Repeats

FISH with the insect TTAGG probe produced no signals on chromosome spreads suggesting telomeres in *C. lectularius* to be of some other molecular composition. For telomere composition to be revealed, dot-blot hybridization of *C. lectularius* gDNA with telomeric probes of different groups of animals and plants, namely, ciliate $(TTTTGGGG)_n$ and $(TTGGGG)_n$, nematode $(TTAGGC)_n$, shrimp $(TAACC)_n$, vertebrate $(TTAGGG)_n$, and plant $(TTTAGGG)_n$, was performed. All these probes yielded likewise negative results.

8.3 DISCUSSION

8.3.1 Diversity in Ribosomal Genes Location within the Heteroptera

This is the first physical mapping effort reported for *C. lectularius* genome using the FISH technique. Physical location of genes remains very poorly studied in true bugs. Out of more than 40,000 described species (Weirauch and Schuh 2011), only 94 species (two of the species are presented by three subspecies each) have been investigated in this respect (Table 8.2) and only the rRNA genes are mapped (Papeschi et al. 2003; Cattani and Papeschi 2004; Cattani et al. 2004; Severi-Aguiar and Azeredo-Oliveira 2005; Severi-Aguiar et al. 2006; Papeschi and Bressa 2006; Morielle-Souza and Azeredo-Oliveira 2007; Bressa et al. 2008, 2009; Grozeva et al. 2010, 2011; Bardella et al. 2010, 2013; Kuznetsova et al. 2012; Panzera et al. 2012; Pita et al. 2013).

TABLE 8.2

Distribution of Fluorescence In Situ Hybridization-Ribosomal DNA Sites in the Heteroptera

Taxon	2n	rDNA Location in Karyotype	DNA Location in the Chromosomes	References
Nepomorpha				
Belostomatidae, Belostomatinae				
Belostoma dentatum	$26 + X_1X_2Y$	A (a medium pair)	T	Chirino et al. 2013
B. elegans	$26 + X_1X_2Y$	A	T	Papeschi and Bressa 2006
B. elongatum	$26 + X_1X_2Y$	A (a medium pair)	T	W
B. gestroi	$26 + X_1X_2Y$	A (a medium pair)	T	Chirino et al. 2013
B. micantulum	$14 + XY$	X and Y	T	Papeschi and Bressa 2006
B. oxyurum	$6 + XY$	X and Y	T	Papeschi and Bressa 2006
Belostomatidae, Lethocerinae				
Lethocerus patruelis	$22 + 2m + XY$	X and Y	T	Kuznetsova et al. 2012
Cimicomorpha				
Cimicidae				
Cimex lectularius	$26 + X_1X_2Y$	X_1 and Y	T	Grozeva et al. 2011
Miridae				
Deraeocoris ruber	$30 + 2m + XY$	X and Y	T	Grozeva et al. 2011
D. rutilus	$30 + 2m + XY$	X and Y	T	Grozeva et al. 2011
Megaloceroea recticornis	$30 + XY$	X and Y	T	Grozeva et al. 2011
Reduviidae, Hammacerinae				
Microtomus lunifer	$28A + X_1X_2Y$	A (largest pair)		Poggio et al. 2011
Reduviidae, Triatominae				
Dipetalogaster maxima	$20A + XY$	X	T	Panzera et al. 2012

TABLE 8.2 (*Continued*)
Distribution of Fluorescence In Situ Hybridization-Ribosomal DNA Sites in the Heteroptera

Taxon	2n	rDNA Location in Karyotype	DNA Location in the Chromosomes	References
Eratyrus cuspidatus	$20A + X_1X_2Y$	X_1 and Y	T	Panzera et al. 2012
Mepraia gajardoi	$20A + X_1X_2Y$	X_1	T (?)	Panzera et al. 2012
M. spinolai	$20A + X_1X_2Y$	X_1	T	Panzera et al. 2012
Panstrongylus chinai	$20A + X_1X_2Y$	A (largest pair)	T	Panzera et al. 2012
P. lignarius	$20A + X_1X_2Y$	A (largest pair)	T	Panzera et al. 2012
P. megistus	$18A + X_1X_2Y$	A (largest pair) 1 sex chromosome	T	Morielle-Souza and Azeredo-Oliveira 2007, Panzera et al. 2012
Psammolestes tertius	20A + XY	X and Y	T	Panzera et al. 2012
Rhodnius colombiensis	20A + XY	X	T	Panzera et al. 2012
R. domesticus	20A + XY	X and Y	T	Panzera et al. 2012
R. ecuadoriensis	20A + XY	X and Y	T	Pita et al. 2013
R. milesi	20A + XY	X and Y	T	Pita et al. 2013
R. nasutus	20A + XY	X	T	Pita et al. 2013
R. neglecticus	20A + XY	X and Y	T	Pita et al. 2013
R. neivai	20A + XY	X and Y	T	Pita et al. 2013
R. pallescens	20A + XY	X and Y	T	Morielle-Souza and Azeredo-Oliveira 2007, Panzera et al. 2012
R. pictipes	20A + XY	X	T	Pita et al. 2013
R. prolixus	20A + XY	X	T	Panzera et al. 2012
R. robustus	20A + XY	X	T	Pita et al. 2013
R. stali	20A + XY	X and Y	T	Pita et al. 2013
Triatoma boliviana	20A + XY	X	I	Panzera et al. 2012
T. brasiliensis	20A + XY	A (largest pair)		Bardella et al. 2010, Panzera et al. 2012
T. carcavalloi	20A + XY	A (largest pair)	T	Panzera et al. 2012
T. carrioni	20A + XY	X	I	Panzera et al. 2012
T. delpontei	20A + XY	A (largest pair) X	I	Panzera et al. 2012
T. dimidiata capitata	$20A + X_1X_2Y$	A (a large pair)	T	Panzera et al. 2012
T. dimidiata dimidiata	$20A + X_1X_2Y$	A (a large pair)	T	
T. dimidiata maculipennis	$20A + X_1X_2Y$	A (a large pair)	T	
T. flavida	$20A + X_1X_2Y$	A (a large pair)		Panzera et al. 2012
T. garciabesi	20A + XY	X	T	Panzera et al. 2012
T. infestans Andean group	20A + XY	A (a large pair)	T	Panzera et al. 2012
T. infestans Non-Andean group	20A + XY	X	T	Morielle-Souza and Azeredo-Oliveira 2007, Panzera et al. 2012
T. infestans melanosoma	20A + XY	X	T	Bardella et al. 2010, Panzera et al. 2012
T. lecticularia	20A + XY	A (a large pair)	T	Panzera et al. 2012
T. maculata	20A + XY	X and Y	T	Panzera et al. 2012
T. mattogrossensis	20A + XY	X and Y	T	Bardella et al. 2010, Panzera et al. 2012

(Continued)

TABLE 8.2 (Continued)

Distribution of Fluorescence In Situ Hybridization-Ribosomal DNA Sites in the Heteroptera

Taxon	2n	rDNA Location in Karyotype	DNA Location in the Chromosomes	References
T. mazzottii	$20A + X_1X_2Y$	A (an autosomal pair)		Panzera et al. 2012
T. nitida	$18A + X_1X_2Y$	A (a large pair)	T	Panzera et al. 2012
T. pallidipennis	$20A + X_1X_2Y$	A (a large pair)		Panzera et al. 2012
T. phyllosoma	$20A + X_1X_2Y$	A (a large pair)	T	Panzera et al. 2012
T. platensis	$20A + XY$	X	T	Panzera et al. 2012
T. protracta	$20A + X_1X_2Y$	A	T	Severi-Aguiar and Azeredo-Oliveira 2005, Panzera et al. 2012
T. pseudomaculata	$20A + XY$	A (a large pair)	T	Panzera et al. 2012
T. rubrovaria	$20A + XY$	A (largest pair)	T	Bardella et al. 2010, Panzera et al. 2012
T. sherlocki	$20A + XY$	A (largest pair)		Panzera et al. 2012
T. sordida	$20A + XY$	X	T	Panzera et al. 2012
T. tibiamaculata	$20A + X_1X_2Y$	A	T	Severi-Aguiar and Azeredo-Oliveira 2005, Panzera et al. 2012
T. vandae	$20A + XY$	X and Y	T	Panzera et al. 2012
T. vitticeps	$20A + X_1X_2X_3Y$	X_2 and X_3	T	Severi-Aguiar , Azeredo- Oliveira 2006, Panzera et al. 2012
T. wigodzinsky	$20A + XY$	A (largest pair)		Panzera et al. 2012
Pentatomomorpha				
Coreidae				
Acanonicus hahni	$18 + X$	A (largest pair)	T	Bardella et al. 2013
Althos obscurator	$22 + 2m + X$	A (a medium pair)	T	Bardella et al. 2013
Athaumastus haematicus	$18 + 2m + X$	A (largest pair)	T	Bardella et al. 2013
Camptischium clavipes	$18 + 2m + X$	A (largest pair)	ST	Cattani et al. 2004
Cebrenis sp.	$20 + 2m + X$	A (largest pair)	T	Bardella et al. 2013
Chariesterus armatus	$22 + 2m + X$	A (a medium pair)	T	Bardella et al. 2013
Holhymenia histrio	$24 + 2m + X$	A (largest pair)	T	Bardella et al. 2013
*H. rubiginosa**	$24 + 2m + X$	A (largest pair)	T	Bressa et al. 2008
Hypselonotus fulvus	$16 + 2m + X$	A (a medium pair)	T	Bardella et al. 2013
H. interruptus	$16 + 2m + X$	A (a medium pair)	T	Bardella et al. 2013
Leptoglossus gonagra	$18 + 2m + X$	M	T	Bardella et al. 2013
L. zonatus	$18 + 2m + X$	A (a medium pair)	T	Bardella et al. 2013
L. occidentalis	$18 + 2m + X$	A (largest pair)	ST	Present study
Pachylis argentinus	$12 + 2m + X$	A (largest pair)	ST	Papeschi et al. 2003
Phthia picta	$18 + 2m + X$	A (a medium pair)	T	Bardella et al. 2013
Spartocera cf. *fusca*	$20 + 2m + X$	A (largest pair)	T	Cattani and Papeschi 2004, Bardella et al. 2013
Zicca annulata	$20 + 2m + X$	A (largest pair)	T	Bardella et al. 2013
Largidae				
Euryophthalmus rufipennis	$12 + X$	X	T	Bardella et al. 2013

TABLE 8.2 (*Continued*)

Distribution of Fluorescence In Situ Hybridization-Ribosomal DNA Sites in the Heteroptera

Taxon	2n	rDNA Location in Karyotype	DNA Location in the Chromosomes	References
Lygaeidae				
Oxycarenus lavaterae	14 + 2m + XY	A (largest pair)	T	Grozeva et al. 2011
Pentatomidae				
Antiteuchus tripterus	12 + XY	A (largest pair)	T	Bardella et al. 2013
Arvelius albopunctatus	12 + XY	A (a medium pair)	T	Bardella et al. 2013
Edessa impura	12 + XY	A (largest pair)	T	Bardella et al. 2013
*E. meditabunda**	12 + XY	A (largest pair)	T	Bardella et al. 2013
E. rufomarginata	12 + XY	A (largest pair)	T	Bardella et al. 2013
Eurydema oleracea	12 + XY	X and Y	T	Grozeva et al. 2011
Euschistus cornutus	12 + XY	A (a medium pair)	T	Bardella et al. 2013
E. heros	12 + XY	A (a medium pair)	I	Bardella et al. 2013
Graphosoma lineatum (as *G. italicum*)	12 + XY	X	T	Grozeva et al. 2011, González-García et al. 1996
Nezara viridula	12 + XY	A	I	Papeschi et al. 2003
Pyrrhocoridae				
Dysdercus albofasciatus	10 + neo-X neo-Y	neo-X	T and I	Bressa et al. 2009
D. chaquensis	12 + X	A	T	Bressa et al, 2009
D. fulvoniger	12 + X	A	T	Bardella et al. 2013
D. imitator	12 + X	A	T	Bardella et al. 2013
D. ruficollis	12 + X	A	T	Bressa et al. 2009, Bardella et al. 2013
Pyrrhocoris apterus	22 + X	A (largest pair)	I	Grozeva et al. 2011
Rhopalidae				
Harmosthes prolixus	10 + 2m + X	A (largest pair)	T	Bardella et al. 2013

A, autosomes; T, telomere position; ST, subtelomere position; I, interstitial position.

* polymorphism for signals.

The species studied belong to 38 genera, 10 families, and 3 (out of 8) infraorders including Nepomorpha (Belostomatidae), Cimicomorpha (Cimicidae, Largidae, Miridae, Reduviidae, and Rhopalidae), and Pentatomomorpha (Coreidae, Lygaeidae, Pentatomidae, and Pyrrhocoridae). The species are widely diversified in terms of karyotypes with autosome numbers in males varying from 6 to 26 (including a pair of m-chromosomes when presents) and sex determining systems of the four types—X (0), XY, X_1X_2Y, $X_1X_2X_3Y$, and $X_1X_2X_3X_4Y$. The sites for rDNA at a rate of 1, 2, 3, or 4 (per diploid genome) are positioned variously in different species: either in autosomes (the largest or one of the medium-sized pairs whereas two pairs in the case of polymorphism); or in m-chromosomes; or in sex chromosomes (X or two Xs when a multiple sex system is present or in both X and Y chromosomes); or both in a pair of autosomes and the X chromosome (Figures 8.3 through 8.10). In *Dysdercus albofasciatus* Berg 1878, two signals are observed in the neo-X chromosome (Figure 8.10).

The autosomal location seems to predominate (found in half of the species studied), at least in the X (0) species, this pattern being observed in species differing in chromosome numbers and sex chromosome systems. For example, in the family Coreidae, which is one of the best explored families characterized by the presence of an X (0) system and m-chromosomes, all of the species have rDNA sites in autosomes (Figure 8.9). In one species, *Leptoglossus gonagra* (Coreidae), a pair of m-chromosomes was found to bear rDNA sites (Bardella et al. 2013). Within the

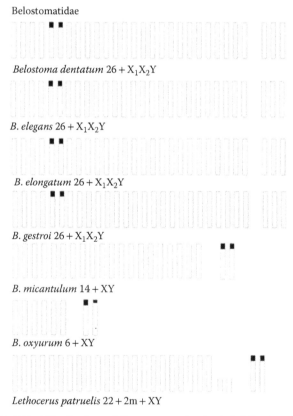

FIGURE 8.3 Distribution of fluorescence in situ hybridization-ribosomal DNA sites in species from the family Belostomatidae. The autosomes are of the same size, because in the original publications, information about their comparative sizes in a karyotype is absent.

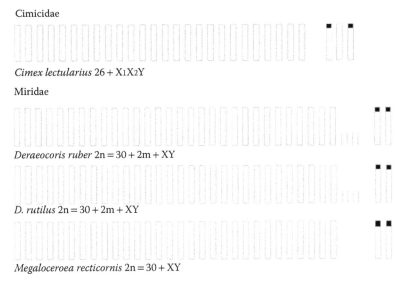

FIGURE 8.4 Distribution of fluorescence in situ hybridization-ribosomal DNA sites in species from the families Cimicidae and Miridae (Cimicomorpha). The autosomes are of the same size, because in the original publications, information about their comparative sizes in a karyotype is absent.

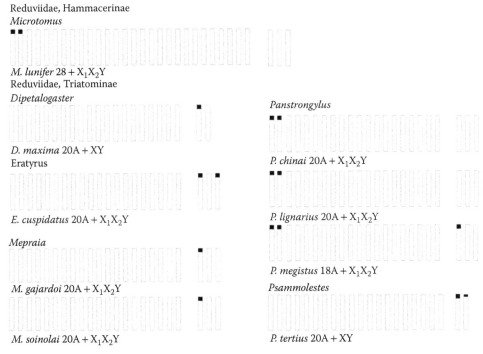

FIGURE 8.5 Distribution of fluorescence in situ hybridization-ribosomal DNA sites in species from different genera of the subfamily Triatominae (Cimicomorpha, Reduviidae). The autosomes are of the same size, because in the original publications, information about their comparative sizes in a karyotype is absent.

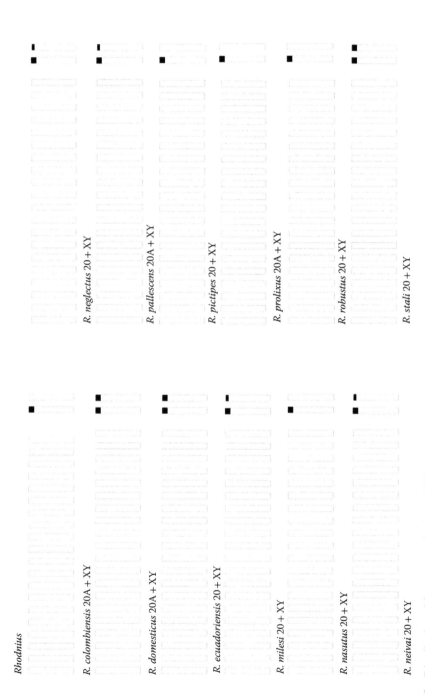

FIGURE 8.6 Distribution of fluorescence in situ hybridization-ribosomal DNA sites in species from genus *Rhodnius* (Cimicomorpha, Reduviidae). The autosomes are of the same size, because in the original publications, information about their comparative sizes in a karyotype is absent.

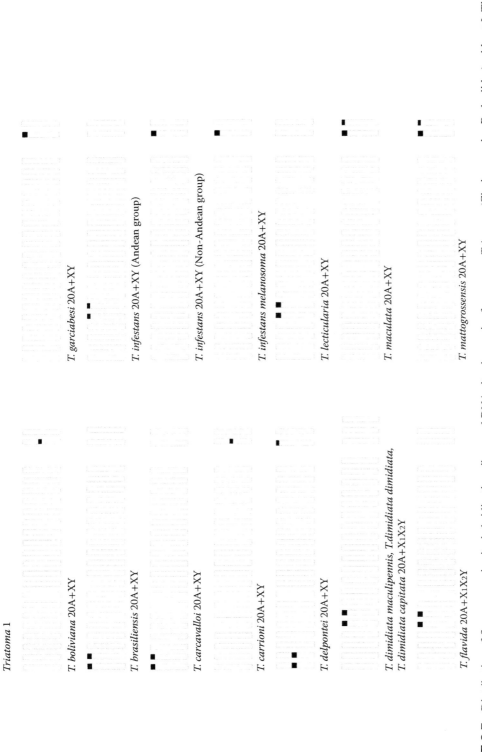

FIGURE 8.7 Distribution of fluorescence in situ hybridization-ribosomal DNA sites in species from genus *Triatoma* (Cimicomorpha, Reduviidae) without I. The autosomes are of the same size, because in the original publications, information about their comparative sizes in a karyotype is absent.

FIGURE 8.8   Distribution of fluorescence in situ hybridization-ribosomal DNA sites in species from genus *Triatoma* (Cimicomorpha, Reduviidae) without II, to be: The autosomes are of the same size, because in the original publications, information about their comparative sizes in a karyotype is absent.

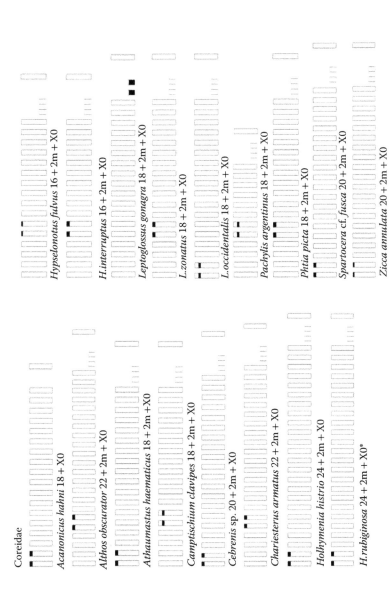

FIGURE 8.9 Distribution of fluorescence in situ hybridization-ribosomal DNA sites in species from the family Coreidae (Pentatomomorpha). The autosomes are of the same size, because in the original publications, information about their comparative sizes in a karyotype is absent.

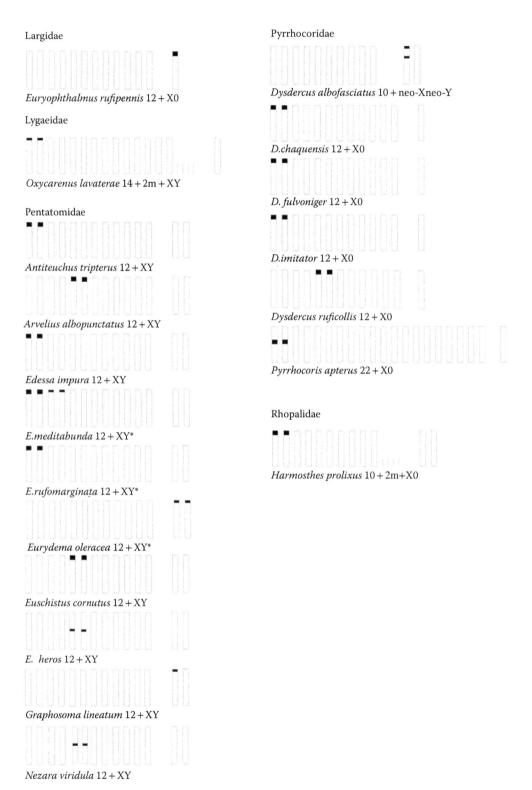

FIGURE 8.10 Distribution of fluorescence in situ hybridization-ribosomal DNA sites in species from the families Lygaeidae, Pentatomidae, Pyrrhocoridae, and Rhopalidae (Pentatomomorpha). The autosomes are of the same size, because in the original publications, information about their comparative sizes in a karyotype is absent.

Heteroptera, this is the first finding to show that ribosomal genes can be positioned in m-chromosomes. The m-chromosomes are a pair of very small chromosomes that occurs in many heteropteran species and behaves differently from both autosomes and sex chromosomes. The origin and function of m-chromosomes in heteropteran genomes are still unclear. They are known to be negatively heteropycnotic and are thought to be achiasmate in male meiosis (Ueshima 1979); however, Nokkala (1986) argued for chiasmate m-chromosomes in males of *Coreus marginatus*. The finding of rRNA genes in m-chromosomes opens new perspectives for understanding the molecular composition of m-chromosomes in the Heteroptera.

In species with a simple XY system, there is a wide variety of ribosomal loci locations, the differences being observed even between closely related species and even between subspecies of a single species. The genus *Triatoma* (Reduviidae) is a very good case in point (Figures 8.7 and 8.8). In this group, 17 species studied so far have $2n = 20 + XY$, one of these species, *Triatoma infestans*, being represented by three forms. Some of these species show rDNA sites in the X chromosome or in both sex chromosomes, some species have rDNA sites in the largest pair or in one of the medium-sized pairs of autosomes whereas one species has them both in X and a medium-sized pair of autosomes. Moreover, in *T. infestans melanosoma* and *T. infestans* (non-Andean group) rDNA signals are observed in the X whereas in *T. infestans* (Andean group) in the largest autosome pair. A considerable amount of variety is also observed among the species with multiple sex chromosomes (Table 8.2). Among such species of the genus *Triatoma*, rDNA sites are situated either in autosomes or in two of the three Xs (Table 8.2) (Figures 8.7 and 8.8).

In the majority of studied heteropteran species, rDNA sites are situated in the largest autosome pair whereas in some species autosomes bearing rDNA sites are referred to a "large" or a "larger" or a "medium" pair in original papers (see Table 8.2). Because of such uncertainty, in all of the species this pair is here arbitrarily taken to be that number 3 (Figures 8.3 through 8.10). We suggest, however, that those rDNA medium-sized autosomes can in fact be different in different species; however, new chromosomal markers are needed to test this hypothesis.

The majority of sites mapped to date in heteropterans showed a terminal localization; however, in rare cases they are positioned interstitially in chromosomes, for example, in *Pachylis argentinus* Berg, 1879, *Laptoglossus occidentalis,* and *Camptischium clavipes* (Coreidae), *Nezara veridula Triatoma delpontei* (Reduviidae), Linnaeus, 1758 (Pentatomidae), and in *Pyrrhocoris apterus*

(Pyrrhocoridae). Moreover, the cogeneric species sometimes differ in chromosomal location of the 18S rDNA genes. For example, *Euschistus heros* and *E. cornutus* (Pentatomidae) have ribosomal sites in a medium-sized pair of autotomes, however, in the first species they are located terminally whereas in the second one interstitially. This difference is probably due to an inversion or unequal crossover. All these findings attest that extensive chromosome rearrangements such as translocations, inversions, and transposition of some rDNA sequences to new places in the same or different chromosomes have occurred during the evolution of heteropterans. One more feature to be mentioned is the presence of a visible difference in intensity between hybridization signals in the X and Y chromosomes. Among other species, this is the case in *Belostoma oxyurum*, *Rhodnius,* and *Psammolestes* (Table 8.2). Interestingly, in each of these species ribosomal signal in the X chromosome is more intense, indicating that the ribosomal copy number is significantly higher in this chromosome compared with that in the Y chromosome. This phenomenon was recently discussed by Pita et al. (2013) in a publication dealing with triatomine bugs. In their opinion, the low copy number inherent in the Y chromosomes is consistent with the mechanism proposed by Dubcovsky and Dvorak (1995), which suggests the dispersion of a single rDNA copy and the successive amplification of the copy number, assuming an ancestral location of rRNA genes in the X chromosome. In many organisms, rDNA clusters may suddenly change their position without any other alteration in the remaining chromosome marks, that is, behave as mobile genetic elements due to the presence of transposable elements adjacent to the ribosomal genes (Schubert and Wobus 1985; Zhang et al. 2008). Several transposable elements have been described in Triatominae (Gilbert et al. 2010), including a nonlong terminal repeat retrotransposon inserted inside the 28S rDNA sequence in *Rhodnius prolixus* (Jakubczak et al. 1991). An rDNA cluster with associated transposable elements was suggested to move through triatomine genomes, as previously postulated for other insects (Cabrero and Camacho 2008; Nguyen et al. 2010; Cabral-de-Mello et al. 2011).

8.3.2 Diversity of Telomeric Motives in the Heteroptera

The telomeres are specific nucleoprotein complexes that terminate eukaryotic chromosomes and are responsible for the stability of chromosomes. DNA of the telomeres consists of short nucleotide motifs repeated thousands and millions of times. Although telomeric sequences may vary in composition in various groups of eukaryotes, they are strictly conserved in some higher rank taxonomic groups (Traut 1999). In animals, three main types of telomeric repeats are known: TTAGGG, TTAGGC, and TTAGG. Motif TTAGGG is typical of all multicellular animals, except roundworms and arthropods, and is probably an ancestral trait in Metazoa (Traut et al. 2007). Motif TTAGG, which is a derivative of TTAGGG, occurs in all main lineages of arthropods, supporting thus their origin from a common ancestor (Traut et al. 2007). It was originally believed that the TTAGG repeat is present in all arthropods. However, more detailed studies showed that this repeat is lacking in some crustaceans (Pelliccia et al. 1994), some spiders (Vítková et al. 2005), and some insects (Frydrychová et al. 2004). In hemimetabolous insects from the order group Paraneoptera (Psocoptera, Phthiraptera, Thysanoptera, and Hemiptera), the TTAGG motif is prevailing and probably initial (Frydrychová et al. 2004; Lukhtanov and Kuznetsova 2010). Among the Hemiptera, aphids, coccids, and aleyrodids (suborder Sternorrhyncha) and cycads (suborder Auchenorrhyncha) have the canonical insect motif $(TTAGG)_n$ (Frydrychová et al. 2004; Mohan et al. 2011; Monti et al. 2011; Maryańska-Nadachowska et al. 2013), however, each of these groups still remains very poorly studied. Until the recent time, the true bugs (suborder Heteroptera) were considered as a group in which the motif TTAGG is absent/lost (Sahara et al. 1999; Frydrychová et al. 2004; Lukhtanov and Kuznetsova 2010; Grozeva et al. 2011; Kuznetsova et al. 2011). This inference was based on the study of seven species from the families Pentatomidae, Pyrrhocoridae, Miridae, and Cimicidae: *Halyomorpha halys, Eurydema oleracea*, *Graphosoma lineatum*, *Pyrrhocoris apterus*, *Deraeocoris rutilus*, *Megaloceroea recticornis*, and *C. lectularius* (Okazaki et al. 1993; Sahara et al. 1999; Grozeva et al. 2010, 2011). The absence of $(TTAGG)_n$ telomeric sequence in these species was documented by different techniques including PCR, southern and/or dot-blot hybridization, and FISH.

It is important that all the previously listed families are classified within the evolutionarily advanced true bug infraorders Pentatomomorpha (Pentatomidae and Pyrrhocoridae) and Cimicomorpha (Miridae and Cimicidae). However, our recent finding of the insect-type $(TTAGG)_n$ telomeric motif in *Lethocerus patruelis* (Figure 8.11) from the family Belostomatidae (infraorder Nepomorpha or true water bugs) is clearly indicative of the heterogeneity of the Heteroptera in telomere organization (Kuznetsova et al. 2012). It can be assumed that Nepomorpha preserved the plesiomorphic telomere structure, whereas Cimicomorpha and Pentatomomorpha have the apomorphic state of this character. This assumption is consistent with the generally accepted opinion that Cimicomorpha

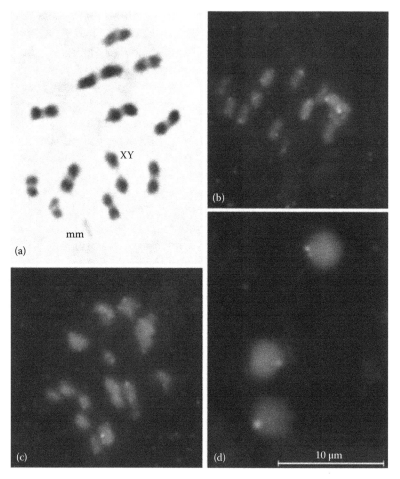

FIGURE 8.11   **(See color insert.)** (a–d): Meiotic chromosomes of *Lethocerus patruelis* after standard staining (a) and fluorescence in situ hybridization (b–d). (a)—metaphase I showing $n = 11AA + mm + XY$; (b–d)—representative fluorescence in situ hybridization images of metaphase I, (b, c) hybridized with 18S rDNA and TTAGG probes and spermatids, (d) hybridized with TTAGG probe. Ribosomal clusters (green) in X and Y chromosomes (b, c), and TTAGG repeats (red) at the ends of chromosomes (b, c) and clustered at the periphery of spermatid nuclei (d). (From Kuznetsova, V. G. et al., *Comp Cytogen*, 6(4), 341–346, 2012.)

and Pentatomomorpha represent a monophyletic lineage, and Nepomorpha has a basal position within the Heteroptera (see Kuznetsova et al. 2012 for references). However, the telomere structure in true bugs other than *L. patruelis* remains unclear. The fact that DNA from pentatomorphan and cimicomorphan species (including in *C. lectularius*) did not cross-hybridize with TTTTGGGG, TTGGGG, TTAGGC, TAACC, TTAGGG, and TTTAGGG telomeric probes using dot-blotting (Grozeva et al. 2011) raises the question of whether these true bugs have evolved an yet unknown telomeric repeat motif or some different mechanism of telomere maintenance. In summary, much more research is needed to decide this problem.

8.4 CONCLUSIONS

Thus, using FISH coupled with a number of standard cytogenetic techniques we have revealed that the common bedbug *C. lectularius* is characterized with the following traits: (1) holokinetic chromosomes, (2) $2n = 26 + X_1X_2Y$ in males, (3) the two rDNA loci located in sex chromosomes X_1 and Y, (4) the absence of TTAGG telomere repeat, (5) achiasmate meiosis of a collochore type in males, and (6) the sex chromosome postreduction in male meiosis. Comparative analysis of the data available in the literature has shown that these characters present to different extent in other true bugs being found in all species (1), in all species with the only exception (4), in the majority of species (6), and in some species and superspecies groups (5). Although multiple sex chromosome systems evolved many times in the Heteroptera, *C. lectularius* is unique to show the outstanding variation in the number of X chromosomes (from 3 to 20 extra Xs) in different populations. The significance of this extraordinary variation and underlying mechanism(s) as well as the origin of multiple sex chromosome systems in the whole Heteroptera (if these systems are a result of transverse fragmentations of the original X chromosome or of X-autosomal rearrangements or of some other causes) remain unclear. *C. lectularius* possessing an rDNA locus as a marker of the X chromosome represents a very promising species to investigate this problem using chromosome mapping with DNA specific probes.

ACKNOWLEDGMENTS

The study was supported by the project "Molecular cytogenetics and phylogeny of the order Hemiptera (Insecta)" within the research agreement between Russian Academy of Sciences and Bulgarian Academy of Sciences (to SG and VK), and by grants from the Russian Foundation for Basic Research (14-04-01051a) and programs of the Presidium of RAS "Gene Pools and Genetic Diversity" and "Origin of the Biosphere and Evolution of Geobiological Systems" (to BA and VK).

REFERENCES

Bai, X., P. Mamidala, S. P. Rajarapu, S. C. Jones, and O. Mittapalli. 2011. Transcriptomics of the bed bug (*Cimex lectularius*). *PLoSOne* 6 (1):e16336. doi: 10.1371/journal.pone.0016336.

Bardella, V. B., T. Fernandes, and A. L. F. Vanzela. 2013. The conservation of number and location of 18S sites indicates the relative stability of rDNA in species of Pentatomomorpha (Heteroptera). *Genome* 56(7): 425–429.

Bardella, V. B., M. L. Gaeta, A. L. L. Vanzela, and M. T. V. Azeredo-Oliveira. 2010. Chromosomal location of heterochromatin and 45S rDNA sites in four South American triatomines (Heteroptera: Reduviidae). *Comp Cytogen* 4 (2):141–149. doi: 10.3897/compcytogen.v4i2.50.

Bressa, M. J., M. J. Franco, M. Y. Toscani, and A. G. Papeschi. 2008. Heterochromatin heteromorphism in Holhymenia rubiginosa (Heteroptera: Coreidae). *Eur J Entomol* 105:65–72. doi: 10.14411/eje.2008.009.

Bressa, M., A. Papeschi, M. Vítková, S. Kubíčková, I. Fuková, M. Pigozzi, and F. Marec. 2009. Sex chromosome evolution in cotton stainers of the genus *Dysdercus* (Heteroptera: Pyrrhocoridae). *Cytogen Genome Res* 125:292–305. doi: 10.1159/000235936.

Busch, H. and K. Smetana. 1970. *The Nucleolus*. Academic Press, New York.

Cabral-de-Mello, D. C., R. C. Moura, and C. Martins. 2011. Cytogenetic mapping of rRNAs and histone H3 genes in 14 species of *Dichotomius* (Coleoptera, Scarabaeidae, Scarabaeinae) beetles. *Cytogen Genome Res* 134:127–135. doi: 10.1159/000326803.

Cabrero, J. and J. P. M. Camacho. 2008. Location and expression of ribosomal RNA genes in grasshoppers: Abundance of silent and cryptic loci. *Chromosome Res* 16:595–607. doi: 10.1007/s10577-008-1214-x.

Cattani, M. V., E. J. Greizerstein, and A. G. Papeschi. 2004. Male meiotic behavior and nucleolus organizer regions in *Camptischium clavipes* (Fabr) (Coreidae, Heteroptera) analyzed by fluorescent banding and in situ hybridization. *Caryologia* 57(3):267–273. doi: 10.1080/00087114.2004.10589403.

Cattani, M. V. and A. G. Papeschi. 2004. Nucleolus organizing regions and semipersistent nucleolus during meiosis in *Spartocera fusca* (Thunberg) (Coreidae, Heteroptera). *Hereditas* 140(2):105–111. doi: 10.1111/j.1601-5223.2004.01752.x.

Chirino, M. G., A. G. Papeschi, and M. J. Bressa. 2013. The significance of cytogenetics for the study of karyotype evolution and taxonomy of water bugs (Heteroptera, Belostomatidae) native to Argentina. *Comp Cytogen* 7 (2):111–129. doi: 10.3897/CompCytogen.v7i2.4462.

Darlington, C. D. 1939. The genetical and mechanical properties of the sex chromosomes. V. *Cimex* and Heteroptera. *J Genet* 39:101–138.

Dubcovsky, J. and J. Dvorak. 1995. Ribosomal RNA multigene loci: Nomads of the Triticeae genomes. *Genetics* 140:1367–1377.

Frydrychová, R., P. Grossmann, P. Trubac, M. Vítková, and F. Marec. 2004. Phylogenetic distribution of TTAGG telomeric repeats in insects. *Genome* 47:163–78.

Gilbert, C., S. Schaack, J. K. Pace II, P. J. Brindley, and C. Feschotte. 2010. A role for host-parasite interactions in the horizontal transfer of transposons across phyla. *Nature* 464:1347–1350. doi: 10.1038/nature08939.

Goddard, J. and R. deShazo.2009. Bed bugs (*Cimex lectularius*) and clinical consequences of their bites. *J Amer Med Ass (JAMA)* 301(13):1358–66. doi: 10.1001/jama.2009.405.

González-Garcia J. M., C. Antonio, J. A. Suja, and J. S. Rufas. 1996. Meiosis in holocentric chromosomes: kinetic activity is randomly restricted to the chromatid ends of sex univalents in *Graphosoma italicum (Heteroptera)*. *Chromosome Res*. 4:124–132.

Grozeva, S., V. Kuznetsova, and B. Anokhin. 2010. Bed bug cytogenetics: Karyotype, sex chromosome system, FISH mapping of 18S rDNA, and male meiosis in *Cimex lectularius* Linnaeus, 1758 (Heteroptera: Cimicidae). *Comp Cytogen* 4(2) doi: 10.3897/compcytogen.v4i2.36.

Grozeva, S., V. Kuznetsova, and B. Anokhin. 2011. Karyotypes, male meiosis and comparative FISH mapping of 18S ribosomal DNA and telomeric (TTAGG)$_n$ repeat in seven species of true bugs (Hemiptera: Heteroptera). *Comp Cytogen* 5(4) doi: 10.3897/CompCytogen.v5i4.2307.

Grozeva, S., V. G. Kuznetsova, and S. Nokkala. 2004. Patterns of chromosome banding in four nabid species (Heteroptera, Cimicomorpha, Nabidae) with high chromosome number karyotypes. *Hereditas* 140(2):99–104. http://dx.doi.org/10.1111/j.1601-5223.2004.01782.x

Grozeva, S. and S. Nokkala. 2002. Achiasmatic male meiosis in *Cimex* sp. (Heteroptera, Cimicidae). *Caryologia* 55:189–192.

Haynes, K. F., M. H. Goodman, and M. F. Potter. 2010. Bed bug deterrence. *BMC Biology* 8:117. http://www.biomedcentral.com/content/pdf/1741-7007-8-117.pdf

Henking, H. 1891. Untersuchungen uber die ersten Entwicklungsvorgange in den Eiern der Insekten. II: Uber spermatogenese und deren Beziehung zur Eientwicklung bei *Pyrrhocoris apterus* L. *Z Wissen Zool* 51:685–736.

Henry, T. J. 2009. Biodiversity of Heteroptera. In *Insect Biodiversity: Science and Society*, eds. R. G. Foottit and P. H. Adler, 223–263. Blackwell Publishing, Chichester, UK.

Hubbell, H. R. 1985. Silver staining as an indicator of active ribosomal genes. *Stain Technol* 60:284–294.

Hwang, S. W., T. J. Svoboda, I. J. De Jong, K. J. Kabasele, and E. Gogosis. 2005. Bed bug infestations in an urban environment. *Emerg Infect Dis* 11:533–538. doi: 10.1371/journal.ppat.1003462.

Jakubczak, J. L., W. D. Burke, and T. H. Eickbush. 1991. Retrotransposable elements RI and R2 interrupt the rRNA genes of most insects. *Proc Natl Acad Sci USA* 88:3295–3299. doi:10.1073/pnas.88.8.3295.

Krinsky, W. 2002. True bugs. In *Medical and Veterinary Entomology*, eds. G. Mullen and L. Durden, 67–86. Academic Press, Orlando, FL.

Kuznetsova, V. G. and S. Grozeva. 2010. Achiasmatic meiosis: A review. *H Vavilov Soc Genet Breed Sci* 14:79–88. (In Russian).

Kuznetsova, V. G., S. Grozeva, and B. A. Anokhin. 2012. The first finding of (TTAGG) n telomeric repeat in chromosomes of true bugs (Heteroptera, Belostomatidae). *Comp Cytogen* 6(4):341–346. doi: 10.3897/CompCytogen.v6i4.4058.

Kuznetsova, V., S. Grozeva, S. Nokkala, and C. Nokkala. 2011. Cytogenetics of the true bug infraorder Cimicomorpha (Hemiptera, Heteroptera): A review. *Zookeys* 154:31–70. doi: 10.3897/zookeys.154.1953.

Kuznetsova, V., S. Grozeva, J. N. Sewlal, and S. Nokkala. 2007. Cytogenetic characterization of the endemic of Trinidad, *Arachnocoris trinitatus* Bergroth: The first data for the tribe Arachnocorini (Heteroptera: Cimicomorpha: Nabidae). *Fol Biol (Kraków)* 55:17–26.

Kuznetsova, V. A. Maryańska-Nadachowska, and T. Karamysheva (2014). Chromosome analysis and FISH mapping of 18S rDNA and (TTAGG)$_n$ telomere sequences in the spittlebug genus *Philaenus* Stål (Hemiptera, Auchenorrhyncha, Aphrophoridae) In: *Protocols for Chromosome Mapping of Arthropod Genomes*, CRC Press, Taylor & Francis, Boca Raton, FL.

Lukhtanov, V. A. and V. G. Kuznetsova. 2010. What genes and chromosomes say about the origin and evolution of insects and other arthropods. *Russ J Genet* 46:1115–1121.

Maryańska-Nadachowska A., Kuznetsova V. G., and Karamysheva T. V. 2013. Chromosomal location of rDNA clusters and TTAGG telomeric repeats in eight species of the spittlebug genus *Philaenus* (Hemiptera: Auchenorrhyncha: Aphrophoridae). *Eur J Entomol* 110(3):411–418. doi: 10.14411/eje.2013.055.

Melters, D. P., L. V. Paliulis, I. F. Korf, and S. W. L. Chan. 2012. Holocentric chromosomes: Convergent evolution, meiotic adaptations, and genomic analysis. *Chromosome Res* 20:579–594. doi: 10.1007/s10577-012-9292-1.

Mohan, K. N., B. S. Rani, P. S. Kulashreshta, and J. S. Kadandale. 2011. Characterization of TTAGG telomeric repeats, their interstitial occurrence and constitutively active telomerase in the mealybug *Planococcus lilacinus* (Homoptera; Coccoidea). *Chromosoma* 120(2):165–175. doi: 10.1007/s00412-010-0299-0.

Monti, V., M. Giusti, D. Bizzaro, G. C. Manicardi, and M. Mandrioli. 2011. Presence of a functional (TTAGG)(n) telomere-telomerase system in aphids. *Chromosome Res* 19:625–633. doi: 10.1007/s10577-011-9222-7.

Morielle-Souza A. and M. T. V. Azeredo-Oliveira. 2007. Differential characterization of holocentric chromosomes in triatomines (Heteroptera, Triatominae) using different staining techniques and fluorescence in situ hybridization. *Genet Mol Res* 6:713–720.

Nguyen, P., K. Sahara, A. Yoshido, and F. Marec. 2010. Evolutionary dynamics of rDNA clusters on chromosomes of moths and butterflies (Lepidoptera). *Genetica* 138:343–354. doi: 10.1007/s10709-009-9424-5.

Nokkala, S. 1986. The mechanisms behind the regular segregation of the m-chromosomes in *Coreus marginatus* L. (Coreidae, Hemiptera). *Hereditas* 105:73–85. doi: 10.1111/j.1601-5223.1986.tb00662.x.

Nokkala, S. and C. Nokkala. 1986. Achiasmatic male meiosis of collochore type in the heteropteran family Miridae. *Hereditas* 105:193–197. http://onlinelibrary.wiley.com/doi/10.1111/j.1601-5223.1986.tb00661.x/pdf doi: 10.1111/j.1601-5223.

Okazaki, S., K. Tsuchida, H. Maekawa, H. Ishikawa, and H. Fujiwara. 1993. Identification of a pentanucleotide telomeric sequence, (TTAGG)$_n$, in the silkworm *Bombyx mori* and in other insects. *Mol Cell Biol* 13:1424–1432.

Panagiotakopulu, E. and P. C. Buckland. 1999. *Cimex lectularius* L., the common bed bug from Pharaonic Egypt. *Antiquity* 73(282):908–911.

Pandey, N. 2005. Good night, sleep tight, don't let the bed bugs bite. *Br J Gen Prac* 55(520):887.

Panzera F., R. Pérez, Y. Panzera, I. Ferrandis, M. J. Ferreiro, and L. Calleros. 2010. Cytogenetics and genome evolution in the subfamily Triatominae (Hemiptera, Reduviidae). *Cytogen Gen Res* 128:77–87. doi:10.1159/000298824.

Panzera, Y., S. Pita, M. J. Ferreiro, I. Ferrandis, C. Lages, R. Pérez, A. E. Silva, M. Guerra, and F. Panzera. 2012. High dynamics of rDNA cluster location in kissing bug holocentric chromosomes (Triatominae, Heteroptera). *Cytogen Gen Res* 138:56–67. doi:10.1159/000341888.

Papeschi, A. G. and M. J. Bressa. 2006. Evolutionary cytogenetics in Heteroptera. *J Biol Res* 5:3–21.

Papeschi, A. G., M. L. Mola, M. J. Bressa, E. J. Greizerstein, V. Lía, and L. Poggio. 2003. Behaviour of ring bivalents in holocentric systems: Alternative sites of spindle attachment in *Pachylis argentinus* and *Nezara viruda* (Heteroptera). *Chromosome Res* 11:725–733.

Pelliccia, F., E. V. Volpi, L. Lanza, L. Gaddini, A. Baldini, and A. Rocchi. 1994. Telomeric sequences of *Asellus aquaticus* (Crustacea, Isopoda). *Heredity* 72:78–80. doi:10.1038/hdy.1994.9.

Pérez R., L. Calleros, V. Rose, M. Lorca, and F. Panzera. 2004. Cytogenetic studies on *Mepraia gajardoi* (Heteroptera: Reduviidae). Chromosome behaviour in a spontaneous translocation mutant. *Eur J Entomol* 101:211–218. doi: 10.14411/eje.2004.027.

Pita, S., F. Panzera, I. Ferrandis, C. Galvão, A. Gómez-Palacio, and Y. Panzera. 2013. Chromosomal divergence and evolutionary inferences in Rhodniini based on the chromosomal location of ribosomal genes. *Mem Inst Oswaldo Cruz* 108(3). pii: S0074-02762013000300376. doi: 10.1590/ S0074-02762013000300017.

Poggio, M. G., M. J. Bressa, and A. G. Papeschi. 2011. Cytogenetics male meiosis, heterochromatin characterization and chromosomal location of rDNA in *Microtomus lunifer* (Berg, 1900) (Hemiptera: Reduviidae: Hammacerinae). *Comp Cytogen* 5(1):1–22. doi:10.3897/compcytogen. v5i1.1143.

Poggio, M. G., M. J. Bressa, A. G. Papeschi, O. D. Iorio, and P. Turienzo. 2009. Insects found in birds' nests from Argentina: Cytogenetic studies in Cimicidae (Hemiptera) and its taxonomical and phylogenetic implications. *Zootaxa* 2315:39–46.

Potter, M. F., B. Rosenberg, and M. Henriksen. 2010. Bugs without borders: Defining the global bed bug resurgence. (Cover Story). *Pest World*: 8–20.

Reinhardt, K. and M. T. Siva-Jothy. 2007. Biology of the bed bugs (Cimicidae). *Ann Rev Ent* 52:351–374. doi: 10.1146/annurev.ento.52.040306.133913.

Romero, A., M. F. Potter, D. A. Potter, and K. F. Haynes. 2007. Insecticide resistance in the bed bug: A factor in the pest's sudden resurgence? *J Med Entomol* 44:175–178. doi: http://dx.doi.org/10.1603/0022-2585(2007)44[175:IRIT BB]2.0.CO;2.

Sadílek, D., F. Šťáhlavský, J. Vilímová, and J. Zima. 2013. Extensive fragmentation of the X chromosome in the bed bug *Cimex lectularius* (Heteroptera: Cimicidae): A population survey across Europe. *Comp Cytogen* 7(4):253–269. doi: 10.3897/CompCytogen.v7i4.6012.

Sahara, K., F. Marec, and W. Traut. 1999. TTAGG telomeric repeats in chromosomes of some insects and other arthropods. *Chrom Res* 7:449–460.

Sailer, R. I. 1952. A review of the stink bugs of the genus Mecidea. *Proc US Nat MUS* 102:471–505.

Schrader, F. 1947. The role of kinetochore in the chromosomal evolution of the Hemiptera and Homoptera. *Evolution* 1:134–142.

Schubert, I. and U. Wobus. 1985. In situ hybridization confirms jumping nucleolus organizing regions in *Allium. Chromosoma* 92:143–148.

Schuh, R. T. and J. A. Slater. 1995. *True Bugs of the World (Hemiptera: Heteroptera). Classification and Natural History.* Cornell University Press XII, Ithaca, New York.

Schwarzacher, T. and P. Heslop-Harrison. 2000. *Practical In Situ Hybridization.* BIOS Scienti?c Publishers Ltd, Oxford, UK.

Severi-Aguiar, G. D. C. and M. T. V. Azeredo-Oliveira. 2005. Localization of rDNA sites in holocentric chromosomes of three species of triatomines (Heteroptera, Triatominae). *Genet Mol Res* 4:704–709.

Severi-Aguiar, G. D., L. B. Lourenco, H. E. M. C. Bicudo, and M. T. V. Azeredo-Oliveira. 2006. Meiosis aspects and nucleolar activity in *Triatoma vitticeps* (Triatominae, Heteroptera). *Genetica* 126:141–151. doi: 10.1007/ s10709-005-1443-2.

Simov, N., T. Ivanova, and I. Schunger. 2006. Bat-parasitic *Cimex* species (Hemiptera: Cimicidae) on the Balkan Peninsula, with zoogeographical remarks on *Cimex lectularius* Linnaeus. *Zootaxa* 1190:59–68.

Slack, H. D. 1938. Chromosome numbers in *Cimex. Nature* 142:358.

Slack, H. D. 1939a. The chromosomes of *Cimex. Nature* 143:78.

Slack, H. D. 1939b. Structural hybridity in *Cimex* L. *Chromosoma* 1:104–118.

Stutt, A. and M. Siva-Jothy. 2001. Traumatic insemination and sexual conflict in the bed bug *Cimex lectularius. Proc Natl Acad Sci USA* 98:5683–5687.

Traut, W. 1999. The evolution of sex chromosomes in insects: Differentiation of sex chromosomes in flies and moths. *Eur J Entomol* 96:227–235.

Traut, W., M. Szczepanowski, M. Vitková, C. Opitz, F. Marec, and J. Zrzavy. 2007. The telomere repeat motif of basal Metazoa. *Chrom Res* 15(3):-371–382. doi: 10.1007/s10577-007-1132-3.

Ueshima, N. 1966. Cytology and cytogenetics. In *Monograph of Cimicidae (Hemiptera-Heteroptera)*, ed. R. L. Usinger, 183–237. Thomas Say Foundation, New York.

Ueshima, N. 1967. Supernumerary chromosomes in the human bed bug, Cimex lectularius L. (Cimicidae: Hemiptera). *Chromosoma* 20:311–331.

Ueshima, N. 1979. Hemiptera II: Heteroptera. In *Animal Cytogenetics, Vol. 3, Insecta 6*, ed. B. John. Gebrüder Borntraeger, Berlin-Stuttgart, Germany.

Usinger, R. 1966. *Monograph of Cimicidae, Vol 7*. Entomological Society of America 7. Thomas Say Foundation, College Park, MD.

Vítková, M., J. Kral, W. Traut, J. Zrzavy, and F. Marec. 2005. The evolutionary origin of insect telomeric repeats, $(TTAGG)_n$. *Chrom Res* 13: 145–156.

Wang, C., K. Saltzmann, E. Chin, G. W. Bennett, and T. Gibb. 2010. Characteristics of *Cimex lectularius* (Hemiptera: Cimicidae), infestation and dispersal in a high-rise apartment building. *J Econom Ent* 103:172–177. doi: http://dx.doi.org/10.1603/EC09230.

Weatherston, J. and J. E. Percy. 1978. Venoms of Rhyncota (Hemiptera). In *Athropod venoms. Handbook of experimental pharmacology*, ed. S. Bettini, 489–509. Springer-Verlag. Berlin, Germany.

Weirauch, C. and R. T. Schuh. 2011. Systematics and evolution of Heteroptera: 25 years of progress. *Annu Rev Entomol* 56:487–510.

White, M. J. D. 1973. *Animal Cytogenetics and Evolution*. Cambridge University Press, Cambridge, UK.

Zhang, X., M. T. Eickbush, and T. H. Eickbush. 2008. Role of recombination in the long-term retention of transposable elements in rRNA gene loci. *Genetics* 180:1617–1626.

9

Aphids (Hemiptera)

Mauro Mandrioli and Gian Carlo Manicardi

CONTENTS

LIST OF ABBREVIATIONS

AFLP, amplification fragment length polymorphism
DABCO, 1,4-diazabicyclo[2.2.2]octane
FISH, fluorescent In Situ hybridization
FITC, fluorescein isothiocyanate
rDNA, ribosomal DNA (DNA sequence coding for ribosomal
RNA)
SSC, saline-sodium citrate buffer

9.1   INTRODUCTION

9.1.1   Aphid Biology

Aphids (Hemiptera: Aphididae) are ancient insects, whose fossils
go back to the Triassic, about 220–210 million years (Grimaldi and
Engel 2005). To date, they are present in most of the world's biomes,
including the tropics and subarctic regions, where they infest a huge
range of plants (Loxdale 2009) causing several damages due not
only to their direct parasitic action against crops, but also because
they are active vectors of numerous crop viruses (Blackman and
Eastop 2000, 2006, 2007; van Emden and Harrington 2007).

Aphids (Figure 9.1a through c) have a quite unusual biology,
involving apomictic cyclical parthenogenesis (a form of reproduction
whereby adult females give birth to female progeny in the absence of
male fertilization), complex life cycles, environmentally determined
polymorphism, and telescoping generations, which means that aphid
parthenogenetic/viviparous females have daughters within them,
who are already parthenogenetically producing their own daughter
(Srinivasan and Brisson 2012) (Figure 9.1f). These uncommon features made aphids intriguing for a long time for entomologists, but
progress in genetic studies of aphids was impeded for many years
due to difficulties in breeding them through the sexual phase.

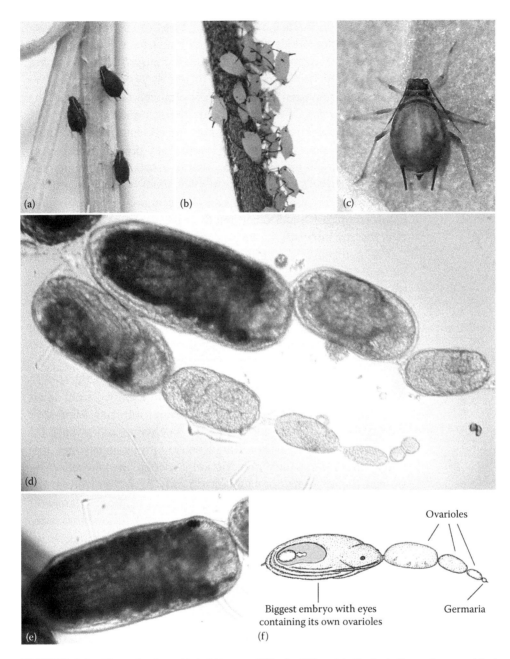

FIGURE 9.1 **(See color insert.)** Aphids may differ in different traits, including color, as evident comparing *Uroleucon grossum* (a), *Aphis nerii* (b), and *Toxoptera aurantii* (c). Dissecting parthenogenetic females, it is possible to easily isolate embryos of different developmental stages and sizes that represent a useful source for mitotic cells, with the exception of the biggest embryos that are generally discarded (marked with arrowheads in (d, e)).

9.1.2 From the Cytogenetic to the Genomic Analyses

The sequencing of the pea aphid *Acyrthosiphon pisum* genome allowed a better understanding of some of the unique functional aspects of aphids (such as the genetic basis of phenotypic plasticity and the interactions with endosymbionts), but unfortunately the chromosomal mapping of the identified contigs/scaffolds has not been included in the main goals of the *A. pisum* genome project so that some peculiar features of aphid genetics, including the presence/absence of highly syntenic regions and/or hot spots of recombination, have been so far not studied (International Aphid Genomics Consortium 2010). However, at present, the use of cytogenetic tools in aphids is becoming crucial for mapping and positional cloning of orthologous genes identified in the genome project of the pea aphid *A. pisum* (International Aphid Genome Consortium 2010) and other aphid species (such as *Myzus persicae*) that are currently in the process of genome annotation.

For cytogenetics, aphids also provide an interesting model because they possess holocentric chromosomes, showing centromeric activity along the whole chromosomal axis (Hughes-Schrader and Schrader 1961; Blackman 1987), so that during mitotic anaphase, chromatids move apart in parallel and do not form the classical V-shaped figures usually observed during the movement of monocentric chromosomes (Figure 9.2).

Aphids could be particularly useful for cytogenetic studies because mitotic chromosomes can easily be obtained from aphid embryonic tissues (Blackman 1987; Mandrioli and Manicardi 2012). However, holocentrism has also been a great drawback for karyotype analyses, because holocentric chromosomes lack primary and/or secondary constrictions, so that in conventionally stained preparations homologues can only be recognized on the basis of their size (Blackman 1987; Hales et al. 1997; Manicardi et al. 2002). Moreover, G-banding produced results that were not clear or consistent enough for use in species comparisons (Blackman 1985) so that only changes affecting the number of chromosomes or their size (such as nonreciprocal translocations of portions of chromosomes) could be detected. Other types of common chromosome rearrangements, such as inversions and duplications, almost certainly occur in aphids, but their presence cannot be detected in conventionally stained preparations because they do not alter the gross chromosome morphology.

Despite this limitation, chromosomal rearrangements have been reported in aphids mainly involving autosomal fusions and dissociations (Blackman 1980a,b; Monti et al. 2012; Rivi et al. 2012) and it has been suggested that they played a large part in aphid evolution since they affected, for instance, the host choice, as reported in the corn leaf aphid *Rhopalosiphum maidis* (Fitch) feeding in barley and sorghum (Brown and Blackman 1988).

The chromosome number and the structure of the karyotypes of aphids are generally stable within genera, although karyotype

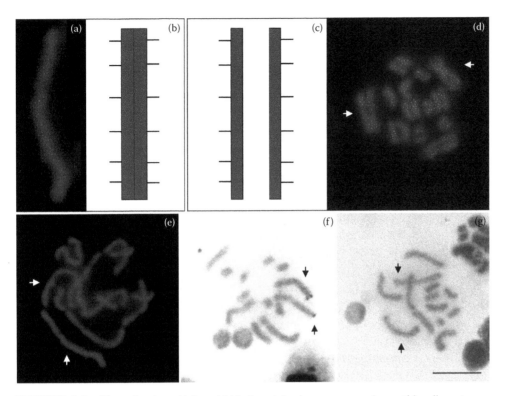

FIGURE 9.2 (See color insert.) In aphid holocentric chromosomes, chromatids adhere to one another without any prominent structure detectable between them, as evident after staining with propidium iodide (a) and in the schematic representation (b). During mitotic anaphase (c and d), chromatids move apart in parallel and do not form the classical V-shaped figures typical of mono-centric chromosomes. Aphid chromosomes may greatly differ in their condensation level, as evident comparing panels (d) and (e) and they can be stained with conventional staining, such as silver + Giemsa stainings (f) and Giemsa staining alone (g) to quickly verify the quality of the chromosome slides. Arrows indicated X chromosomes. Bar corresponds to 10 μm.

variations are relatively common within some species. Exceptional findings of chromosome polymorphisms have been reported in the genus *Amphorophora*, where species with similar morphology and biology vary in female karyotype from $2n = 4$ to $2n = 72$ (Blackman 1980a; Blackman et al. 2000), in *Trama maritime* (Eastop), where specimens of the same species have different karyotypes (interindividual polymorphisms) (Blackman et al. 2000) and in *M. persicae*, where extensive intraclonal and intraindividual chromosomal mosaicisms were observed in some clones (Monti et al. 2012). Furthermore, differences due to recombination of the ribosomal DNA genes between the two X chromosomes were reported in diverse aphid species, showing that karyotypic changes can occur rapidly within aphid populations (Mandrioli et al. 1999a,b).

Despite these data, some aphid genera have remarkably stable karyotypes. In the large genus *Aphis*, for instance, a very stable chromosome number has been observed in different species, such as *A. gossypii*, *A. verbasci*, *A. spiraecola*, *A. affinis*, *A. clematidis*, *A. sambuci*, *A. pomi*, *A. solanella*, and *A. fabae* making this taxon intriguing in term of karyotype constancy (Khuda-Bukhsh and Pal 1985; Criniti et al. 2005; Rivi et al. 2009).

The frequent occurrence of different chromosome number in aphids is related to their holocentric chromosomes, because chromosomal fragments can contact the microtubules and move properly in the daughter cells during cell division (Blackman 1980a). In contrast, fragments of monocentric chromosomes may be lost during mitosis and meiosis in the absence of centromeric activity in the chromosome fragment.

The publication of the pea aphid genome sequence by the International Aphid Genomics Consortium in 2010 made *A. pisum* the aphid species of greatest genomic and transcriptomic interest, but the knowledge about the structure of its chromosomes is still scarce. On the contrary, the peach potato aphid *M. persicae* (whose genome project is ongoing) is a good experimental model for the study of chromosome rearrangements in aphids, because several variations, mainly due to chromosomal translocations and occasionally to fragmentations, have been reported in the chromosome number and structure (Blackman 1980; Lauritzen 1982). Several populations of *M. persicae* were, for example, heterozygous for a translocation between autosomes 1 and 3 and this particular rearrangement has been shown to be involved in resistance to organophosphate and carbamate insecticides (Spence and Blackman 1998). *M. persicae* populations with 13 chromosomes have also been identified in different countries as the result of an autosome 3 fission (Blackman 1980). At least two independent and diverse fragmentations of the autosome 3 were reported (Blackman 1980; Lauritzen 1982) suggesting that different naturally occurring rearrangements of the same chromosome may be observed in the aphid karyotype (Blackman 1980; Lauritzen 1982; Monti et al. 2012). Finally, some *M. persicae* populations possessed a further fission of autosome 2 giving karyotype consisting of $2n = 14$ chromosomes (Blackman 1980; Lauritzen 1982).

9.1.3   From Genome Size to Genome Structure

The genome size of aphids is quite variable because it ranges from 0.36 to 1.77 pg, with the genome size less variable in the family Adelgidae (Finston et al. 1995). According to literature data, the *A. pisum* genome size was about 300 Mb, whereas the haploid genome of the *A. pisum* LSR1 clone used for the genome project was estimated by flow cytometry to be 517 Mb clearly supporting the idea that genome size could be highly variable among clones of the same species (International Aphid Genome Consortium 2010).

The complete *A. pisum* genome contains 72,844 contigs and has a total length of 464 Mb. According to gene prediction, more than 34,604 genes could be present, even if predictions also include unsupported *ab initio* models and partial gene models so that this estimate is likely to exceed the true number of protein-coding genes (International Aphid Genome Consortium 2010). This unusually high gene number is due to an extensive set of gene duplications occurring in more than 2000 gene families mainly involved in chromatin modification, microRNA synthesis, and sugar transport. Gene losses have been also observed involving several genes functioning in the aphid immune system, in the selenoprotein utilization, purine salvage, and the urea cycle (International Aphid Genome Consortium 2010).

Nucleotide divergences for sequenced open reading frames of orthologous genes range from 5% to 10% in comparisons of *A. pisum* to other aphid species belonging to tribe Macrosiphini and up to 15% for comparisons to other Aphidinae (Moran et al. 1999; Von Dohlen and Teulon 2003). Assuming that rates of nucleotide divergence and genomic rearrangements are similarly correlated in aphids and other animals, these values indicate that *A. pisum* will show substantial synteny of gene order and orientation with other Aphidinae with excellent prospects for being able to extend genomic information from *A. pisum* to other aphid species.

Few studies have been focused on the construction of genetic maps in aphids (Brisson and Davis 2008). The first densest pea aphid genetic map has been developed by Hawthorne and Via (2001) to study the aphid host plant specialization. They developed a linkage map of 173 dominant amplified fragment length polymorphism (AFLP) markers grouped into four linkage groups. Successively, Braendle et al. (2005) developed additional seven AFLP markers on the X chromosome. Similarly, Smith and MacKay (1989) hypothesized that the winged state of males is determined by a locus on the X chromosome and this hypothesis was successively supported by Caillaud et al. (2002), who showed that the trait is determined by a single locus and segregates accordingly in the F2 generation of a mapping panel produced from an initial cross between a clone that only produced winged males, and a clone that only produced unwinged males. They further confirmed that the locus was on the X chromosome by showing that the trait cosegregated with three X-linked microsatellite markers. Braendle et al. (2005) then detected AFLP markers flanking this locus and named the locus *aphicarus*.

From a cytogenetic point of view, aphid chromosomes have been studied mainly to identify cytogenetic markers that could be useful for taxonomic identification, as well as for the analysis of karyotype evolution (Manicardi et al. 2002). In particular, most papers focused on the karyotyping of parthenogenetic XX females, whereas few studies regarded the X0 males. Aphid males are produced by parthenogenetic females as a consequence of an X chromosome loss occurring in the course of a single maturation division (Orlando 1974; Blackman 1987; Blackman and Spence 1996).

The aphid X chromosomes can be easily identified in aphids because they generally are the longest chromosomes in the complement and they present a large rDNA cluster located at one telomere (with the unique exception of *Schoutedenia ralumensis* and *Maculachnus submacula* that present autosomal nucleolar organizing regions [NORs] and *Amphorophora idaei* showing interstitial NORs at X chromosomes) (Mandrioli et al. 1999a,b; Manicardi et al. 2002).

At present, few genes have been located on chromosomes in aphids, other than the 28S rDNA genes located at NOR, and they include the 5S rDNA (Bizzaro et al. 2000) and histone genes (Mandrioli and Manicardi 2013) in both *A. pisum* and *M. persicae* and the esterase E4 coding genes in *M. persicae* only (Blackman et al. 1995). Ten satellite DNAs have been also identified and localized on chromosomes in five aphid species: one in *Megoura viciae* (Bizzaro et al. 1996), one in *Rhopalosiphon padi* (Monti et al. 2010), two in *M. persicae* (Spence et al. 1998; Mandrioli et al. 1999a), two in *Amphorophora tuberculata* (Spence and Blackman 1998), and four in *A. nerii* (Mandrioli et al. 2011). The chromosomal localization of these satellites, investigated by fluorescent in situ hybridization (FISH), mostly corresponded to C-positive heterochromatic areas on the X chromosome. However, a 169 bp tandem repeat DNA marker for subtelomeric heterochromatin on all chromosomes has been described in aphids of the *M. persicae* group by Spence et al. (1998).

Despite the few mapped genes, several methods are already available for the study of aphids at a cytogenetic level not only to determine the localization of genes and repeated DNA sequences, but also to gain a better knowledge of several relevant biological processes that involve aphid chromosomes, such as the sex determination, the host choice, and the resistance to insecticides.

9.2   PROTOCOLS

9.2.1   Laboratory Equipment

Cytogenetic analyses can be performed using hybridizers, hands-free co-denaturation and hybridization instruments designed for slide-based fluorescence (FISH) and chromogenic in situ hybridization, or using more common (and less costly) laboratory equipments. The main advantage of the hybridizer is that this system reduces the manual steps and improves the efficiency, through-put, and precision compared to manually performed conventional hybridization procedures. However, optimal results can be also obtained using conventional lab tools, including a programmable temperature-controlled heating block (normally based on a poly-merase chain reaction [PCR] machine), a water bath (with vari-able temperature and, preferably, shaking) for denaturation of the probe and posthybridization washes, and an incubator for over-night hybridization.

In both cases, an epifluorescence microscope equipped with different filter set and a charge-coupled device (CCD) camera is necessary for the chromosome analysis.

9.2.2 Aphid Breeding

To have high-quality chromosome slides enriched in mitotic plates, it is essential to have healthy aphid strains. If you plan to work with laboratory populations, it is essential to maintain aphids on fresh plants at 20°C with a light-dark regime of 16 hours of light and 8 hours of darkness. It could be optimal to put partheno-genetic aphid females on new plants 1 or 2 days before dissection. Because aphid populations generally consist of parthenogenetic females only, if you plan to study male chromosomal plates, male induction can be obtained by exposing parthenogenetic females to short photoperiods (8 hours light: 16 hours dark) at 20°C.

The analysis of chromosomes from natural populations of aphids collected in the field has to be performed few hours after aphid collecting. The permanence of aphids on cut leaves or tree branches strongly reduces the number of mitoses observed in the chromosome slides.

9.2.3 Source of Chromosomes and Chromosome Slide Preparations

Aphids are ideal sources of mitotic chromosomes because parthe-nogenetic females contain multiple embryos developing sequen-tially within each ovariole (Figure 9.1d through f). In particular, chromosome preparations from parthenogenetic females can be made by spreading embryo cells and by squash preparation of sin-gle embryos. The main advantage of the spreading method is that it brings to high-quality slides a reduced background noise due to cytoplasm residue or to portions of cuticle, but it needs several adult females to have a sufficient amount of material on each slide. The squashing approach can be applied on single aphid (includ-ing single embryo), but slides contain also a high background and chromosomes cannot be properly separated in the mitotic plates.

9.2.3.1 Spreading Preparation

1. Put a single female aphid in a drop of a freshly made 0.8% hypotonic solution of sodium citrate onto a clean glass slide and dissect embryos by using microdissecting needles (or entomological needles) under a stereomicroscope. Using a Pasteur pipet, collect small embryos only (they have a high number of mitosis) in a Bovery dish and discard the adult body and larger embryos (with evident red eyes). Dissect at least 20–25 aphids and leave the embryos in the sodium citrate solution for 45 minutes at room temperature before starting the spreading procedure.

2. Transfer embryos to a minitube and centrifuge them at 350 g for 3 minutes to remove the sodium citrate solution. Then add to the pellet 500 µL of a freshly mixed mixture of 3 parts glacial acetic acid and 1 part methanol and flow up and down the acetic acid/methanol fixative solution for 3 minutes through a needle of a 1 mL hypodermic syringe to break embryos.

3. Centrifuge the embryos at 350 g for 3 minutes to remove the fixative solution containing cytoplasm residues and the cell debris and resuspend the pellet in 200 µL fresh fixative, then flow up and down the acetic acid/methanol solution for 3 minutes.

4. Under a chemical fume hood, drop 20 µL of the cellular suspension onto a clean slide from a height of about 20 cm and repeat dropping again to have 10 chromosome slides that have to be air-dried before use.

5. Store chromosome slide fully desiccated for up to 1 month at room temperature for FISH or indefinitely for conventional Giemsa, propidium iodide, or silver stainings.

9.2.3.2 Squash Preparation

1. Dissect embryos from a single aphid by using a stereomicroscope and microdissecting needles (or entomological needles) in two drops of a freshly made 0.8% hypotonic solution of sodium citrate. Discard the adult body and larger embryos if they have evident red eyes to collect only small embryos with a high number of mitosis. Leave the embryos in the sodium citrate solution for 15 minutes at room temperature.

2. Remove the citrate solution with a piece of blotting paper and replace it with two drops of a freshly mixed mixture of 2 parts glacial acetic acid and 1 part methanol and leave for 3 minutes at room temperature.

3. Apply a cover glass and gently squash the preparation by gently tapping it with a blunt instrument (such as a needle

or flat back of a pencil) or by finger (using the thumb with a pressure that just turns the nail white).

4. After an initial squashing, take up the excess acid with a piece of blotting paper and then flick off the cover slip with a razor blade.

5. Air-dry preparations at room temperature and store them fully desiccated for up to 3 months at 4°C for FISH experiments or banding or at room temperature for conventional Giemsa, propidium iodide, or silver stainings.

   Note: After flicking off the cover slip, chromosomes and interphase material will be present not only on the glass slide, but also on the cover slip so that you may preserve both for further applications. If you prefer to preserve all the material on the glass slides only, immediately after squashing submerge slides into liquid nitrogen until frozen (about 30 seconds) and then flick off the cover slip with a razor blade.

The quality of the chromosomal slides (in terms of quantity and quality of metaphases) can be checked directly under a phase-contrast microscope at a 40× magnification. If quality needs to be further checked to evaluate the quality of the chromosome morphology, chromosome slides can be stained with Giemsa or with propidium iodide.

9.2.3.3 Giemsa Staining

1. Stain slides with a 50 mL Giemsa solution (4% Giemsa in phosphate-buffered saline [PBS] 7.2–7.4) in a glass staining jar for 15 minutes at room temperature and then shortly wash (about 30 seconds) in distilled water or PBS to remove the Giemsa excess.

2. Carefully dry the slides at room temperature and mount cover slips in 2–3 drops of distrene plasticizer xylene (DPX) under a chemical fume hood and let DPX dry for 16 hours. Alternatively, slides can be examined directly without any coverslip at 40× magnification or mounted with two drops of distilled water if higher magnifications are necessary. The main advantage of DPX is that slides are permanently mounted and stained.

9.2.3.4 Propidium Iodide Staining

1. Stain slides with 100 µL of a 150 ng/mL propidium iodide solution in a dark moist chamber at room for 15 minutes.

2. Wash in distilled water or PBS 7.2–7.4 for 1 minute with a 50 mL water/PBS volume in a glass staining jar.

3. Mount slides with cover slips in 40 μL of a mixture of buffered glycerol and 1,4-diazabicyclo[2.2.2]octane (DABCO) (1% DABCO in 90% glycerol and 10% PBS).

9.2.4 Types of Probes and Labeling Method

At present, probes used for gene mapping in aphids mainly consisted in DNA fragments isolated by restriction enzymes or by PCR products that have been labeled by random primer or PCR.

In both cases, the nonradioactive digoxigenin (DIG) system has been used because DIG, a steroid hapten, can be used for both FISH (using fluorescein labeled anti-DIG antibodies for the detection) and Southern blotting by detecting the DIG-labeled probe with antibodies conjugated to an alkaline phosphatase that allows a subsequent color- or luminescence detection.

The DIG is coupled to deoxyuridine nucleotide triphosphate (dUTP) via an alkali-labile ester bond. The labeled dUTP can be easily incorporated by enzymatic nucleic acid synthesis using DNA polymerases. The combination of nonradioactive labeling with PCR is a powerful tool for the analysis of PCR products, and also for the preparation of labeled probes from small amounts of a respective target sequence.

9.2.4.1 Digoxigenin Labeling by Polymerase Chain Reaction

Most of the probes used in aphid cytogenetics were obtained using the "PCR DIG Probe Synthesis Kit" according to the Roche protocol using specific oligonucleotide primers. Using this approach, the thermostable polymerase routinely used for PCR reactions incorporates the DIG-dUTP as it amplifies a specific region of the template DNA. The result is a highly labeled, very specific, and very sensitive hybridization probe.

The PCR DIG Probe Synthesis Kit requires less optimization than most labeling methods and it is necessary just to optimize PCR amplification parameters (cycling conditions, template concentration, primer sequence, and primer concentration) for each template and primer set in the absence of DIG-dUTP before attempting incorporation of DIG. For most templates, no optimization of $MgCl_2$ concentration is required, but cheap thermostable polymerase can produce a reduced yield of labeled probe because some DNA templates (especially those with high AT content or longer templates) could be not efficiently amplified. This is due to the steric hindrance of DIG molecules that could prevent the synthesis of long T_n arrays.

To label the probe, it is necessary only to substitute the ordinary deoxyribose nucleoside triphosphates with those furnished in the PCR DIG Probe Synthesis Kit (Roche Diagnostics, Mannheim, Germany).

9.2.4.2 Random Primed Labeling of DNA

The "random primed" DNA labeling method is based on the hybridization of oligonucleotides of all possible sequences to the denatured DNA to be labeled. The input DNA serves as a template for the synthesis of labeled DNA, and it is not degraded during the reaction, making it possible to label minimal amounts of DNA (10 ng) with this method. The complementary DNA strand is synthesized by Klenow polymerase using the 3′-OH termini of the random oligonucleotides as primers. Alkali-labile DIG-dUTP nucleotides, labeled with DIG, are present in the reaction so that they are incorporated into the newly synthesized complementary DNA strand. Random primed labeling can be used for templates of almost any length, but because different six-base primers bind simultaneously the template, the labeled probe product will actually be a collection of fragments of variable length. The labeled probe will therefore appear as a smear, rather than a unique band on a gel. The size distribution of the labeled probe depends on the length of the original template.

Following steps are to be performed to label the probe by random priming (according to the Roche protocol):

1. Add 300 ng of template DNA to 16 μL of autoclaved and double-distilled water in a 1.5 mL reaction vial.
2. Denature DNA by heating in boiling water for 7 minutes and quickly chill in an ice/water bath.
3. Add 4 μL of *DIG-High prime* mix (Roche), thoroughly mix, centrifuge briefly, and then incubate for 20 hours at 37°C.
4. Stop the reaction by adding 2 μL of 0.2 M EDTA (pH 8.0) or by heating to 65°C for 10 minutes.

9.2.5 Hybridization and Detection Methods

FISH has been repeatedly used in aphid cytogenetics to localize different gene families (such as the 28S and 5S rDNA arrays and the histone coding genes), and repeated DNA sequences. At present, two different methods are currently used and they differ mainly for the denaturation step of the chromosomal DNA so that the choice of the method can be evaluated in view of the available laboratory equipments.

9.2.5.1 Fluorescent In Situ Hybridization with High Temperature Denaturation

Chromosome preparations made by the spreading method are generally used for gene mapping in aphids. Because of the denaturation at 85°C, slides need to be at least 48–72 hours old to preserve chromosome morphology after FISH. A pretreatment with ribonuclease (RNase)/pepsin and a refixation are also required. Pretreatment of chromosome preparations with RNase and pepsin solutions is useful to remove RNA, cytoplasm, and other cellular material, and to make the cells and chromatin permeable to the probe. Refixation of the material with a paraformaldehyde solution serves to maintain chromosomal morphology, and prevent loss of material during subsequent steps.

The following steps are to be performed to make FISH:

1. Wash spreading slides in a PBS pH 7.2–7.4 solution for 5 minutes at room temperature in a plastic (or glass) staining jar to remove the fixative solution.
2. Pretreat chromosomal slides with 100 µL of RNase A solution (200 µg/mL in 2× SSC) for 30 minutes at 37°C in a moist chamber (any kind of box can be used with two layers of wet 3-mm Whatman paper on the bottom; use plastic bars to hold the slides).
3. Incubate slides in pepsin solution (5 µg/mL in 0.01 M HCl) for 5 minutes at 37°C in a moist chamber.
4. Refix chromosome preparations with a freshly depolymerized paraformaldehyde solution (4% in phosphate buffer 0.1 M) for 15 minutes at room temperature in a plastic (or glass) staining jar and then wash in fresh water for 15 minutes.
5. Immediately after refixing, transfer the slides into an ethanol series (70%, 80%, 90%, 95%, and 100%, 5 minutes each at room temperature) and air-dry slides.
6. Prepare in a 1.5 mL minitube 30 µL of hybridization mixture (hyb mix) for each slide. The hyb mix contains 50% formamide, 0.25% sodium dodecyl sulfate (SDS), 2× SSC, 10% dextran sulfate, and 1 ng/µL of labeled probe. Care should be taken that the solution is well mixed, because the dextran sulfate solution is very sticky. Denature the hybridization mix by boiling for 5 minutes and immediately cool on ice.
7. Immediately after probe cooling, put 30 µL of denatured hyb mix on each slide, seal slides under a cover slip, and place them quickly in a prewarmed humidified chamber.
8. Denature at 85°C for 10 minutes and then move slide to 37°C for an overnight hybridization. Most protocols involve an overnight hybridization step of about 16 hours, which is a sufficient time for most probes to find homologous sequences on the chromosomes. However, 6 hours of hybridization work well for repeated sequences allowing a single-day FISH experiment.

9. After hybridization, slides need to be washed to remove unbound or loosely bound DNA probes. Wash twice, for 15 minutes, in 0.1× SSC at room temperature and twice in 0.1× SSC for 15 minutes at 42°C, and finally in the TNA solution (consisting of 0.1 M Tris-HCl and 0.2 M NaCl) at room temperature for 10 minutes. All washes are performed in 50 mL volumes in a plastic (or glass) staining jar.

10. Place slides in a moist chamber and preincubate them in 100 μL TNA plus 0.5% blocking reagent (Roche) and 0.2% Tween 20 at 37°C for 30 minutes. Preincubation avoids noisy background due to not specific binding of the antibodies.

11. Incubate slides at 37°C with the fluorescein isothiocyanate (FITC)-conjugated anti-DIG antibody, diluted 1:400 in PBS pH 7.2–7.4, for 45 minutes in a humid chamber, then wash slides once in TNA with 0.2% Tween 20 added for 5 minutes and twice in TNA alone, for 5 minutes each.

12. Counterstain chromosomes with 100 μL of a 100 ng/mL 4',6-diamidino-2-phenylindole (DAPI) solution or 150 ng/mL propidium iodide and mount slides in 40 μL of a mixture of buffered glycerol and DABCO (1% DABCO in 90% glycerol and 10% PBS). Thus, use of DABCO is useful to avoid a quick fade of the FITC fluorescence.

13. Store slides in the dark at least for 3 hours at 4°C before collecting FISH images.

9.2.5.2   Fluorescent In Situ Hybridization with Denaturation in Formamide Solution

The denaturation with a formamide solution at 70°C preserves chromosomes better than the previous method involving high temperature denaturation, so that 1-day-old chromosome can be used. As reported in the Section 9.2.5.1, a pretreatment with RNase/pepsin and a refixation are required. To make FISH, the following steps are to be followed:

1. Wash slides in a PBS solution for 5 minutes at room temperature in a plastic (or glass) staining jar to remove the fixative solution.

2. Pretreat chromosomal slides with 100 μL of RNase A solution (200 μg/mL in 2× SSC) for 30 minutes at 37°C in a moist chamber (any kind of box can be used with two layers of wet 3-mm Whatman paper on the bottom; use plastic bars to hold the slides).

3. Incubate slides in pepsin solution (5 μg/mL in 0.01 M HCl) for 5 minutes at 37°C in a moist chamber.

4. Refix chromosome preparations with a freshly depolymerized paraformaldehyde solution (4% in phosphate buffer 0.1 M) for 15 minutes at room temperature in a plastic (or glass) staining jar and then wash in fresh water for 15 minutes.

5. After refixing, place the slides (by using a slide holder that can handle 10 slides) in a prewarmed 70% formamide/2× SSC solution at 70°C for 3 minutes to denature chromosomal DNA. This solution can be used several times for denaturation, but to preserve its efficiency, it must be stored at 4°C after use. The quality of the formamide is important, so that a molecular biology grade formamide should be preferred.

6. After denaturation, immediately transfer slides into a cold (–20°C) ethanol series (70%, 80%, 95%, and 100%, 5 minutes each) and allow them to air-dry.

7. When slides are dried, prepare in a 1.5 mL minitube 30 μL of hyb mix for each slide. The hyb mix contains 50% formamide, 0.25% SDS, 2× SSC, 10% dextran sulfate, and 1 ng/μL of labeled probe. Care should be taken that the solution is well mixed, because the dextran sulfate solution is very sticky. Denature the hyb mix by boiling for 5 minutes and immediately cool on ice.

8. Put 30 μL of denatured hyb mix on each slide, seal slides under a coverslip, and place them quickly in a prewarmed humidified chamber to 37°C for an overnight hybridization. As previously reported, most protocols involve an overnight hybridization step of about 16 hours.

9. After hybridization, slides need to be washed to remove unbound or loosely bound DNA probes. Wash twice, for 15 minutes, in 0.1× SSC at room temperature and twice in 0.1× SSC for 15 minutes at 42°C, and finally in the PBS pH 7.2–7.4 at room temperature for 10 minutes. All washes are performed in 50 mL volume in a plastic (or glass) staining jar.

10. Place slides in a moist chamber and preincubate them in 100 μL PBS pH 7.2–7.4 plus 0.5% blocking reagent (Roche) and 0.2% Tween 20 at 37°C for 30 minutes. Preincubation avoids noisy background due to not specific binding of the antibodies.

11. Incubate slides at 37°C with the FITC-conjugated anti-DIG antibody, diluted 1:400 in PBS pH 7.2–7.4, for 45 minutes in a humid chamber, then wash slides twice in PBS pH 7.2–7.4 for 5 minutes.

12. Counterstain chromosomes with 100 μL of a 100 ng/mL DAPI solution or 150 ng/mL propidium iodide and mount

slides in 40 µL of a mixture of buffered glycerol and DABCO (1% DABCO in 90% glycerol and 10% PBS).

13. Store slides at least 3 hours in the dark at 4°C before collecting FISH images.

9.2.5.3 Fiber Fluorescent In Situ Hybridization

The term fiber FISH refers to the common practice of FISH conducted on preparations of artificially extended chromatin fibers on a slide glass. Differently from standard FISH applied to metaphase chromosomes that provides a mapping resolution of 1–3 Mb, fiber FISH has a resolution range of at least 1–400 kb so that it provides a precious mapping tool also useful to generate "color barcodes" for specific chromosomal regions, which can be used to study suspected chromosome rearrangements or the fine structures of some chromosomal portion, such as the telomeric or subtelomeric regions.

Slides with aphid DNA fibers can be obtained from chromosomes fixed onto slides by spreading, as reported in a previous section. Following steps are to be followed to obtain DNA fibers:

1. Make fresh chromosome spreads, and as soon as an iridescent halo appeared on the slide drying surface, wash in PBS pH 7.2–7.4 for 2 minutes in a glass staining jar at room temperature.

2. Drop onto the slides 60 µL NaOH/ethanol solution (0.07 N NaOH/absolute ethanol 5:2) and smear the solution on the slide with a cover slip.

3. Put two drops of methanol, and after 20 seconds, two further methanol drops placing the slide in vertical position for draining them.

4. Add further four drops of methanol, place the slide in vertical position for draining them, and air-dry slides.

5. When dried, examine slides by phase-contrast microscopy and dehydrate them in an alcohol hydration series (70%, 80%, 90%, 95%, and 100%) at room temperature before FISH experiments.

6. Fiber FISH hybridization has to be performed according to previously reported protocol with the formamide denaturation.

7. Counterstain DNA fiber with 150 ng/mL propidium iodide and mount slides in 40 µL of a mixture of buffered glycerol and DABCO (1% DABCO in 90% glycerol and 10% PBS).

8. Observe fiber FISH slides using a laser-scanning confocal microscope.

9.2.6   Image Acquisition and Analysis

FISH slides are generally observed using an epifluorescence microscope equipped with a 100 W mercury light source. The microscope needs to be equipped with a filter set for propidium and FITC separately or for simultaneous observation of propidium iodide/FITC. A common filter set for the simultaneous observation of propidium iodide/FITC consists of the following filters: excitation BP 485/20, beam splitter FT 51, emission LP 520.

Images have to be collected using a cooled CCD camera and the successive image processing and printing can be done using commercial image processing software, such as Adobe Photoshop (Adobe Systems; San Jose, CA).

Fiber FISH slides have been observed using a laser-scanning confocal microscope and images have to be collected using the software supplied with the confocal microscope.

9.2.7   Troubleshooting

The main problems in aphid cytogenetics are generally related to the number of metaphase plates on slides and on the chromosomal denaturation during FISH.

9.2.7.1   Problem 1: Low Number of Metaphases on Slides

To improve the number of mitotic plates on slides, use healthy parthenogenetic females that contain multiple embryos developing sequentially within each ovariole. An optimal result can be obtained moving aphids on a new plant 2 days before dissection.

9.2.7.2   Problem 2: Poor Quality of Chromosomes after Fluorescent In Situ Hybridization

Hybridization procedures, in particular if involving denaturation at high temperature, can affect the morphology of the hybridized chromosomes showing a sort of C banding on chromosomes. Frequently, overdenatured chromosomes appear "fluffy" because heat denaturation may have a very destructive influence on chromosomal morphology. To overcome this problem, it is important to avoid freshly made chromosomes and to avoid an unusually long permanence of chromosomes in the acetic acid–methanol solution during the spreading procedure. Two- to 3-day-old slides give optimal results not only in terms of chromosome morphology, but also allowing a proper denaturation. Overdenaturation can also result from an incomplete drying of the slides prior to denaturation.

9.2.7.3   Problem 3: Absent or Weak Hybridization Signals

Weak or absent signals after FISH can result from different causes including small probe size, low quality of the DNA probe, and insufficient denaturation of the slides. For repetitive DNA, in particular if clustered, short probes (less than 100 bp) can be used, but larger probes (at least 1,000 bp) are requested for moderately repeated DNA targets. Probes longer that 10,000 bp have been used to localize single copy genes, but an optimal localization can be obtained with cosmids or bacterial artificial chromosome.

The quality of the probe is related to an adequate probe labeling that is particularly relevant if FITC-conjugated probes are used due to the progressive fading of the FITC fluorescence.

Slide denaturation time varies depending on the age of the slides so that old slides (more than 5–7 days) generally request longer denaturation time and/or long pretreatment with pepsin before denaturation.

9.3   DISCUSSION

9.3.1   Mapping Aphid Genes

Despite the availability of several methods for gene mapping, few papers addressed the study of the distribution and localization of genes on aphid chromosomes. Among them, the X chromosomes are the most studied in the aphid complement and hybridization with 28S probes revealed that 18S, 5.8S, and 28S rDNA genes are usually arranged as tandemly repeated clusters at one telomere of each X chromosome, making these genes a specific marker for the Xs (Figure 9.3) (Blackman and Spence 1996; Mandrioli et al. 2011). Indeed, the majority of the aphid species has X-linked rDNA genes with few exceptions including the interstitial position of rDNA genes in *Amphorophora idaei* (Fenton et al. 1994) and the autosomal localization of NORs in *Schoutedenia lutea* (Hales 1989). FISH evidenced also a certain amount of heterogeneity between homologous NORs indicating that there are a variable number of ribosomal genes clustered at each X telomere due to mitotic unequal crossing over occurring in parthenogenetic females (Mandrioli et al. 1999a,b).

FISH experiments with 5S rDNA genes as probes showed that they are located in a single cluster on autosome 1 in *A. nerii* (Mandrioli et al. 2011) and in two interstitial clusters on the X chromosomes of *A. pisum* (Bizzaro et al. 2000), so that, differently from what reported for the 28S rDNA genes, 5S localization on aphid chromosomes may vary between species. At the same time, 5S heteromorphism has never been observed in aphids suggesting that only part of the aphid genome may have a high rate of recombination.

In situ hybridization has been particularly useful for the study of the composition of aphid heterochromatin (Figure 9.3). Indeed, even if several papers analyzed the distribution of constitutive (C-) heterochromatin in numerous aphid species in the last three decades, few of them analyzed its composition. FISH localization of repeated DNAs confirmed that they constitute a considerable portion of the aphid genome and represent the major components of heterochromatin (Mandrioli et al. 1999a; Manicardi et al. 2002). In particular, the chromosomal localization of the satellite DNAs, isolated in *M. viciae* (Bizzaro et al. 1996), *R. padi* (Monti et al. 2010), *M. persicae* (Spence et al. 1998; Mandrioli et al. 1999a), *A. tuberculata* (Spence and Blackman 1998), and *A. nerii* (Mandrioli et al. 2011), mostly corresponded to C-positive heterochromatic areas on the X chromosome. Interestingly, FISH with the 169 bp tandem repeat DNA showed that in the peach potato aphid *M. persicae* it occurs at both ends of all autosomes of the standard chromosome complement and at one end of the X chromosome, but is absent from the NOR-bearing end (Spence et al. 1998), making this repeated DNA

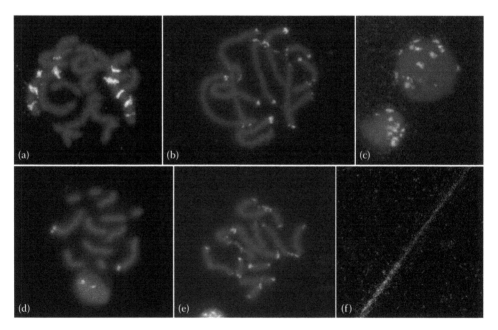

FIGURE 9.3 (See color insert.) Fluorescent in situ hybridization has been frequently used to localize satellite DNAs, such as Hind200 (a) and the subtelomeric DNA repeat of the peach potato aphid *M. persicae* not only in chromosomes (b), but also in the interphase nuclei (c). At the same time, fluorescent in situ hybridization gave a good chromosomal mapping of different gene families, such as the major rDNA array (d), and other repeated sequences, including the telomeric $(TTAGG)_n$ sequence (e) in *M. persicae*. A fine mapping has been also obtained at higher resolution by fiber fluorescent in situ hybridization mapping of specific DNA sequences, such as the green FITC-labeled subtelomeric DNA repeat in the red propidium-stained fiber in *M. persicae* (f).

a useful marker for the study of chromosome rearrangements in *M. persicae*. At this regard, FISH with the 169 bp tandem repeat DNA assessed the presence of recurrent fragmentations of chromosomes X, 1, and 3 in several *M. persicae* clones (Rivi et al. 2012) showing that, contrarily to what generally reported in literature, aphid X chromosomes are frequently involved in fragmentations, in particular at their telomeric ends opposite to the NORs.

In situ hybridization has been also useful for the study of both the telomeric and subtelomeric regions. At present, the occurrence of the (TTAGG)$_n$ repeat has been reported in some aphid species, where FISH experiments clearly showed a hybridization signal on each telomere of all the aphid chromosomes (Spence et al. 1998; Bizzaro et al. 2000; Monti et al. 2011). In aphid nuclei (Monti et al. 2011), telomeres appeared clustered into few foci and were not located mainly near the nuclear periphery, as reported in other insects such as *Drosophila melanogaster* (Hochstrasser et al. 1986) and the cabbage moth, *Mamestra brassicae* (Mandrioli 2002), clearly assessing that FISH could be extremely intriguing and also to better understand the fine architecture of the aphid interphase nuclei.

9.3.2 Scientific Benefits of Genome Mapping in Aphid Genetics and Genomics

Gene mapping has not been included as one of the main goals of the pea aphid genome project, but the availability of well-established cytogenetic methods could favor their use for the mapping and positional cloning of orthologous genes identified not only as a result of this genome project, but also in other aphid species that are currently in the process of genome annotation. These resources will provide also unprecedented opportunities for investigating many features of the aphid genome, including the presence of syntenic regions and the occurrence of fragile sites and/or hot spots of recombination. Finally, the availability of cytogenetic maps will be useful for a better understanding of the effects of chromosome rearrangements in the plant host choice and in the development of insecticide resistance in aphids.

ACKNOWLEDGMENTS

We are greatly indebted to Giuseppe Calabrese (Università degli studi "G. D'Annunzio," Chieti, Italy) for his precious suggestions about the fiber FISH protocol. This work is supported by the grant "Experimental approach to the study of evolution" from the Department of Animal Biology of the University of Modena and Reggio Emilia (MM).

REFERENCES

Bizzaro, D., Manicardi, G. C., and U. Bianchi. 1996. Chromosomal localization of a highly repeated *Eco*RI DNA fragment in *Megoura viciae* (Homoptera, Aphididae) by nick translation and FISH. *Chromosome Res* 4:392–396.

Bizzaro, D., Mandrioli, M., Zanotti, M., Giusti, M., and G. C. Manicardi. 2000. Chromosome analysis and molecular characterization of highly repeated DNAs in the aphid Acyrthosiphon pisum (Aphididae, Hemiptera). *Genetica* 108:197–202.

Blackman, R. L. 1980a. Chromosome numbers in the Aphididae and their taxonomic significance. *Syst Entomol* 5:7–25.

Blackman, R. L. 1980b. Chromosomes and parthenogenesis in aphids. *Symp Roy Entomol Soc Lond* 10:133–148.

Blackman, R. L. 1985. Aphid cytology and genetics (a review). In *Evolution and Biosystematics of Aphids*, Szelegiewicz H. (Editor), pp. 171–237, Ossolineum, Wroclaw, Poland.

Blackman, R. L. 1987. Reproduction, cytogenetics and development. In *Aphids, Their Biology, Natural Enemies and Control*, Minks, A. K. and Harrewijn, P. (Editors) Volume 2A, pp. 163–195, Elsevier, Amsterdam, The Netherlands.

Blackman, R. L. and V. F. Eastop. 2000. *Aphids on the World's Crops* (2nd Edn). Wiley, Chichester, United Kingdom.

Blackman, R. L. and V. F. Eastop. 2006. *Aphids on the World's Herbaceous Plants and Shrubs*. Wiley, Chichester, United Kingdom.

Blackman, R. L. and V. F. Eastop. 2007. *Aphids as Crop Pests*. CABI, London, United Kingdom.

Blackman, R. L. and J. M. Spence. 1996. Ribosomal DNA is frequently concentrated on only one X chromosome in permanently apomictic aphids, but this does not inhibit male determination. *Chromosome Res* 4:314–320.

Blackman, R. L., Spence, J. M., Field L. M., and A. L. Devonshire. 1995. Chromosomal localization of the amplified esterase genes conferring resistance to insecticides in the aphid Myzus persicae. *Heredity* 75:297–302.

Blackman, R. L., Spence, J. M., and B. B. Normark. 2000. High diversity of structurally heterozygous karyotypes and rDNA arrays in parthenogenetic aphids of the genus Trama. *Heredity* 84:254–260.

Braendle, C., Caillaud, M. C., and D. L. Stern. 2005. Genetic mapping of aphicarus: A sex-linked locus controlling a wing polymorphism in the pea aphid (Acyrthosiphon pisum). *Heredity* 94:435–442.

Brisson, J. A. and G. K. Davis. 2008. Pea aphid. In *Genome Mapping and Genomics in Arthropods*, Hunter W. and Kole C. (Editors), Volume 1, pp. 59–67, Springer-Verlag, Berlin and Heidelberg, Germany.

Brown, G. and R. L. Blackman. 1988. Karyotype variation in the corn leaf aphid, Rophalosiphon maidis (Fitch), species complex (Hemiptera, Aphididae) in relation to host plant and morphology. *Bull Entomol Res* 78:351–363.

Caillaud, M. C., Boutin, M., Braendle, C., and J. C. Simon. 2002. A sex-linked locus controls wing polymorphism in males of the pea aphid, Acyrthosiphon pisum (Harris). *Heredity* 89:346–352.

Criniti, A., Simonazzi, G., Cassanelli, S., Ferrari, M., Bizzaro, D., and G. C. Manicardi. 2005. X-linked heterochromatin distribution in the holocentric chromosomes of the green apple aphid Aphis pomi. *Genetica* 124:93–98.

Fenton, B., Birch, A. N. E., Malloch, G., Woodford, J. A. T., and C. Gonzalez. 1994. Molecular analysis of ribosomal DNA from the aphid Amphorophora idaei and an associated fungal organism. *Insect Mol Biol* 3:183–189.

Field, L. M. and A. L. Devonshire. 1998. Evidence that the E4 and FE4 esterase genes responsible for insecticide resistance in the aphid Myzus persicae (Sulzer) are part of a gene family. *Biochem J* 330:169–173.

Finston, T. L., Hebert, P. D. N., and R. B. Foottit. 1995. Genome size variation in aphids. *Insect* Biochem Mol Biol 25:189–196.

Grimaldi, D. and M. S. Engel. 2005. *Evolution of the Insects*. Cambridge University Press, Cambridge, United Kingdom.

Hales, D. F. 1989. The chromosomes of Schoutedenia lutea (Homoptera, Aphidoidea, Greenideinae), with an account of meiosis in the male. *Chromosoma* 98:295–300.

Hales, D. F., Tomiuk, J., Wohrmann, K., and P. Sunnucks. 1997. Evolutionary and genetic aspects of aphid biology: A review. *European J Entomol* 94:1–55.

Hawthorne, D. J. and S. Via. 2011. Genetic linkage of ecological specialization and reproductive isolation in pea aphids. *Nature* 412:904–907.

Hochstrasser, M., Mathog, D., Gruenbaum Y., Saumweber, H., and J. W. Sedat. 1986. Spatial organization of chromosomes in the salivary gland nuclei of Drosophila melanogaster. *J Cell Biol* 102:112–115.

Hughes-Schrader, S. and F. Schrader. 1961. The kinetochore of the Hemiptera. *Chromosoma* 12:327–350.

International Aphid Genomics Consortium. 2010. Genome sequence of the pea aphid Acyrthosiphon pisum. *PLoS Biol* 8:e1000313.

Khuda-Bukhsh, A. R. and N. B. Pal. 1985. Cytogenetic studies on aphids (Homoptera: Aphididae) from India: Karyomorphology of eight species of Aphis. *Entomol* 10:171–177.

Lauritzen, M. 1982. Q- and G- band identification of two chromosomal rearrangements in the peach-potato aphids Myzus persicae (Sulzer), resistant to insecticides. *Hereditas* 97:95–102.

Loxdale, H. D. 2009. What's in a clone: the rapid evolution of aphid asexual lineages in relation to geography, host plant adaptation and resistance to pesticides. In *Lost Sex: The Evolutionary Biology of Parthenogenesis*, Schon I., Martens K., and van Dijk P. (Editors), pp. 535–557, Springer, Heidelberg, Germany.

Mandrioli, M. 2002. Cytogenetic characterization of telomeres in the holocentric chromosomes of the lepidopteran Mamestra brassicae. *Chromosome Res* 9:279–286.

Mandrioli, M., Azzoni, P., Lombardo, G., and G. C. Manicardi. 2011. Composition and epigenetic markers of heterochromatin in the aphid Aphis nerii (Hemiptera: Aphididae). *Cytogenet Genome Res* 133:67–77.

Mandrioli, M., Bizzaro, D., Gionghi, D., Bassoli, L., Manicardi, G. C., and U. Bianchi. 1999a. Molecular cytogenetic characterization of a highly repeated DNA sequence in the peach potato aphid Myzus persicae. *Chromosoma* 108:436–442.

Mandrioli, M. and G. C. Manicardi. 2012. Unlocking holocentric chromosomes: New perspectives from comparative and functional genomics? *Curr Genom* 13:343–349.

Mandrioli, M. and Manicardi, G.C. 2013. Chromosomal mapping reveals a dynamic organization of the histone genes in aphids (Hemiptera: Aphididae). Entomologia 1:e2.

Mandrioli, M., Manicardi, G. C., Bizzaro, D., and U. Bianchi. 1999b. NORs heteromorphism within a parthenogenetic lineage of the aphid Megoura viciae. *Chrom Res* 7:157–162.

Manicardi, G. C., Mandrioli, M., Bizzaro, D., and U. Bianchi. 2002. Cytogenetic and molecular analysis of heterochromatic areas in the holocentric chromosomes of different aphid species. In *Some Aspects of Chromosome Structure and Function*, Sobti R. G., Obe G., and Athwal R. S. (Editors), pp. 47–56, Narosa Publishing House, New Delhi, India.

Monti, V., Giusti, M., Bizzaro, D., Manicardi, G. C., and M. Mandrioli. 2011. Presence of a functional (TTAGG) n telomere-telomerase system in aphids. *Chrom Res* 19:625–633.

Monti, V., Mandrioli, M., Rivi, M., and G. C. Manicardi. 2012. The vanishing clone: Karyotypic evidence for extensive intraclonal genetic variation in the peach potato aphid, Myzus persicae (Hemiptera: Aphididae). *Biol J Linnean Soc* 105:350–358.

Monti, V., Manicardi, G. C., and M. Mandrioli. 2010. Distribution and molecular composition of heterochromatin in the holocentric chromosomes of the aphid Rhopalosiphum padi (Hemiptera: Aphididae). Genetica 138:1077–1084.

Moran, N. A., Kaplan, M. E., Gelsey, M. J., Murphy, T. G., and E. A. Scholes. 1999. Phylogenetics and evolution of the aphid genus Uroleucon based on mitochondrial and nuclear DNA sequences. *Sys Entomol* 24:85–93.

Orlando, E. 1974. Sex determination in Megoura viciae Bukton (Homoptera, Aphididae). *Monit Zool* 8:61–70.

Rivi, M., Cassanelli, S., Mazzoni, E., Bizzaro, D., and G. C. Manicardi. 2009. Heterochromatin and rDNA localization on the holocentric chromosomes of black bean aphid, Aphis fabae Scop. (Hemipetra, Aphididae). *Caryologia* 62:341–346.

Rivi, M., Monti, V., Mazzoni, E., Cassanelli, S., Panini M., Bizzaro D., Mandrioli M., G. C. Manicardi. 2012. Karyotype variations in Italian populations of the peach-potato aphid Myzus persicae (Hemiptera: Aphididae). *Bull Entomol Res* 102:1–9.

Smith, M. A. H. and P. A. MacKay. 1989. Genetic variation in male alary dimorphism in populations of the pea aphid, Acyrthosiphon pisum. *Entomol Exp Appl* 51:125–132.

Spence, J. M. and R. L. Blackman. 1998. Orientation of the stretched univalent X chromosome during the unequal first meiotic division in male aphids. *Chrom Res* 6:177–181.

Spence, J. M., Blackman, R. L., Testa, J. M., and P. D. Ready. 1998. A 169 bp tandem repeat DNA marker for subtelomeric heterochromatin and chromosomal re-arrangement in aphids of the Myzus persicae group. *Chrom Res* 6:167–175.

Srinivasan, D. G. and J. A. Brisson. 2012. Aphids: A model for polyphenism and epigenetics. *Genet Res Int*, ID 431531.

van Emden, H. F. and R. Harrington. 2007. Aphids as Crop Pests. CABI, Wallingford, United Kingdom.

Von Dohlen, C. D. and D. A. J. Teulon. 2003. Phylogeny and historical biogeography of New Zealand indigenous Aphidini aphids (Hemiptera, Aphididae): An hypothesis. *Ann Entomol Soc Am* 96:107–116.

10

Spittlebugs (Hemiptera)

Valentina G. Kuznetsova, Anna Maryańska-Nadachowska, and Tatyana Karamysheva

CONTENTS

LIST OF ABBREVIATIONS

AgNOR, silver-binding nucleolus organizer region
AT, adenine–thymine
Bp, base pairs
C-banding, staining of heterochromatic regions
CCD-camera, charge-coupled device camera
CGH, comparative genomic hybridization
CMA3, chromomycin A3
COI, cytochrome c oxidase subunit I
CytB, cytochrome b
DAPI, 4′,6-diamidino-2-phenylindole
dATP, deoxyadenosine triphosphate
dCTP, deoxycytidine triphosphate
2D, two-dimensional
ddH$_2$O, double-distilled water
dGTP, deoxyguanosine triphosphate
DNA, deoxyribonucleic acid
dNTP, deoxyribonucleotide triphosphate
dTTP, deoxythymidine triphosphate
EDTA, ethylene diaminetetraacetic acid
FISH, fluorescence in situ hybridization
FITC, fluorescein isothiocyanate
GISH, genomic in situ hybridization

HCl, hydrochloric acid
ITS2, internal transcribed spacer
mt-genome, mitochondrial genome
NORs, nucleolus organizing regions
PBS, phosphate-buffered saline
PCR, polymerase chain reaction
rDNA, ribosomal DNA
RNase A, ribonuclease A
SAT-chromosome, satellite chromosome
SSC, saline sodium citrate
TE buffer, tris-ethylene diaminetetraacetic acid (EDTA) buffer
TTAGG, thymine–thymine–adenine–guanine–guanine

10.1   INTRODUCTION

10.1.1   Taxonomy and Importance of *Philaenus* spp

The hemipteran (homopteran) suborder Auchenorrhyncha involves
five superfamilies including the Fulgoroidea (planthoppers),
Cicadoidea (cicadas), Membracoidea (leafhoppers and treehoppers), Myerslopioidea stat. nov. (ground-dwelling leafhoppers),
and Cercopoidea (spittlebugs or froghoppers). The object of the
present study is the spittlebug genus *Philaenus* Stål, 1864 belonging to the cercopoid family Aphrophoridae.

Olli Halkka, one of the earliest and most known researchers
of chromosomes in Auchenorrhyncha, believed that they "are a
group well suited for comparative karyological work. Technically,
this group presents no special difficulties. The numbers of the
chromosomes are relatively low and the chromosomes themselves
are fairly large" (Halkka 1959). By now, approximately 820
auchenorrhynchan species (just 2% of the 42,000 extant species described) are known from a cytogenetic viewpoint; these
species represent 482 genera and 30 families from all but one
(the Myerslopioidea stat. nov.) superfamilies of Auchenorrhyncha.
Cytological studies in the Auchenorrhyncha have been carried
out using conventional cytogenetic techniques and predominantly
restricted to chromosome counts and sex chromosome systems
in the species addressed. Several recent studies used some differential chromosome-staining techniques, such as C-banding,
silver-binding nucleolar organizer region (AgNOR)-banding, and
base-specific fluorochrome banding, that led to a better characterization of the auchenorrhynchan chromosomes and their content
(Kuznetsova and Aguin-Pombo 2014).

In the last few decades, the ability to identify the individual chromosomes in a karyotype has been markedly improved by the development of molecular cytogenetic techniques. There are a number of

such techniques to suit different purposes such as fluorescence in situ hybridization (FISH) to locate positions of specific DNA sequences and genes on chromosomes, comparative genomic hybridization (CGH) for analyses of genome homology, genomic in situ hybridization (GISH) to identify alien chromosomes or segments, immunofluorescence to detect location and relative abundance of the proteins, and others. Some of these techniques are being exploited in a number of agriculturally or medically important hemipterans such as aphids (Hales et al. 2000; Manicardi et al. 2002; Mandrioli and Borsatti 2007; Monti et al. 2013), the Chagas disease–transmitting triatomine bugs (Panzera et al. 2012), and bedbugs (Grozeva et al. 2010, 2011, 2014; Bai et al. 2011). However, Auchenorrhyncha remain hardly characterized in this respect because only two relevant publications are currently available (Frydrychová et al. 2004; Maryańska-Nadachowska et al. 2013). In the first, the Southern hybridization technique was used to reveal molecular structure of telomeres in *Calligypona pellucida* (Delphacidae) (Frydrychová et al. 2004). In the second, FISH with telomere (TTAGG)$_n$ and ribosomal 18S rDNA as probes was implemented to study the representatives of the spittlebug genus *Philaenus* (Maryańska-Nadachowska et al. 2013). The results obtained in this pioneered work have important implications for cytogenetic evolution of this highly diverse insect group.

During the last 50 years, the genus *Philaenus* has attracted the particular interest of both geneticists and taxonomists because of its outstanding color polymorphism (e.g., Halkka and Halkka 1990; Drosopoulos et al. 2010). The nature and origin of this polymorphism and its possible contribution to the evolution of reproductive isolation and sympatric speciation have been extensively documented for the meadow spittlebug *Philaenus spumarius* (Linnaeus, 1758) (Halkka and Halkka 1990; Stewart and Lees 1996; Drosopoulos 2003; Drosopoulos et al. 2010). This species is widely distributed throughout the temperate zones of both northern and southern hemispheres although its original distribution was restricted to the Palaearctic region *P. spumarius* is a highly polyphagous species inhabiting grass and herb habitats with evident preferences for feeding on nitrogen-fixing plants (Thompson 1994). The species has become a pest of fodder plants and strawberries in areas where it is not a native species (Halkka et al. 1967; Zając and Wilson 1984).

Because of high polymorphism for the dorsal body surface color pattern, more than 50 synonyms have been given to *P. spumarius* (Nast 1972). Until the late 1980s, only three *Philaenus* species were recognized in the western half of the Palaearctic: the Holarctic *P. spumarius* and the Mediterranean *P. signatus* Melichar, 1896 (inhabiting the Balkans and Middle East), and *P. tesselatus* Melichar, 1889 (Southern Iberia and Maghreb).

At the turn of the twenty-first century, intensive purposeful investigations of the genus in the Mediterranean region have led to the discovery of five more species differing in the structure of genitalia and anal tube of the male: *P. loukasi* Drosopoulos et Asche, 1991; *P. arslani* Abdul-Nour et Lahoud, 1995; *P. italosignus* Drosopoulos et Remane, 2000; *P. maghresignus* Drosopoulos et Remane, 2000; and *P. tarifa* Remane et Drosopoulos, 2001 (Abdul-Nour and Lahoud 1995; Drosopoulos and Remane 2000; Remane and Drosopoulos 2001; Drosopoulos and Quartau 2002). Although the Mediterranean species are sympatric with *P. spumarius*, they are partially allopatric with one another (Figure 10.1). *P. signatus* is suggested to distribute beyond the Mediterranean being recorded also from Iraq, Iran, and Afghanistan; however, these records need verification (Drosopoulos and Remane 2000; Tishechkin 2013). Recently, two new *Philaenus* species, *P. elbursianus* and *P. iranicus*, were described by Tishechkin (2013) from Iran.

The current taxonomy of the genus accepts its division into two groups based on morphological similarities in the male anal tube. The *Philaenus spumarius* species group embraces *P. spumarius*, *P. tesselatus*, *P. loukasi*, and *P. arslani*, whereas the *Philaenus signatus* species group comprises *P. signatus*, *P. italosignus*, *P. maghresignus*, *P. tarifa* (Drosopoulos and Remane 2000), and *P. elbursianus* Tishechkin sp. n. (Tishechkin 2013). *P. iranicus* Tishechkin sp. n. sufficiently differs from all the congeners in the morphology of genitalia and anal tube, which warranted its separation in the subgenus *Gyrurus* Tishechkin subgen. n.; all other species being thus united into the subgenus *Philaenus* s. str. (Tishechkin 2013).

Furthermore, based on the larval food plant preferences, the genus is classified into three groups: (1) the lily, *Asphodelus aestivus*

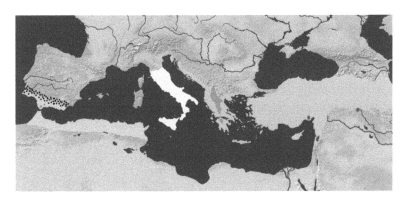

FIGURE 10.1   (**See color insert.**) Map showing distribution of *Philaenus* species in the Mediterranean region: light blue—*P. maghresignus*; red—*P. tarifa*; yellow—*P. italosignus*; pink—*P. loukasi*; brown—*P. arslani*; orange—*P. signatus*; black dots—*P. tesselatus*. *P. spumarius* is sympatrically distributed with other species.

(= *A. microcarpus*) (*P. signatus*, *P. italosignus*, *P. maghresignus*, and *P. tarifa*), (2) xerophilic plants (*P. loukasi* and *P. arslani*), and (3) various dicotyledonous and monocotyledonous plants (*P. spumarius* and *P. tesselatus*) (Drosopoulos 2003). Unfortunately, Tishechkin (2013) did not provide information on the food plant of new species from Iran. The results of recent molecular phylogenetic studies using nucleotide sequences from *COI* and *CytB* genes and *ITS2* nuclear region (Maryańska-Nadachowska et al. 2010, 2012), and cytogenetic analyses using different approaches (Kuznetsova et al. 2003; Maryańska-Nadachowska et al. 2008, 2012) including FISH-mapping of 18S ribosomal and TTAGG telomere sequences (Maryańska-Nadachowska et al. 2013) carried out on all (excepting *P. elbursianus* sp. nov. and *P. iranicus* sp. nov.) species of *Philaenus* are essentially congruent with morphological and food plant preference classifications.

10.1.2 Classical Cytogenetics of *Philaenus* spp

10.1.2.1 Standard Karyotypes

10.1.2.1.1 *Karyotype of Philaenus spumarius*

$2n = 23$ (22 + X); Figure 10.2a and b

Boring (1913) was first to describe the male karyotype of *P. spumarius* as consisting of $2n = 22 + X$. Hereafter, Kurokawa (1953) confirmed this karyotype for males, and then Kuznetsova et al. (2003) recorded $2n = 22 + X(0)$ for males and $2n = 22 + XX$ for females of this species. The largest autosome pair is about 10 μm long and displays a subterminal gap in every homologue; the remaining autosomes form a graded size series, and the X is close in size to autosomes number 3. In the first meiotic division in males, 11 bivalents and the univalent X chromosome are present (Kuznetsova et al. 2003).

10.1.2.1.2 *Karyotype of Philaenus tesselatus*

$2n = 23$ (22 + X); Figure 10.2c and d

In chromosome number and sex chromosome system, and in gross morphology of karyotype in terms of size of chromosomes and the presence of gaps in the first pair of autosomes, *P. tesselatus* resembles *P. spumarius* (Maryańska-Nadachowska et al. 2012).

10.1.2.1.3 *Karyotype of Philaenus arslani*

$2n = 20$ (18 + XY); Figure 10.2e

Karyotype includes 18 autosomes and X and Y chromosomes in males, both sex chromosomes being considered the neo-chromosomes. Autosomes show a size gradient from large to small. This chromosome complement is suggested to have been derived from the karyotype inherent in *P. spumarius* by means

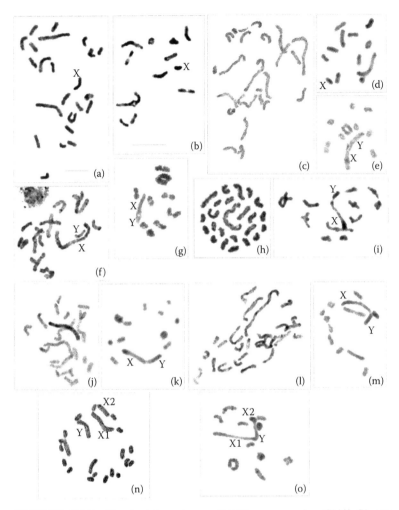

FIGURE 10.2 Standard karyotypes of *Philaenus* species (Shiff–Giemsa staining). (a, b)—*P. spumarius*: spermatogonial metaphase, $2n = 23$ (a) and spermatocyte diakinesis, $n = 11 + X$ (b); (c, d)—*P. tesselatus*: spermatogonial metaphase, $2n = 23$ (c) and spermatocyte diakinesis, $n = 11+X$ (d); (e)—*P. arslani*: spermatocyte diakinesis, $n = 9 +$ neo-XY; (f, g)—*P. loukasi*: spermatogonial metaphase, $2n = 20$ (f) and spermatocyte diakinesis, $n = 9 +$ neo-XY (g); (h, i)—*P. signatus*: spermatogonial metaphase, $2n = 22$ (h) and spermatocyte diakinesis, $n = 10 +$ neo-XY (i); (j, k)—*P. tarifa*: spermatogonial metaphase, $2n = 24$ (j) and spermatocyte diakinesis, $n = 11 +$ neo-XY (k); (l, m)—*P. maghresignus*: spermatogonial metaphase, $2n = 24$ (l) and spermatocyte diakinesis, $n = 11 +$ neo-XY (m); (n, o)—*P. italosignus*: spermatogonial metaphase, $2n = 23$ (n) and spermatocyte diakinesis, $n = 11 + X1X2Y$ (o). Bar = 10 µm. (From Kuznetsova, V. G. et al., *Folia Biol-Krakow*, 51 (1–2), 33–40, 2003.)

of two fusions: the first between two pairs of autosomes and the second between an autosome and the X chromosome. In the first meiotic division in males, nine autosome bivalents and a large pseudo-bivalent of sex chromosomes are present. The X and Y chromosomes are of approximately the same size (Maryańska-Nadachowska et al. 2008).

10.1.2.1.4   Karyotype of Philaenus loukasi

$2n = 20$ (18 + XY); Figure 10.2f and g

In chromosome number and the presence of a neo-XY sex chromosome system, the species resembles *P. arslani*. However, in *P. loukasi*, X chromosome is approximately three times longer than the Y (Maryańska-Nadachowska et al. 2012).

10.1.2.1.5   Karyotype of Philaenus signatus

$2n = 24$ (22 + XY); Figure 10.2h and i

Karyotype includes 22 autosomes and X and Y sex chromosomes in males, and the sex determination system is suggested to be of a neo-XY type. Like in *P. loukasi*, X chromosome is approximately three times longer than the Y. The remaining chromosomes show a size gradient from large to small (Maryańska-Nadachowska et al. 2012).

10.1.2.1.6   Karyotype of Philaenus tarifa

$2n = 24$ (22 + XY); Figure 10.2j and k

By chromosome number and the presence of a neo-XY sex chromosome system, the species resembles *P. signatus*. However, in *P. tarifa*, X chromosome is nearly twice as large as the Y, thus difference in size between sex chromosomes are not as marked as in *P. signatus* and *P. loukasi* (Maryańska-Nadachowska et al. 2012).

10.1.2.1.7   Karyotype of Philaenus maghresignus

$2n = 24$ (22 + XY); Figure 10.2l and m

By chromosome number and the presence of a neo-XY sex chromosome system, the species resembles *P. signatus* and *P. tarifa*. However, in the relative size of X and Y chromosomes, *P. maghresignus* bears similarity with *P. tarifa* rather than with *P. signatus* (Maryańska-Nadachowska et al. 2012).

10.1.2.1.8   Karyotype of Philaenus italosignus

$2n = 23$ (20 + X_1X_2Y); Figures 10.2n and o, and 10.3

Karyotype consists of 23 chromosomes, including 20 autosomes and 3 sex chromosomes suggesting the presence of multiple sex chromosome system of the neo-neo-X_1X_2Y type. Sex chromosomes are different in size, with X_1 and Y being the largest chromosomes of the set and X_2 smaller than the largest pair of autosomes. The X_1 is about twice as long as X_2, and the latter is about 1.5 times smaller than the Y (Maryańska-Nadachowska et al. 2012). Figure 10.3 represents the assumed mechanism of the origin of neo-neo-X_1X_2Y system in *P. italosignus*.

10.1.2.2   Nucleolus Organizer Region Banding

This technique reveals the nucleolus organizer regions (NORs) containing the genes that code for ribosomal RNA. The application

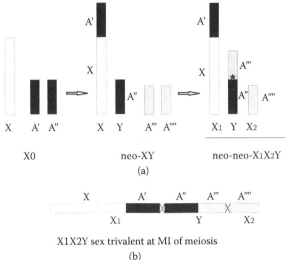

FIGURE 10.3 The presumed origin of the X_1X_2Y sex chromosome system (a) and the trivalent X_1X_2Y at metaphase I (b) in *Philaenus italosignus*. * Ribosomal DNA-fluorescence in situ hybridization loci.

of NOR-banding (silver staining) to seven out of the eight (except *P. maghresignus*) species of *Philaenus* detected differences in the number of NOR-bearing chromosomes and position of NORs between the species including those displaying the same chromosome complement. It is worth noting that Ag-positive sites could be seen and identified only in the extended prophase chromosomes. At mitotic and meiotic prophases, argentophilic material was separated into interconnected granules grouped more commonly around particular, presumably satellite chromosomes (SAT-chromosomes). After silver staining, in *P. italosignus* NORs were detected on the sex chromosomes but in all other species on the autosomes (Figure 10.4a through c). *P. spumarius* was found to have two NOR-bearing pairs of autosomes, the largest one and one of the middle-sized pairs, whereas in *P. tarifa*, *P. signatus*, *P. loukasi*, and *P. arslani* NORs were revealed on the largest pair of autosomes only. In *P. tesselatus*, the variability was observed: generally, AgNORs were present on the first autosome pair while occasionally one more very small silver-positive site was found on one of the small bivalents (Kuznetsova et al. 2003; Maryańska-Nadachowska et al. 2012).

10.1.2.3 C-Banding

This technique reveals the extent and location of heterochromatic segments (C-bands) that contain highly condensed, repetitive, and largely transcriptionally silent DNA. Conventional opinion holds that holokinetic chromosomes contain a small amount of

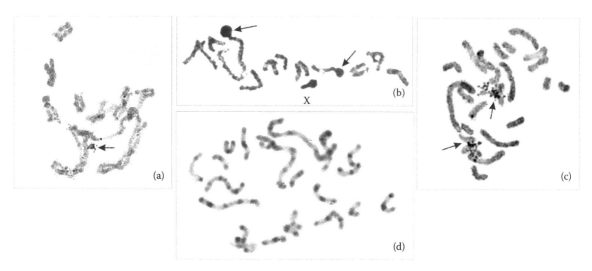

FIGURE 10.4   Silver-binding nucleolar organizer region banding (a–c) in *Philaenus italosignus* (a), *P. spumarius* (b), and *P. loukasi* (c) and C-banding (d) in *P. italosignus*. Arrows indicate nucleolus organizer regions in the sex trivalent (a), in the first and in a medium-sized bivalents (b), and in a medium-sized pair of autosomes (c). Bar = 10 µm. (From Kuznetsova, V. G. et al., *Folia Biol-Krakow*, 51 (1–2), 33–40, 2003 (b); Maryańska-Nadachowska, A. et al., *J Insect Sci* 12(54), 1–17, 2012 (a,c,d).)

constitutive heterochromatin, which is generally located on chromosome ends or in their vicinities (Blackman 1987). However, *Philaenus* species showed both terminal and interstitial C-bands on autosomes and sex chromosomes (Kuznetsova et al. 2003; Maryańska-Nadachowska et al. 2008, 2012). The greatest amount of C-heterochromatin was found in *P. italosignus* in which prominent C-bands were numerous and variably located along the complement, allowing the majority of homologous chromosomes to be identified (Figure 10.4d).

10.1.2.4   Meiosis

Unlike other Hemiptera in which meiosis is either highly specific (e.g., in true bugs) (Kuznetsova et al. 2011) or very often aberrant (e.g., in whiteflies and coccids) (White 1973), this process is essentially simple and follows the classical scheme in the Auchenorrhyncha (Kuznetsova and Aguin-Pombo 2014). This is also true of *Philaenus* spp. in which homologous chromosomes during first meiotic division undergo pairing, synapsis, recombination, and segregation at anaphase I; during second division, sister chromatids separate apart and migrate to opposite poles at anaphase II creating haploid daughter cells. Bivalents typically form one-two chiasmata only. The low number of chiasmata is a common pattern in the Auchenorrhyncha (Halkka 1964; Kuznetsova et al. 2009a,b, 2010) being most likely inherent in holokinetic bivalents as such. It has been demonstrated for the occurrence of fundamental differences between monocentric and holokinetic chromosomes in the condensation processes or underlying structural elements that act indirectly as a restrictive factor on the number of chiasmata formed in holokinetic bivalents (Nokkala et al. 2004).

10.1.2.5 Karyotype Transformations in Evolution

Differences in chromosome number between *Philaenus* species are caused by variation in the number of autosomes and sex chromosomes, due to the existence of three autosome numbers (18, 20, 22) and three sex systems (X(0), XY, and X_1X_2Y) in males. As has been mentioned previously, the within-genus karyotype diversity is largely in agreement with the recognized grouping proposed from morphology (Drosopoulos and Remane 2000) and food plant relationships (Drosopoulos 2003). Thus, the four representatives of the *signatus* group have similar karyotypes, $2n = 22 + XY$ in *P. signatus*, *P. maghresignus*, and *P. tarifa*, and $2n = 20 + X_1X_2Y$ in *P. italosignus*. The last karyotype might have arisen from $2n = 22 + XY$ via a fusion between the ancestral Y chromosome and an autosome pair (Maryańska-Nadachowska et al. 2013). All of these species are similar to one another in the morphology of male genitalia and feeding on *Asphodelus aestivus* (Asphodelaceae). Within the *spumarius* species group, *P. loukasi* and *P. arslani* whose larvae develop on plants growing in arid conditions share $2n = 18 + XY$, whereas the two polyphagous species, *P. spumarius* and *P. tesselatus*, feeding on a wide range of dicotyledonous plants possess $2n = 22 + X(0)$.

Plant-feeding insects nearly always mate on their food plant, and this assortative mating restricts gene flow between populations of a species and increases the likelihood of adaptation to different food plants. Certain host-specific phytophagous insect species are suggested to arise in the absence of geographic isolation (i.e., in sympatry) in the process of shifting and adapting to new plants (e.g., Borghuis et al. 2009). We speculate that in *Philaenus*, host switching has generated selection for chromosome rearrangements and divergence in chromosomal complements and that karyotype changes have occurred independently several times in the evolution of the genus (Figure 10.5).

10.1.3 Complete Mitochondrial Genome Sequence of *Philaenus spumarius*

Mitochondrial DNA is a useful and particularly popular marker in molecular ecology, population genetics, evolutionary biology, as well as in phylogeographic and phylogenetic studies of animals and insects (Stewart and Beckenbach 2005; Hahn et al. 2013). High mutation rate, lack of recombination, maternal inheritance,

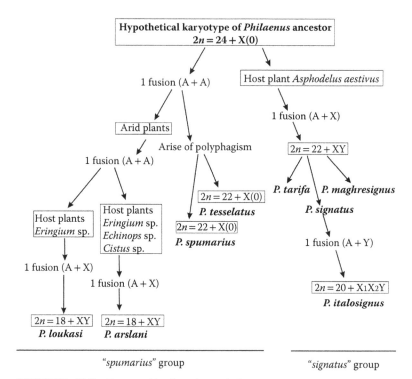

FIGURE 10.5   Presumable directions of chromosome rearrangements during karyotype evolution and changes of host plant preferences in *Philaenus* species.

high copy number, and therefore relatively easy accessibility often make mitochondrial DNA the molecular marker of choice. In the early years, only relatively short mitochondrial regions were targeted, but with improving methodology, sequencing of complete mitochondrial genomes (mt-genomes) became more common, even when exploring difficult templates such as ancient DNA. Complete mt-genomes are particularly useful when attempting to answer long-standing questions of evolutionary histories and reconstruct phylogenies. The development of next-generation sequencing instruments has led to the complete mt-genome sequence of *P. spumarius* (Stewart and Beckenbach 2005) (http://www.bch .umontreal.ca/ogmp/projects/other/mt_list.html). To date, only five species of the insect cohort Paraneoptera representing, however, all of its orders, Psocoptera, Phthiraptera, Thysanoptera, and Hemiptera, have their complete mitochondrial DNA sequences available in the sequence databases (see for references Stewart and Beckenbach 2005). Besides *P. spumarius*, these are the heteropteran species *Triatoma dimidiata* (the kissing bug), the phthirapteran species *Heterodoxus macropus* (the wallaby louse), the thysanopteran species *Thrips imaginis* (the plague thrips), and a psocopteran species (a booklouse identified as lepidopsocid RS-2001). In *P. spumarius*, the mt-genome is a circular molecule of 16,324 bp with a total A + T content of 77.0% and 76.7% for coding regions only (Stewart and Beckenbach 2005). This genome is relatively conservative in terms of gene organization and nucleotide composition. The genome organization is the same as observed in *Drosophila yakuba* (Clary and Wolstenholme 1984) and the hypothesized ancestral arthropod genome arrangement (Crease 1999) with, however, the addition of a relatively A + T-deficient repeat region located within the A + T-rich region. This conservation of genome structure is consistent with the other mt-genome example observed within the Hemiptera but varied from the other species sampled from the remaining paraneopteran orders. The nucleotide composition and patterns of nucleotide strand biases are more similar to those observed in other insects than in the other hemipteran, the kissing bug (Stewart and Beckenbach 2005).

10.1.4   Molecular Cytogenetics of *Philaenus* spp

Cytogenetics entered the molecular era with the introduction of in situ hybridization, a procedure that allows researchers to locate the positions of specific DNA sequences and genes on chromosomes. FISH is a very straightforward technique that involves hybridizing a DNA probe to its complementary sequence on chromosomal preparations. Using this technique, it is possible to integrate the molecular information of DNA sequences to their physical location along chromosomes of a species. Recently, we adapted FISH for the purpose of characterizing *Philaenus* karyotypes (Maryańska-Nadachowska et al. 2013). This is the first application of FISH technique in the Auchenorrhyncha. Ribosomal 18S rDNA and telomere $(TTAGG)_n$ probes were hybridized to chromosomes of eight species in an effort to identify additional chromosome markers and improve understanding karyotype transformations in the process of evolution of the genus *Philaenus*.

10.2   FLUORESCENCE IN SITU HYBRIDIZATION PROTOCOLS

10.2.1   Materials and Supplies

Prepare all solutions using ultrapure water (prepared by purifying deionized water to attain a sensitivity of 18 MΩ·cm at 25°C) and analytical grade reagents. Prepare and store all reagents at room temperature (RT) (unless indicated otherwise).

Standard solutions (such as ethanol, methanol/glacial acetic acid 3:1, 20× saline sodium citrate [SSC, Roche Applied

Science, Mannheim, Germany] phosphate-buffered saline [PBS, Sigma-Aldrich, Steinheim, Germany], formamide, Sigma-Aldrich, Steinheim, Germany, formaldehyde, F8775, Sigma-Aldrich, Steinheim, Germany) are used. Note that some of the chemicals below are environmental toxins (e.g., formaldehyde and formamide). All work with these solutions must be done in a chemical fume hood. Please ensure that these substances are collected and treated as hazardous waste after use.

1. Formamide (Cat. No. F7503, Sigma-Aldrich, Steinheim, Germany).
2. Deionized Formamide for hybridization buffer (Cat. No. F9037, Sigma-Aldrich, Steinheim, Germany).
3. Prolong Gold Antifade reagent with 4′,6-diamidino-2-phenylindole (DAPI) (Cat. No. P36931, VECTOR Laboratories, Burlingame).
4. PBS, pH 7.4 (without Ca^{2+} and Mg^{2+}) (phosphate buffered saline—Cat. No. P3813, Sigma, Steinheim, Germany).
5. 20× SSC (175.3 g NaCl, 88.2 g sodium citrate, water to 1000 mL), adjust to pH 7.0 (Product No. 11666681001, Roche Applied Science, Mannheim, Germany).
6. Salmon Sperm DNA (stock ~10 mg/mL) (Cat. No. D7656, Sigma-Aldrich, Steinheim, Germany).
7. Ribonuclease A (RNase A) Solution from bovine pancreas for molecular biology deoxyribonuclease (DNase) free (Cat. No. R4642, Sigma-Aldrich, Steinheim, Germany).
8. Rubber cement: Fixogum™ (Marabu, Tamm, Germany).
9. Hybridization buffer: dissolve 2 g dextran sulfate in 10 mL 50% deionized formamide/2× SSC, 0.05% NP-40 (USB, Cleveland, USA) for 3 hours at 70°C (final pH ~7.0). Aliquot and store at −20°C.
10. 1N HCl (mix 10 mL 36% HCl with 90 mL water, store in glass bottle).
11. 0.01 N HCl for pepsin digestion (mix 0.5 mL 1 N HCl in 49.5 mL water).
12. Pepsin stock (10%): dissolve 1 g pepsin (Sigma, Steinheim, Germany) in 10 mL H_2O at 37°C; store in aliquots at −20°C. Pepsin solution: add 1 mL of 1 M HCl to 99 mL distilled water and incubate at 37°C for 20 minutes; then add 500 μL of pepsin stock solution 10% (w/v) (Cat. No. P-7012, Sigma, Steinheim, Germany) and leave coupling jar at 37°C; make fresh.
13. Postfix solution: add 3 mL acid-free formaldehyde 37% (Sigma, Cat. No. F8775), 2.5 mL 1 M $MgCl_2$ to 100 mL of 1× PBS.

14. RNaseA stock solution (10 mg/mL) for slide pretreatment. Dissolve 10 mg RNaseA powder (Sigma, DNase free) in 1 mL water. Boil tube for 5 minutes, and then store indefinitely in a freezer.

15. 20× SSC: 175.3 g sodium chloride, 88.2 g sodium citrate. Add 1 L of double-distilled water (ddH$_2$O), pH 7.0, autoclave before storage at RT.

16. Washing buffer (diluted from stock 20× SSC): 1× SSC (mix 50 mL 20× SSC, 950 mL water); 4× SSC, 0.05% NP-40 (mix 200 mL 20× SSC, 800 mL water, 0.5 mL NP-40).

17. Modified labeled nucleotide at 1 mM, such as digoxigenin-11-dUTP and biotin-16-dUTP (Cat. No. 11 093 088 910 and Cat. No. 11 093 070 910, respectively, Roche Diagnostics GmbH, Germany).

18. Taq DNA polymerase with 10× Taq buffer (Cat. No. EP0402, Fermentas Life Science, St. Leon-Rot, Germany).

19. Deoxyribonucleotide triphosphate (dNTP): 10 mM stock solutions of deoxyadenosine triphosphate, deoxycytidine triphosphate, deoxyguanosine triphosphate, and deoxythymidine triphosphate (Cat. No. D6920, D7045, D7170, T7791, Sigma, Steinheim, Germany).

20. Tris-ethylene diaminetetraacetic acid (EDTA) buffer (TE buffer) (10 mM Tris, pH 7.0; 1 mM EDTA, pH 8.0).

10.2.2 Equipment

For FISH, the standard molecular cytogenetic equipment is needed. Apart from the standard equipment, the following more specialized items are needed: CCD (charge-coupled device) camera with image capture and hard- and software; epifluorescence microscope; hot plate with digital temperature control for slide warming (see Section 10.2.6 for more details).

1. Speed vacuum Concentrator 5301 (Eppendorf, Hamburg, Germany)

2. Thermomixer compact (Eppendorf, Hamburg, Germany)

3. Mastercycler personal (Eppendorf, Hamburg, Germany)

4. Centrifuge 5415C (Eppendorf, Hamburg, Germany)

5. Mini agarose gel electrophoresis apparatus (Bio-Rad Laboratories, Inc, Hercules, CA)

6. UV transilluminator (Bio-Rad Laboratories, Inc., Hercules, CA)

7. Phase-contrast microscope (Leica, Wetzlar, Germany, DM4000 B microscope with a Leica DFC350 FX camera and a Leica Application Suite 2.8.1. software with an Image Overlay module)

10.2.3   Chromosome Preparation

Fixative solution: 3:1 ethanol/glacial acetic acid

Superfrost™ microscope slides (Article Number AA00008032E, Menzel, Braunschweig, Germany)

Microscope coverslips (24 × 24 mm) (Article Number BB024024A1, Menzel, Braunschweig, Germany)

Slide storage boxes (Kartell, Milano, Italy)

10.2.4   Probe Preparation

10.2.4.1   Labeling of Ribosomal 18S rDNA Probe

For ribosomal DNA (rDNA), chromosome sequences detection by FISH, a 1.8 kb fragment of 18S rDNA was generated by polymerase chain reaction (PCR) using genomic DNA isolated from the specimens of *Pyrrocoris apterus* as a template.

For rDNA-probe, prepare 25 µL PCR reaction mixture containing 1.5 mM $MgCl_2$, 2.5 mM dNTPs, 10 µM of the two primers the 18Sai forward primer (5′-CCT GAG AAA CGG CTA CCA CAT C-3′) and the 18Sbi reverse primer (5′-GAG TCT CGT TCG TTA TCG GA-3′) (Whiting et al. 1997), 100 ng template DNA, and 5 U Taq DNA polymerase (Qiagen, Hilden, Germany). Perform PCR with the thermal conditions: an initial denaturation period of 3 minutes at 94°C was followed by 30 cycles of 60 seconds at 94°C, annealing for 60 seconds at 51°C, a 1.5 minutes extension step at 72°C, and concluded with a final extension step of 10 minutes at 72°C. The probe was labeled by nick translation with biotin-16-dUTP (Roche Diagnostics GmbH, Mannheim, Germany) according to the manufacturer's instructions (Invitrogen, NY).

10.2.4.2   Labeling of Telomeric (TTAGG)$_n$ Probe

To visualize clusters of telomeric repeats, the pentanucleotide (TTAGG)$_n$-specific telomeric DNA probe was generated by PCR using a modified version of López-Fernández technique (López-Fernández et al. 2004). A nontemplate PCR was carried out in 50 µL of reaction mixture containing 1.5 mM $MgCl_2$, 0.2 mM of each dNTP, 0.5 µM of each of the two primers (5′-GGTTA-GGTTA-GGTTA-GGTTA-GG-3′ and 5′-TAACC-TAACC-TAACC-TAACC-TAA-3′), and 2 U Taq DNA polymerase. The PCR was performed with an initial cycle of 90 seconds at 94°C, followed by 30 cycles of 45 seconds at 94°C, 30 seconds at 40°C, and 60 seconds at 72°C, and a final extension step of 10 minutes at 72°C. The PCR product was then labeled with digoxigenin-11-dUTP during additional PCR cycles to produce PCR telomeric probe. Using 5 µL of PCR product, check the quality of amplified DNA by agarose gel electrophoresis. A smear of DNA in size range between 100 and 1000 bp should be visible.

10.2.5　Hybridization and Detection

10.2.5.1　Postlabeling Treatment of the Probe

Before use for FISH, the PCR-labeled DNA should be prepared as follows for each slide to be hybridized: precipitate 200–400 ng of the digoxigenin or biotin labeled probes together with 5 mg of salmon sperm DNA with 2.5 volumes of ethanol (100%, –20°C) and 0.1 volume of sodium acetate (3 M, pH 5.2). Precipitation can be done for either 20 minutes at –80°C or 12–20 hours at –20°C. Pellet the DNA by centrifugation at 15,000 rpm at 4°C for 30 minutes. After centrifugation, a white pellet should be seen; discard the supernatant. Untreated dNTP is then removed by washing the DNA pellet with 2.5 volumes of ice-cold 70% ethanol. Centrifuge at 15,000 rpm at 4°C for 15 minutes and discard the supernatant without removing the pellet.

Allow to air-dry at 37°C in a heat block or using a speed vacuum. Subsequently, the probe mixture has been resuspended in hybridization buffer (see recipe). To improve resuspension, place the tube(s) containing the DNA pellet and 15 μL of hybridization buffer in Thermomixer Eppendorf at 45°C and vortex them for at least 1 hour, until the pellet dissolves thoroughly. Also, the pellet can be resuspended in 15 μL of hybridization buffer (per slide) for at least 2–3 hours at RT. The DNA is stable at this stage and can be left at RT for use on the same day. Alternatively, it can be kept at –20°C for long periods of time (Garimberti and Tosi 2010).

10.2.5.2　Slide Pretreatment and Denaturation

10.2.5.2.1　Pretreatment with Ribonuclease A

Prepare RNase A:

1. Dilute 10 μL (10 mg/mL) RNase (Sigma) in 990 μL 2× SSC (pH 7.0–7.4) and prewarm to 37°C.
2. Apply 100 μL per slide and incubate for 1 hour at 37°C under coverslip in a humid chamber.
3. Rinse slides 3× 5 minutes with 2× SSC (pH 7.0–7.4) at RT.
4. Dehydrate the slides through an alcohol series (make up fresh): 3 minutes in 70% ethanol, 3 minutes in 80% ethanol, and 3 minutes in 100% ethanol.
5. Allow to air-dry for at least 10 minutes.

10.2.5.2.2　Pretreatment with Pepsin

In a conventional FISH approach, pretreatment of the slides with pepsin followed by postfixation with formalin buffer is required to reduce the background.

1. Put slides for 15 minutes in pepsin solution at 37°C in a coupling jar.

2. Wash slides in 100 mL 1× PBS (RT) for 5 minutes. Once with PBS containing 50 mM MgCl$_2$.
3. Postfix for 10 minutes at RT with a postfix solution of PBS containing 50 mM MgCl$_2$ and 1% formaldehyde.
4. Incubate slides in 100 mL 2 × PBS (RT) for 5 minutes.
5. Dehydrate slides in an ethanol series (70%, 80%, 100%, 3 minutes each) and air-dry.

10.2.5.3 Fluorescence In Situ Hybridization

1. Add 15 µL of probe in hybridization mixture directly onto each slide, put a 24 × 24 mm coverslip (avoiding bubbles) on the drop, and seal with rubber cement.
2. Incubate slides on a warming plate for 5 minutes at 75°C. The amount of probe/probe solution must be reduced according to the coverslip size.
3. Incubate slides for overnight at 42°C in a moisten chamber, humidified with wet paper towels and placed inside a standard thermostat overnight.
4. Take the slides out of the 42°C chamber. Gently remove the rubber cement with forceps avoiding movements of the coverslips that can result in damage to chromosomes and nuclei.

10.2.5.4 Slide Washing and Mounting

1. To remove the unbound or nonspecifically bound probe fragments, the slides will be washed under stringent conditions.
2. Then place the slides for 5 minutes in a prewarmed coupling jar containing 1× SSC for 5 minutes at 60°C without agitation.
3. Transfer the slides in a coupling jar containing 4× SSC/0.2% NP-40 (100 mL, 45°C) for 10 minutes at 45°C on a shaking platform.
4. Biotin- and digoxigenin-labeled probes were visualized with avidin—Alexa 488 (Invitrogen) and mouse antidigoxigenin antibodies conjugated to Cy3 (Sigma-Aldrich, Tokyo, Japan), respectively.
5. Wash the slides (2× 5 minutes) with 4× SSC/0.2% NP-40 at 45°C on a shaking platform.
6. Put the slides in PBS solution (RT) for 5 minutes.
7. Wash the slides briefly with water for a few seconds.
8. Dehydrate slide in ethanol (70%, 80%, 100%, 4°C, 3 minutes each) and air-dry in dark place.
9. Add 15 µL DAPI/Antifade solution, cover with a coverslip, and look at the results under a fluorescence microscope.

10.2.6   Visualization and Mapping

Visualization of fluorescent hybridization signals requires the use
of an epifluorescence microscope equipped with appropriate filter
set specific for the fluorochromes to be viewed. In our laboratory,
the system in use for two-dimensional-FISH analysis is a Zeiss
microscope (Axioskop 2 PLUS, Carl Zeiss Jena GmbH, Germany)
equipped with $100 \times$ objective and a CCD-camera (CV M300; JAI
Corporation, Yokohama, Japan). Appropriate filters are mounted
in a six-position computerized filter wheel connected with the
microscope. Often used filter sets are No. 49 (Zeiss, Germany),
SP101 fluorescein isothiocyanate (CHROMA, Lake Forest, CA),
and SP103v1 Cy3tmv1 (CHROMA, Lake Forest, CA); software used for acquisition and storage of FISH analysis is ISIS4
(METASystems GmbH, Altlussheim, Germany).

10.2.7   Troubleshooting

1. If no or weak signal is observed, then amplify signal.

   For most FISH solutions, it is necessary to autoclave
   ddH_2O for 20 minutes to destroy any DNase activity and
   ensure sterility. The water quality in these procedures
   has a stronger effect on the signal-to-noise ratio than the
   purity of the chemicals (Garimberti and Tosi 2010).

   The quality of the preparation is absolutely crucial
   if good hybridization results are to be obtained. The
   preparation should be well spread, flat, and have plenty
   of chromosomes with good morphology. In addition, the
   chromosomes should be free from cytoplasmic remains
   and other cellular material. Each slide should be carefully
   checked under a phase-contrast microscope immediately
   after the spread.

   RNase/pepsin pretreatment is recommended. Without
   pretreatment, such metaphase preparations would give
   poor FISH results, because the target DNA is inaccessible to the probe, aggravated by the autofluorescence of
   the cytoplasm. Normally, the pretreatment with RNase A
   is done for between 30 minutes and 1 hour. If it is done
   for less than 30 minutes, the digestion is not complete; on
   the other hand, RNase A may begin to influence the chromosome morphology with treatments that are longer than
   1 hour.

   Check the size of the labeled probe. In our laboratory,
   we use PCR product with labeled DNA fragments in the
   desired range (200–600 bp) and not more than 1000 bp.
   Larger DNA fragments would result in bright fluorescent

signals all over the slide. If the size of labeled DNA is larger than 1000 bp, it should be recut using DNase I. When labeled DNA fragments are too small (<200 bp), they might not hybridize efficiently to chromosomal DNA; hence, FISH signals would not be visible.

2. If there is starry background fluorescence, then spin the detection solution 4 minutes at 14,000 rpm and take only supernatant.

   Remove the supernatant and wash the pellet of DNA probe twice with ethanol. Dry the DNA pellet on a Speedy-vac sample concentrator. The DNA usually becomes completely transparent when dry. However, overdrying the DNA pellet might make it difficult to resuspend it in hybridization buffer. Alternatively, the pellet can be resuspended in 0.5–1 μL water or TE buffer at 50°C with repeated vortexing for 5–10 minutes. After the pellet is hydrated, add 15 μL of hybridization buffer and mix well for 1 hour at 45°C, spin down before dropping. The use of high temperatures should be avoided to prevent the denaturation of the probe at this stage.

3. If there is a general strong staining of chromosomes and nuclei, then check the size of the labeled probe on an agarose minigel with a suitable size marker. If the probe is too small, relabel the probe.

10.2.8   Representative Results

10.2.8.1   Detection of a Tandem Telomere Repeat Sequence in *Philaenus* spp by Fluorescence In Situ Hybridization with the (TTAGG)$_n$ Probe

In each of the eight species tested, bright hybridization signals were revealed at the ends of spermatocyte chromosomes whereas no signals were found in nontelomeric locations (Figure 10.6a through h). This finding indicates conclusively that the telomeres in *Philaenus* species are composed of the simple (TTAGG)$_n$ nucleotide sequence.

10.2.8.2   Detection of Ribosomal Sites in *Philaenus* spp by Fluorescence In Situ Hybridization with the 18S rDNA Probe

In different *Philaenus* species, FISH with 18S rRNA-targeted probe yielded hybridization signals either on autosomes or on sex chromosomes or both on autosomes and sex chromosomes. Location patterns showed no evident variation within a species. The three species, *P. tesselatus*, *P. arslani*, and *P. signatus*, had a single hybridization locus (Figure 10.6a through c), whereas *P. spumarius*,

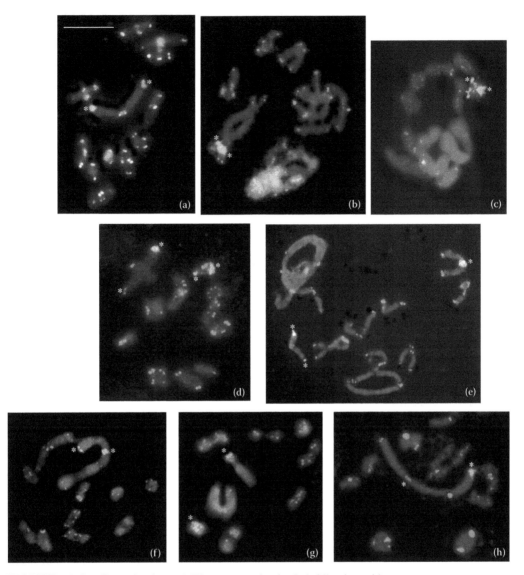

FIGURE 10.6 **(See color insert.)** Fluorescence in situ hybridization with 18S ribosomal DNA and TTAGG telomere repeats as probes in *Philaenus* species: *P. tesselatus* (a), *P. arslani* (b), *P. signatus* (c), *P. spumarius* (d), *P. tarifa* (e), *P. italosignus* (f), *P. loukasi* (g), and *P. maghresignus* (h). *, 18S rDNA arrays in chromosomes. (From Maryańska-Nadachowska, A. et al., *Eur J Entomol* 110 (3), 411–8, 2013.)

P. tarifa, *P. italosignus*, *P. loukasi*, and *P. maghresignus* had two such loci (Figure 10.6d through h). Figure 10.7 summarizes the most representative results from FISH analysis.

In *P. tesselatus* and *P. arslani*, positive hybridization signals were present on the largest pair of autosomes being located terminally in the first species while clearly subterminally in the second one. In contrast to these one-locus species, *P. signatus* showed fluorescent bright spots at the ends of homologues of the pair 6. In two-locus

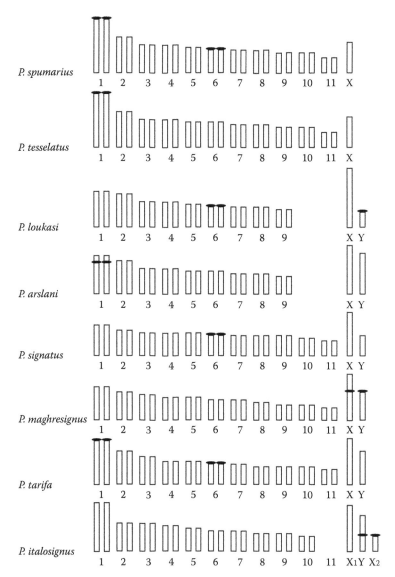

FIGURE 10.7   Idiograms of *Philaenus* species karyotypes showing the physical location of the 18S rDNA sites (black lenticular bodies). (From Maryańska-Nadachowska, A. et al., *Eur J Entomol* 110 (3), 411–8, 2013.)

species, hybridization sites were located differently: in the terminal position on the largest and one of the medium-sized (probably 6) pairs of autosomes in *P. spumarius* and *P. tarifa*; terminally on one of the medium-sized pairs of autosomes (probably 6) and on the Y chromosome in *P. loukasi*; and on the X and Y chromosomes in *P. maghresignus* and *P. italosignus*. In *P. maghresignus*, signals were interstitial on the X chromosome while terminal on the Y chromosome, whereas in *P. italosignus*, they were interstitial on the Y while terminal on the X_2 (Figure 10.6).

10.3 DISCUSSION

10.3.1 Chromosome Organization and Evolution in *Philaenus*

The genus *Philaenus*, like all other Auchenorrhyncha and Hemiptera as a whole (Hallka 1959; White 1973), is characterized by holokinetic chromosomes. In contrast to monocentric chromosomes, holokinetic chromosomes have no localized centromere. This is considered to be nonlocalized or diffuse and is formed by a large kinetochore plate extending along all or most of holokinetic chromosome length (Schrader 1947; Wolf 1996). The large kinetochore plate is suggested to encourage rapid karyotype evolution via occasional fusion/fission events. First, fusion of holokinetic chromosomes would not create the problems characteristic of a dicentric chromosome in monocentric organisms. Second, fission of a holokinetic chromosome should create chromosome fragments that exhibit a part of the kinetochore plate and can attach themselves to the spindle fibers at cell divisions. As a result, chromosome fragments that would be acentric (lacking a centromere) and hence lost in organisms with monocentric chromosomes may be inherited in a Mendelian fashion in holokinetic organisms. The gametes harboring chromosome fragments are consequently expected to be viable (Hipp et al. 2010). Fusion/fission rearrangements are therefore conventionally accepted as the most common mechanisms of chromosome evolution in holokinetic groups (White 1973), fusions being probably more common in holokinetic groups including Auchenorrhyncha (Halkka 1959, 1964; Kuznetsova and Aguin-Pombo 2014). The point is that a chromosome, whether holokinetic or monocentric, has to display two functional telomeres to survive a mitotic cycle. A fusion chromosome always displays two functional telomeres originated from the two ancestral chromosomes, whereas fission chromosomes have to be able to develop functional telomeres de novo (Nokkala et al. 2007). However, more studies are indubitably needed to better appreciate the mechanisms of evolution of holokinetic chromosomes, because, for example, Hipp et al. (2010) have shown that both fusions and fissions restrict gene flow in plants with holokinetic chromosomes.

10.3.2 Molecular Organization of Telomeres in *Philaenus*

Telomeres are defined as regions of the chromosomal ends that are required for complete replication, meiotic pairing, and stability of a chromosome (Zakian 2012). The molecular structure of telomeres is characterized by a tandem repeat of a short DNA sequence, which is synthesized by the reverse transcriptase activity of telomerase and is diversely differentiated in eukaryotes (Blackburn 1991). Comparative analysis of these repeats (motifs) in various groups of organisms showed that they were evolutionarily stable, and, having once appeared during evolution, defined taxa and phylogenetic branches of high rank (Traut et al. 2007). Quite recently, Frydrychová et al. (2004) assembled and analyzed all the data available on the telomere motifs in Insecta, and, together with some original observations, they interpreted these character data in a phylogenetic framework. The majority of insect species was demonstrated to share the telomeres composed of the pentanucleotide TTAGG repeat. Because the same telomere composition is characteristic of the vast majority of evolutionary lineages in Arthropoda, the $(TTAGG)_n$ telomere sequence was considered an ancestral motif in the Insecta. Many higher level insect groups preserved this motif; however, several orders, for example, Dermaptera, Diptera, and some others, are suggested to have lost this telomere sequence during their evolution (Sahara et al. 1999; Frydrychova et al. 2004; Vitkova et al. 2005; Lukhtanov and Kuznetsova 2010).

To date, a single attempt has been made to detect molecular organization of telomeres in Auchenorrhyncha. By using single-primer PCR and southern hybridization of genomic DNA, Frydrychová et al. (2004) suggested the $(TTAGG)_n$ sequence to be characteristic of the telomeres in *Calligypona pellucida* from the fulgoroid family Delphacidae. However, these techniques are known to be a rank below FISH in terms of accuracy of telomere repeat detection because they reveal only the presence of a sequence in the genome, but not its location on the chromosome. The data coming from our pioneering FISH (Maryańska-Nadachowska et al. 2013) confirmed that the chromosome ends of *Philaneus* species were composed of the $(TTAGG)_n$ sequence. It seems probable that this telomere motif has been conserved in the Auchenorrhyncha; however, evidence for the latter remains still scarce. For example, recently it has been shown that in heteropterans, the canonical insect $(TTAGG)_n$ motif is absent in the evolutionarily advanced families Pyrrhocoridae, Miridae, Cimicidae, and Pentatomidae (Grozeva et al. 2011) but is present in the basal family Belostomatidae (Kuznetsova et al. 2012).

10.3.3 Mapping a Ribosomal 18S rDNA Cluster to *Philaenus* spp Chromosomes

In most eukaryotes, rDNA consists of tandemly repeated arrays of three genes (18S, 5.8S, and 28S) encoding nuclear rRNA. These arrays make up the NORs and can be located on one or a few chromosomes in different species. Knowledge of the relative physical locations and number of rDNA loci is important and useful in the construction of physical maps of chromosomes and in phylogenetic studies. Members of the *Philaenus* showed wide variation in the number and location of 18S rDNA sites: eight species were examined and six patterns of rDNA location were observed. Figure 10.7 shows the locations of 18S rDNA sites on chromosomes of each of the species tested. Although the numbers of loci were low (one or two in different species), they were visualized variously in different species: either on autosomes (*P. spumarius*, *P. signatus*, *P. tarifa*, *P. tesselatus*, and *P. arslani*) or on sex chromosomes (*P. italosignus* and *P. maghresignus*), or both on autosomes and sex chromosomes (*P. loukasi*). On autosomes, the location of 18S rDNA sites did not vary randomly, occurring preferentially on the largest and/or on one of the medium-sized pairs. It is very difficult to identify chromosome pairs and to identify homeologies between species; however, a medium-sized pair with the 18S rDNA site seems to be the same pair number 6 in each of the species. We observed also that rDNA sites occupied preferentially the terminal regions of chromosomes, both in karyotypes with single site and in karyotypes with two sites, and both in the case of autosomes and in the case of sex chromosomes. This observation allows suggestion that there seems to be a strong positive selection favoring the location of 18S rDNA sites at the terminal region. However, in some cases, hybridization spots were located subterminally as in *P. arslani* (on the first pair of autosomes), or clearly interstitially as in *P. maghresignus* (on the X chromosome) and in *P. italosignus* (on the Y chromosome). The variation in the chromosomal location of rDNA sites in species with similar karyotypes (in *P. tesselatus* and *P. spumarius* with $2n = 22 + X$; in *P. loukasi* and *P. arslani* with $2n = 18 + XY$; in *P. signatus*, *P. maghresignus*, and *P. tarifa* with $2n = 22 + XY$) shows that their rDNA is mobile allowing suggestion that such mobility may be an evolutionary feature of the *Philaenus*. Although we were unable to infer with confidence which location is the ancestral state in the genus, one 18S rDNA-bearing pair of autosomes, most likely the largest seems to correspond to the ancestral condition because such a location of NORs has been observed in all auchenorrhynchan species so far examined using AgNOR and fluorochrome chromomycin A3 (CMA$_3$) techniques (Kuznetsova et al. 2009b, 2010).

Several authors doubted the distinctness of *P. tesselatus* considering it as a subspecies (Wagner 1959), or a geographical and morphological form (Maryańska-Nadachowska et al. 2013), or a synonym of *P. spumarius* (Nast 1972). However, Drosopoulos and Quartau (2002) argued for a validity of *P. tesselatus* due to that it possesses constant species-level characters. Cytogenetic data on

P. tesselatus in terms of AgNOR and FISH-18S rDNA patterns are in conflict. *P. tesselatus* differs from *P. spumarius* in having a single hybridization site of 18S rDNA located on the first pair of chromosomes, whereas *P. spumarius* displays two FISH-rDNA arrays, one on the first pair of autosomes and the other on the autosomal pair 6 (Figure 10.7). This difference would favor the view that *P. tesselatus* is a valid species; however, after AgNOR-banding performed in this species, an additional very weak silver-positive site was occasionally revealed on the bivalent number 6 (Maryańska-Nadachowska et al. 2013). Possibly, *P. tesselatus* has in fact two 18S rDNA sites, but if copy number falls below the threshold for in situ hybridization, we may not be able to reliably detect the site. Noteworthy also is a lack of perfect correlation between results obtained through FISH, silver staining, and CMA_3 fluorochrome staining in determining the number of SAT-chromosomes in some other *Philaenus* species. As in the case of *P. spumarius* and *P. tesselatus*, the number of major rDNA revealed by FISH was always lower than the number of AgNORs (Maryańska-Nadachowska et al. 2013). The noncorrespondence between the number of FISH-rDNA signals and the number of Ag-positive signals has been reported in many animal species (see for review Gromicho et al. 2005), including insects such as ants (Lorite et al. 1997), beetles (Colomba et al. 2000), and grasshoppers (Vitturi et al. 2008).

10.4   CONCLUSIONS

Numerous studies, each with a radically different approach, have been performed on the spittlebug genus *Philaenus* to date. They have been focused on a wide range of aspects, including cytogenetic characters (Kuznetsova et al. 2003; Maryańska-Nadachowska et al. 2008, 2012, 2013). The FISH technique being applied for the first time to the genus *Philaenus* and to the Auchenorrhyncha as a whole takes the cytogenetics of this large group of Hemiptera a step further. The application of FISH with ribosomal rDNA and telomere $(TTAGG)_n$ as probes has shown that (1) *Philaenus* species underwent an extensive reorganization of their genomes: the ribosomal genes changed repeatedly their relative position along the chromosomes indicating that large-scale chromosomal alterations, that is, autosomal and sex-autosomal translocations, did occurred in the evolution of this group and (2) a pentanucleotide sequence repeat $(TTAGG)_n$ is inherent in *Philaenus* species allowing thus suggestion that this canonical and ancestral motif of insect telomeres was conserved in the Auchenorrhyncha as a whole.

It can be anticipated that data generated by FISH with some other nucleotide probes will provide more definitive insights into the origin and mechanisms of *Philaenus* karyotypic diversity.

ACKNOWLEDGMENTS

The study was financially supported by the Russian Scientific Foundation (grant 14-14-00541 to VK) and by the Russian Foundation for Basic Research (grant 14-04-00086a to TK). Work on this project was facilitated by the research agreement between the Russian Academy of Sciences and the Polish Academy of Sciences (to VK and AM-N)

REFERENCES

Abdul-Nour, H. and L. Lahoud. 1995. Révision du genere *Philaenus* Stål, 1864 au Liban, avec la description d'une nouvelle espèce: *P. arslani*, n.sp. (Homoptera, Auchenorrhyncha, Cercopidae). *Nouv Rev Entomol (NS)* 12:297–303.

Bai, X., P. Mamidala, S. P. Rajarapu, S. C. Jones, and O. Mittapalli. 2011. Transcriptomics of the bed bug (Cimex lectularius). *PLoS One* 6(1):e16336. doi:10.1371/journal.pone.0016336.

Blackburn, E. H. 1991. Structure and function of telomeres. *Nature* 350:569–73.

Blackman, R. L. 1987. Reproduction, cytogenetics and development. In *Aphids: Their Biology Natural Enemies and Control. World Crop Pests*, eds. A. K. Minks and P. Harrewijn, 2A: 164–95, Elsevier, Amsterdam, The Netherlands.

Borghuis, A., J. van Groenendael, O. Madsen, and J. Ouborg. 2009. Phylogenetic analyses of the leaf beetle genus Galerucella: Evidence for host switching at speciation? *Mol Phylogenet Evol* 53:361–7.

Boring, A. M. 1913. The chromosomes of the Cercopidae. *Biol Bull* 24:133–47.

Clary, D. O. and D. R. Wolstenholme. 1984. The Drosophila mitochondrial genome. *Oxf Surv Eukaryotic Genes* 1:1–35.

Colomba, M. S., R. Vitturi, and M. Zunin. 2000. Chromosome analysis and rDNA FISH in the stag beetle Dorcus parallelipipedus L. (Coleoptera: Scarabaeoidea: Lucanidae). *Hereditas* 133:249–53.

Crease, T. J. 1999. The complete sequence of the mitochondrial genome of Daphnia pulex (Cladocera: Crustacea). *Gene* 233:89–99.

Drosopoulos, S. 2003. New data on the nature and origin of colour polymorphism in the spittlebug genus Philaenus (Hemiptera: Aphrophoridae). *Ann Soc Entomol Fr* 39:31–42.

Drosopoulos, S., A. Maryańska-Nadachowska, and V. G. Kuznetsova. 2010. The Mediterranean: Area of origin of polymorphism and speciation in the spittlebug Philaenus (Hemiptera, Aphrophoridae). *Zoosyst Evol* 86:125–8.

Drosopoulos, S. and J. A. Quartau. 2002. The spittlebug Philaenus tesselatus Melichar, 1899 (Hemiptera, Auchenorrhyncha, Cercopidae) is a distinct species. *Zootaxa* 68:1–8.

Drosopoulos, S. and R. Remane. 2000. Biogeographic studies on the spittlebug species group Philaenus signatus with the description of two new allopatric species. *Ann Soc Entomol Fr* 36:269–77.

Frydrychová, R., P. Grossmann, P. Truba, M. Vítková, and F. Marec. 2004. Phylogenetic distribution of TTAGG telomeric repeats in insects. *Genome* 47:163–78.

Garimberti, E. and S. Tosi. 2010. Chapter 1. Fluorescence in situ hybridization (FISH), basic principles and methodology. In *Fluorescence In Situ Hybridization (FISH): Protocols and Applications, Methods in Molecular Biology*, eds. J. M. Bridger and E. V. Volpi. vol. 659:3–21. Springer Science+Business Media, Berlin, Germany. doi: 10.1007/978-1-60761-789-1_1.

Gromicho, M., C. Ozout-Costaz, and M. J. Collares-Pereira. 2005. Lack of correspondence between CMA3-Ag-positive signals and 28S rDNA loci in two Iberian minnows (Teleostei, Cyprinidae) evidenced by sequential banding. *Cytogenet Genome Res* 109:507–11.

Grozeva, S., B. Anokhin, and V. Kuznetsova. 2014. Chapter 10. Bed bugs (Hemiptera). In: *Protocols for Chromosome Mapping of Arthropod Genomes*, CRC Press, Taylor & Francis, Boca Raton, FL.

Grozeva, S., V. Kuznetsova, and B. Anokhin. 2010. Bed bug cytogenetics: Karyotype, sex chromosome system, FISH mapping of 18S rDNA, and male meiosis in Cimex lectularius Linnaeus, 1758 (Heteroptera: Cimicidae). *Comp Cytogenet* 4:151–60. doi: 10.3897/compcytogen.v4i2.36.

Grozeva, S., V. Kuznetsova, and B. Anokhin. 2011. Karyotypes, male meiosis and comparative FISH mapping of 18S ribosomal DNA and telomeric (TTAGG) n repeat in eight species of true bugs (Hemiptera, Heteroptera). *Comp Cytogenet* 5:355–74. doi: 10.3897/compcytogen. v5i4.2307.

Hahn, C., L. Bachmann, and B. Chevreux. 2013. Reconstructing mitochondrial genomes directly from genomic next-generation sequencing reads - a baiting and iterative mapping approach. *Nucleic Acids Res* (July) 41(13): e129. Published online May 9, 2013. doi: 10.1093/nar/gkt371.

Hales, D., A. C. C. Wilson, J. M. Spence, and R. L. Blackman. 2000. Confirmation that Myzus antirrhinii (Macchiati) (Hemiptera: Aphididae) occurs in Australia, using morphometrics, microsatellite typing and analysis of novel karyotypes by fluorescence in situ hybridization. *Aust J Entomol* 39:123–29.

Halkka, O. 1959. Chromosome studies on the Hemiptera, Homoptera, Auchenorrhyncha. *Ann Acad Sci Fen* (A IV) 43:1–71.

Halkka, O. 1964. Recombination in six homopterous families. *Evolution* 18:81–8.

Halkka, O. and L. Halkka. 1990. Population genetics of the polymorphic spittle-bug, Philaenus spumarius (L.). *Evol Biol* 24:149–91.

Halkka, O., M. I. Raatikainen, and J. Vilbaste. 1967. Ecology and ecological genetics of Philaenus spumarius (L.) (Homoptera). *Ann Zool Fenn* 4: 1–18.

Hipp, A. L, P. E. Rothrock, R. Whitkus, and J. A. Weber. 2010. Chromosomes tell half of the story: The correlation between karyotype rearrangements and genetic diversity in sedges, a group with holocentric chromosomes. *Mol Ecol* 19:3124–38. doi: 10.1111/j.1365-294X.2010.04741.x.

Kurokawa, H. A. 1953. A studies of the chromosomes in some species of the Cicadidae and the Cercopidae. *Jpn J Genet* 28:1–5.

Kuznetsova, V. G. and D. Aguin-Pombo. 2014. Comparative cytogenetics of Auchenorrhyncha: A review. In *Leafhoppers of the World and Their Relatives*, eds. J. Badmin and M. Webb London, UK.

Kuznetsova, V. G., S. Grozeva, and B. A. Anokhin. 2012. The first finding of (TTAGG)n telomeric repeat in chromosomes of true bugs (Heteroptera, Belostomatidae). *Comp Cytogent* 6(4):341–46. doi: 10.3897/ CompCytogen.v6i4.4058.

Kuznetsova, V. G., S. Grozeva, S. Nokkala, and C. Nokkala. 2011. Cytogenetics of the true bug infraorder Cimicomorpha (Hemiptera, Heteroptera): A review. *ZooKeys* 154:31–70. doi: 10.3897/zookeys.154.1953.

Kuznetsova, V. G., A. Maryańska-Nadachowska, and A. F. Emeljanov. 2009a. A contribution to the karyosystematics of the planthopper families Dictyopharidae and Fulgoridae (Hemiptera: Auchenorrhyncha). *Eur J Entomol* 106:159–70. doi: 10.14411/eje.2009.019.

Kuznetsova, V. G., A. Maryańska-Nadachowska, and V. M. Gnezdilov. 2010. Meiotic karyotypes and testis structure of 14 species of the planthopper tribe Issini (Hemiptera: Fulgoroidea, Issidae). *Eur J Entomol* 107:465–80. doi: 10.14411/eje.2010.055.

Kuznetsova, V. G., A. Maryańska-Nadachowska, and S. Nokkala. 2003. A new approach to the Auchenorrhyncha (Hemiptera, Insecta) cytogenetics: Chromosomes of the meadow spittlebug Philaenus spumarius (L.) examined using various chromosome banding techniques. *Folia Biol-Krakow* 51 (1–2):33–40.

Kuznetsova, V. G., A. Maryańska-Nadachowska, and S. Nokkala. 2009b. Karyotype characterization of planthopper species Hysteropterum albaceticum Dlabola, 1983 and Agalmatium bilobum (Fieber, 1877) (Homoptera: Auchenorrhyncha: Issidae) using AgNOR-, C- and DAPI/CMA3 –banding techniques. *Comp Cytogenet* 3(2):111–23. doi: 10.3897/compcytogen.v3i2.18.

López-Fernández, C., E. Pradillo, M. Zabal-Augirre, J. L. Fernandez, C. Garcia de la Vega, and J. Gisalvez. 2004: Telomeric and interstitial telomeric-like DNA sequence in Orthoptera genomes. *Genome* 47:757–63.

Lorite, P., E. Aranega, F. Luque, and T. Palomeque. 1997. Analysis of the nucleolar organizing regions in the ant Tapinoma nigerrinum (Hymenoptera, Formicidae). *Heredity* 78:578–82.

Lukhtanov, V. A. and V. G. Kuznetsova. 2010. What genes and chromosomes say about the origin and evolution of insects and other arthropods. *Russ J Genet* 46(9):1115–21. doi: 10.1134/S1022795410090279.

Mandrioli, M. and F. Borsatti. 2007. Analysis of heterochromatic epigenic markers in the holocentric chromosomes of the aphid Acyrthosiphon pisum. *Chromosome Res* 15:1015–22.

Manicardi, G. C., M. Mandrioli, D. Bizzaro, and U. Bianchi. 2002. Cytogenetic and molecular analysis of heterochromatic areas in the holocentric chromosomes of different aphid species. In *Some Aspects of Chromosome Structure and Functions*, eds. R. G. Sobti, G. Obe, and R. S. Athwal, 47–56. Narosa Publishing House, New Delhi, India.

Maryańska-Nadachowska, A., S. Drosopoulos, D. Lachowska, Ł. Kajtoch, and V. G. Kuznetsova. 2010. Molecular phylogeny of the Mediterranean species of Philaenus (Hemiptera: Auchenorrhyncha: Aphrophoridae) using mitochondrial and nuclear DNA sequences. *Syst Entomol* 35:318–28. doi: 10.1111/j.1365-3113.2009.00510.x.

Maryańska-Nadachowska, A., V. G. Kuznetsova, and H. Abdul-Nour. 2008. A chromosomal study on a Lebanese spittlebug Philaenus arslani (Hemiptera: Auchenorrhyncha: Aphrophoridae). *Eur J Entomol* 105:205–10. doi: 10.14411/eje.2008.029.

Maryańska-Nadachowska, A., V. G. Kuznetsova, and T. V. Karamysheva. 2013. Chromosomal location of rDNA clusters and TTAGG telomeric repeats in eight species of the spittlebug genus Philaenus (Hemiptera: Auchenorrhyncha: Aphrophoridae). *Eur J Entomol* 110(3):411–8. doi: 10.14411/eje.2013.055.

Maryańska-Nadachowska, A., V. G. Kuznetsova, D. Lachowska, and S. Drosopoulos. 2012. Mediterranean species of the spittlebug genus Philaenus: Modes of chromosome evolution. *J Insect Sci* 12(54):1–17. http://www.insectscience.org/12.54.

Monti, V., C. Serafini, G. C. Manicardi, and M. Mandrioli. 2013. Characterization of non-LTR retrotransposable TRAS elements in the aphids Acyrthosiphon pisum and Myzus persicae (Aphididae, Hemiptera). *J Hered* 104(4):554–64. doi: 10.1093/jhered/est017.

Nast, J. 1972. *Palaearctic Auchenorrhyncha (Homoptera): An Annotated Check List*. Polish Scientific Publishers, Warszawa, Poland.

Nokkala, C., V. Kuznetsova, S. Grozeva, and S. Nokkala. 2007. Direction of karyotype evolution in the bug family Nabidae (Heteroptera): New evidence from 18S rDNA analysis. *Eur J Entomol* 104:661–5. doi: 10.14411/eje.2007.083.

Nokkala, S., V. G. Kuznetsova, A. Maryańska-Nadachowska, and C. Nokkala. 2004. Holocentric chromosomes in meiosis. I. Restriction of the number of chiasmata in bivalents. *Chromosome Res* 12:733–9. doi: 10.1023/B:C HRO.0000045797.74375.70.

Panzera, Y., S. Pita, M. J. Ferreiro, I. Ferrandis, C. Lages, R. Perez, A. E. Silva, M. Guerra, F. Panzera. 2012. High dynamics of rDNA cluster location in kissing bug holocentric chromosomes (Triatominae, Heteroptera). *Cytogenet Genome Res* 138(1):56–67. doi: 10.1159/000341888. Epub 2012 Aug 18.

Remane, R. and S. Drosopoulos. 2001. Philaenus tarifa nov. sp. – an additional spittlebug from Southern Spain (Homoptera – Cercopidae). *Deut Entomol Z* 48:277–79.

Sahara, K., F. Marec, and W. Traut. 1999. TTAGG telomeric repeats in chromosomes of some insects and other arthropods. *Chromosome Res* 7:449–60. doi: 10.1023/A:1009297729547.

Schrader, F. 1947. The role of the kinetochore in the chromosomal evolution of the Heteroptera. *Evolution* 1:134–42.

Stewart, A. J. A. and D. R. Lees. 1996. The coulor/pattern polymorphism of Philaenus spumarius (L.) (Homoptera: Cercopidae) in England and Wales. *Philos Trans R Soc London B* 351:69–89.

Stewart, J. B. and A. T. Beckenbach. 2005. Insect mitochondrial genomics: The complete mitochondrial genome sequence of the meadow spittlebug Philaenus spumarius (Hemiptera: Auchenorrhyncha: Cercopoidae). *Genome* 48:46–54. doi: 10.1139/g04-090.

Thompson, V. 1994. Spittlebug indicators of nitrogen–fixing plants. *Ecol Entomol* 19:391–98.

Tishechkin, D. Y. 2013. Two new species of the genus Philaenus (Homoptera, Aphrophoridae) from Iran. *Entomol Rev* 93(1):73–6. doi: 10.1134/S0013873813010119.

Traut, W., M. Szczepanowski, M. Vítková, C. Opitz, F. Marec, and J. Zrzavý. 2007. The telomere repeat motif of basal Metazoa. *Chromosome Res* 15:371–82. doi: 10.1007/s10577-007-1132-3.

Vitkova, M., J. Kral, W. Traut, J. Zrzavy, and F. Marec. 2005. The evolutionary origin of insect telomeric repeats (TTAGG)n. *Chromosome Res* 13:145–56. doi: 10.1007/s10577-005-7721-0.

Vitturi, R., A. Lannino, C. Mansueto, V. Mansueto, and M. Colomba. 2008. Silvernegative NORs in Pamphagus ortolaniae (Orthoptera: Pamphagidae). *Eur J Entomol* 105:35–9. doi:10.14411/eje.2008.004.

Wagner, W. 1959. Zoologische Studien in Westgriechenland. IX. Tail Homoptera. Sitzungsberichte der Sterreichisen Akademie der Wissenschaften, Mathematisch–Naturwissenschaftliche Klasse, Abteilung 168: 583–605.

White, M. J. D. 1973. *Animal Cytology and Evolution*. 3rd ed. Cambridge University Press, London.

Whiting, M. F., J. C. Carpenter, Q. D. Wheeler, and W. C. Wheeler. 1997. The Strepsiptera problem: Phylogeny of the holometabolous insect orders inferred from 18S and 28S ribosomal DNA sequences and morphology. *System Biol* 46:1–68. doi:10.1093/sysbio/46.1.1.

Wolf, K. W. 1996. The structure of condensed chromosomes in mitosis and meiosis of insects. *Int J Insect Morphol Embryol* 25:37–62.

Zając, M. A. and M. C. Wilson. 1984. The effect of nymphal feeding by the meadow spittlebug, Philaenus spumarius (L.) on strawberry yield and quality. *Crop Prot* 3:167–75.

Zakian, V. A. 2012. Telomeres: The beginnings and ends of eukaryotic chromosomes. *Exp Cell Res* 318:1456–60. doi: 10.1016/j.yexcr.2012.02.015.

11

Grasshoppers (Orthoptera)

Juan Pedro M. Camacho, Josefa Cabrero,
Maria Dolores López-León, Diogo C. Cabral-de-Mello,
and Francisco J. Ruiz-Ruano

CONTENTS

LIST OF ABBREVIATIONS

AFLP, amplified fragment length polymorphism
BSA, bovine serum albumin
CMA$_3$, chromomycin A$_3$
Cot, concentration of DNA and time
DA, distamycin A
DAPI, 4′,6 -diamidino-2-phenylindole
dATP, deoxyadenosine triphosphate
dCTP, deoxycytidine triphosphate
dGTP, deoxyguanosine triphosphate
DNA, deoxyribonucleic acid
DNase, deoxyribonuclease
dNTPs, deoxynucleotide triphosphates

DOP, degenerate oligonucleotide-primed
DPX, distrene 80, plasticizer, xylene
DTT, dithiothreitol
dTTP, deoxythymidine triphosphate
dUTP, 2′-deoxyuridine 5′-triphosphate
EDTA, ethylenediaminetetraacetic acid
FIAD, Feulgen image analysis densitometry
FISH, fluorescence in situ hybridization
IOD, integrated optical density
ITS, internal transcribed spacers
JPEG, joint photographic experts group
LTRs, long terminal repeats
NOR, nucleolus organizer region
PBS, phosphate buffered saline
PBT, phosphate buffered Tween 20
PCR, polymerase chain reaction
RAPD, random amplified polymorphism DNA
rDNA, ribosomal DNA
RNase, ribonuclease
RPM, revolutions per minute
SDS, sodium dodecyl sulfate
SSC, saline sodium citrate
TAE, Tris-acetate
Taq, Thermus aquaticus
TE, Tris-EDTA
TIFF, tagged image file format
WGA, whole genome amplification

11.1   INTRODUCTION

11.1.1   Taxonomy

Grasshoppers belong to the insect order Orthoptera, which also includes locusts and crickets. This order comprises more than 26,000 species (http://orthoptera.speciesfile.org) with global distribution and higher diversity in the tropics. The name of the order is derived from "orthos," meaning "straight," and "pteron," meaning "wing." Most orthopteran species are included in two main suborders: Ensifera, including the long-horned grasshoppers (superfamily Tettigonioidea) and several types of crickets (Grylloidea and Gryllacridoidea), and Caelifera, including the short-horned grasshoppers (Acridoidea), the grouse locusts (Tetrigoidea) and the pigmy mole crickets (Tridactyloidea).

Although most protocols described here were developed for caeliferan insects, many of them are also useful for orthopterans and other insects. Approximately half of the known orthopteran species belong to the superfamily Acridoidea (more than 7600 species), with the Acrididae being the most diverse family. Although Acridoidea is considered monophyletic, the internal relationships between families are not well understood (Song 2010), but recently, Leavitt et al. (2013) have contributed to elucidate this topic by using the entire mitogenome for phylogenetic analysis. Acridid grasshoppers are phytophagous, although many species can be omnivores (even cannibalistic) when cultured in the laboratory. Females lay several clutches of eggs (egg pods) in the ground during their life (approximately one pod per week, depending on the species). Grasshoppers usually show cryptic coloration (Rentz 1991), and their most noticeable characteristic is the ability to jump. As part of the courtship ritual, males in many species "sing" by via stridulation of their rear legs and forewings.

11.1.2   Importance of Grasshopper Species

Approximately 20 acridid species in several different subfamilies exhibit gregarious behavior and migrate in dense swarms; these severe pests generate massive damage to crops. The most well known of these species are the desert locust (*Schistocerca gregaria*) and the migratory locust (*Locusta migratoria*) in Africa and the Middle East; *Schistocerca piceifrons* in tropical Mexico and Central America; *Melanoplus bivittatus*, *M. femurrubrum*, *M. differentialis*, and *Camnula pellucida* in North America; *Romalea guttata*, *Brachystola magna*, and *Sphenarium purpurascens* in northern and central Mexico; some species of *Rhammatocerus* in South America; and *Oedaleus senegalensis* and *Zonocerus variegatus* in Africa.

Grasshoppers constitute part of the diet in some African, American, and Asian countries, as they are a source of protein and fat. A recent FAO Forestry paper (Van Huis et al. 2013) indicates that approximately 80 grasshopper species are consumed worldwide. Most grasshopper species are edible, so they, and other insects, can be used for food. Locusts are particularly easy to harvest when they swarm. In Africa, the desert locust, the migratory locust, the red locust (*Nomadacris septemfasciata*), and the brown locust (*Locustana pardalina*) are eaten, although the insecticide treatment of these pests calls for caution in their consumption. This problem does not apply to nonswarming grasshoppers that are easily captured or cultured. Some countries with long traditions of using grasshoppers and/or crickets for food are Mexico, Niger, Thailand, the Lao People's Democratic Republic, and Cambodia, where these insects are farmed.

11.1.3 Karyotype

The large size and low number of grasshopper chromosomes have significantly contributed to the general understanding of chromosome structure and function during mitosis and meiosis. No other organism provides such a convenient and complete collection of stages showing the complete course of meiosis, that is, the most complex type of cell division. The pioneer work of McClung (1902) describing male grasshopper meiosis opened a long series of meiotic studies using grasshoppers as the preferred material. Even today, male grasshopper meiosis is visualized in student practical activities in many universities. Key topics of chromosome biology, such as chromosome structure, condensation, pairing, movement, chiasma formation, and chromosome rearrangements, have been elucidated using grasshopper cytogenetic materials. Grasshopper meiosis can also be analyzed in vivo in short-term cultures of spermatocytes (Nicklas 1961; Rebollo and Arana 1995; Rebollo et al. 1998), and grasshoppers are one of the few animals in which female meiosis has been analyzed in detail (Hewitt 1976; Henriques-Gil et al. 1987; Cano and Santos 1989).

Most grasshopper species, especially those in the family Acrididae, have 23 chromosomes in males and 24 in females. This difference is due to their X0/XX sex chromosome determinism. The karyotype composed from $2n = 23$, X0♂/24, XX♀ is considered atavistic, at least for Caelifera representatives (White 1973; Hewitt 1979). Between species, variation in chromosome number and derived sex systems, such as neo-XY and neo-X_1X_2Y, occurs in some Acrid groups (Hewitt 1979, Castillo et al. 2010), and there are also cases of extensive intraspecific variation caused by polymorphic chromosome rearrangements, supernumerary segments, and supernumerary (B) chromosomes (Hewitt 1979; Jones and Rees 1982; Camacho et al. 2000; Camacho 2004, 2005).

Many grasshopper species have an acro/telocentric chromosome morphology, with most chromosomes appearing to have a single arm with the centromere placed close to one end. The karyotypes are composed of a continuous series of chromosomes gradually decreasing in size, which frequently complicates their identification. However, grasshopper autosomes are classified into three size groups, that is, long (L), medium (M), and short (S). In many cases, the frontier between the L and M autosomes is marked by the size of the X chromosome, whereas between the M and S autosomes is marked by the so-called "megameric bivalent." This autosomal bivalent shows positive heteropyknosis (high condensation) during meiotic prophase (Corey 1938). It is usually the ninth autosomal bivalent in size in species with $2n♂ = 23$ ($n = 11 + X0$) and the sixth one in species with $2n♂ = 17$ ($n = 8 + X0$). As a borderline bivalent, the megameric bivalent is named M9 (or M6) in some species but S9 (or S6) in others. Similarly, the X chromosome is similar in size to the L chromosomes in some species but similar to the M chromosomes in others (Camacho 1980).

Chiasma frequency in grasshoppers has profusely been used in comparisons between sexes (Fletcher and Hewitt 1980; Cano et al. 1987; Cano and Santos 1990), B chromosome effects (John and Hewitt 1965; Cano and Santos 1988; Camacho et al. 2002), temperature or x-ray effects (Church and Wimber 1969), and changes associated with locust phase transformation (Dearn 1974).

The C-banding technique (Sumner 1972) allowed the characterization of heterochromatin distribution in many species of grasshoppers (King and John 1980; Santos et al. 1983; Cabrero and Camacho 1986a), and the silver impregnation technique (Goodpasture and Bloom 1975; Rufas et al. 1982) revealed the localization of the active nucleolus organizer regions (Cabrero and Camacho 1986b). Similarly, triple CMA_3-DA-DAPI staining (Schweizer 1980) provided information on the chromosome location of chromatin regions that are rich in A+T or G+C, as these regions preferentially bind to the DAPI or CMA_3 fluorochromes, respectively (Schweizer et al. 1983; John et al. 1985; Camacho et al. 1991).

More recently, fluorescence in situ hybridization (FISH) has opened the door to the physical mapping of several repetitive DNA families, such as 45S ribosomal DNA (rDNA) (Cabrero and Camacho 2008), histone genes (Cabrero et al. 2009), and 5S rDNA (Cabral de Mello et al. 2011a,b). In addition, the mapping of satellite DNAs and transposable elements has elucidated B chromosome evolution in the species *Eyprepocnemis plorans* (Cabrero et al. 2003; Montiel et al. 2012).

11.1.4   Genome Size

Grasshopper genomes are among the largest genomes in insects. For instance, the migratory locust has a C value (6 pg) that is double that of human beings, and it is even higher in the grasshopper *Podisma pedestris* (16.93 pg) (Westerman et al. 1987). The mass of total genomic DNA is known in only 39 grasshopper species (Animal Genome Size Database: http://www.genomesize.com) (Hanrahan and Johnston 2011), and most of these estimations were evaluated by microdensitometry. Currently, flow cytometry (Geraci et al. 2007), Feulgen Image Analysis Densitometry (FIAD) (Hardie et al. 2002), and real-time PCR (Wilhelm et al. 2003) are the most common techniques for measuring genomic DNA. Given the excellent correlation between genome size measurements performed by flow cytometry and FIAD (Dolezel et al. 1998), the simplicity, ease, accuracy, and cost-effectiveness of the latter (Hardie et al. 2002) make it the best choice for most cytogenetic laboratories.

11.1.5 Genome Sequencing Projects

Because of their huge size, grasshopper genomes constitute a challenge for full sequencing projects. In spite of that, the decreasing costs of high-throughput sequencing methods have recently allowed the publication of the first complete draft sequence in the migratory locust (*L. migratoria*), which is the largest animal genome hitherto sequenced (Wang et al. 2014). It is thus presumable that the genomes of other species nominated for sequencing in the i5k initiative, including some *Schistocerca* and *Chorthippus* species (accessed on July 31, 2013, at http://arthropodgenomes.org/wiki/i5K_nominations), will be promptly sequenced.

The 6.3 Gb *L. migratoria* draft genome sequence published has uncovered some interesting characteristics of this genome, compared to the genomes of other insects. For instance, whereas there is no difference in the length of coding regions, compared with *Drosophila melanogaster*, the *L. migratoria* genome shows much longer introns and intergenic regions, presumably because of the proliferation of mobile elements combined with slow rates of loss for these elements (Wang et al. 2014). These authors have performed the most complete genomic analysis ever published in a single paper, by also performing methylome and transcriptome analyses. This has revealed complex regulatory mechanisms involved in microtubule dynamic-mediated synapse plasticity during phase change, and expansion of gene families associated with energy consumption and detoxification, the latter being consistent with long-distance flight and phytophagy characteristics of gregarian locusts. Remarkably, these authors have found in this genome hundreds of potential insecticide target genes, thus offering new insights into the biology and sustainable management of this pest species.

11.2 PROTOCOLS

11.2.1 Biological Materials and Grasshopper Culture

The easiest way to visualize grasshopper chromosomes is to analyze male meiosis because no colchicine pretreatment is necessary. Female meiosis is also amenable to analysis, but the technique is rather complex, and the resulting cells cannot be used for chromosome-banding techniques (Hewitt 1976; Henriques-Gil et al. 1987). The best mitotic chromosomes are obtained from embryos, especially from the neuroblast cells. In adult females, mitosis can be visualized in cells from ovariole walls or gastric caeca.

When collecting grasshoppers in the field, the first challenge for beginners is to distinguish males from females. In all grasshopper species, females are larger than males, and the end of the abdomen is pointed in females but rounded in males (http://keys.lucidcentral .org/keys/grasshopper/nonkey/html/Gender/Gender.htm)

Field-collected males and females can be prepared for cytological analysis or maintained in the laboratory for controlled crosses, obtaining embryos for cytological analysis or next-generation adults. Culture conditions are simple, as grasshoppers have scarce requirements: 27°C–30°C with 30% humidity and a 12:12 photoperiod (Figure 11.1a and b), although these conditions can vary among species. They can be fed almost anything, but lettuce, cabbage, and bran work well for most species.

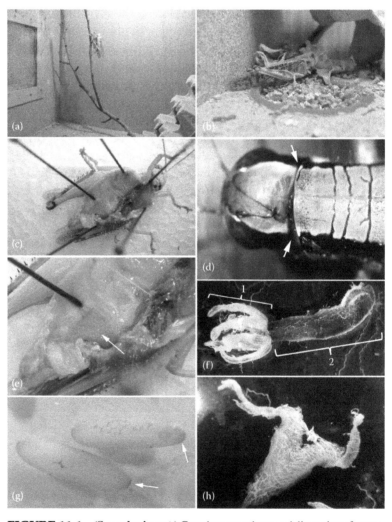

FIGURE 11.1   (**See color insert.**) Grasshopper culture and dissection of appropriate tissues for chromosome analysis. (a) Mating pair of *Locusta migratoria*; (b) laying female of *L. migratoria*; (c) dissected male of *Eyprepocnemis plorans* showing a yellowish mass (indicated by an arrow in e) corresponding to the testes; (d) dorsal view of the head and pronotum of the grasshopper *Parascopas sanguineus* showing where to cut (arrows) for a rapid dissection of gastric caeca (1 in f) and gizzard (2 in f); (g) eggs showing the micropyle end (arrows); (h) ovaries.

11.2.2 Equipment, Materials, and Reagents

A laboratory used to perform the protocols included in this chapter should be equipped with the following:

Equipment
- Fluorescence microscope (Cat. No. 909, Olympus BX41, Olympus, Tokyo, Japan) with a digital camera (Olympus DP70) and appropriate filter set
- Zoom Stereomicroscope (Cat. No. SMZ-1000, Nikon, Tokyo, Japan)
- Thermal cycler Mastercycler ep gradient S (Cat. No 13038553, Eppendorf, Hamburg, Germany)
- Freezers (−20°C and −80°C) (Cat. No. 365GTL, AEG; Cat. No. 14230-102, VWR, Radnor, PA)
- Refrigerators Samsung no frost (Cat. No. RL58GEGSW1, Samsung, Seoul, South Korea)
- SW22 Shaking Water bath (Cat. No. 9550322, Julabo, Seelbach, Germany)
- Biological safety cabinet ESCO Class II (Cat. No. SC2-4A1, ESCO, Singapore)
- Mini-sub cell GT, Electrophoresis apparatus; PowerPac 3000 (Cat. No. 166-4288EDU, Cat. No. 165-5056, Bio-rad, CA)
- UVP Visi-Blue transilluminator (Cat. No. UV95-0461-01, Fischer Scientific, Hampton, NH)
- BioPhotometer plus (Cat. No. 6132 000.008, Eppendorf)
- Universal Precision Ovens (Cat. No. 2005151, Selecta, Barcelona, Spain)
- TransferMan NK 2, Micromanipulator (Cat. No. 920000011, Eppendorf) coupled to an inverted microscope (Axiovert 200, Zeiss, Jena, Germany) for chromosome microdissection
- Hot plate X5 (Cat. No. 23-PC800, Bio-Optica, Milan, Italy)
- Microcentrifuge 5415D (Cat. No. 022621408, Eppendorf)
- Autoclave (Cat. No. AHS-75 N, Raypa, Barcelona, Spain)
- Shaking platform MVH-40 (Cat. No. 2063MVH40, ICT, SL, La Rioja, Spain, Lardero, La Rioja, Spain)

Materials
- Coplin jar (Cat. No. 12954000, Endo glassware, Beijing, China)
- Coverslips (Cat. No. BB018018A1, Menzel-Gläser, Braunschweig, Germany)
- Dissecting scissors (Cat. No. 72940, Dumont, Montignez, Switzerland)
- Dissecting tweezers (Cat. No. 72873D, Dumont)

- Eppendorf micropestle for 1.2- to 2-mL tubes (Cat. No. Z317314, Sigma-Aldrich, St. Louis, MO)
- Filter paper (Cat. No. 1305, Filtros Anoia, Barcelona, Spain)
- Homogenizer (Cat. No. 6102, Kartell, Melbourne, Australia)
- Laboratory film (Cat. No. PM996, Parafilm, Pechiney Plastic Packaging Company, Chicago, IL)
- Micropipette different volumes Nichipet (Cat. No. NPX-2, NPX-20, NPX-200, NPX-1000, Nichiryo, Tokyo, Japan)
- Microcentrifuge tube Eppendorf (Cat. No. 175508N, Daslab, Barcelona, Spain)
- Petri dish (Cat. No. P 9.0-720, Soria Genlab, Madrid, Spain)
- Razor blade (Cat. No. 61204100, Nahita, Auxilab SL, Beriáin, Navarra, Spain)
- Microscope Slides (Cat. No. AB00000112E, Menzel-Gläser)

Reagents

- Acetic acid glacial (Cat. No.131008.1211, Panreac, Barcelona, Spain)
- Acridine orange solution (Cat. No. A9231, Sigma-Aldrich)
- Agarose (Cat. No. A9539, Sigma)
- Anti-digoxigenin-rhodamine (Cat. No. 11207750910, Roche, Basilea, Switzerland)
- Barium hydroxide octahydrate (Cat. No. 101737, Merck, Hunterdon County, NJ)
- BioNick DNA Labeling System (Cat. No. 18247-015, Invitrogen, Life Technologies, Carlsbad, CA)
- BSA (bovine serum albumin) (Cat. No. A3294, Sigma-Aldrich)
- Calcium chloride dihydrate ($CaCl_2.2H_2O$) (Cat. No. 131232, Panreac)
- Chloroform (Cat. No. EC 200-663-8, Amresco, Solon, OH)
- Citric acid (Cat. No. 131808, Panreac)
- Colchicine (Cat. No. C9754, Sigma-Aldrich)
- Chromomycin A (Cat. No. C2659, Sigma-Aldrich)
- DAPI (4′, 6′-diamidino-2-phenylindole) (Cat. No. D9542, Sigma-Aldrich)
- Dextran sulfate (Cat. No. D8906, Sigma-Aldrich)
- DIG-nick translation mix (Cat. No. 11 745 816 910, Roche)
- Disodium phosphate (Na_2HPO_4) (Cat. No. 141655.1210, Panreac)

- Distamycin A (Cat. No. D6135, Sigma-Aldrich)
- Dithiothreitol (Cat. No. 43815, Sigma-Aldrich)
- DNA polymerase I/DNase I (Cat. No. 18162-016, Invitrogen)
- dNTPs set (Cat. No. DNTP10, Sigma-Aldrich)
- DPX (mountant for microscopy) (Cat. No. 36029, BDH, VWR)
- Ethanol (Cat. No. 121086.1211, Panreac)
- Formamide (Cat. No. F7503, Sigma-Aldrich)
- Formaldehyde (Cat. No. F8775, Sigma-Aldrich)
- Formic acid (Cat. No. 131030, Panreac)
- GenElute PCR Clean-Up Kit (Cat. No. NA1020, Sigma-Aldrich)
- GenomePlex WGA Reamplification Kit (Cat. No. WGA3, Sigma-Aldrich)
- GenomePlex Single Cell Whole Genome Amplification Kit (Cat. No. WGA4, Sigma-Aldrich)
- Giemsa (Cat No. 1.09204, Merck)
- Glycogen (Cat. No. 10 901 393 001, Roche)
- Hydrochloric acid (HCl) (Cat. No. 20 252 290, BDH, VWR)
- Hyperladder DNA marker (Cat. No. BIO-33039, Bioline, London, UK)
- Illustra GenomiPhi V2 DNA Amplification Kit (Cat. No. 25 6600 30, GE Healthcare Life Sciences, Little Chalfont, UK)
- Isoamyl acohol (Cat. No. 121372.1611, Panreac)
- Labeled nucleotide, Fluorescein 12-dUTP (Cat. No. 11 373 242 910) and Tetramethylrhodamine-5-dUTP (Cat. No. 11 534 378 910, Roche)
- Lactic acid (Cat. No. 141034.1211, Panreac)
- Magnesium chloride ($MgCl_2$) (Cat. No. M1028, Sigma-Aldrich)
- Monopotassium phosphate (KH_2PO_4) (Cat. No. 131509, Panreac)
- Orcein (Cat. No. 251324.1604, Panreac)
- Paraformaldehyde (Cat. No. P6148, Sigma-Aldrich)
- Pepsin (Cat. No. P6887, Sigma-Aldrich)
- Phenol (Cat. No. P4682, Sigma-Aldrich)
- Potassium chloride (KCl) (Cat. No. 131494, Panreac)
- Potassium ferricyanide ($K_3[Fe(CN)_6]$) (Cat. No. P4066, Sigma-Aldrich)
- Propionic acid (Cat. No. P1386, Sigma-Aldrich)
- Proteinase K (Cat. No. P2308, Sigma-Aldrich)
- RNase A (Cat. No. R6513, Sigma-Aldrich)
- Salmon testes DNA (Cat. No. D7656, Sigma-Aldrich)
- S1 nuclease (Cat. No. 18001-016, Invitrogen)

- Schiff's reagent (Cat. No. 3952016, Sigma-Aldrich)
- Silver nitrate (Cat. No. 101512, Merck)
- Sodium acetate (Cat. No. 131633.1210, Panreac)
- Sodium bicarbonate ($NaHCO_3$) (Cat. No. S5761, Sigma-Aldrich)
- Sodium chloride (NaCl) (Cat. No. 121659.1211, Panreac)
- Tri-sodium citrate (C6H5Na3O7) (Cat. No. 141655.1210, Panreac)
- SDS (sodium dodecyl sulfate) (Cat. No. L5750, Sigma-Aldrich)
- Sodium hydroxide (NaOH) (Cat. No. 131687.1210, Panreac)
- Sodium metabisulfite (Cat. No. 13459, Sigma-Aldrich)
- Sodium thiosulfate ($Na_2S_2O_3$) (Cat. No. S7026, Sigma-Aldrich)
- SpectrumOrange dUTP (Cat. No. 02N33 050, Vysis, Abbott Molecular, Abbott Park, IL)
- Streptavidin-Alexa Fluor 488 Conjugate (Cat. No. S11223, Life Technologies, TermoFisher Scientific)
- SYBR safe DNA gel stain (Cat. No. S33102, Invitrogen)
- *Taq* DNA polymerase (Cat. No. P0023, Canvax, Córdoba, Spain)
- Tris (Cat. No. A7455, AppliChem, Darmstadt, Germany)
- Tween 20 (Cat. No. P5927, Sigma-Aldrich)
- Ultrapure water (Cat. No. W4502, Sigma-Aldrich)
- VECTASHIELD mounting medium (Cat. No. H-1000, Vector Laboratories, Burlingame, CA)

11.2.3 Sources of Chromosomes

In adults, the best mitotic cells are obtained from gastric caeca in males and females and from female ovarioles. Testes do not require colchicine pretreatment and provide convenient mitotic metaphases from pre-meiotic spermatogonial cell divisions. In the remaining cases, colchicine treatment is necessary to increase the proportion of cells in mitotic metaphase, which is the best stage to visualize chromosomes for karyotyping and physical mapping. Undoubtedly, the best source of mitotic chromosomes is embryos obtained from eggs incubated in the laboratory.

11.2.4 Tissue Extraction and Fixation

11.2.4.1 Mitotic Chromosomes in Adults

The following protocol is quite simple and useful for obtaining mitotic plates from ovarioles and gastric caeca, which avoids the requirement of laboratory strain maintenance. Although it is possible to obtain good metaphase plates using this approach, the number of cells could be reduced, depending on the animal.

1. Dissolve colchicine in insect saline solution to a final concentration of 0.05% and inject it into the abdomen in amounts corresponding to body size. For an average grasshopper, 0.1 mL may be enough.
2. After 6–8 hours, anesthetize the animal in ethyl acetate vapors and remove the desired organ.
3. With scissors, make a longitudinal and ventral cut of the abdomen (Figure 11.1c), open it with entomological pins, and, with forceps, extract the testes (Figure 11.1e) or the ovarioles (Figure 11.1h), which are located under the digestive tube. Gastric caeca can also be obtained this way or via a dorsal cut between the head and the thorax (Figure 11.1d and f). Immediately fix all the materials in freshly prepared 3:1 solution of absolute ethanol–acetic acid.
4. Alternatively, place the material in potassium chloride (KCl) hypotonic solution (0.75%) for 45–60 minutes before fixation to improve chromosome spreading.
5. Then, immerse the materials in fresh 3:1 absolute ethanol–acetic acid for 1 hour at room temperature (RT) and store them at 4°C or −20°C.
6. In ovarioles, the interesting part is the terminal filament. Under a stereomicroscope, dissect the ovarioles and eliminate the developing egg (if it is large). Similarly, clean gastric caeca (in glacial acetic acid) to remove digestive remains.

In some cases, we want to obtain colchicine-treated mitotic metaphase cells and use body remains without the effect of colchicine, for example, to extract RNA. In this case, we dissect the animal, extract the desired organ for chromosome analysis (e.g., ovarioles and/or gastric caeca) and immerse it in 2% colchicine in insect saline for 2–6 hours. They are then fixed in 3:1 ethanol/acetic acid and, after a 1-hour fixation at RT, are stored at 4°C until study. Body remains are immediately frozen in liquid nitrogen and stored at −80°C.

11.2.4.2 Mitotic Chromosomes in Embryos

Embryos are excellent for chromosome studies because many mitotic metaphases are obtained from a single embryo, and the chromosomes from these cells are easy to spread, providing high-quality material for chromosome banding and physical mapping. In addition, embryo neuroblasts are the very best cells for karyotyping, banding, and mapping. These cells have very large chromosomes because of a special low-condensation state, allowing the detection of very thin bands that are difficult to detect in other types of cells. To obtain embryos and fix them for cytological analysis, we proceed as follows:

1. Maintain gravid females in culture cages with humid vermiculite (or sand) to facilitate laying (Figure 11.1b). Monitor vermiculite every day for egg pods.
2. Place egg pods in a petri dish with 10% humid vermiculite and incubate at 25°C–28°C, with readjustments to humidity every 4–5 days. Fix mitosis-rich embryos before they enter diapause, typically before embryo eyes are pigmented. Depending on the species, the appropriate incubation period to obtain an optimum number of mitotic metaphase cells varies, for example, from 6 days in *L. migratoria* to 10 days in *E. plorans* or even 15 days in *Chorthippus jacobsi*.
3. At the end of the incubation period, immerse egg pods in insect saline solution and dissect them to separate the eggs and extract the embryo from each egg. For this purpose, perform a transversal cut close to the micropyle end, which has a crown of minute holes (Figure 11.1g).
4. Immerse embryos in 2 mL 0.05% colchicine in insect saline solution for 2 hours and in 2 mL distilled water (hypotonic treatment) for 10–15 minutes, depending on the species.
5. Immerse embryos in 3:1 absolute ethanol–acetic acid for 1 hour at RT and store them at 4°C.

11.2.4.3 Meiotic Chromosomes

Meiosis is easy to observe in adult males. No colchicine treatment is necessary, and it is preferable not using it. Female meiosis is more difficult to observe, but Henriques-Gil et al. (1987) developed a protocol for observing primary and secondary oocytes from metaphase I onward.

Male meiosis is analyzed in testis tubules of adult individuals previously anaesthetized with ethyl acetate vapors to extract the testis mass (a yellowish mass dorsally placed in the abdomen and including both testes together) (Figure 11.1e).

Testes are immersed in freshly prepared 3:1 absolute ethanol–acetic acid in a tube, which is vigorously shaken to separate the testis tubules (if necessary, the tubules in the fixative could be separated with two needles under a stereomicroscope). The fixative should be exchanged with fresh solution several times. After approximately 1 hour at RT, the fixed material should be stored at 4°C or at −20°C.

Alternatively, a small amount of 0.05% colchicine dissolved in insect saline solution could be injected in the abdomen of the animal 5–8 hours before fixation and follicle dissection. In addition, the testis follicles could be placed in potassium chloride (KCl) hypotonic solution (0.75%) for 45–60 minutes.

11.2.5 Chromosome Preparations

11.2.5.1 Squashing

In all types of adult tissues (testis tubules, ovarioles, and gastric caeca), the cytological preparations can be performed similarly, except that gastric caeca should be immersed for 1–2 minutes in glacial acetic acid to eliminate digestive remains. The preparations are performed by squashing as follows:

1. On a clean slide, immerse two testis tubules (or ovarioles) in a drop of 45%–50% acetic acid.
2. Crush the material with the flat end of an appropriate macerator to separate the tissue into individual cells. Place a coverslip over the material, and holding the coverslip in one corner with filter paper, gently push with a needle to eliminate air drops and spread the cells between the slide and the coverslip. Finally, remove excess fixative and squash by placing new filter paper on the preparation and pressing strongly with the thumb.
3. After 10 minutes, immerse the preparation in liquid nitrogen for 1 minute. Separate the coverslip with a razor blade and dry the slide for 15 minutes at RT.

11.2.5.2 Spreading

Alternatively, preparations can be made with the following protocol:

1. Place 2–3 testis follicles (or ovarioles) on a slide, add a drop of 50% acetic acid and macerate the tissue to separate cells.
2. Add another drop of 50% acetic acid and spread the solution over the slide.
3. Place the slide onto a hot plate at 45°C–50°C to dry the solution. Tilt the slide to facilitate the spreading of the cell solution. Avoid higher temperatures that may degrade DNA or chromosomes.

Spreading preparations can also be made as follows (modified from Castillo et al. 2011):

1. Cut a small piece of gastric caecum and place it in a small petri dish containing 700 μL 50% acetic acid. Disperse the caecum by pipetting with a Pasteur pipette.

2. To improve the quality of the preparation, use a hot plate at 45°C–50°C and spread cells with a pipette.
3. Place a dry and clean slide onto the hot plate. Transfer 100 μL of disaggregated caecum to the slide. Aspirate the solution and place it again on another region of the slide. Repeat this step as necessary until several preparations are obtained, with several drops of cell suspension on each slide (as explained in more detail for embryo preparations).
4. Air-dry the slides at RT.

Although embryo preparations can be made by squashing (see Section 11.2.5.1), the best results are obtained with Meredith's technique for mammal meiosis (Meredith 1969), with slight modifications as follows:

1. Place the embryo in an Eppendorf tube with 20 μL of 75% acetic acid. Wait for 3 minutes to soften the material and then perform repeated pipetting with a micropipette to separate embryo cells.
2. Pipette 20 μL with the micropipette and slowly place a drop of the cell suspension on a slide previously warmed on a hot plate at 60°C. Repeat this process with the same micropipette to transfer 6–8 nonoverlapping drops per slide. Dry the slide on the warm plate.

Embryo preparations can also be produced by spreading, as described by Crozier (1968). An embryo is crushed in a drop of 60% acetic acid and one or two drops of 3:1 methanol/acetic acid is added to facilitate cell spreading. The preparations are air-dried. Preparations from embryo or adult materials can be stored at −20°C for 2 days or dehydrated in an alcohol series (3 minutes in 70%, 5 minutes in 90%, and 8 minutes in absolute ethanol) and frozen at −80°C.

11.2.5.3 Chromosome Microdissection

Few studies of chromosome painting have been performed in grasshopper species, with the only exceptions being those performed on the X and B chromosomes of *L. migratoria* (Teruel et al. 2009a) and *E. plorans* (Teruel et al. 2009b) and the B chromosomes of *Podisma kanoi* (Bugrov et al. 2007). However, this interesting assay elucidates chromosome evolution, and it provides precious information about the molecular content of specific chromosomes or chromosome regions.

Chromosome microdissection can be performed from testis tubules, ovarioles, or embryo cells. To minimize DNA damage, the material needs to be fixed in 3:1 absolute ethanol–acetic acid for 10 minutes and stored in 70% ethanol at −20°C until use (for months).

Before making the preparation, the coverslips can be subjected to a salt treatment, which facilitates separation of the chromosomes with the microdissection needle, but the treatment is not strictly necessary. This treatment is performed by incubating them overnight in 10× SSC (saline sodium citrate). Excess salt is removed by washing with hot running water and distilled water. The coverslips are then air-dried and stored until use. The salts form a film on the coverslip, which subsequently facilitates the separation of chromosomes from the glass.

Immediately before microdissection, preparations are made in 50% acetic acid on a 24 × 60-mm coverslip at 27°C on a warm plate, following Meredith's method (see Section 11.2.5.2). A single drop is placed in the center of the coverslip because the interesting cells usually occupy the periphery of the drop, and they can be repeatedly localized by their position in "hours," as on a round clock face. Two or more drops can also be placed, but if different drops overlap, it is difficult to establish cell localization.

Chromosome microdissection is carried out with glass needles in an inverted microscope coupled to an electronic micromanipulator. These needles are manually made from 2-mm-diameter glass capillaries with a horizontal pipette puller and are steps of UV sterilized twice. Appropriate chromosomes for microdissection can be obtained from pachytene, diplotene, diakinesis or metaphase I cells, where the desired chromosome or bivalent is completely separated from the remaining chromosomes to avoid contamination. The microdissected chromosomes are collected in 0.2-mL tubes with 20 μL of 1× PCR buffer (Roche) (10 mM Tris-HCl, 1.5 mM MgCl$_2$, 50 mM KCl, pH 8.3) when they are going to be amplified by degenerate oligonucleotide-primed PCR (DOP-PCR) or in 9 μL DNase-free ultrapure water for the whole genome amplification (WGA) method (GenomePlex, Sigma).

11.2.5.4 Chromosomes for Fiber-FISH

The material analyzed is the cerebral ganglion of adult individuals. Insect ganglia are dissected under a stereomicroscope and immersed in saline solution, and proceeded as described in the following protocol:

1. Immerse the cerebral ganglion in 250 μL of 60% acetic acid for 1 minute.
2. Set the ganglion in a homogenizer in 500 μL of 60% acetic acid and homogenize 10–15 times.
3. Pick up the homogenate with a micropipette and transfer it to an Eppendorf tube.

4. For very soft materials, it is sufficient to homogenize the tissue in an Eppendorf tube with 750 μL of 60% acetic acid and pipetting until complete disintegration.
5. Centrifuge for 10 minutes at 1000 rpm and discard the supernatant.
6. Resuspend in 500–750 μL fixative (3:1 absolute ethanol–acetic acid).

To make preparations:

1. Set a clean slide on a warm plate at 50°C.
2. With a 200-μL micropipette, place several drops of the material on the slide.
3. Wait for the fixative to evaporate and immediately immerse the slide in 1× PBS (phosphate buffered saline) for 1 minute.
4. Pour 200 μL of 0.05 M NaOH (in 30% ethanol) on one end of the slide. Move the end of another slide, slightly inclined, along the entire slide. Discard the latter slide.
5. Add a few drops of absolute ethanol (⊠500 μL) on the slide, keeping it inclined.
6. Air-dry the slides. Select the best preparations under a microscope.
7. Dehydrate in an ethanol series for 3 minutes (70%), 5 minutes (90%), and 8 minutes (absolute).
8. Freeze at −20°C or at −80°C.

11.2.6   Staining Protocols

11.2.6.1   Conventional Staining

Pour a small drop of 2% acetic (or lactopropionic) orcein on the center of a clean slide and immerse one or two testis tubules or ovarioles. The material is crushed with the flat end of a macerator to disintegrate the tissue and separate cells. Place a coverslip on the orcein. Using a needle or lancet and filter paper, remove air bubbles and excess orcein. Place several pieces of filter paper on the coverslip and firmly press the preparation (see examples in Figures 11.2 and 11.3a and b).

11.2.6.2   C-Banding

The purpose of this technique is to visualize constitutive heterochromatin (C-positive blocks) (Figure 11.3c). Preparations obtained by the methods mentioned above are treated as follows:

1. Hydrolyze in 0.2 N HCl at 28°C for 30 minutes.
2. Wash preparations in tap water for 2 minutes.

3. Immerse the preparations in a saturated and filtered solution of 5% barium hydroxide at 28°C for 3–10 minutes, depending on the type of material and fixation.
4. Wash vigorously in tap water for 2 minutes.
5. Wash briefly in 0.2 N HCl to eliminate barium hydroxide remains.
6. Wash vigorously in tap water for 2 minutes.
7. Immerse the preparations in 2× SSC at 60°C for 1 hour.
8. Wash vigorously in tap water for 2 minutes.
9. Stain with 5% Giemsa in phosphate buffer for 1–5 minutes, depending on the material.
10. Mount the air-dried preparations in DPX (distrene 80, plasticizer, xylene).

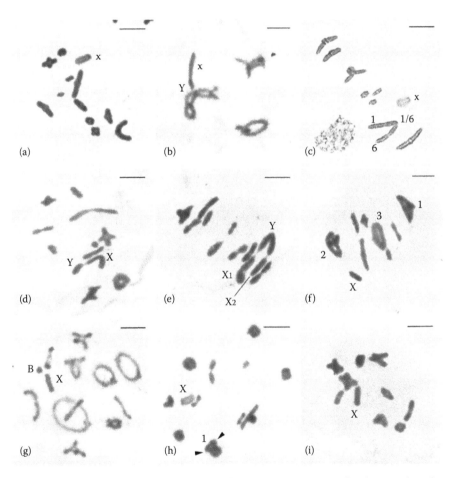

FIGURE 11.2 Conventional staining of meiotic cells at metaphase I in nine species of grasshoppers from three distinct families showing the chromosomal diversity observed in the group. (a) *Abracris flavomileata*, $2n = 23$, X0; (b) *Dichroplus silveiraguidoi*, $2n = 8$, neo-XY; (c) *Dichroplus pratensis*, $2n = 18$, X0 with heterozygote fusion between chromosomes 1 and 6; (d) *Ronderosia bergi*, $2n = 22$, neo-XY; (e) *Dichromatos lilloanus*, $2n = 21$, neo-X_1X_2Y; (f) *Chorthippus nevadensis*, $2n = 17$; (g) *Eyprepocnemis plorans*, $2n = 23$, X0 plus one B chromosome; (h) *Ommexecha virens*, $2n = 23$, X0; (i) *Stiphra robusta*, $2n = 19$, X0. Arrowheads in (h) show centromere position in the largest bivalent with metacentric morphology. Bar = 5 μm.

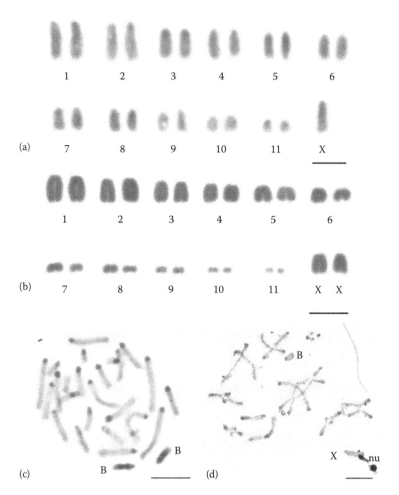

FIGURE 11.3 (**See color insert.**) Typical grasshopper karyotypes (a and b) and C-banded (c) and silver stained (d) chromosomes. (a) Male $2n = 23$, X0 karyotype obtained from an embryo cell of *Eyprepocnemis plorans*; (b) female $2n = 24$, XX karyotype obtained from a gastric caecum cell of *Adimantos ornatissinus*; (c) C-banded embryo mitotic metaphase cell from *E. plorans*; (d) silver-stained diplotene cell from an *E. plorans* male. nu = nucleolus, Bar = 5 μm.

11.2.6.3 Silver Impregnation

In grasshoppers, silver impregnation stains the nucleolus but not the nucleolar organizer regions (NORs) (Figure 11.3d). However, we can infer the chromosome localization of NORs from the chromosome regions that are closely associated with nucleoli during the first meiotic prophase (especially pachytene and diplotene). The best material for this technique is the testis, and a very simple version of this technique was developed by Rufas et al. (1982).

1. Wash the preparations with a formic acid solution prepared by adding a few drops of formic acid (pH 3–3.5) to 200 mL deionized water and dry the preparations with warm air.
2. Prepare a solution of 0.5 g silver nitrate and 0.5 mL above-mentioned formic acid solution.
3. Place a drop of this solution on the slide and place a coverslip.
4. Incubate the preparations in a humid chamber, in the dark, at 60°C.
5. Wash with distilled water.
6. Dry and mount in DPX.

11.2.6.4 Double Silver Impregnation and Fluorescence In Situ Hybridization

This technique reveals the physical location of NORs (by FISH) and nucleoli attached to the chromosomes (by silver impregnation), thus allowing to ascertain which NOR was active in every cell. The following sequential staining technique is based on the one described by Zurita et al. (1998):

1. Perform the conventional silver staining technique for testis preparations (Section 11.2.6.3).
2. Place a drop of distilled water and a coverslip on the slide and observe it under the microscope. Select and photograph interesting cells and write cell coordinates in the preparation.
3. To eliminate silver nitrate, immerse the preparations in 7.5% potassium ferricyanide for 4 minutes and then immediately immerse them in 20% sodium thiosulfate for 5 minutes.
4. Wash with distilled water.
5. Dry the preparations and perform FISH, as described in Section 11.2.8.
6. It is important to permeabilize the material before performing FISH (see Section 11.2.8).
7. Search for the same cells previously photographed after silver impregnation, photograph them with the appropriate fluorescence filters and then merge the images with the appropriate software.

11.2.6.5 Triple Fluorescent CMA$_3$-DA-DAPI Staining

Triple fluorescent staining with CMA$_3$, DA, and DAPI is based on the procedure described by Schweizer (1980, 1981). This technique reveals two types of chromosome bands: those containing G+C–rich chromatin (which are CMA$_3$+) and those containing A+T–rich chromatin (DAPI+). The procedure is performed in the dark as follows:

1. Pour approximately 100 μL CMA$_3$ (0.5 mg/mL) in McIlvaine's buffer on the slide and add a paraffin coverslip. Incubate for 1 hour at 37°C in the dark.
2. Wash with tap water removing the coverslip.
3. Wash with distilled water.
4. Dry the preparation with warm air.
5. Pour 100 μL DA (0.05–0.1 mg/mL) in McIlvaine's buffer on the preparation, place a paraffin coverslip and incubate in the dark at 37°C for 15 minutes.
6. Immerse the preparation in a mixture containing 1:1 McIlvaine's buffer and distilled water to separate the coverslip.
7. Pour 100 μL DAPI (1 μL /mL) in McIlvaine's buffer on the slide and incubate it for 40 minutes at RT in the dark.
8. Wash with distilled water or PBS.
9. Mount in VECTASHIELD.
10. Store the preparations at 37°C in the dark for a minimum of 48 hours before microscopy.

11.2.6.6 Acridine Orange Staining

This technique is used for N banding in grasshoppers (Fox and Santos 1985). After comparing it with FISH for several repetitive DNA probes, it was observed that the N bands include repetitive G+C–rich DNAs, such as rDNA and histone genes. The best material for this technique is mitosis in embryos. It is performed with the following steps:

1. Immerse the slides in absolute ethanol for 2–5 minutes.
2. Dry them with warm air.
3. Immerse the slides in a 1:1 mixture of formamide and 2× SSC at 60°C for 1 hour.
4. Wash the slides vigorously with tap water.
5. Without drying them, immerse the slides in a 0.1% acridine orange solution for 1 minute.
6. Immerse the slides in a decoloring buffer (pH = 7.0) for 15 minutes. Repeat three times.
7. Wash in abundant water.
8. Mount the preparation with a drop of buffer and observe under the fluorescence microscope.

The chromosomes will appear light green, whereas the N bands will be dark green. Alternatively, staining in 5–10% Giemsa for 2–5 minutes can improve contrast. In all cases, dry the preparations and mount them in DPX.

To make the decoloring buffer, prepare solution 1 (28.392 g/L Na$_2$HPO$_4$) and solution 2 (21.01 g/L citric acid) and mix 82.35 mL solution 1 and 17.65 mL solution 2.

11.2.7   Molecular Techniques

11.2.7.1   Genomic DNA Isolation

The manual extraction using phenol/chloroform/isoamyl alcohol (25:24:1) works well for DNA extraction in grasshoppers, although some commercial kits are also useful. DNA can be extracted from all body parts, but the muscle from the posterior leg provides high amount of good-quality DNA for subsequent procedures, such as probe generation. For each sample, follow this procedure:

1. Prepare 500 μL of the following solution in a 1.5 mL microtube

   1. 5 M Sodium chloride (NaCl)   10 μL
   2. 1 M Tris-HCl, pH 8.00   5 μL
   3. 0.5 M EDTA, pH 8.00   25 μL
   4. 10% SDS   25 μL
   5. 10 mg/mL Proteinase K   10 μL
   6. Distilled H_2O 425 μL

2. Drain the ethanol from the tissue and macerate the material in the microtube with the solution described above using scissors or a pestle.
3. Incubate the microtube in a water bath at 45°C for 90–120 minutes or until the tissue dissolves. Homogenize periodically during this period.
4. Add 500 μL phenol/chloroform/isoamyl alcohol (25:24:1) and homogenize with circular rotation for 15 minutes.
5. Centrifuge at 15,000 rpm for 15 minutes at 4°C.
6. Transfer the upper aqueous phase to a clean 1.5 mL microtube.
7. To precipitate DNA, add 0.2× volumes 1 M NaCl and 2× volumes absolute cold ethanol and mix by inversion.
8. Centrifuge at 15,000 rpm for 15 minutes at 4°C.
9. Discard the supernatant and add 375 μL of 70% cold ethanol, without agitation.
10. Centrifuge at 15,000 rpm for 15 minutes at 4°C.
11. Discard the supernatant and incubate at 37°C or at RT to dry the pellet.
12. Rehydrate the DNA in 100 μL ultrapure water for 1 hour.
13. To check DNA quality, load 5 μL sample on a 0.8% agarose gel and measure the concentration using a spectrophotometer.
14. If necessary, treat each 100 μL extracted DNA with 1 μL RNase (1 mg/mL) and incubate for 60 minutes at 37°C.

11.2.7.2   DNA Amplification from Microdissected Chromosomes

There are several methods to amplify the DNA obtained from microdissected chromosomes. We will describe three of them, namely DOP-PCR, GenomePlex, and Genomiphi methods.

The DOP-PCR technique was developed by Telenius et al. (1992), and it is commonly used to amplify DNA in chromosomes obtained by microdissection or flow sorting. The primers are partially degenerate oligonucleotides, which are used in two rounds of amplification. The first primer has a low hybridization temperature to facilitate primer binding to many genomic regions, whereas the second primer is more specific and has a higher hybridization temperature. The second round may also serve to mark probes for chromosome painting.

The microdissected chromosomes are placed in 0.2-mL tubes containing 20 μL of 1× Roche buffer (10 mM Tris-HCl, 1.5 mM $MgCl_2$, 50 mM KCl, pH 8.3).

The first round PCR is performed as follows:

1. Add 1 μM oligo-DOP, 200 μM dNTPs, 100 pg/μL BSA, 15 mM $MgCl_2$, 2 U DNA polymerase and perform the following PCR program: initial denaturation at 94°C for 5 minutes, 8 cycles of denaturation at 94°C for 1 minute, hybridization at 45°C for 1 minute and extension at 72°C for 3 minutes, 28 cycles of denaturation at 94°C for 1 minute, hybridization at 56°C for 1minute and extension at 72°C for 3 minutes and a final extension at 72°C for 5 minutes. The product should be maintained at 4°C.
2. Visualize the amplification product (5 μL) on a 2% agarose gel with 5 μL of 1× SYBR safe or an adequate quantity of another DNA stain.

The second PCR is done as follows:

1. Add 1× PCR buffer (100 mM Tris-HCl, 15 mM $MgCl_2$, 500 mM HCl pH 8.3), 4 μM oligo-DOP, 200 μM dNTPs, 20 μM SpectrumOrange-dUTP (Vysis), 2 U DNA polymerase to 5 μL of the first PCR product.
2. Perform a PCR program consisting of an initial denaturation at 94°C for 5 minutes, 5 cycles of denaturation at 94°C for 30 seconds, hybridization at 45°C for 30 seconds and extension at 72°C for 90 seconds, 28 cycles of denaturation at 94°C for 30 seconds, hybridization at 56°C for 30 seconds and extension at 72°C for 90 seconds and a final extension at 72°C for 7 minutes.

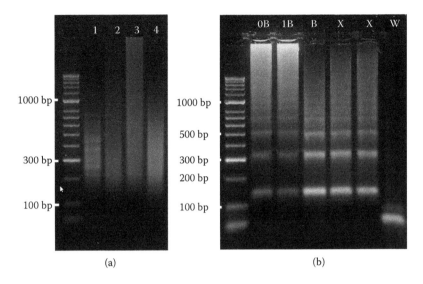

(a)   (b)

FIGURE 11.4 PCR amplification of (a) microdissected B and X chromosomes or (b) satellite DNA in *E. plorans*. (a) Agarose gel (1.5%) showing the fragments obtained after DOP-PCR (lane 1) or GenomePlex (lanes 2–4) amplification. Lanes 1 and 2 correspond to the microdissected B_{24} and lanes 3 and 4 correspond to the X chromosome; (b) PCR amplification of the satellite DNA in 0B and 1B individuals as well as in microdissected DNA from B and X chromosomes. w = white.

3. To verify the result, run 5 μL of the reaction on a 2% agarose gel with 5 μL of 1× SYBR safe or an adequate quantity of another DNA stain (Figure 11.4a).

The GenomePlex method performs WGA by LA-PCR (linker-adapted PCR). Random genome fragmentation is followed by ligation of the OmniPlex Library adaptors and PCR amplification with primers specific for the adaptors.

The amplification of chromosome DNA with the GenomePlex Single Cell kit is performed following the manufacturer's (Sigma) recommendations with no modifications.

1. Place the microdissected chromosomes in 9 μL DNase-free ultrapure water and fragment DNA by adding 1 μL Lysis and Fragmentation-K solution (32 μL of 10× Lysis and Fragmentation plus 1 μL proteinase K). Incubate the mixture for 1 hour at 55°C, heat shock it at 99°C for 4 minutes to inactivate the enzyme and place it on ice.

2. Ligate the OmniPlex Library primers by adding 2 μL 1× Library Preparation and 1 μL Stabilization Solution to the product of the former reaction. Incubate the mixture for 2 minutes at 95°C and place it on ice.

3. Add 1 μL Enzyme Library to generate fragments with the following program: 20 minutes at 16°C, 20 minutes

at 24°C, 20 minutes at 37°C, and 20 minutes at 75°C. Maintain the product at 4°C.

4. Finally, amplify the generated fragments in a final volume of 70 μL by adding the following reagents and to 14 μL of the previous product: 48.5 μL DNase-free ultrapure water, 7.5 μL of 10× Amplification Master Mix, and 5 μL WGA DNA polymerase.

5. Apply a PCR program consisting of an initial denaturation at 95°C for 3 minutes, 25 cycles of denaturation at 94°C for 30 seconds and hybridization/extension at 65°C for 5 minutes and a final step maintaining the product at 4°C.

6. Visualize the product (5 μL) on a 2% agarose gel with 5 μL of 1× SYBR safe or an adequate quantity of another DNA stain (Figure 11.4a).

7. Purify the amplified DNA with the GenElute PCR Clean-Up Kit (Sigma). Quantify the DNA visualization on a 2% agarose gel, as described in step 6 and store it at −20°C.

The chromosome DNA obtained as described in Section 11.2.7.1 should be reamplified to obtain several working aliquots. Reamplification is performed following the protocol of the GenomePlex WGA Reamplification Kit (Sigma) with some modifications.

1. Perform the amplification reaction in a final volume of 70 μL by adding the following reagents to 5 μL of the previously amplified DNA: 52.5 μL DNase-free ultrapure water, 7.5 μL of 10× Amplification Master Mix and 5 μL WGA DNA polymerase.

2. Use a PCR program consisting of an initial denaturation at 95°C for 3 minutes, 15 cycles of denaturation at 94°C for 25 seconds and hybridization/extension at 65°C for 5 minutes and maintenance at 4°C.

3. Purify the amplified DNA with the GenElute PCR Clean-Up Kit (Sigma), quantify it by running 5 μL on a 2% agarose gel (described earlier) and store it at −20°C.

The GenomiPhi (Φ29) method uses the Φ29 DNA polymerase and random hexamer primers for unbiased WGA. We used the illustra GenomiPhi V2 DNA Amplification Kit (GE Healthcare Life Sciences). It contains all components necessary for minis-cale WGA by isothermal strand displacement. This kit yields 4–7 μg of representative genomic DNA (gDNA) in 1.5 hours from 1–10 ng input gDNA, with no amplification in nontemplate negative controls. The high processivity and fidelity of the Φ29

DNA polymerase allows highly uniform amplification across the genome. The average amplification is 10,000-fold with an average product length >10 kb.

We have successfully used two different amplification programs:

Program 1: 95°C treatment for 3 minutes and keep at 4°C.
Program 2: 30°C for 2 hours (it can be extended to 3 hours if the starting amount of DNA is low), 65°C for 10 minutes, and keep at 4°C.

To perform this type of DNA amplification, do the following:

1. In a 0.2-mL Eppendorf tube, add 9 μL sample buffer to the microdissected chromosome. Pulse spin. Run Program 1 to denature DNA. Pulse spin and store the Eppendorf tubes on ice to anneal random hexamers.
2. Add 9 μL reaction buffer and 1 μL Φ29 enzyme to the tubes on ice. Run Program 2. The enzyme will be inactivated during the final step at 65°C. Store the samples at −20°C.
3. Purify the product with a purification kit (this step can be omitted if the initial amount of DNA is low).
4. Test 1 μL product on a 1% agarose gel.

To obtain several working aliquots, we can reamplify the chromosomal DNA through the following procedure:

1. Amplify 1 μL chromosomal DNA with the Φ29 DNA polymerase and dilute 1:10 in DNase-free ultrapure water.
2. To quantify the resulting DNA, run 1 μL DNA on a 1% agarose gel. Store the DNA at −20°C until use.

11.2.8 Fluorescence In Situ Hybridization

Since its first description by Gall and Pardue (1969), in situ hybridization is the technique of choice for the physical mapping of chromosomes in plant and animal species. Currently, FISH, using one or a pool of fluorescent probes, is widely used in cytogenetic laboratories because of its quickness, high resolution, and versatility. These qualities are especially true with direct fluorophore probe labeling because no further detection steps are needed and nonspecific background is avoided. Probes made of biotin- or digoxigenin-conjugated oligonucleotides need additional antibody-based detection procedures but provide greater sensitivity. Typical FISH experiments use DNA or RNA as probes, which may be composed of a specific sequence, a fraction of genomic DNA or amplified DNA from either an entire microdissected chromosome or a specific chromosomal region. The most common strategies for probe labeling are PCR and nick translation. Here, we describe three FISH protocols—specially focused on grasshopper chromosomes—that routinely give the best results in our labs.

11.2.8.1   Probe Generation

Probe sequences can be isolated directly from genomic DNA, or they can be cloned into plasmids to amplify DNA before labeling. If the sequence of interest is known, it can be amplified by PCR, or synthetic oligonucleotides can be purchased commercially. Some examples of commonly used probes in grasshoppers and the method for obtaining these probes are presented.

11.2.8.1.1 C_0t-1 Repetitive DNA

The isolation of highly and moderately repetitive DNA is based on the reassociation kinetics of genomic DNA, as described by Zwick et al. (1997), and the process is listed as follows:

1. Dilute the genomic DNA to 100−500 ng/μL in 0.3 M NaCl in a 1.5-mL microtube. It is important to use nondegraded genomic DNA.
2. Fragment the DNA by autoclaving at 1.4 atm/120°C or using DNase I. The time used for DNA fragmentation is variable, and it is useful to test a range of times to obtain an optimal result.
3. Run 3 μL autoclaved or digested DNA on a 1% agarose gel to check the size of the fragments. The recommended size of DNA fragments ranges from 100 to 1000 bp.
4. Denature at least three samples (tubes 1, 2, and 3) of 50 μL fragmented DNA using a thermocycler or water bath at 95°C for 10 minutes.
5. Place the tubes on ice for 10 seconds. Add S1 nuclease enzyme to tube 1 and incubate at 37°C for 8 minutes. After 10 seconds on ice, immediately transfer tubes 2 and 3 to a water bath/thermocycler at 65°C to renature the DNA.
6. After 1 minute, add S1 nuclease enzyme to tube 2, and after 5 minutes, add S1 nuclease enzyme to tube 3. Incubate at 37°C for 8 minutes. Other times could be tested to obtain a large amount of repetitive DNA. Use 1 U of S1 nuclease enzyme and 5.5 μL of 10× nuclease buffer for each 1 μg DNA.
7. Add an equal volume of phenol/chloroform/isoamyl alcohol (25:24:1) and rotate the tubes.
8. Centrifuge for 5 minutes at 13,000 rpm and transfer the supernatant (aqueous phase) to a clean 1.5-mL microtube.
9. Add 2.5 volumes of cold absolute ethanol to precipitate the DNA and place the tube in a −70°C deep freezer for 30 minutes.

10. Centrifuge for 15 minutes at 15,000 rpm at 4°C.
11. Dry the pellet at RT and add 30–50 µL ultrapure water.
12. Check the fragment sizes on a 0.8% agarose gel. The fragments should be 50–500 bp.
13. Quantify the DNA.

To use the C_0t-1 repetitive DNA in chromosomal mapping, label the probe by a nick translation reaction with the appropriate amount of DNA (see Section 11.2.8.2).

11.2.8.1.2   Telomere Repeat

The telomere motif in grasshoppers is TTAGG and the self-annealing primers F (TTAGG)₅ and R (CCTAA)₅ should be used to obtain this type of probe, as suggested by Ijdo et al. (1991).

The reaction is done with the following products:

1. 5 µL *Taq* DNA polymerase enzyme buffer (10×)
2. 0.5 µL MgCl₂ (50 mM)
3. 2 µL F primer (10 mM)
4. 2 µL R primer (10 mM)
5. 1 µL dATP (2 mM)
6. 1 µL dCTP (2 mM)
7. 1 µL dGTP (2 mM)
8. 0.7 µL dTTP (2 mM)
9. 0.6 µL labeled dUTP (1 mM)
10. 0.4 µL *Taq* DNA polymerase (5 U/µL)
11. Sterile ultrapure water (up to 50 µL)

And PCR cycles are performed as indicated in Table 11.1.

Check the amplification on a 1% agarose gel. The fragments should be a smear between 100 and 1000 bp, but if they are higher, use DNase I to cut the fragments. Alternatively, the reaction can be adjusted with higher primer concentrations. Note that for telomeric probe generation it is not necessary to use a DNA template.

Telomeric probes can also be obtained from commercial suppliers as 3′-, 5′-, or both end-labeled synthetic oligonucleotides ([TTAGG]₇, [CCTAA]₇). See examples of FISH with telomeric probes in Figure 11.5a and b.

11.2.8.1.3   Multigene Families, Transposable
Elements, and Satellite DNA

Moderate and highly repetitive sequences, such as distinct families of satellite DNA (Figures 11.4b and 11.5c) and transposable elements (Figure 11.5d), can also be identified and isolated by standard cloning protocols or next-generation sequencing (NGS) approaches. Amplification is performed by PCR using specific

TABLE 11.1

PCR Program for Amplifying Telomeric DNA

Step	Temperature	Time
Initial denaturation	95°C	5 minutes
10 Cycles	95°C	1 minute
	55°C	30 seconds
	72°C	1 minute
35 Cycles	95°C	1 minute
	60°C	30 seconds
	72°C	1 minute and
		30 seconds
Final extension	72°C	5 minutes
Hold	4°C	

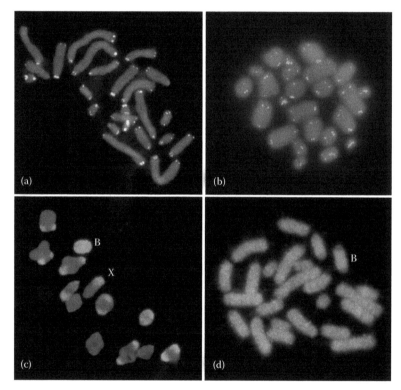

FIGURE 11.5 **(See color insert.)** FISH mapping of tandem (a, b, c) and scattered (d) repetitive DNA sequences in L. migratoria (a), Adimantos arnatissimus (b) and E. plorans (c, d). (a, b) Telomeric DNA, (c) a 180-bp satDNA, (d) Gypsy transposable element. (a, d) Embryonic mitotic cells, (c) first meiotic metaphase cell, (b) mitotic cell obtained from gastric caeca. Note the presence of B chromosomes in c and d.

primers. Some examples of primers used for the amplification of these repetitive sequences in the grasshopper, *E. plorans,* are listed in Table 11.2.

Multigene families of DNA sequences (Figure 11.6) are amplified through PCR using specific or universal primers. Some

TABLE 11.2

Primers Used for PCR Amplification of Some Repetitive DNAs in the Grasshopper
Eyprepocnemis plorans, **Such as a 180-bp Satellite DNA and Three Mobile Elements**

Target	Primer Sequences	Estimated Fragment Size	Reference
180-bp satellite DNA	5′ GCACTGCTTTCCAGATATCACACTAAAATG 3′ 5′ CGCATTTCTGCCGCCTGTGGCGCTACATT 3′	148 bp	Teruel (2009)
Gypsy, LTR retrotransposon	F 5′ GTKTTIKTIGAYACIGGIKC 3′ R 5′ GCGTTIKTIAGICCRAAIGGCAT 3′	1107 bp	Miller et al. (1999)
RTE, non-LTR retrotransposon	F 5′ CTGGACGCAGAGAARGCVTTYGAC 3′ R 5′ CGGGCTCAGNGGWCANCCYTGA 3′	1242 bp	Montiel et al. (2012)
Mariner transposon	F 5′CGCGCATGAATGGATTAACG 3′ F 5′AAGAGCCGACATCGAAGGATC 3′	1112 bp	Burke et al. (1993)

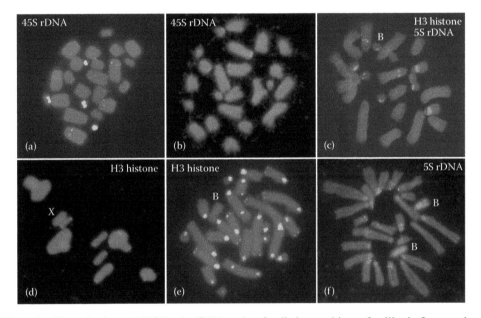

FIGURE 11.6   (**See color insert.**) FISH using DNA probes for distinct multigene families in five grasshopper species, (a) *Adimantos arnatissimus,* (b and c) *Locusta migratoria,* (d) *Chorthippus jacobsi,* (e) *Abracris flavolineata,* and (f) *Eyprepocnemis plorans.* (a and e) Mitotic cells obtained from gastric caeca; (b, c, f) embryo mitotic cells, (d) meiotic metaphase I cell. The probes used are indicated in the cells. Note the presence of B chromosomes in c, e, f.

primers currently used to obtain these sequences in grasshoppers from distinct families are listed in Table 11.3.

The PCR reaction should be performed using the parameters shown in Table 11.4.

It is important to sequence the obtained fragment to perform the FISH technique, to be sure of fragment identity. Test different temperatures within the indicated range. The theoretical optimal temperature is provided by the primer manufacturer, but it should be empirically determined. Check the fragment sizes on a 1% agarose gel.

TABLE 11.3

Primers Used for PCR Amplification of Some Multigene Families in Grasshoppers

Target Gene	Primer Sequences	Estimated Fragment Size	Reference
18S rDNA	F 5′ CCCCGTAATCGGAATGAGTA 3′ R 5′ GAGGTTTCCCGTGTTGAGTC 3′	822 bp	Cabral-de-Mello et al. (2010)
5S rDNA	F 5′ AACGACCATACCACGCTGAA 3′ R 5′ AAGCGGTCCCCCATCTAAGT 3′	92 bp	Loreto et al. (2008)
U1 snDNA	F 5′ CTTACCTGGCGTRGRGGWY 3′ R 5′ CAKTCCCRRCTACCAAAAATT 3′	127 bp	Cabral-de-Mello et al. (2012)
U2 snDNA	F 5′ ATCGCTTCTCGGCCTTATG 3′ R 5′ TCCCGGCGGTACTGCAATA 3′	178 bp	Bueno et al. (2013)
H3 histone	F 5′ ATGGCTCGTACCAAGCAGACVGC 3′ R 5′ATATCCTTRGGCATRATRGTGAC 3′	370 bp	Colgan et al. (1998)
H4 histone	F 5′ TSCGIGAYAACATYCAGGGIATCAC 3′ R 5′ CKYTTIAGIGCRTAIACCACRTCCAT 3′	210 bp	Pineau et al. (2005)

TABLE 11.4

PCR Program for Amplifying Multigene Families

Step	Temperature	Time
Initial denaturation	95°C	5 minutes
30 Cycles	95°C	1 minute
	(45°C–60°C)*	30 seconds
	72°C	1 minute
Final extension	72°C	5 minutes
Hold	4°C	

*Distinc temperatures should be tested in the indicated range

Double FISH can be performed by combining different probes and colors (Figure 11.6c), which is especially indicated for the fiber-FISH technique (Figure 11.7a).

11.2.8.2   Probe Labeling

PCR and nick translation are the most frequently used strategies for probe labeling in FISH experiments. The choice of labeling method depends on the type of DNA probe.

PCR labeling is useful for small DNA sequences. For DNA fragments larger than 600 bp, the PCR product should be fragmented with DNase I to allow the probe to access the target chromosomal DNA. This fragmentation prevents the occurrence of background after posthybridization washes. The probe-labeling PCR is identical to a regular PCR, except that the nucleotide concentrations are modified. The ratio of regular to modified nucleotides (e.g., Dig-11-dUTP or Bio-16-dUTP) should be approximately 70% to 30%. A reaction example follows:

1. 2.5 μL *Taq* DNA polymerase enzyme buffer (10×)
2. 0.25 μL MgCl$_2$ (50 mM)

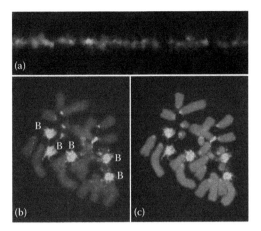

FIGURE 11.7 **(See color insert.)** (a) Double FISH for 45S rDNA (red) and a sequence-characterized amplified region (SCAR) marker specific to B chromosomes in *E. plorans* was performed on chromatin fiber. Note the alternating arrangement of the two sequences; chromosome painting on an embryonic mitotic cell of *Locusta migratoria* using a B-chromosome DNA probe obtained through microdissection, (b) paint probe signal, (c) DAPI pattern in gray scale merged with paint probe signal.

3. 1 μL F primer (10 mM)
4. 1 μL R primer (10 mM)
5. 0.5 μL dATP (2 mM)
6. 0.5 μL dCTP (2 mM)
7. 0.5 μL dGTP (2 mM)
8. 0.35 μL dTTP (2 mM)
9. 0.3 μL labeled dUTP (1 mM)
10. 0.1 μL *Taq* DNA polymerase (5 U/μL)
11. 2 μL genomic DNA (50–100 ng/μL)
12. Sterile ultrapure water (up to 25 μL)

Nick translation labeling is recommended for DNA fragments larger than 600 bp, but it is also suitable for shorter probes. Indirect probe labeling with biotin or digoxigenin can be performed with commercially available kits, such as the BioNick Labeling System (Cat. no. 18247-015, Invitrogen, Carlsbad, CA) and the DIG Nick Translation Mix (Roche). Nick translation can also be performed with a mix of DNA polymerase I and DNase I.

For direct probe fluorophore labeling using DNA polymerase I/ DNase I, follow this procedure:

1. Add the following to an Eppendorf tube on ice and mix:

10× nick translation buffer	5 μL
0.2 mM unlabeled ACG nucleotide mixture	5 μL
100 mM DTT	1 μL

0.05 mM dTTP	1 μL
DNA template	1 μg
1 mM fluorophore-conjugated dUTP	1 μL
Ultrapure water	up to 45 μL
DNA polymerase I/DNase I (0.4 U/μL)	5 μL
Total volume	50 μL

2. Incubate for 2–3 hours at 15°C.
3. Stop reaction with 5 μL 0.5 M EDTA (pH 8.0).

The reaction provides sufficient labeled probe for four to seven FISH reactions. For labeling of microdissected chromosomes, see Section 11.2.8.5.

For probe precipitation:

1. Add 5 μL 3 M sodium acetate (pH 5.2) and 150 μL chilled absolute ethanol for probe precipitation. Mix well.
2. Keep on ice for 15 minutes.
3. Store at −20°C overnight or at −80°C for 2 hours.
4. Centrifuge at 16,000 rpm for 30 minutes in a microcentrifuge.
5. Discard the supernatant.
6. Wash the labeled probe in chilled 70% ethanol.
7. Dry the probe.
8. Resuspend in 20 μL distilled water or TE (pH 8.0).
9. Store at −20°C until use.

11.2.8.3 FISH with Direct Fluorophore Labeling and Detection

The following protocol is modified from the one by Schwarzacher and Heslop-Harrison (2000). It is a 3-day-long procedure that includes the first slide pretreatment and dehydration (day 1), additional slide pretreatments and hybridization reactions (day 2), and slide posthybridization washing (day 3). Detection steps are not needed after posthybridization washes.

Day 1

For optimum results, slides are pretreated with pepsin, which permeabilizes the cell membrane and eliminates cytoplasm, thus facilitating probe access to target chromosomes. RNA must also be removed to avoid background hybridization, and a paraformaldehyde fixative is used to preserve the material during in situ hybridization.

Pepsin digestion is necessary for meiotic preparations where removing cytoplasm is essential for efficient probe binding. For this purpose:

1. Add 100 μL 50 μg/mL pepsin solution in 0.01 N HCl to the preparation and cover with a parafilm coverslip to avoid drying.
2. Incubate the slides 2–5 minutes at 37°C in a humid chamber. Avoid chromosome degradation by stopping the reaction once the cytoplasm is removed.
3. Wash the preparation thoroughly in distilled water at RT and air-dry. This step can be omitted for embryonic mitotic preparations with hypotonic treatments.
4. Dehydrate chromosomes in a 70%, 90%, and absolute ethanol series for 3, 3, and 5 minutes, respectively, and incubate at 60°C overnight.

Day 2

1. Add 200 μL RNase (100 μg/mL in 2× SSC) to each preparation, cover with a parafilm coverslip and incubate for 90–120 minutes at 37°C in a humid chamber.
2. Wash three times in 2× SSC at RT for 5 minutes.
3. Place the slides in a Coplin jar with 100 mL freshly prepared 4% paraformaldehyde and incubate them for 10 minutes in a fume hood.
4. Wash three times with shaking for 5 minutes each in 2× SSC at RT.
5. Dehydrate preparations in a 70% (3 minutes), 90% (3 minutes), and absolute (5 minutes) ethanol series.
6. Prepare a 30 μL hybridization reaction mix by adding the following reagents to an Eppendorf tube:

Formamide	12 μL
Dextran sulfate (50%)	6 μL
20× SSC	1.5 μL
10% SDS	0.5 μL
Salmon sperm DNA (5 μg/μL)	1 μL
Probe	100–250 ng per slide
Ultrapure water	up to 30 μL

7. Denature the probe in a water bath or thermocycler at 70°C for 10 minutes.
8. Place on ice for 5 minutes.
9. Apply the hybridization mixture to slides and cover with a parafilm coverslip. Avoid bubbles and cover the entire preparation.
10. Denature the chromosomal DNA by placing the slides with the hybridization mixture on a hot plate for 6 minutes at 80°C.
11. Incubate slides overnight at 37°C in a humid chamber.

Day 3

Slides must be protected from light during washing and detection procedures.

1. Carefully remove the coverslips from slides.
2. Put the slides in a Coplin jar and wash them twice with 2× SSC at 37°C, 5 minutes per wash, with gentle shaking in a water bath.
3. Wash slides with fresh 2× SSC at RT by vigorously shaking on a shaking platform.
4. Wash with 4× SSC/0.2% Tween 20 solution for 5 minutes.
5. Counterstain slides with 100 µL DAPI solution (2 µg/mL) in McIlvaine's buffer for 15 minutes with a parafilm coverslip. Protect the slides from light.
6. Wash briefly in 4× SSC/0.2% Tween 20 solution.
7. Add one drop of antifading solution (VECTASHIELD) and place glass coverslips on slides, avoiding bubbles.
8. Remove the excess antifading solution with filter paper. Repeat this step twice or until the filter paper remains clean after pressing.
9. Store slides horizontally in a box at 4°C in the dark. Visualize on an epifluorescence microscope coupled to appropriate filters.

11.2.8.4   FISH with Indirect Labeling and Antibody Detection

This protocol for chromosome in situ hybridization is divided into three main stages: (1) slide pretreatment (day 1), (2) DNA denaturation/hybridization (day 1) and (3) washing/probe detection (day 2), as modified from that of Cabral-de-Mello et al. (2010). Freshly made or stored (at −20°C) slides can be used for FISH experiments.

Day 1
1. Dehydrate slides in an ethanol series (70%, 85%, and absolute) for 5 minutes each at RT and air-dry at 37°C.
2. Incubate the preparation in 100 µg/mL RNase solution in 2× SSC under a parafilm coverslip for 1 hour at 37°C.
3. Wash three times in 2× SSC for 5 minutes each at RT.
4. Incubate the preparation in 10 µg/mL pepsin solution in 0.1 N HCl under a parafilm coverslip for 20 minutes at 37°C (optional).
5. Wash three times in 2× SSC for 5 minutes each at RT.
6. Place the slide in a Coplin jar with 3.7% formaldehyde diluted in wash-blocking buffer for 10 minutes.
7. Wash three times in 2× SSC for 5 minutes each at RT.

8. Dehydrate the slides in an ethanol series (70%, 85%, and absolute) for 5 minutes each and air-dry at 37°C.
9. Prepare the hybridization probe mixture. In a microtube, add at least 100 ng labeled DNA, formamide (final concentration 50%), SSC (final concentration 2×), and dextran sulfate (final concentration 10%); see an example below.

   1 6 µL Labeled DNA (at least 100 ng)
   2 15 µL of 100% Formamide
   3 6 µL of 50% Dextran sulfate
   4 3 µL of 20× SSC

10. Denature the hybridization probe mixture at 95°C for 10 minutes and immediately place the tube on ice for 5 minutes.
11. Place the hybridization probe mixture on the slide and cover with a glass coverslip, avoiding bubbles. The quantity of hybridization mixture will determine the coverslip size.
12. Incubate the slides with hybridization mixture at 75°C using a metal plate in a water bath or directly in a thermocycler for 5 minutes.
13. Incubate the slides overnight in a humid chamber at 37°C.

Day 2
1. Remove the coverslip and incubate the slides in a Coplin jar with 2× SSC for 5 minutes at RT.
2. Wash slides twice in 2× SSC at 42°C for 5 minutes each.
3. Wash slides twice in 0.1× SSC at 42°C for 5 minutes each.
4. Wash slides once in 2× SSC at 42°C for 5 minutes.
5. Place slides in 2× SSC at RT for 10 minutes.
6. Transfer the slides to a Coplin jar containing wash-blocking buffer.
7. Dilute the streptavidin-Alexa Fluor 488 conjugate (Life Technologies) for detecting probes labeled with biotin or the anti-digoxigenin-rhodamine (Roche, Basilea, Switzerland) for detecting probes labeled with digoxigenin in wash-blocking buffer, as follows:

   1. 1:100 µL for streptavidin-Alexa Fluor 488 conjugate (initial concentration 2 mg/mL): wash-blocking buffer
   2. 0.5:100 µL for anti-digoxigenin-rhodamine (initial concentration 200 g/mL): wash-blocking buffer.

Note: For two-color FISH experiments, dilute 0.5 µL anti-digoxigenin-rhodamine and 1 µL streptavidin-Alexa Fluor 488 conjugate in 100 µL wash-blocking buffer solution.

8. Add the solution on the slide and cover with a parafilm coverslip. Incubate at 37°C for 1 hour.
9. Wash the slide three times in wash-blocking buffer at 45°C for 5 minutes each.
10. Mount the slide with 0.5 μL DAPI (0.2 mg/mL) mixed in 15 μL VECTASHIELD antifade solution. Use an appropriately sized glass coverslip.
11. Store the slides in the dark at 4°C until the analysis.
12. Analyze the chromosome preparations under an epifluorescence microscope coupled to an adequate filter set.

11.2.8.5   Chromosome Painting

Chromosome painting is a FISH variation in which a specific chromosome probe homologous to part or the entire length of a particular chromosome is used on metaphase or interphase cells. Since it was developed several decades ago (Cremer et al. 1988; Lichter et al. 1988; Pinkel et al. 1988), the number of applications has increased over time, revealing that FISH procedure is useful for the identification of numerical and structural chromosome aberrations or for the establishment of evolutionary relationships between chromosomes from the same or different species (Figure 11.7b and c). The protocol described here is based on the work by Marchal et al. (2004).

The DNA probe is obtained by chromosome microdissection following the protocols described earlier. Amplified microdissected chromosome DNA is labeled using different approaches depending on the amplification methods used.

1. For chromosomal DNA amplified by DOP-PCR, use the second PCR described in Section 11.2.7.2.
2. For labeling chromosomal DNA amplified by GenomePlex (Sigma-Aldrich), use the standard nick translation procedure.

For probe precipitation, follow the steps:

1. Add the following to an Eppendorf tube and mix:

Distilled water	6.5 μL
3 M sodium acetate (pH 5.2)	4 μL
Salmon sperm (50 ng/μL)	1 μL
Glycogen	0.5 μL
Labeled probe	8 μL
Absolute ethanol	20 μL
Total volume	40 μL

2. Precipitate overnight at −20°C.

3. Centrifuge at 16,000 rpm at 4°C for 30 minutes in a microcentrifuge.
4. Discard the supernatant.
5. Wash the probe with 100 µL chilled 70% ethanol.
6. Centrifuge at 14,000 rpm at 4°C for 30 minutes.
7. Discard the supernatant.
8. Dry the probe.
9. Resuspend the probe in the following solution:

Ultrapure water	9 µL
20× SSC	3 µL
Formamide	15 µL
Dextran sulfate (50%)	3 µL
Total volume	30 µL

10. Mix and vortex for several seconds.
11. Incubate for 3 h at 37°C.
12. Store at −20°C until use.

The slides should be pretreated as follows:

1. Incubate slides at 37°C for at least 24 hours.
2. Dehydrate slides in a 70%, 90%, and absolute ethanol series for 5 minutes each.

Perform the hybridization the following way:

1. Denature chromosomal DNA in 70% formamide in 2× SSC for 2.5 minutes.
2. Put slides in 2× SSC at RT for 1 minute.
3. Dehydrate slides in a 70%, 90%, and absolute ethanol series for 5 minutes each.
4. Denature the probe for 6 minutes at 73°C in a water bath.
5. Place the denatured probe on ice for 5 minutes.
6. Add the probe mix to slides, cover with a parafilm coverslip and incubate at 37°C for 16 hours in a humidified chamber with formamide/2× SSC.

For posthybridization washing and chromosome painting detection, do the following:

1. Wash slides in 0.4× SSC/0.3% Tween 20 at 70°C for 2 minutes.
2. Wash in 2× SSC/0.1% Tween 20 at RT for 30 seconds.
3. Counterstain slides with 100 µL DAPI (20 µg/mL) in McIlvaine's buffer for 15 minutes with a parafilm coverslip in the dark.

4. Wash in PBT (phosphate buffered Tween 20), and mount slides with antifading VECTASHIELD (Vector) using a glass coverslip.
5. Remove the excess antifading solution by gently pressing with filter paper.
6. Store at 4°C several days before analysis under an epifluorescence microscope.

11.2.9 Measuring DNA by Feulgen Image Analysis Densitometry

The following protocol is used to measure the haploid DNA amount (C-value) or even the DNA content of single chromosomes. This method is useful for whole genome sequencing projects, coverage estimates, or simply for estimating the differences in chromosome size caused by chromosome polymorphisms. For instance, we recently estimated the size of three B chromosome variants in *E. plorans* to be 0.51, 0.54, and 0.64 pg, whereas the B chromosome in *L. migratoria* is only 0.15 pg (Ruiz-Ruano et al. 2011).

11.2.9.1 Sample Preparation

Testis tubules fixed in 3:1 absolute ethanol–acetic acid are a good choice for this technique, as they supply spermatids, for measuring the C-value, and spermatocytes, for measuring chromosome size. For C-value measurements, we need to include in the analysis an additional species with a known C-value to be used as a standard, which will allow to estimate the absolute DNA amount in picograms. It is convenient if both species' materials (the sample and the standard) were fixed at the same time in the same conditions (Hardie et al. 2002).

Preparations are made by squashing 2–3 testis tubules from the sample in 50% acetic acid on the left half of a slide and the same amount for the standard on the right half of the same slide. Thus, the Feulgen staining is performed in the same conditions for both the sample and the standard. Given that most grasshoppers show XX/X0 sex chromosome determinism, males are expected to contain two types of spermatids, which differ in DNA amount because of the presence or absence of the X chromosome. The size of the X chromosome is inferred from the difference between +X and −X spermatids (Ruiz-Ruano et al. 2011). To measure chromosome size, we use the X chromosome of spermatocytes as an internal standard and measure the amount of DNA in the autosomes.

11.2.9.2 Feulgen Reaction

Feulgen reaction is a DNA-specific stoichiometric staining performed as follows:

1. Hydrolyze the preparations in 5 N HCl for 20 minutes at RT to depurinize DNA and generate free aldehyde groups. Then rinse the preparations in 0.1 N HCl. Adjust the hydrolysis time for each type of material. Perform the next steps in the dark.
2. Stain with Schiff's reagent (Sigma-Aldrich) for 90–120 minutes at RT. This reagent binds DNA and thus gives color to chromatin. Fresh Schiff's reagent is recommended, although it can be reused several times if stored at 4°C.
3. Remove the unbound stain with three 5-minute shaking washes in sulfurous water (300 mL containing 1.5 g sodium or potassium metabisulfite and 15 mL of 1 N HCl in distilled water). Wash in running tap water and rinse in distilled water.
4. Air-dry the slides, mount them in DPX, and store them at 4°C in the dark.

11.2.9.3   Image Capture

If image analysis is being performed for the first time, some tests are recommended to estimate the reliability of the microscope and camera. The linearity of the camera response can be tested with density filters, with uniformity testing so that the entire field is homogeneously captured by the camera, with no change in light intensity.

1. Turn on the microscope 10–20 minutes before image capture with the 100× objective to ensure stable measurements. During this time, test Köhler illumination and setup the camera.
2. Exposure time, resolution, and picture format must be the same for all captures. We use 1/120 seg, 1360 × 1024 pixels, and TIFF (tagged image file format), respectively. For higher sensitivity, 16-bit TIFF images are recommended. Although 8-bit images can be used, it is better to save them in the TIFF format. Do not use JPG format because information is lost during compression.
3. Select a region in the slide without materials and lacking dark spots. Open the condenser to an average value, such that the green channel does not increase, and then adjust the light source until the maximum pixel value is approximately 200 to avoid overexposure of the camera cells. This point is critical for good quantification.
4. Find a region of the preparation with groups of nonoverlapping spermatids. The best results are obtained in spermatids that are starting to elongate because they show a more homogeneously stained nucleus (i.e., discard round spermatids). Assure that the spermatid compaction is similar in the sample and the standard, as inferred from a similar

shape. Capture 50–100 spermatids. For measuring selected chromosomes, capture images of 10–20 complete diplotene or metaphases I cells where chromosomes do not overlap.

11.2.9.4   Image Analysis

1. Open the images in ImageJ software (Magelhaes et al. 2004). For 16-bit TIFF images, move the bottom bar to the middle to select the green channel. Trim each object, but include a small part of the background surrounding it. Paste it into a new file (File >> New >> Internal Clipboard), and save it as text with a txt extension. If you use 8-bit TIFF or JPG files, split the channels (Image >> Color >> Split channels), trim objects from the green channel, and save them in the same format (i.e., with the tif or jpg extension). Files from each sample and the standard need to be saved in different folders. Indicate identifiable chromosomes in the file name.

2. Integrated Optical Density (IOD) measures are calculated with a threshold set by Otsu's method and applying the Beer–Lambert's law with our open-source Python-written pyFIA software (Ruiz-Ruano et al. 2011), which can be downloaded from http://code.google.com/p/pyfia/. pyFIA has been tested in GNU/Linux Debian and Ubuntu distributions. Instructions for installing and running the program are found on the Web site.

3. In each folder, pyFIA creates a text file called "output" with the IOD values (for analysis) and the average of background values (which is subtracted from the optical density value for each stained pixel).

   The format of the output file is optimized for the open-source spreadsheet Gnumeric, although it is compatible with other software of this type.

   For spermatid analysis, build a histogram series with different interval amplitudes (Statistics >> Descriptive Statistics >> Frequency Tables >> Histogram), and choose one that gives the best representation of the data. In the best conditions, a bimodal distribution is observed (Figure 11.8) where the peak with lower IOD corresponds to spermatids lacking the X chromosome. The peak with the higher IOD, however, corresponds to spermatids carrying the X chromosome and represents the C-value. Logically, the difference between the two peaks represents the DNA content of the X chromosome. We then copy the data on the histogram to the open-source software Qtiplot and make a graph (Plots >> Columns).

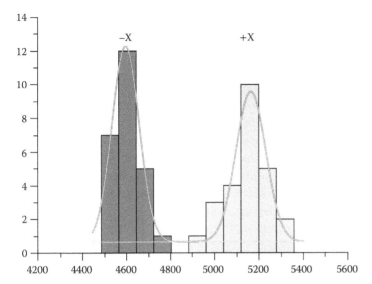

FIGURE 11.8 **(See color insert.)** Histogram of IOD values obtained from 50 spermatids of an *E. plorans* male. A bimodal distribution shows that one peak belongs to spermatids without (left) or with (right) the X chromosome. The latter peak corresponds to the C-value of the species, and the difference between the two peaks is equivalent to the DNA content of the X chromosome.

With the histogram window selected, we select Analysis >> Fit Multi-peak >> Gaussian >> 2 peaks. Select the bar at the center of each distribution and calculate the mean of each peak by adjusting to a Gaussian curve. These values are the IOD for the analysis. To measure each object (i.e., the difference between −X and +X spermatids or the biggest chromosome) in the same individual, a coefficient of variation of up to 10% is permitted (Hardie et al. 2002).

4. Finally, the DNA amount of the haploid set (C-value) or the X chromosome can be calculated, in picograms, from the IOD values in the sample and the standard by the following equations:

$$C = \frac{IOD_C \times C_S}{IOD_s}$$

$$X = \frac{IOD_X \times C}{IOD_C}$$

where C = C-value for the sample (picogram); Cs = C-value for the standard (picogram); IOD_C = IOD for the +X peak in the sample; IOD_S = IOD for the +X peak in the standard; X = DNA amount of the X chromosome in the sample (picogram); IOD_X = Difference in IOD between the +X and −X peaks in the sample.

5. To measure selected chromosomes, it is better to measure only the autosomal bivalents because the X chromosome usually shows different condensation. Arrange the autosomal IODs from high to low, sum them, and calculate the relative amount for each autosome by dividing their sum. Calculate the size of each autosome in picograms by multiplying the obtained proportion by the C-value minus X chromosome size in picograms. Arrange all chromosome measurements in picograms, including the X chromosome, from larger to smaller values. This result will indicate the size order of the X chromosome, in respect to the autosomes, avoiding the problem of differential condensation. If the X chromosome size in picograms is unknown because the analysis of spermatids did not exhibit two peaks, the only option is to include the X chromosome in IOD measurements and double it. This method will solve the problem in primary spermatocytes of bivalent autosomes and univalent X chromosome.

6. As DNA amount is usually expressed in picograms, we can easily convert it to bp because 1 pg of DNA is equivalent to 0.978×10^9 bp (Dolezel et al. 2003).

11.2.10   Reagents and Solutions

1. BSA: Dilute in PBS at the required concentration. Make aliquot and store at $-20°C$.
2. CMA_3: Dissolve 5 mg chromomycin A_3 in 10 mL solution of 1:1 McIlvaine buffer pH 7.0 and distilled water. Add 10 µL of 5 M $MgCl_2$.
3. DAPI 20 µg/mL: Dilute in McIlvaine's buffer from a stock solution of 100 µg/mL in water. Store at $-20°C$.
4. Dextran sulfate (50%): Mix the solution in distilled water by heating at $70°C$ until dissolved. Filter to sterilize.
5. 1 M DTT: Dissolve 30.9 g DTT in 20 mL 0.01 M sodium acetate (pH 5.2).
6. 100 mM DTT solution: Mix 100 µL DTT 1 M, 3.3 µL sodium acetate 3 M, pH 5.2 and 897 µL ultrapure water.
7. EDTA (ethylenediaminetetraacetic acid): Dissolve 0.5 M EDTA solution in distilled water, pH 8.0.
8. Glycogen: 20 mg/mL in distilled water.
9. Insect saline solution: Dissolve 9 g NaCl, 0.42 g KCl, 0.33 g $CaCl_2.2H_2O$, 0.2 g $NaHCO_3$ in 1000 mL distilled water.
10. McIlvaine's buffer: Prepare two separate solutions of 200 mM PO_4HNa_2 and 100 mM citric acid. Add 100 mM

citric acid solution (18 mL) to the 200 mM PO_4HNa_2 solution (82 mL) for a 100 mL solution, pH 7.0.

11. 10× Nick translation buffer: 0.5 M Tris-HCl pH 7.8, 50 mM $MgCl_2$, 5 mg/mL BSA.

12. Orcein (lactopropionic): Dilute 4 g orcein in 100 mL propionic acid and 100 mL lactic acid. Filter twice.

13. Paraformaldehyde 4% solution: Add 4 g paraformaldehyde to 80 mL water and stir while heating at 60°C in a fume hood until translucid. Then add 0.5 mL 4M NaOH to make the solution transparent. Cool the solution and complete to 100 mL.

14. PBS: 137 mM NaCl, 2.7 mM KCl, 10 mM Na_2HPO_4, 2 mM KH_2PO_4, pH 7.4.

15. PBT: Add 10 mL of 10× PBS and 200 μL Tween 20 to 90 mL distilled water.

16. Pepsin 50 μg/mL in HCl 0.01N: Dilute a 5 mg/mL pepsin stock solution in HCl 0.01 N.

17. Phosphate buffer: 34 mM KH_2PO_4, 36 mM Na_2PO_4, pH 6.8

18. RNase stock solution: 10 mg/mL RNase in 10 mM Tris-HCl (pH 7.5), 15 mM NaCl. Heat the solution at 100°C for 15 seconds and allow to cool. Make aliquot and store at −20°C.

19. Salmon sperm: Dilute in distilled water from a stock solution.

20. 10% SDS: Make the solution in distilled water and filter it to sterilize.

21. 3 M Sodium acetate: Adjust pH with acetic acid and filter to sterilize.

22. 2× SSC: Dilute from a 20× SSC stock solution.

23. 20× SSC: 3 M NaCl, 0.3 M sodium citrate, pH 7.0.

24. 0.4× SSC/0.3% Tween 20: Dilute from 20× SSC and add 0.3% (v/v) of Tween 20.

25. 2× SSC/0.1% Tween 20: Dilute from 20× SSC and add 0.1% (v/v) of Tween 20

26. 4× SSC/0.2% Tween 20: Dilute from 20× SSC and add 0.2% (v/v) of Tween 20.

27. TE (Tris-EDTA): 10 mM Tris of the desired pH, 1 mM EDTA, pH 8.0.

28. 1 M Tris-HCl: disolve 121.1 g Tris-base in 800 mL distilled water. Adjust pH adding HCl until required. Complete to a final 1000 mL volume.

29. Unlabeled ACG nucleotide mixture: dATP, dCTP, and dGTP, 0.2 mM each in 100 mM Tris-HCl (pH 7.5) or ultrapure water.

30. Wash-blocking buffer: 0.4× SSC, 0.1% Triton X, 1% BSA or skimmed milk.

11.2.11   Troubleshooting

During the realization of the various techniques described earlier, some problems can occur impoverishing the final result and some solutions are provided in Table 11.5.

TABLE 11.5

Technical Problems and Possible Solutions

Technique	Problem	Solution
Colchicine treatment	Scarce mitotic metaphase cells	Increase concentration or time of colchicine treatment.
Squash chromosome preparations	Scarce cells	Use a lower volume of acetic acid and increase time lapse till coverslip separation.
	Cells too aggregated and mounted on one another	Stronger squashing
		5 minutes treatment in 50% acetic acid just before squashing will soften the material.
Spreading chromosome preparations	Opened chromatids	Shorter time of osmotic shock or colchicine treatment.
Chromosome microdissection	Difficulty to identify the chromosome to be microdissected	2% Giemsa staining for 1 minute
Chromosomes for Fiber-FISH	Aggregated cells	Improve cell homogenization
C-Banding	No bands and chromosomes with normal appearance	Add fresh fixative for 1 hour at RT.
		Increase time or temperature of barium hydroxide treatment
	No bands and faint chromosomes	Decrease time or temperature of barium hydroxide treatment
Silver impregnation	No visible nucleoli	Immerse the material in fresh fixative for 1 hour at RT.
		Check that oven temperature is at 60°C, and that solution pH is appropriate.
	Nucleoli are stained but chromosomes are scarcely visible	2% Giemsa for 1 minute after silver impregnation.
Triple fluorescent CMA$_3$-DA-DAPI Staining	No fluorescence	Be sure that preparations are not too old and were stored in the dark
		Increase DA counterstaining time
DNA amplification from microdissected chromosomes	Low amount of DNA obtained	Increase the number of chromosomes microdissected

TABLE 11.5 (*Continued*)
Technical Problems and Possible Solutions

Technique	Problem	Solution
		Increase the amplification time (Φ29)
FISH	No or scarce material in the slide after FISH	Take a look to the preparation under the microscope, before FISH, to be sure of the presence of enough cells
		Reduce denaturation temperature.
	No or faint fluorescence signal	Check probe fluorescence under the microscope before using it
		Check probe concentration and use a positive control slide
		Increase the time of pepsin treatment
		Reduce washing temperature
		Try a different antifading batch
	Background	Increase FISH stringency
		Reduce probe concentration in the hybridization reaction
		Improve slide preparation protocol and the post-pepsin washings
		Use new autoclaved solutions.
Feulgen reaction and image analysis	Scarcely stained cells	Adjust time of hydrolysis
		Be sure that Schiff's reagent is freshly prepared and stored in the dark
	Absence of repeatability in the measurements	Be sure that fixation and image capture were done in the same conditions for both the sample and the standard
		Check the linearity, uniformity, and stability of camera of microscope

11.3 DISCUSSION

11.3.1 Integration of Cytogenetic, Linkage, and Physical Maps and Genome Sequences

Currently, the available techniques have provided rather biased information on grasshopper genomes, with extensive data at the chromosome level, rather scarce information at molecular and genomic levels, and no information for genetic maps obtained from linkage analysis. Grasshopper genomes are the material of choice for cytogenetic studies because of their accessibility and low cost. But, the huge size of grasshopper genomes poses additional difficulty to the implementation of most molecular techniques. For instance, we tried to develop AFLP (amplified fragment length polymorphism) markers for population studies in *E. plorans*, but after 2 years, we had to abandon the project because we were unable to reproduce the observed AFLP patterns. The large genome of this species (10^{10} bp) (Ruiz-Ruano et al. 2011) and the presumed presence of many uncontrolled pseudogenes may have contributed to the lack of repeatability. However, we were more fortunate in developing inter-simple sequence repeat (ISSR) markers in this species, which showed much better reproducibility and allowed us to analyze population genetic structure and gene flow (Manrique-Poyato et al. 2013). Other authors have successfully used random amplified polymorphism DNAs (Sesarini and Remis 2008). This situation will change in the few next years, as NGS becomes increasingly cheap and accessible.

An appropriate combination of many of the techniques described here could advance some genome sequencing projects. The complete assembly of a giant genome, which is 2–5 times larger than the human genome, is a very difficult task because grasshopper genomes contain many copies of mobile elements (Montiel et al. 2012), satellite DNAs, and other repetitive elements (Cabrero et al. 2003; Cabrero and Camacho 2008; Cabrero et al. 2009; Cabral-de-Mello et al. 2011a,b). Many of these paralogous copies are scattered over the entire genome and especially abundant in euchromatin. Thus, the assembly of any given chromosome has many uncertainties. However, we can combine microdissection with PCR amplification of selected DNA sequences to map genes (or DNA sequences) into chromosomes. This type of approach is especially easy in grasshoppers because of their large meiotic chromosomes, which are excellent for microdissection. In this way, we inferred that the B chromosome in *L. migratoria* most likely arose 750,000 years ago (Teruel et al. 2010).

Physical maps of several repetitive DNAs, such as 45S rDNA (Cabrero and Camacho 2008), histone genes (Cabrero et al. 2009), 5S rDNA (Cabral-de-Mello et al. 2011a,b), or mobile elements such as *Gypsy, RTE,* and *Mariner* (Montiel et al. 2012), have been obtained in grasshoppers. Phylogenetic tracing of the physical maps should be used to compare species and to infer possible evolutionary patterns at the level of different subfamilies. This analysis could detect phylogenetic incongruences that suggest horizontal transfer, which is especially probable for mobile elements.

The physical maps will be helpful for completing genomic studies because they identify chromosomes. This task is difficult because some chromosomes are nearly the same size, but the mapping of all markers to chromosomes will solve the problem of distinguishing nonhomologous chromosomes of very similar size. Microdissection PCR mapping will help in chromosome identification, provided that one has sequence information.

11.3.2   Chromosome and Genome Organization

Chromosome organization in grasshoppers is rather uniform, at least in the Acrididae family, with two main patterns of karyotypes. The immense majority have 23 acro/telocentric chromosomes in males (24 in females); part of the Gomphocerinae subfamily has 17 (18) by the fixed occurrence of three centric fusions, yielding three long meta-submetacentric pairs (Hewitt 1979). Molecular phylogenetic studies in grasshoppers have shown that the ancestral condition in the Gomphocerinae is the 23 acro/telocentric pattern typically found in *Dociostaurus* and other genera (Bugrov et al. 2006; Contreras and Chapco 2006).

One interesting question is whether all 17 karyotypes in gomphocerine grasshoppers from several different genera (e.g., *Chorthippus, Omocestus, Stenobothrus*) are monophyletic. In this case, we expect similar gene content in each of the long meta-centric pairs from different species; the three centric fusions may have occurred as separate events to yield, polyphyletically, long metacentric chromosome pairs that are not homologous between species. Many of these species carry 45S rDNA genes in the L_2 and L_3 chromosomes at similar interstitial locations, which support the monophyletic hypothesis. Similarly, the chromosome location of histone H3 and H4 genes supports this hypothesis. In species with 23 chromosomes, the location of these genes is highly conserved; they are interstitially located in the eighth autosome, in order of decreasing size. It is also highly conserved in species with 17 chromosomes, where it is located in the short arm of the smallest metacentric autosomes (L_3). This change is most parsimoniously explained by common ancestry, that is, the involvement of the H3–H4-carrying acrocentric (M_8) chromosome in the centric fusion created the L_3 metacentric autosome (Cabrero et al. 2009). However, synteny for other markers should be analyzed in these species to test the monophyletic hypothesis. Today, we have appropriate tools for the tests: microdissection of separate chromosomes and molecular analysis of the gene content of each chromosome in different species. This approach can be extremely useful in groups where chromosome numbers have experienced dramatic changes. For instance, whereas most acridid grasshoppers show $2n\male = 22 + X0$ chromosomes, the genus *Dichroplus* includes species with $2n\male = 18 + X0$, such as *Dichroplus pratensis* (Bidau et al. 1991), and even with $2n\male = 6 + XY$, such as *Dichroplus silveiraguidoi* (Cardoso et al. 1974). Unveiling how syntenic relationships among genes have changed in parallel with the complex chromosome rearrangements taking place in this and similar cases, will be an interesting topic to investigate in next years, by using the methodology described earlier.

11.3.3   Chromosome and Genome Evolution

The interplay between chromosome and genome evolution is, for the moment, a black box in grasshoppers. We know that acridid chromosomes are rather conserved because most species within a subfamily have very similar karyotypes. This picture changes a little when the location of heterochromatin (Cabrero and Camacho 1986a) or nucleolus organizer regions (Cabrero and Camacho 1986b) are considered, and it changes even more after physical mapping of a comprehensive number of species (Cabrero and Camacho 2008; Cabrero et al. 2009; Cabral-de-Mello et al. 2011a,b). These kinds of studies have revealed a remarkable evolutionary trend in grasshopper genomes for the two families of ribosomal RNA genes (i.e., 45S and 5S). In many species, both families have experienced an intragenomic spread to reach most chromosomes. This spread has been shown at both the intra- and interspecific levels. For instance, the eastern populations of *E. plorans* in Dagestan (Caucasus, Russia) carry 45S rDNA only in the S_9 and S_{11} chromosomes, whereas Spanish and Moroccan populations carry it in almost all chromosomes (López-León et al. 2008). At the interspecific level, a similar pattern is observed with a broad range of species carrying this gene family, from a single chromosome pair to copies in all chromosomes (Cabrero and Camacho 2008). Similarly, the 5S rRNA gene family is located in a single chromosome pair in some species but in all chromosomes in others (Cabral-de-Mello et al. 2011b). This huge variation in chromosome location could be due to an inherent mobility of rDNA (Schubert 1984; Schubert and Wobus 1985). Remarkably, *Pezotettix giornae* and *Oedipoda caerulescens* carry a single cluster for 45S and 5S rDNA located at two different chromosome pairs. In contrast, *Omocestus bolivari* carries both types of rDNA in all chromosomes, which suggests that they have common mechanisms for intragenomic mobility (e.g., association with the same type of mobile element). However, species where all chromosomes carry one of the rDNA types but only one carries the other type (e.g., *Stauroderus scalaris* carries 45S rDNA in all chromosome pairs but 5S rDNA in a single chromosome pair, whereas *Chorthippus nevadensis* shows the opposite pattern) (Cabral-de-Mello et al. 2011b) indicate that both rDNA types show independent mobility.

Even with conventional staining, numerous polymorphisms have been described for centric shifts (White 1973; Hewitt 1979), supernumerary segments (Camacho and Cabrero 1982), and B chromosomes (Hewitt 1979; Camacho 2005). We are only beginning to understand—with very scarce and partial details—the relationship between variation at chromosome level and genome evolution. For instance, B chromosomes in *E. plorans* are among the most polymorphic in any animal or plant (Camacho 2005). More than 50 variants had been described by López-León et al. (1993), but B chromosome classification in this species has not been continued. New types appear in nearly every new population that is exhaustively sampled. However, there are few widespread B variants; B_1 is the most common in the Iberian Peninsula, Balearic Islands, Morocco, Tunisia, and Sicily, suggesting that this was the ancestor variant for the entire western Mediterranean region (Cabrero et al. 2014). The majority of B chromosome variants in this species carry rDNA. In eastern populations (Dagestan, Armenia, Turkey, and Greece), B chromosome variants have the repetitive DNA and small amounts of 180-bp DNA tandem repeats (satDNA). Whereas rDNA is also the major component of the B_1 variant, all other variants in some Spanish populations (e.g., B_2, B_5, and B_{24}) carry more satDNA than rDNA, suggesting that B_1 was replaced by variants with higher relative amount of satDNA and lower amount of rDNA. The possibility that this difference could have influenced the replacement of B_1 for the other variants was noted by Cabrero et al. (1999).

A sequence-characterized amplified region (SCAR) marker found in *E. plorans* is specific to the B chromosomes because it is amplified only from B-carrying individuals (Muñoz-Pajares et al. 2011). The 1510 bp sequence of this marker is remarkably similar in B-carrying individuals from Spain, Morocco, Greece, Turkey, and Armenia. Because B chromosomes are dispensable (i.e., B-lacking individuals survive without them), it is unlikely that sequence conservation is due to selective constraints. The high similarity of the SCAR sequence between so distant regions thus suggests that B chromosomes are very young in this species. This example shows how the joint analysis of chromosome and genome evolution can allow hypothesis testing about the origin of certain genomic compartments, such as a B chromosome. A similar approach, based on the comparison of the internal transcribed spacers 45S rDNA sequences, suggests that the B chromosome arose from the smallest autosome (S_{11}) (Teruel et al. 2014). Similarly, we inferred that the B chromosome in *L. migratoria* could have been derived from the M_8 autosome because the H3 and H4 histone genes mapped to only these two chromosomes (Teruel et al. 2010). As in *L. migratoria*, the mapping of a repetitive DNA (U2 snDNA) indicates the possible origin of the B chromosome from the longest autosome pair in *Abracris flavolineata*. This hypothesis is based on isochromosome formation because of the distribution of the U2 snDNA clusters in both arms of the B chromosome (Bueno et al. 2013). Sex chromosomes have also been recently analyzed by FISH mapping. These data suggested an independent origin of sex-derived systems in related Melanoplinae genera, that is, *Eurotettix* and *Dichromatos*, and a common origin and subsequent differential accumulation of multigene families in neo-X_1X_2Y *Dichromatos* sex chromosomes (Palacios-Gimenez et al. 2013).

These examples show the information that can be obtained by combining cytogenetics with molecular tools. Undoubtedly, there will be new developments that allow further testing of interesting hypotheses on the evolution of different genomic compartments—chromosomes, chromosome segments, mobile genomic elements, repetitive DNAs, and so forth—that lead to the first complete grasshopper genome sequence.

11.4   CONCLUSIONS

The recent publication of the first draft genome for a grasshopper species (*L. migratoria*) by Wang et al. (2014) will represent a qualitative change in the kind of molecular approaches that can be applied at all levels. The availability of a reference genome will undoubtedly allow designing primers for PCR amplification of many genomic regions in other species, with higher easiness for conserved regions, and higher availability for close relative species such as those belonging to the Oedipodinae subfamily. The locust genome is not yet complete because, as in other genomes, the regions being rich in repetitive DNA need additional work, a subject where chromosome studies as those reported here will be of great aid.

With this reference genome, a variety of comparative studies to infer evolutionary changes of many genomic regions can now be performed, thus opening new avenues for cytogenetical work determining the correspondence between physical maps and genome location. In addition, multitude of phylogenetic studies can now be carried out to unveil the evolutionary history of grasshoppers by using many nuclear markers. So far, phylogenomic studies in grasshoppers have been performed only with mitochondrial genes, but next years will surely witness the inclusion of multiple nuclear genes, as soon as other genomes are available.

ACKNOWLEDGMENTS

This study was supported by grants from the Spanish Ministerio de Ciencia e Innovación (CGL2009-11917), Plan Andaluz de Investigación (CVI-6649), Fundação de Amparo a Pesquisa do Estado de São Paulo-FAPESP (2011/19481-3), Conselho Nacional de Desenvolvimento Científico e Tecnológico-CNPq (475308/2011-5), PROPE/UNESP, and was partially performed by FEDER funds. The authors are grateful to Dr. Dardo A. Martí and Dr. Maria José de Souza for providing the material of *Dichroplus pratensis* and *Dichroplus silveiraguidoi*, respectively.

REFERENCES

Bidau, C. J., C. Belinco, P. Mirol, and D. Tosto. 1991. The complex Robertsonian system of *Dichroplus pratensis* (Melanoplinae, Acrididae). I. Geographic distribution of fusion polymorphisms. *Genet Sel Evol* 23: 353–370.

Bueno, D., O. M. Palacios-Gimenez, and D. C. Cabral-de-Mello. 2013. Chromosomal mapping of repetitive DNAs in *Abracris flavolineata* reveal possible ancestry for the B chromosome and surprisingly H3 histone spreading. *PLoS ONE* 8: e66532.

Bugrov, A. G., T. V. Karamysheva, E. A. Perepelov, E. A. Elisaphenko, D. N. Rubtsov, E. Warchałowska-Sliwa, H. Tatsuta, and N. B. Rubtsov. 2007. DNA content of the B chromosomes in grasshopper *Podisma kanoi* Storozh. (Orthoptera, Acrididae). *Chromosome Res* 15: 315–325.

Bugrov, A., O. Novikova, V. Mayorov, L. Adkison, and A. Blinov. 2006. Molecular phylogeny of Palaearctic genera of Gomphocerinae grasshoppers (Orthoptera, Acrididae). *Syst Entomol* 31: 362–368.

Burke, W. D., D. G. Eickbush, Y. Xiong, J. Jakubczak, and T. H. Eickbush. 1993. Sequence relationship of retrotransposable elements R1 and R2 within and between divergent insect species. *Mol Biol Evol* 10: 163–185.

Cabral-de-Mello, D. C., J. Cabrero, M. D. López-León, and J. P. M. Camacho. 2011a. Evolutionary dynamics of 5S rDNA location in acridid grasshoppers and its relationship with H3 histone gene and 45S rDNA location. *Genetica* 139: 921–931.

Cabral-de-Mello, D. C., C. Martins, M. J. Souza, and R. C. Moura. 2011b. Cytogenetic mapping of 5S and 18S rRNAs and H3 histone genes in 4 ancient Proscopiidae grasshopper species: Contribution to understanding the evolutionary dynamics of multigene families. *Cytogenet Genome Res* 132: 89–93.

Cabral-de-Mello, D. C., R. C. Moura, and C. Martins. 2010. Chromosomal mapping of repetitive DNAs in the beetle *Dichotomius geminatus* provides the first evidence for an association of 5S rRNA and histone H3 genes in insects, and repetitive DNA similarity between the B chromosome and A complement. *Heredity* 104: 393–400.

Cabral-de-Mello, D. C., G. T. Valente, R. T. Nakajima, and C. Martins. 2012. Genomic organization and comparative chromosome mapping of the U1 snRNA gene in cichlid fish, with an emphasis in *Oreochromis niloticus*. *Chromosome Res* 20: 279–292.

Cabrero, J. and J. P. M. Camacho. 1986a. Cytogenetic studies in gomphocerine grasshoppers. I. Comparative analysis of chromosome C-banding pattern. *Heredity* 56: 365–372.

Cabrero, J. and J. P. M. Camacho. 1986b. Cytogenetic studies in gomphocerine grasshoppers. II. Chromosomal location of active nucleolar organizing regions. *Can J Genet Cytol* 28: 540–544.

Cabrero, J. and J. P. M. Camacho. 2008. Location and expression of ribosomal RNA genes in grasshoppers: Abundance of silent and cryptic loci. *Chromosome Res* 16: 595–607.

Cabrero, J., M. D. López-León, M. Bakkali, and J. P. M. Camacho. 1999. Common origin of B chromosome variants in the grasshopper *Eyprepocnemis plorans*. *Heredity* 83: 435–439.

Cabrero, J., M. D. López-León, M. Ruíz-Estévez, R. Gómez, E. Petitpierre, J. S. Rufas, B. Massa, M. K. Ben Halima, and J. P. M. Camacho. 2014. B$_1$ was the ancestor B chromosome variant in the western Mediterranean area in the grasshopper *Eyprepocnemis plorans*. *Cytogenet Genome Res* 142: 52–58.

Cabrero, J., M. D. López-León, M. Teruel, and J. P. M. Camacho. 2009. Chromosome mapping of H3 and H4 histone gene clusters in 35 species of acridid grasshoppers. *Chromosome Res* 17: 397–404.

Cabrero, J., F. Perfectti, R. Gómez, J. P. M. Camacho, and M. D. López-León. 2003. Population variation in the A chromosome distribution of satellite DNA and ribosomal DNA in the grasshopper *Eyprepocnemis plorans*. *Chromosome Res* 11: 375–381.

Camacho, J. P. M. 1980. Variabilidad cromosómica en poblaciones naturales de Tettigonioidea, Pamphagoidea y Acridoidea. PhD Thesis. Granada, Spain: Universidad de Granada.

Camacho, J. P. M. 2004. *B Chromosomes in the Eukaryote Genome*. Basel, Switzerland: Karger.

Camacho, J. P. M. 2005. B chromosomes. In *The Evolution of the Genome*, edited by T. R. Gregory, pp. 223–286. San Diego, CA: Elsevier.

Camacho, J. P. M., M. Bakkali, J. M. Corral, et al. 2002. Host recombination is dependent on the degree of parasitism. *Proc Biol Sci* 269: 2173–2177.

Camacho, J. P. M., and J. Cabrero. 1982. Supernumerary segments in five species of grasshoppers (Orthoptera: Acridoidea). *Genetica* 59: 113–117.

Camacho, J. P. M., J. Cabrero, E. Viseras, M. D. López-León, J. Navas-Castillo, and J. D. Alché. 1991. G-banding in two species of grasshoppers and its relationship to C, N, and fluorescence banding techniques. *Genome* 34: 638–643.

Camacho, J. P. M., T. F. Sharbel, and L. W. Beukeboom. 2000. B-chromosome evolution. *Phil Trans R Soc Lond B* 355: 163–178.

Cano, M. I., G. H. Jones, and J. L. Santos. 1987. Sex differences in chiasma frequency and distribution in natural populations of *Eyprepocnemis plorans* containing B-chromosomes. *Heredity* 59: 237–243.

Cano, M. I. and J. L. Santos. 1988. B chromosomes of the grasshopper *Heteracris littoralis*: Meiotic behaviour and endophenotypic effects in both sexes. *Genome* 30: 797–801.

Cano, M. I. and J. L. Santos. 1989. Cytological basis of the B chromosome accumulation mechanism in the grasshopper *Heteracris littoralis* (Ramb). *Heredity* 62: 91–95.

Cano, M. I. and J. L. Santos. 1990. Chiasma frequencies and distributions in gomphocerine grasshoppers: A comparative study between sexes. *Heredity* 64: 17–23.

Cardoso H., F. A. Saez, and N. Brum-Zorrilla. 1974. Location, structure and behaviour of C-heterochromatin during meiosis in *Dichroplus silveiraguidoi* (Acrididae-Orthoptera). *Chromosoma* 48: 51–64.

Castillo, E. R., D. A. Martí, and C. J. Bidau. 2010. Sex and neo-sex chromosomes in Orthoptera: A review. *J Orthoptera Res* 19: 213–231.

Castillo, E. R., A. Taffarel, and D. A. Martí. 2011. Una técnica alternativa para le caritipado mitótico en saltamontes: bandeo C y Fluorescente en *Adimantus ornatissimus* (Orthoptera: Acrididae). *Rev Cienc Tecnol* 16: 31–35.

Church, K. and D. E. Wimber. 1969. Meiosis in the grasshopper: Chiasma frequency after elevated temperature and x-rays. *Can J Genet Cytol* 11: 209–216.

Colgan, D. J., A. McLauchlan, G. D. F. Wilson et al. 1998. Histone H3 and U2 snRNA DNA sequences and arthropod molecular evolution. *Austral J Zool* 46: 419–437.

Contreras, D. and W. Chapco. 2006. Molecular phylogenetic evidence for multiple dispersal events in gomphocerine grasshoppers. *J Orthoptera Res* 15: 91–98.

Corey, H. I. 1938. Heteropycnotic elements of orthopteran chromosomes. *Arch Biol* 49: 159–172.

Cremer, T., P. Lichter, J. Borden, D. C. Ward, and L. Manuelidis. 1988. Detection of chromosome aberrations in metaphase and interphase tumor cells by in situ hybridization using chromosome specific library probes. *Hum Genet* 80: 235–246.

Crozier, R. 1968. An acetic acid dissociation air-drying technique for insect chromosomes, with aceto–lactic orcein staining. *Stain Technol* 43: 171–173.

Dearn, J. M. 1974. Phase transformation and chiasma frequency variation in locusts. II. *Locusta migratoria*. *Chromosoma* 45: 339–352.

Dolezel, J., J. Bartos, H. Voglmayr, and J. Greilhuber. 2003. Nuclear DNA content and genome size of trout and human. *Cytometry A* 51: 127–128.

Dolezel, J., J. Greilhuber, S. Lucretti et al. 1998. Plant genome size estimation by flow cytometry: Inter-laboratory comparison. *Annals Bot* 82 (Suppl A): 17–26.

Fletcher, H. L. and G. M. Hewitt. 1980. A comparison of chiasma frequency and distribution between sexes in three species of grasshoppers. *Chromosoma* 77: 129–144.

Fox, D. P. and J. L. Santos. 1985. N-bands and nucleolus expression in *Schistocerca gregaria* and *Locusta migratoria*. *Heredity* 54: 333–341.

Gall, J. and M. L. Pardue. 1969. Formation and detection of RNA-DNA hybrid molecules in cytological preparations. *Proc Natl Acad Sci USA* 63: 378–383.

Geraci, N. S., J. Spencer Johnston, J. Paul Robinson, S. K. Wikel, and C. A. Hill. 2007. Variation in genome size of argasid and ixodid ticks. *Insect Biochem Mol Biol* 37: 399–408.

Goodpasture, C. and S. E. Bloom. 1975. Visualization of nucleolar organizer regions in mammalian chromosomes using silver staining. *Chromosoma* 53: 37–50.

Hanrahan, S. J. and J. S. Johnston. 2011. New genome size estimates of 134 species of arthropods. *Chromosome Res* 19: 809–823.

Hardie, D. C., T. R. Gregory, and P. D. Hebert. 2002. From pixels to picograms: A beginners' guide to genome quantification by Feulgen image analysis densitometry. *J Histochem Cytochem* 50: 735–749.

Henriques-Gil, N., G. H. Jones, M. I. Cano, P. Arana, and J. L. Santos. 1987. Female meiosis during oocyte maturation in *Eyprepocnemis plorans* (Orthoptera: Acrididae). *Can J Genet Cytol* 28: 84–87.

Hewitt, G. M. 1976. Meiotic drive for B-chromosomes in the primary oocytes of *Myrmeleotettix maculatus* (Orthoptera: Acrididae). *Chromosoma* 56: 381–391.

Hewitt, G. M. 1979. Grasshopper and crickets. In *Animal Cytogenetics*, Insecta 1 Orthoptera, Vol. 3, edited by B. John. Berlin, Germany: Gebruder Borntraeger.

Ijdo, J. W., R. A. Wells, A. Baldini, and S. T. Reeders. 1991. Improved telomere detection using a telomere repeat probe (TTAGGG)*n* generated by PCR. *Nucleic Acids Res* 19: 4780.

John, B. and G. M. Hewitt. 1965. The B-chromosome system of *Myrmeleotettix maculatus* (Thunb.). I. The mechanics. *Chromosoma* 16: 548–578.

John, B., M. King, D. Schweizer, and M. Mendelak. 1985. Equilocality of heterochromatin distribution and heterogeneity in acridid grasshoppers. *Chromosoma* 91: 185–200.

Jones, R. N. and H. Rees. 1982. *B Chromosomes*. New York: Academic Press.

King, M. and B. John. 1980. Regularities and restrictions governing C-band variation in acridoid grasshoppers. *Chromosoma* 76: 123–150.

Leavitt, J. R., K. D. Hiatt, M. F. Whiting, and H. Song. 2013. Searching for the optimal data partitioning strategy in mitochondrial phylogenomics: A phylogeny of Acridoidea (Insecta: Orthoptera: Caelifera) as a case study. *Mol Phylogenet Evol* 67: 494–508.

Lichter, P, T. Cremer, J. Borden, L. Manuelidis, and D. C. Ward. 1988. Delineation of individual human chromosomes in metaphase and interphase cells by in situ suppression hybridization using chromosome specific library probes. *Hum Genet* 80: 224–234.

López-León, M. D., J. Cabrero, V. V. Dzyubenko, A. G. Bugrov, T. V. Karamysheva, N. B. Rubtsov, and J. P. M. Camacho. 2008. Differences in ribosomal DNA distribution on A and B chromosomes between eastern and western populations of the grasshopper *Eyprepocnemis plorans plorans*. *Cytogenet Genome Res* 121: 260–265.

López-León, M. D., J. Cabrero, M. C. Pardo, E. Viseras, J. P. M. Camacho, and J. L. Santos. 1993. Generating high variability of B chromosomes in *Eyprepocnemis plorans* (grasshopper). *Heredity* 71: 352–362.

Loreto, V., J. Cabrero, M. D. López-León, J. P. M. Camacho, and M. J. Souza. 2008. Possible autosomal origin of macro B chromosomes in two grasshopper species. *Chromosome Res* 16: 233–241.

Magelhaes, P. J., S. J. Ram, and M. D. Abramoff. 2004. Image processing with ImageJ. *Biophotonics Int* 11: 36–42.

Manrique-Poyato, M. I., M. D. López-León, R. Gómez, F. Perfectti, and J. P. M. Camacho. 2013. Population genetic structure of the grasshopper *Eyprepocnemis plorans* in the south and east of the Iberian Peninsula. *PloS One* 8: e59041.

Marchal, J. A., M. J. Acosta, H. Nietzel et al. 2004. X chromosome painting in Microtus: Origin and evolution of giant sex chromosomes. *Chromosome Res* 12: 767–776.

McClung, C. E. 1902. The accessory chromosome—sex determinant? *Biol Bull* 3: 43–84.

Meredith, R. 1969. A simple method for preparing meiotic chromosomes from mammalian testis. *Chromosoma* 26: 254–258.

Miller, K., C. Lynch, J. Martin, E. Herniou, and M. Tristem. 1999. Identification of multiple Gypsy LTR-retrotransposon lineages in vertebrate genomes. *J Mol Evol* 49: 358–366.

Montiel, E. E., J. Cabrero, J. P. M. Camacho, and M. D. López-León. 2012. Gypsy, RTE and Mariner transposable elements populate *Eyprepocnemis plorans* genome. *Genetica* 140: 365–374.

Muñoz-Pajares, A. J., L. Martínez-Rodríguez, M. Teruel, J. Cabrero, J. P. M. Camacho, and F. Perfectti. 2011. A single, recent origin of the accessory B chromosome of the grasshopper *Eyprepocnemis plorans*. *Genetics* 187: 853–863.

Nicklas, R. B. 1961. Recurrent pole-to-pole movements of the sex chromosome during prometaphase I in *Melanoplus differentialis* spermatocytes. *Chromosoma* 12: 97–115.

Palacios-Gimenez, O. M, E. R. Castillo, D. A. Martí, and D. C. Cabral-de-Mello. 2013. Tracking the evolution of sex chromosome systems in Melanoplinae grasshoppers through chromosomal mapping of repetitive DNA sequences. *BMC Evol Biol* 13: 167.

Pineau, P., M. Henry, R. Suspène et al. 2005. A universal primer set for PCR amplification of nuclear histone H4 genes from all animal species. *Mol Biol Evol* 22: 582–588.

Pinkel, D., J. Landegent, C. Collins et al. 1988. Fluorescence in situ hybridization with human chromosome specific libraries: Detection of trisomy 21 and translocations of chromosome 4. *Proc Natl Acad Sci USA* 85: 9138–9142.

Rebollo, E. and P. Arana. 1995. A comparative study of orientation at behavior of univalent in living grasshopper spermatocytes. *Chromosoma* 104: 56–67.

Rebollo, E., S. Martín, S. Manzanero, and P. Arana. 1998. Chromosomal strategies for adaptation to univalency. *Chromosome Res* 6: 515–531.

Rentz, D. C. F. 1991. Orthoptera. In *Insects of Australia,* edited by CSIRO. Melbourne, Australia: Melbourne University Press.

Rufas, J. S., P. Iturra, W. De Souza, and P. Esponda. 1982. Simple silver staining procedures for the localization of nucleolus and nucleolar organizers under light and electron microscopy. *Arch Biol* 93: 267–274.

Ruiz-Ruano, F. J., M. Ruiz-Estévez, J. Rodríguez-Pérez, J. L. López-Pino, J. Cabrero, and J. P. M. Camacho. 2011. DNA amount of X and B chromosomes in the grasshoppers *Eyprepocnemis plorans* and *Locusta migratoria*. *Cytogenet Genome Res* 134: 120–126.

Santos, J. L., P. Arana, and R. Giraldez. 1983. Chromosome C-banding patterns in Spanish Acridoidea. *Genetica* 61: 65–74.

Schubert, I. 1984. Mobile nucleolus organizing regions (NORs) in *Allium* (Liliaceae s. lat.)?—Inferences from the specifity of silver staining. *Plant Syst Evol* 144: 291–305.

Schubert, I. and U. Wobus. 1985. In situ hybridization confirms jumping nucleolus organizing regions in *Allium*. *Chromosoma* 92: 143–148.

Schwarzacher, T. and J. S. Heslop-Harrison. 2000. *Practical In Situ Hybridization*. Oxford, United Kingdom: BIOS Scientific.

Schweizer, D. 1980. Simultaneous fluorescent staining of R bands and specific heterochromatic regions (DA-DAPI bands) in human chromosomes. *Cytogenet Cell Genet* 27: 190–193.

Schweizer, D. 1981. Counterstain-enhanced chromosome banding. *Hum Genet* 57: 1–14.

Schweizer, D., M. Mendelak, M. J. D. White, and N. Contreras. 1983. Cytogenetics of the parthenogenetic grasshopper *Warramaba virgo* and its bisexual relatives. X. Patterns of fluorescent banding. *Chromosoma* 88: 227–236.

Sesarini, S. and M. I. Remis. 2008. Molecular and morphometric variation in chromosomally differentiated populations of the grasshopper *Sinipta dalmani* (Orthopthera: Acrididae). *Genetica* 133: 295–306.

Song, H. 2010. Grasshopper systematics: Past, present and future. *J Orthoptera Res* 19: 57–68.

Sumner, A. T. 1972. A simple technique for demonstrating centromeric heterochromatin. *Exp Cell Res* 75: 304–306.

Telenius, H., N. P. Carter, C. E. Bebb, M. Nordenskjold, B. A. Ponder, and A. Tunnacliffe. 1992. Degenerate oligonucleotide-primed PCR: General amplification of target DNA by a single degenerate primer. *Genomics* 13: 718–725.

Teruel, M. 2009. *Origen, expresión y efectos fenotípicos de un parásito genómico*. PhD Thesis. Granada, Spain: Editorial de la Universidad de Granada.

Teruel, M., J. Cabrero, E. E. Montiel, M. J. Acosta, A. Sánchez, and J. P. M. Camacho. 2009a. Microdissection and chromosome painting of X and B chromosomes in *Locusta migratoria*. *Chromosome Res* 17: 11–18.

Teruel, M., J. Cabrero, F. Perfectti, M. J. Acosta, A. Sánchez, and J. P. M. Camacho. 2009b. Microdissection and chromosome painting of X and B chromosomes in the grasshopper *Eyprepocnemis plorans*. *Cytogenet Genome Res* 125: 286–291.

Teruel, M., J. Cabrero, F. Perfectti, and J. P. M. Camacho. 2010. B chromosome ancestry revealed by histone genes in the migratory locust. *Chromosoma* 119: 217–225.

Teruel, M., F. Ruiz-Ruano, J. A. Marchal, A. Sánchez-Baca, J. Cabrero, J. P. M. Camacho, and F. Perfectti. 2014. Disparate molecular evolution of two types of repetitive DNA in the genome of the grasshopper Eyprepocnemis plorans. *Heredity* 112: 531–542.

Van Huis, A., J. Van Itterbeeck, H. Klunder et al. 2013. Edible insects: Future prospects for food and feed security. *FAO Forestry paper* 171.

Wang, X., X. Fang, P. Yang et al. 2014. The locust genome provides insight into swarm formation and long-distance flight. *Nature Commun* 5: 2957.

Westerman, M., N. H. Barton, and G. M. Hewitt. 1987. Differences in DNA content between two chromosomal races of the grasshopper *Podisma pedestris. Heredity* 58: 221–228.

White, M. J. D. 1973. *Animal Cytology and Evolution*, Ed. 3rd . London, United Kingdom: Cambridge University Press.

Wilhelm, J., A. Pingoud, and M. Hahn. 2003. Real-time PCR-based method for the estimation of genome sizes. *Nucleic Acids Res* 31: e56.

Zurita, F., R. Jiménez, M. Burgos, and R. D. de la Guardia. 1998. Sequential silver staining and in situ hybridization reveal a direct association between rDNA levels and the expression of homologous nucleolar organizing regions: A hypothesis for NOR structure and function. *J Cell Sci* 111: 1433–1439.

Zwick M. S., R. E. Hanson, T. D. McKnight, M. Nurul-Islam-Faridi, D. M. Stelly. 1997. A rapid procedure for the isolation of C0t-1 DNA from plants. *Genome* 40: 138–142.

12

Ticks (Ixodida)

Monika Gulia-Nuss, Jason M. Meyer,
and Catherine A. Hill

CONTENTS

LIST OF ABBREVIATIONS

BAC, bacterial artificial chromosome
cDNA, complementary deoxyribonucleic acid
cM, centimorgan
DAPI, 4′,6-diamidino-2-phenylindole
DIG, digoxigenin
DNA, deoxyribonucleic acid
EDTA, ethylenediaminetetraacetic acid
EST, expressed sequence tag
FISH, fluorescence in situ hybridization
Gb, gigabase
gDNA, genomic DNA
HGA, human granulocytic anaplasmosis
ISE18, *Ixodes scapularis* cell line 18
ISR, *Ixodes scapularis* repeat
LD, Lyme disease
Mb, megabase
NGS, next-generation sequencing
NIH, National Institutes of Health
NOR, nucleolar organizing region
PCR, polymerase chain reaction

POW, Powassan virus
RAD, restriction site-associated DNA
RADseq, RAD sequencing
RAPD, random amplified polymorphic DNA
RMR, *Rhipicephalus microplus* repeat
RMSF, Rocky Mountain spotted fever
rRNA, ribosomal ribonucleic acid
RT, room temperature
SNP, single nucleotide polymorphism
SSC, saline sodium citrate
STARI, Southern Tick-Associated Rash Illness
TBE, tris-borate-EDTA acid
TE, transposable element
TR, tandem repeat
USDA, United States Department of Agriculture

12.1 INTRODUCTION

Hard and soft ticks (subphylum: Chelicerata; subclass: Acari; super-order: Parasitiformes; superfamily: Ixodida) are hematophagous ectoparasites of animals. Among arthropods, it is reported that ticks are second only to mosquitoes in terms of their public health impact (Hoogstraal 1956; Sonenshine et al. 2002; Goodman et al. 2005). Many species of ticks are of medical and/or veterinary significance due to their ability to transmit bacteria, protozoa, fungi, and viruses, which cause diseases in humans and animals. Tick attachment and feeding is also associated with physical damage to the host dermis, blood loss, and secondary infections, as well as paralysis caused by toxins transferred in tick saliva. Although the karyotype of multiple hard and soft ticks was reported in the early work of Oliver (1977), cytogenetic studies involving ixodid ticks have been limited. Published protocols for chromosome preparation, staining, and physical mapping of specific DNA sequences are available for species in two genera in the family Ixodidae (hard ticks), *Ixodes* and *Rhipicephalus*. These studies involved use of repetitive DNA as probes and, primarily, localization using fluorescence in situ hybridization (FISH) techniques. The *Ixodes scapularis* (Lyme disease [LD] tick) genome assembly (*Ixodes scapularis* Genome Consortium, unpublished data) is the first available for a tick. Ongoing efforts to integrate the *I. scapularis* sequence, genetic and physical maps will significantly advance understanding of the chromosome biology of this and other tick species, and promote studies to determine the genetic basis of processes involved in tick–host–pathogen interactions that could be exploited to achieve tick and tick-borne disease control. With the advent of next-generation sequencing (NGS) techniques, genome sequence data and assemblies are anticipated for multiple species of ixodid ticks. The demand for cytogenetic techniques to assist the assembly and interpretation of these data within a chromosomal context among members of the superfamily is expected to expand considerably. Here we present an overview of protocols used for cytogenetic work in two species of hard ticks, the prostriate tick *I. scapularis* and the metastriate tick *Rhipicephalus* (*Boophilus*) *microplus,* and discuss the potential application of these approaches for work in other species of ixodid ticks.

12.1.1   Taxonomy and Importance of Ixodid Ticks

Three families are recognized within the superfamily Ixodida, namely the Ixodidae (hard ticks), Argasidae (soft ticks), and Nuttalliellidae. Approximately 80% of the world's tick fauna are ixodid ticks (683 species), whereas 183 species are classified as argasid ticks and a single species is recognized within the Nuttalliellidae (Horak et al. 2002). The family Ixodidae, which is the focus of this chapter, comprises two major linages, the Prostriata (241 species) consisting of the single genus *Ixodes,* and the Metastriata (442 species) inclusive of the genera *Amblyomma, Anomalohimalaya, Boophilus, Bothriocroton, Cosmiomma, Dermacentor, Haemaphysalis, Hyalomma, Margaropus, Nosomma, Rhipicentor,* and *Rhipicephalus* (Klompen et al. 1996). Horak et al. (2002) reclassified five species of *Boophilus* as members of the genus *Rhipicephalus,* and the name *Boophilus* is commonly used as a subgenus when referring to these species (Barker and Murrell 2002; Horak et al. 2002). The family Argasidae includes the genera *Argas, Carios, Ornithodoros,* and *Otobius* (Camicas et al. 1998; Horak et al. 2002).

Ticks classified in the families Ixodidae and Argasidae differ in many aspects of their natural history; however, all life-cycle stages require a blood meal from a vertebrate host to complete development. The larvae, nymphs, and adults of ixodid ticks live primarily off-host in leaf litter and vegetation associated with host habitat. Depending on the species, ixodid ticks typically attach and feed to repletion on their vertebrate host over days or weeks. Molting may occur either on-host as is the case for one-host ticks and immature stages of two-host ticks, or all stages may molt off-host (i.e., three-host ticks). When molting occurs off-host, the subsequent developmental stage must locate and attach to a new host to complete development. In comparison, the nymphs and adults of soft ticks live in close proximity to their vertebrate hosts, usually in nests or burrows of small vertebrates, and each developmental stage can take multiple blood meals, remaining attached to the host for an average of 30–70 minutes during feeding. Female hard ticks produce a single egg batch of several hundred to several thousand eggs, whereas female soft ticks are capable of producing multiple, smaller egg batches (Seraji-Bozorgzad and Tselis 2013).

Hard and soft ticks can significantly impact the health of their vertebrate hosts. Tick attachment and blood feeding can lead to anemia, severe dermatitis, and the development of secondary infections at the feeding wound. High tick burdens may also result in reduced host body weight, and milk and meat production, leading to significant losses in animal production. In addition, many species of ticks are highly competent vectors of bacteria, viruses, and protozoa and can impact the health of the host indirectly via transmission of the causative agents of many human and animal diseases. An extensive review of tick-borne diseases and their associated tick vectors is beyond the scope of this chapter; for further information, the reader is referred to the work of Jongejan and Uilenberg (2004).

Members of the genus *Ixodes* are three-host ticks. The *I. xodes ricinus* species complex is one of the most important affecting public health globally, and includes competent vectors of several types of human pathogens, most notably bacteria in the genus *Borrelia* that cause Lyme borreliosis in North America, Europe, and Asia (Delaye et al. 1997). LD caused by *Borrelia burgdorferi* is the most prevalent vector-borne disease in the United States. Over 33,000 human LD cases were reported to the Centers for Disease Control and Prevention (CDC) in 2011 (CDC 2012), and it is suspected that approximately 10-fold more infections are underreported or misdiagnosed by clinicians (Walker 1998). Recent data collected by the CDC and presented at the 2013 International Conference on Lyme Borreliosis and Other Tick-borne Diseases (http://iclb2013 .com/overview.htm) provide additional support for this estimate. In the eastern and central regions of the United States, LD is transmitted by the black-legged or LD tick, *I. scapularis,* whereas the western black-legged tick, *I. xodes pacificus*, serves as the vector of LD in the western United States. In Europe and Asia, *Borrelia* bacteria are transmitted by *I. ricinus* and *I. xodes persulcatus*, respectively (Jongejan and Uilenberg 2004). Lyme borreliosis has a similar impact on public health in Europe with approximately 100,000 cases reported in 2002 (Lindgren and Jaenson 2006). In North America, *I. scapularis* also transmits the bacterium *Anaplasma phagocytophilum* that causes human granulocytic anaplasmosis, the protozoan *Babesia microti* that causes human babesiosis, and the flavivirus that causes Powassan encephalitis. Investments in the sequencing of the *I. scapularis* genome and complementary genomics research reflect the fact that this tick is arguably one of the most important vectors affecting human health in the United States. The *I. scapularis* assembly (*Ixodes scapularis* Genome Consortium, unpublished data; see discussion below) serves as a valuable reference for genomic work aimed at improving our understanding of the biology of this and other tick vectors of disease.

Species of *Rhipicephalus* are one-host ticks; all life-cycle stages feed and complete development on the same vertebrate host. The tropical or southern cattle tick, *Rhipicephalus (Boophilus) microplus* (henceforth *Rhipicephalus microplus*), is an important pest of livestock and wildlife, particularly in the southern hemisphere. This tick is the vector of the protozoan, *Babesia bovis,* and the bacterium, *Anaplasma marginale*, which cause bovine babesiosis and anaplasmosis, respectively. These diseases reduce milk and beef production in affected cattle and are associated with severe economic losses in livestock production. Feeding wounds caused by *R. microplus* also reduce the value of hides used for leather manufacturing (Jongejan and Uilenberg 2004). Efforts by cattle producers and the United States Department of Agriculture (USDA) to control these diseases and prevent the reestablishment of *Rhipicephalus microplus* in the United States exceed more than $2.5 billion annually (Foil et al. 2004).

12.1.2   Karyotypes

Oliver (1977) pioneered work on the cytogenetics of the Acari and determined the karyotype of 27 tick species. Karyotypes were reported for eight genera of hard ticks, including eight *Ixodes* species and two *Boophilus* species. Among the soft ticks, cytogenetic data are available for the genera *Ornithodoros* (11 species), *Argas* (11 species), and *Otobius* (two species). In all cases where it was possible to recognize the sex chromosomes, the male was invariably the heterogametic sex (Oliver 1977). Most hard ticks studied have an XX–XO (female–male) sex-determination system (Dutt 1954; Kahn 1964; Oliver and Bremner 1968), whereas sex determination is typically XX–XY in soft ticks (Goroschenko 1962). Of the tick species examined by Oliver (1977), the number of somatic chromosomes ranged from 2 to 36 and sex chromosome systems included XX–XY, XX–XO, and $X_1X_1X_2X_2–X_1X_2Y$ variants. Both monokinetic (i.e., one centromere or kinetochore per chromosome) and holokinetic (i.e., centromere or kinetochore activity diffused along the entire length of the chromosome) chromosomes were observed.

Cytogenetic analyses of *I. scapularis* chromosomes revealed a $2n = 28$ karyotype (Oliver et al. 1993) and, interestingly, an XX–XY sex-determination system. This finding was further supported by chromosomal studies using C- and G-banding patterns by both Chen et al. (1994) and Munderloh et al. (1994). Using mitotic chromosomes derived from cell lines, Chen et al. (1994) observed that the X and Y chromosomes were the largest and smallest in the karyotype, respectively, and localized the ribosomal DNA (rDNA) repeat unit associated with the nucleolar organizing regions (NORs) to two pairs of autosomes and the X chromosome. In *R. microplus*, Oliver (1977) reported an XX–XO sex-determination system with 22 diploid chromosomes in females and 21 in males. Preliminary karyotypes based on silver staining and C-banding patterns were developed for *R. microplus* from meiotic chromosome spreads obtained from the testes of newly molted adult males by Hilburn et al. (1989) and Garcia et al. (2002).

More recently, FISH has proved a useful approach for mapping specific DNA sequences to tick chromosomes. For example, Vítková et al. (2005) used FISH to map a $(TTAGG)_n$ telomeric repeat to the terminal heterochromatic regions of *I. ricinus* chromosomes. This finding suggested that the mechanism of telomere maintenance reported for many arthropods (Okazaki et al. 1993; Meyne et al. 1995; Sahara et al. 1999) is also conserved in ixodid ticks. Hill et al. (2009) produced the first FISH-based karyotype for *R. microplus* using the NORs and several classes of sub-telomeric-localizing tandem repeat (TR) units. In brief, the karyotype comprised the X chromosome and three groups of autosomes, which were indistinguishable based on length and hybridization intensity. Subsequently, Meyer et al. (2010) produced the first FISH-based karyotype for *I. scapularis* by localization of multiple families of major TRs and structural features of the *I. scapularis* chromosomes associated with repetitive DNA, including the NORs, telomere, and presumed peri-centromeric region. In addition to the sex chromosomes, the karyotype distinguished four individual autosomes and four groups of autosomes displaying highly similar hybridization patterns. The above-described karyotypes provide an important starting point for studies to understand genome organization and evolution in prostriate and metastriate ticks. Progress toward these goals requires the development of additional markers to permit the identification of each chromosome in the respective karyotypes.

12.1.3 Genome Sequencing Projects

Advances have been made toward the development of complete genome sequence projects for species of mites that have small haploid genomes with a low content of repetitive DNA relative to sequenced arthropods. Notably, the 75 Mb genome of the two-spotted spider mite, *Tetranychus urticae* (superorder Acariformes) was sequenced and assembled, providing the first such genomic resource for a Chelicerate species that facilitated important insight into the biology of this phytophagous mite and major agricultural pest (Grbić et al. 2011). The *T. urticae* assembly provides an important framework for *de novo* genome sequencing efforts involving other acarine species. Other notable efforts include that for the *Varroa destructor* mite parasite of honeybees (Cornman et al. 2010) and the phytoseiid predatory mite, *Metaseiulus occidentalis* (M. A. Hoy, pers. comm., November, 11, 2012). Initial 454 pyrosequencing of *V. destructor* to 4.3-fold genome coverage produced an N_{50} contig length of 2,262 bp, whereas NGS-based transcriptome analysis was undertaken for *M. occidentalis* (Hoy et al. 2013). The *V. destructor* and *M. occidentalis* projects will underpin efforts toward complete assemblies for these and other mite species.

The *I. scapularis* genome sequence, funded by the National Institutes of Health (NIH), is the first genome assembly for a tick and noninsect arthropod. The assembly and automated annotation was produced via a joint effort between the J. Craig Venter Institute, the Broad Institute of Massachusetts Institute of Technology/Harvard, and VectorBase (www.vectorbase.org), and analysis of the genome was undertaken in collaboration with the international tick research community (Hill and Wikel 2005; *Ixodes scapularis* Genome Consortium, unpublished data). Genome sequencing was performed using the Sanger whole genome shotgun method and the assembly and annotation statistics are summarized in Table 12.1. The current IscaW1 assembly, comprising nearly 369,495 scaffolds (N_{50} scaffold length ~52 kb) was produced from more than 17 million trace reads using an

---

TABLE 12.1

Ixodes scapularis **Genome Assembly and Annotation Statistics**

IscaW1 Assembly Statistics

Estimated genome size	2.1 Gb
Total number of sequence reads	17.4 M
Estimated fold coverage of the assembly	3.8-fold
Number of scaffolds	369,495
N_{50} scaffold length	51,551 bp
Total length of combined contigs	1.4 Gb
Total length of combined scaffolds (including gaps)	1.8 Gb

Annotation Release IscaW1.2 Statistics

Total number of genes	24,925
Protein-coding genes	20,486
Mean gene length	10,589 bp
Mean coding sequence length	855 bp

Source: Ixodes scapularis Genome Consortium, unpublished.

innovative Celera assembly algorithm to account for heterogeneity in the donor DNA. Also produced as part of the effort were paired-end sequence reads from more than 180,000 bacterial artificial chromosome (BAC) clones, 45 sequenced and assembled BACs, and approximately 200,000 expressed sequence tags. The assembly represents approximately 3.8-fold coverage of approximately 2.1 Gb *I. scapularis* haploid genome. The IscaW1.2 annotation includes the automated prediction of 20,486 gene models and 4,439 noncoding RNA genes, and revealed expansions of gene families associated with tick–host interactions, and neurological, developmental, and chemoreception processes likely unique to chelicerates.

The *I. scapularis* genome assembly and annotation are an invaluable resource for tick genome research. The assembly fragmentation is reflective of the high repeat content of the genome. Regions of low sequence complexity proved difficult to resolve using existing assembly technologies and were likely collapsed during the assembly process. Consequently, approximately one-third of the *I. scapularis* genome is not represented by assembled sequence. The identification of genome features located within these regions such as heterochromatin, telomeres, and gene models will be problematic, and the lack of large contiguous stretches of assembled sequence (i.e., scaffolds of >1 Mb) is an impediment to studies of genome organization. The greater number of *I. scapularis* gene model predictions as compared to many sequenced arthropods (*Ixodes scapularis* Genome Consortium, unpublished data) likely reflects the existence of split and incomplete gene models. Characterization of the repertoire of *I. scapularis* coding sequences and associated gene products will require further sequencing of the transcriptome and experimental validation via techniques such as polymerase chain reaction (PCR) (Ullmann et al. 2008). To increase the utility of this resource, in 2011, the NIH approved a subsequent initiative to improve the *I. scapularis* assembly and annotation via additional deep sequencing of the nuclear genome and transcriptome (Hill 2010). Also approved were parallel efforts to sample the transcriptomes of the prostriate tick vectors, *I. pacificus, I. ricinus,* and *I. persulcatus*; the metastriate tick vectors, *Amblyomma americanum* and *Dermacentor variabilis*; the soft tick, *Ornithodorus moubata*; and the mite vector of scrub typhus, *Leptotrombidium deliense.* These pilot projects are a first step toward expanding genome research in additional species that are important vectors of human and animal diseases, and that represent the major lineages of acarine species. The development of the first genome sequencing road map for the Ixodida highlights the need for physical mapping protocols to assist assembly efforts and support comparative, genome-level studies among members of the subclass Acari.

Several studies have begun to investigate the genome biology of metastriate ticks. Sequencing of the *R. microplus* genome was first proposed by Guerrero et al. (2006) with the hope of developing this important veterinary pest as a model for genomics work in the metastriate lineage. Recently, Guerrero et al. (2010) used a reassociation kinetics-based method to filter out much of the repetitive DNA comprising the *R. microplus* genome before performing NGS. One run on a GS-FLX machine generated approximately 0.1 coverage of the genome and an N_{50} contig length of 624 bp. Five *R. microplus* BAC clones were entirely sequenced, and both class I and class II transposable elements (TEs) were identified in the BAC assemblies. In addition to this USDA-led initiative, the genome sequencing road map of Hill (2010) calls for the expansion of genome project efforts to include other species of metastriate ticks, most notably the *A. americanum* vector of Southern Tick-Associated Rash Illness and the *D. variabilis* vector of Rocky Mountain Spotted Fever. Collectively, the above-described resources will have great utility for comparative genomics research as well as studies of biology and vector–host–pathogen interactions among species of the subclass Acari.

12.1.4   Genome Size and Organization

In general, ticks have large genomes. The haploid genome sizes of multiple ixodid and argasid ticks have been estimated as greater than 1 Gb (Palmer et al. 1994; Ullmann et al. 2005; Geraci et al. 2007), with *R. microplus* having the largest acarine genome estimated to date, at 7.5 pg (7.1 Gb) (Ullmann et al. 2005). The approximately 2.1 Gb haploid genome size of *I. scapularis*, first estimated via reassociation kinetics and flow cytometry, was confirmed by the assembly effort (*Ixodes scapulatis* Genome Consortium, unpublished data). Of the members in the suborder Ixodida for which genome sizes have been estimated, larger genomes were determined for species comprising the presumably more recently diverged family Ixodidae than in the more basal family Argasidae (Geraci et al. 2007). Furthermore, among the species analyzed, the metastriate ticks have larger genomes than do prostriate species. Together, these findings indicate a possible trend toward larger genome size in more recently diverged tick taxa, presumably reflecting the accumulation of repetitive DNA, but the mechanisms underlying this phenomenon remain unknown (Meyer and Hill 2014). Unfortunately, the large genome sizes of the ticks analyzed to date have important consequences for genome studies of ixodid ticks. High sequencing costs associated with ensuring adequate genome coverage and difficulties with assembling large, repeat-rich tick genomes have created a roadblock to genome sequencing initiatives (Meyer and Hill 2014).

The size of the *I. scapularis* genome reflects the accumulation of classes of highly and moderately repetitive DNA to an extent that is extreme among sequenced arthropods (*Ixodes scapularis* Genome Consortium, unpublished data). Repetitive DNA is estimated to comprise approximately 66% of the *I. scapularis* genome with the highly and moderately repetitive fractions accounting for approximately 27% and 39% of the genome, respectively (Ullmann et al. 2005). Sequencing efforts and cytogenetic studies have shown that the genome is composed of numerous copies of TRs and class I and class II TEs (Meyer et al. 2010; *Ixodes scapularis* Genome Consortium, unpublished data). Collectively, the low complexity *I. scapularis* Repeat families 1–3 (ISR-1–3), which range in size from 90 to 385 bp, account for an estimated 159 Mb (8%) of the genome and the distribution of these repeats on the chromosomes has been revealed using physical mapping (Meyer et al. 2010). The annotation of assembled *I. scapularis* BAC clones as part of the genome effort revealed low coding sequence density (~1.3% of DNA) within the cloned regions, and a similar density is expected across the euchromatin. The majority of BAC clones displayed nonspecific hybridization signals along each *I. scapularis* chromosome suggestive of high repeat content that is dispersed among euchromatic regions.

Like that of its prostriate relative, the genome of *R. microplus* also contains significant amounts of low-complexity DNA. Approximately 69% of the genome is estimated to comprise repetitive DNA, including 31% highly repetitive and 38% moderately repetitive sequences (Ullmann et al. 2005). Some work has been conducted to identify and characterize the complement of *R. microplus* TRs and TEs, although these efforts are hindered by the lack of a genome assembly for this tick. Two major repeat families, the *R. microplus* Repeats 1 and 2 (RMR-1 and RMR-2) were identified by sequencing of a small-insert genomic DNA (gDNA) library and developed as FISH probes to assess their distribution on the chromosomes (Hill et al. 2009). The 149 bp RMR-1 was localized to the subtelomeric regions of autosomes 1–6 and 8–10, whereas the RMR-2 family localized to the subtelomeric regions of all autosomes and the X chromosome, and was composed of three distinct repeat populations, RMR-2a (178 bp), RMR-2b (177 bp), and RMR-2c (216 bp).

The above-mentioned studies suggest that the genomes of both *I. scapularis* and *R. microplus* are permissive to the accumulation and retention of multiple classes of TRs and TEs. Given the estimates of genome size in representative species of ixodid and argasid ticks (Palmer et al. 1994; Ullmann et al. 2005; Geraci et al. 2007), this phenomenon is expected to apply broadly to members of the superfamily Ixodida. Although families of repetitive DNA are an important potential source of genetic markers and probes for chromosome identification, these sequences can complicate genome sequencing and assembly, physical and genetic mapping, and other approaches used to study aspects of genome biology.

12.1.5   Review of Genetic and Physical Maps

Genetic (or linkage) maps are important tools for reverse genetics studies to identify genes associated with traits of interest. Currently, little is known regarding the genetic basis of tick attributes such as vector competence, host preference, and insecticide resistance. Knowledge of the genes and gene products associated with phenotypes could greatly assist efforts to control tick and tick-borne diseases. The only genetic map available for a tick is that of the preliminary *I. scapularis* linkage map produced by Ullmann et al. (2003). The map was constructed based on segregation among 127 loci that were genotyped in 232 F1 offspring from a single female tick. The markers used to produce the map included 84 random amplified polymorphic DNA (RAPD), 32 sequence-tagged RAPD, 5 complementary deoxyribonucleic acid (cDNA), and 5 microsatellite markers. These markers segregated into 14 linkage groups that parallel the described haploid number of chromosomes in *I. scapularis*. The preliminary map of 616 cM included a marker interval of one marker per every 10.8 cM, and the relationship of physical to genetic distance was estimated at approximately 663 kb/cM. This map provides an important foundation for the development of a high-density *I. scapularis* linkage map. Beyond this effort, the development of suitable genetic markers and mapping populations to enable genetic mapping efforts across the superfamily Ixodida has been identified as a high priority for the tick research community (Ullmann et al. 2008; Meyer and Hill 2014). For a comprehensive review of genetic markers available for species of ixodid ticks and associated mapping resources, the reader is referred to the work of Meyer and Hill (2014).

The advent of NGS technologies has greatly facilitated the process of genetic mapping by allowing the rapid generation of dense linkage maps consisting of thousands of sequenced markers. More specifically, the technique of restriction-site-associated DNA sequencing (RADseq) (Baird et al. 2008; Hohenlohe et al. 2010) has been used to develop linkage maps for many eukaryote species based on the discovery of thousands of single nucleotide polymorphism (SNP)-based markers. In an effort funded by the NIH, the

RADseq approach is being used to develop the first high-density genetic map for *I. scapularis* (Gulia-Nuss et al. unpublished data). To date, thousands of SNPs have been identified in the *I. scapularis* Wikel reference strain used to produce the IscaW1 assembly. RADseq provides a cost-effective method for the genotyping of multiple individuals within a mapping population across SNP marker-tagged loci. In addition to genetic mapping, RADseq has facilitated population genomics studies in multiple species, including an ongoing effort led by the authors to evaluate genetic diversity in multiple *I. scapularis* populations collected from across North America (Gulia-Nuss, Meyer, Thimmapuram, and Hill, pers. comm.).

FISH is a powerful cytogenetic application used to determine the chromosomal position of specific DNA and RNA probes of multiple types. FISH-based physical maps are typically constructed by localizing scaffold-linked BAC or other gDNA clone types directly to the chromosomes, thus providing a visible physical position and information regarding the orientation of large stretches of assembled sequence. Such work can then be used to improve the assembly and direct gap closure efforts via a variety of methods that include the PCR and the sequencing of additional clones identified as spanning sequence gaps. DNA clones that are associated with genetic markers provide an opportunity to localize markers along the chromosomes and thus to integrate genetic, sequence, and physical maps (Jiang and Gill 2006). The development of preliminary FISH-based physical maps for *I. scapularis* and *R. microplus* by Hill et al. (2009) and Meyer et al. (2010), are first steps toward map integration for both species.

12.2   PROTOCOL

C-banding has been used to develop karyotypes for both *I. scapularis* and *R.microplus* (Hilburn et al. 1989; Garcia et al. 2002), but chromosome resolution is complicated due to uniformity in autosome length, lack of distinguishing morphological traits, and relatively uniform C-banding patterns. 4',6'-diamidino-2-phenylindole (DAPI) staining of *R. microplus* chromosomes suggested that significant amounts of heterochromatin are associated with the termini of chromosomes, although the distribution was too uniform to permit identification of individual chromosomes (Hill et al. 2009). Consequently, FISH-based mapping approaches were explored by Hill et al. (2009) and Meyer et al. (2010) in an attempt to advance cytogenetics work in ixodid ticks. Here, we review published protocols for FISH-based mapping in the prostriate species, *I. scapularis* (Meyer et al. 2010), and the metastriate tick, *R. microplus* (Hill et al. 2009). We discuss the development of chromosome preparations from mitotic and meiotic cells and the utility of multiple probe types derived from repetitive DNA. We fully expect that these protocols could be adapted for use with other species of ixodid ticks. The success of such efforts will largely depend on two factors: the availability of suitable chromosome preparations and sufficient DNA sequence information for probe design.

12.2.1   Materials

The following is a list of materials that are used routinely for FISH-based labeling of tick chromosomes:

1. Alexa Fluor 488-conjugated anti-biotin (Molecular Probes, Eugene, Oregon; Cat. No. S-32354)
2. Ampicillin (Sigma-Aldrich Corp., St. Louis, Missouri; Cat. No. A9393)
3. Biotin Nick Translation Mix (Roche, Indianapolis, Indiana; Cat. No. 11745824910)
4. Bovine Serum Albumin (BSA) (Sigma-Aldrich; Cat. No. A2153)
5. $CaCl_2$ 5.64 mM (Sigma-Aldrich; Cat. No. C1016)
6. Cover slips (VWR International, Arlington Heights, Illinois; Cat. No. 48393-048)
7. 4′,6-diamidino-2-phenylindole (DAPI) (Sigma-Aldrich; Cat. No. D9542)
8. Demecolcine (Sigma-Aldrich; Cat. No. D7385)
9. Dextran sulfate (Sigma-Aldrich; Cat.No. 42867)
10. DIG Nick Translation Mix (Roche; Cat. No. 11745816910)
11. Ethanol, 200 proof (Koptec, King of Prussia, Pennsylvania; Cat. No. V1016)
12. Formamide (≥99.5%) (Sigma-Aldrich; Cat. No. F7508)
13. Glacial acetic acid, 50% (Sigma-Aldrich; Cat. No. A9967)
14. KCl, 2.16 mM (Sigma-Aldrich; Cat. No. P9541)
15. Luria -Bertani broth (LB) media (Sigma-Aldrich; Cat. No. L3522)
16. $MgCl_2$, 200 mM (Sigma-Aldrich; Cat. No. M8266)
17. Microscope frosted glass slides (VWR International; Cat. No. 89049-670)
18. NaCl, 154 mM (Sigma-Aldrich; Cat. No. S3014)
19. Nick Translation Kit (Roche; Cat. No. 10976776001)
20. Pepsin, 100 mg/mL (Sigma-Aldrich; Cat. No. P7125)
21. Phosphate Buffered Saline (PBS) (Sigma-Aldrich; Cat. No. P5493)
22. QIAprep Spin Miniprep Kit (Qiagen, Valencia, California; Cat.No. 27106)

23. QIAquick Nucleotide Removal Kit (Qiagen; Cat.No. 28304)
24. Rhodomine α-DIG (Molecular Probes; Cat. No. S-6366)
25. RNAse A (Sigma-Aldrich; Cat. No. R4875)
26. Rubber cement (Elmer's Products Inc., Westerville, Ohio; Cat. No. 231)
27. Saline Sodium Citrate (SSC) 2× (Sigma-Aldrich; Cat. No. S6639)
28. Salmon Sperm DNA (ssDNA) Solution UltraPure™ (Invitrogen™, Life Technologies, Green Island, New York; Cat. No. 15632-011)
29. Tris-HCl, 100 mM (Sigma-Aldrich; Cat. No. T3253)
30. Tween 20 (Sigma-Aldrich; Cat. No. P1379)
31. VECTASHIELD with DAPI (Vector Laboratories, Burlingame, California; Cat. No. H-1200)
32. Wheaton Coplin staining jars (Sigma-Aldrich; Cat. No. S6016)

12.2.2 Source of Chromosomes

Polytene chromosomes are considered the best model for cytogenetic studies in insects as they contain thousands of longitudinally bound individual interphase chromosomes (chromatids) that are easily discernible by an optical microscope. Unfortunately, despite analysis of numerous tissues, researchers have failed to reliably identify polytene chromosomes in ticks (Ullmann et al. 2008). Highly condensed metaphase chromosome preparations have limited application for resolving in situ hybridization signals of DNA probes for cytogenetic studies. Chromosomes have been isolated from multiple species of ticks and mites (see Oliver [1977] for a comprehensive list). However, the suitability of these preparations for modern cytogenetics work has not been determined and obtaining sufficient chromosome material from some species may prove problematic, especially in the case where laboratory colonies are not available. Meiotic prophase chromosomes, which typically have less condensed chromatin, a well-defined axial core, and are considerably longer than mitotic metaphase chromosomes, can be used to detect proximal DNA sequences in whole-mount, surface-spread chromosomes (Spyropoulos and Moens 1994). Meiotic chromosome spreads have been prepared from the testes of *R. microplus* (Hill et al. 2009) and have also been observed in testes tissue isolated from *I. scapularis* males, although the percentage of cells undergoing meiosis is insufficient for labeling experiments on any meaningful scale in the latter species. Mitotic chromosome spreads prepared from cell lines have proved a useful alternative to meiotic spreads for initial cytogenetic studies in *I. scapularis* (Chen et al. 1994; Meyer et al. 2010). Cell lines have also been generated for other species of ticks (Bell-Sakyi et al. 2007) and these may serve as valuable resources to broaden cytogenetic initiatives in ixodid ticks.

12.2.2.1 *Ixodes scapularis* Chromosomes

Mitotic chromosomes have been obtained from the *I. scapularis* cell line ISE18 (Munderloh et al. 1994). Demecolcine (0.1 µg/mL) was added to cultures for 6–8 hours to arrest mitosis in metaphase and substantially increase yield of chromosome spreads for FISH; typically this approach can generate hundreds of spreads for analysis. Chromosome preparations were held at −20°C in fixative until use. For slide preparation, 10 µL of cell preparation was dropped onto cold (−20°C) microscope slides (one drop per slide) from a height of approximately 1 ft and the slides were air-dried for at least 24 hours. The density of cells and chromosome spreads on the slides can be adjusted by varying the relative amount of fixative according to the researchers' discretion and should be optimized for each preparation.

12.2.2.2 *Rhipicephalus microplus* Chromosomes

Meiotic chromosome spreads have been prepared from the *R. microplus* Deutsch strain (Hill et al. 2009). To obtain cells undergoing meiosis, the testes from 25 newly molted adult males were dissected under 0.5× Ringer's saline (154 mM NaCl; 5.64 mM $CaCl_2$; 2.16 mM KCl, pH 7.4), transferred to a 3:1 ethanol/glacial acetic acid solution for 5 minutes in a 1.5 mL centrifuge tube, and pelleted by centrifugation at 12×g for 5 minutes. Cells were resuspended in 50% glacial acetic acid and 5–10 µL drops were placed on microscope slides that had been chilled by storage at −20°C and the slides were allowed to air-dry.

Excess cytoplasmic material can complicate chromosome visualization and probe detection. This is particularly problematic when chromosomes are prepared from dissected tissue (e.g., *R. microplus* chromosomes), but is less of an issue when spreads are prepared from cell lines, as is the case for *I. scapularis* chromosomes. Cytoplasmic material was removed from *R. microplus* preparations by a series of washes in 200 µL 2× SSC and 0.5% RNase A at 37°C, followed by treatment with pepsin (100 mg/mL; Sigma-Aldrich) in 85 µL prewarmed 10 mM HCl at 37°C for 2 hours. Following incubation, slides were washed with 1× PBS, 200 mM $MgCl_2$, serially dehydrated in 70%, 90%, and 100% ethanol and air-dried. Slides can be stored at −80°C in a humid chamber up to 6 months.

12.2.3 Types of Probes

Several types of probes derived from classes of repetitive DNA have been used for FISH mapping in *I. scapularis* and/or *R.microplus*. These include (1) multiple classes of short (~100–400 bp) and presumably species-specific telomeric/sub-telomeric TRs, (2) the 28S rDNA repeat associated with the NOR, (3) the telomeric repeat (TTAGG)$_n$ motif that shows conservation among many arthropod species, and (4) the C_ot-1 fraction prepared from the rapidly annealing fold-back or highly repetitive fraction of gDNA. BAC clones containing approximately 120 kb DNA inserts have also been used successfully as probes in both species, and are particularly effective when the insert contains multiple copies of TRs. Centromeric probes prepared from fractionated gDNA have also been used for *I. scapularis*. Unfortunately, probes corresponding to single genes of interest have not proved successful in either species, presumably due to their short length (<10 kb) that falls below the limit of sensitivity for FISH mapping. The preparation of each of these probe types is discussed below.

12.2.3.1 Telomeric/Subtelomeric Tandem Repeat Probes

Small insert (~4 kb) gDNA clone libraries were prepared from sheared *I. scapularis* egg (Wikel strain) (two pulses of sonication for 30 seconds each) and *R. microplus* (Deutsch strain) DNA using the TOPO PCR 4.0 cloning vector (Invitrogen, Carlsbad, California). End sequencing of a 384-well plate from each library was conducted and the sequences are available at GenBank (accession numbers GU318418–GU319109 and FJ223571–FJ223604). Sequences composed of at least 100 bp of tandemly repetitive DNA were identified using Tandem Repeats Finder software (Benson 1999). Three classes of tandem repeats for *I. scapularis* (ISR-1–3; range 90–385 bp) (Meyer et al. 2010) and two for *R. microplus* (RMR 1–2; range 149–216 bp) (Hill et al. 2009) were used for FISH probes.

12.2.3.2 Ribosomal DNA Repeat Probe

rDNA primers designed based on the *R. microplus* 28S rDNA sequence obtained from GenBank (accession number AF200189) (forward primer: 5'-CTC TTG TGG TAG CCA AAT GC-3'; reverse primer: 5'-AAG CGA CGT CGC TAT GAA CG-3') (Hill et al. 2009) were used to PCR-amplify a portion of the 28S rRNA gene from *R.microplus* and *I. scapularis* gDNA. The PCR was performed using 1 μg gDNA and the following conditions: initial denaturation at 94°C for 2 minutes; 94°C for 10 seconds, 53°C for 30 seconds, and 72°C for 1 minute for 30 cycles; final extension at

72°C for 10 minutes. The resulting 749 bp *R. microplus* (Hill et al. 2009) and 629 bp *I. scapularis* (Meyer et al. 2010) amplicons were subcloned and analyzed.

12.2.3.3 (TTAGG)$_n$ Telomeric Repeat Probe

The (TTAGG)$_n$ telomeric sequence was selected as a probe for *I. scapularis* cytogenetics because of its reported conservation among multiple arthropod species, and the successful localization of this repeat to the chromosomes of *I. ricinus* (Vítková et al. 2005). A telomeric-specific probe was generated by the PCR using a method modified from Lorite et al. (2002) and Ijdo et al. (1991). This protocol did not require template DNA; rather, the PCR primers were self-amplifiable because of their overlapping and complementary sequences (Meyer et al. 2010). Labeling results suggest that the (TTAGG)$_n$ probe may have broad application for work with other species of prostriate ticks. The primers Telo_F (5′-TTA GGT TAG GTT AGG TTA GGT TAG G-3′) and Telo_R (5′-TAA CCT AAC CTA ACC TAA CCT AAC C-3′) were used in the PCR. Amplification consisted of 1 minute at 94°C followed by 30 cycles each of 1 minute at 94°C, 30 seconds at 50°C, 1 minute at 72°C, and one final step of 5 minutes at 72°C. PCR-amplified DNA was separated on 0.8% agarose tris-borate-ethylenediaminetetraacetic acid (TBE) gels, stained with ethidium bromide, and visualized under ultraviolet light. The entire reaction was purified and labeled as described in Section 12.2.4 to produce a range of pooled (TTAGG)$_n$-based PCR products for use as a probe.

12.2.3.4 C$_o$t-1 DNA Fraction

The C$_o$t-1 DNA fraction prepared from *I. scapularis* gDNA extracted from embryos has been used to block strong signal from heterochromatic regions of the chromosomes in an attempt to enhance probe detection. C$_o$t-1 DNA can be prepared by DNA reassociation experiments using autoclaving of gDNA in the shearing step or by using the protocol of Zwick et al. (1997), which involves sonication to achieve an average fragment size of approximately 800 bp. The renaturation of the gDNA to specific C$_o$t values can be performed as described by Peterson et al. (2002).

12.2.3.5 Centromeric Probe from Fractionated Genomic DNA

The gDNA fractionation protocol reported by Luo et al. (2004) to identify methylated centromeric DNA in eukaryotes was modified as follows for use in *I. scapularis*. Approximately 2 μg of *I. scapularis* gDNA was digested in a 50 μL reaction mix containing 10 units of the cytosine methylation-sensitive restriction enzyme *Hpa*II and 1× restriction digest buffer I (New England BioLabs Inc., Ipswich, Massachusetts) for 16 hours at 37°C. A control reaction included identical components except for *Hpa*II. After incubation, DNA was analyzed by gel electrophoresis. DNA fragments of interest were excised from the gel and purified with the QIAquick Gel Extraction Kit (Qiagen) as described in Meyer et al. (2010).

12.2.3.6 Bacterial Artificial Chromosome Clone Probes

BAC clones selected from the 10× *I. scapularis* and *R. microplus* BAC libraries have been used successfully as probes (Hill et al. 2009; Meyer et al. 2010). Clones that contain multiple copies of TRs and the rRNA repeat proved particularly effective for chromosome labeling. The authors note that attempts to develop FiberFISH protocols for the localization and orientation of *R. microplus* BAC clones on DNA fibers have proved unsuccessful.

12.2.4 Methods for Labeling Probes

12.2.4.1 Labeling by Nick Translation and Polymerase Chain Reaction

Probes have been prepared from *I. scapularis* and *R. microplus* small insert (<4 kb) gDNA clones and BAC clones containing large inserts (~120 kb), and from *I. scapularis* fractions prepared from gDNA (Hill et al. 2009; Meyer et al. 2010). Small insert clones and BAC clones were grown in 5 mL Luria-Bertani (LB) medium + antibiotic, and DNA was extracted using the QIAprep spin miniprep kit (Qiagen). DNA was labeled by nick translation using either biotin- or digoxigenin (DIG)-conjugated deoxyuridine triphosphate (dUTP) (Roche) according to the manufacturers' recommendations. Briefly, probes were prepared by addition of 1 μg template DNA (small insert clone, BAC clone, or gDNA fraction), 4 μL of the biotin or DIG Nick Translation Mix, and double distilled water (ddH$_2$O) to a final volume of 20 μL. The solution was mixed well, centrifuged briefly, and incubated at 15°C for 90 minutes. Following incubation, 1 μL 500 mM ethylenediaminetetraacetic acid (EDTA) (pH 8.0) was added and the DNA was heated to 65°C for 10 minutes to stop the reaction. PCR products obtained by amplification of repetitive sequence from templates such as plasmid DNA clones, the 28S rRNA gene, and the (TTAGG)$_n$ telomeric repeat can be labeled using either DIG 11-dUTP or biotin-16-dUTP as a partial substitute for deoxythimidine triphosphate (dTTP) in the PCR, according to the instructions provided by the manufacturer (Roche). For all probe types, unincorporated biotin- or DIG-labeled nucleotides were removed using the QIAquick Nucleotide Removal Kit (Qiagen).

The detection of labeled cDNA clones for localization of expressed genes of interest on the *I. scapularis* and *R. microplus* chromosomes has proved problematic. Polytene chromosome preparations have been widely used for this purpose, and serve to greatly enhance the signal intensity of short DNA probes as they represent multiple copies of each chromosome, and hence, each gene. However, to the best of the authors' knowledge, polytene chromosomes have not been observed in ticks.

12.2.5   Hybridization and Detection Methods

12.2.5.1   Slide Preparation

1. Prepare fresh slides as described in Section 12.2.2, or if relabeling slides stored at −80°C, remove coverslips.
2. Serially dehydrate slides by immersing in 70%, 90%, and 100% ethanol solutions in a Coplin jar for 5 minutes per solution at room temperature (RT) on an orbital shaker (gentle shaking). Remove slides, and allow to air-dry.
3. Add 100 μL denaturation solution containing 70% formamide in 2× SSC and cover slides with 22 × 40-mm glass coverslips.
4. Denature DNA on slides using a flat-surface heating block set at 85°C for 90 seconds.
5. Immediately serially dehydrate slides in 70% ethanol pre-chilled at −20°C, and then in 90% and 100% ethanol at RT for 5 minutes each. Allow slides to air-dry.
6. Prepare a hybridization solution containing 1 μL salmon sperm DNA (ssDNA) (nonspecific blocking agent), 5 μL deionized formamide (40%), 1 μL 2× SSC, and 2 μL 50% Dextran sulfate.
7. Add 1 μL biotin and/or 1 μL DIG-labeled DNA to the hybridization solution (4–10 ng DNA/μL final concentration), mix well, denature at 85°C for 15 minutes, and chill solution on ice for 1 minute.

12.2.5.2   Probe Hybridization

1. Add 10 μL hybridization solution/probe mix to each slide and cover with a 22 × 22-mm glass coverslip.
2. Seal the edges of the coverslip to prevent desiccation by applying a thin coating of rubber cement.
3. Transfer slides to a plastic container lined with several moistened paper towels to produce a high relative humidity and prevent desiccation of the slide, and incubate at 37°C for 16 hours (or overnight).
4. Carefully remove the rubber cement with forceps (gently hold the cover slip so that it does not move). Place slides in

2× SSC in a Coplin jar at RT for approximately 5 minutes to allow coverslips to fall off (remove gently if necessary with forceps if coverslips do not fall off).

5. Wash slides with 2× SSC at 42°C for 10 minutes, followed by 2× SSC at RT for 5 minutes, and then 1× PBS at RT for 5 minutes with gentle shaking on an orbital shaker.

12.2.5.3 Probe Detection

Detection can be performed using Alexa Fluor 488-conjugated anti-biotin and rhodamine-conjugated anti-DIG (Molecular Probes). For multilayer detection of biotin-conjugated probes, Alexa Fluor 488-streptavidin can be used for the initial layer of immunodetection, followed by a layer of biotin anti-streptavidin and a second layer of Alexa Fluor 488-streptavidin. For multilayer detection of DIG-conjugated probes, mouse anti-DIG can be used in the first layer of detection, followed by a layer of anti-mouse Alexa Fluor 568 (Molecular Probes).

1. Prepare the antibody mix by the addition of the following reagents for each slide: 20 µL 5× antibody buffer (1% BSA in 1× PBS), 1 µL rhodamine-conjugated anti-DIG and 1 µL of fluorescene isothyocyanate (FITC) or Alexa Fluor 488-conjugated anti-biotin and ddH$_2$O to a final volume of 100 µL.

2. Add 100 µL antibody mix to each slide. Cover slides with a 22 × 40-mm coverslip and incubate at 37°C for 90 minutes in a sealed, light-proof container, lined with moist paper towels to create a humidified chamber.

3. Wash slides three times with 1× TNT buffer (100 mM Tris-HCl, 150 mM NaCl, 0.05% Tween 20, pH 7.5) at RT for 10 minutes per wash in a Coplin jar protected from light.

4. Wash slides with 1× PBS for 10 minutes, on shaker and protected from light.

5. Apply 10 µL mounting medium (VECTASHIELD with DAPI) and cover with a 22 × 22-mm glass coverslip. Press the glass coverslip down firmly, be careful not to cause it to slide and potentially damage the chromosomes/cellular material.

6. Use a Kimwipe to gently remove excess medium from the sides of the coverslip and store each slide in a box at 4°C, protected from light. If stored correctly, it is possible to detect signal for several weeks.

Rehybridization experiments can be performed to localize multiple probes to chromosome spreads. Briefly, subsequent hybridizations were conducted as described earlier. Slides were washed twice for 10 minutes in 1× PBS to remove the glass coverslip and mounting medium between experiments. Slide coordinates for chromosome spreads were recorded so that the identical fields could be reexamined and photographed for each probe tested. Images corresponding to each probe were superimposed onto the original image of the DAPI-stained chromosomes with MetaVue software version 6.3r2 (Molecular Devices, Sunnyvale, California). This compensated for the significant reduction in chromosome quality and increased background associated with successive hybridizations.

12.2.6   Visualization and Mapping

Detection of labeled tick chromosomes was performed on a fluorescent microscope using channels as appropriate for each label used. Digital images of fluorescently labeled *I. scapularis* and *R. microplus* chromosomes were recorded using an ORCA-ER (Hamamatsu, Iwata City, Japan) digital camera mounted to an Olympus BX51 microscope. Digital images were examined using MetaMorph imaging software (Universal Imaging Corp., Downingtown, Pennsylvania).

12.2.7   Representative Results

12.2.7.1   *Ixodes scapularis* Karyotype

Representative results showing the hybridization of fluorescently labeled tandem repeats that include a telomeric $(TTAGG)_n$ motif, the NORs, and the major repeat families ISR-1–3 to mitotic chromosomes prepared from the *I. scapularis* ISE18 cell line are shown in Figure 12.1. Relative hybridization patterns were used to construct the ideogram shown in Figure 12.2. A "two-spot" hybridization pattern at the termini of all sister chromatids was observed with the telomeric probe that supports a telocentric (or acro-centric) chromosome structure, consistent with the original description of ISE18 chromosomes by Chen et al. (1994). The sex chromosomes X (typically the largest chromosome in the karyotype) and Y (usually the smallest chromosome in karyotype), as well as three pairs of chromosomes that hybridize to only ISR-1 (90 bp), ISR-2a (95 bp), and ISR-2a + ISR-3 (385 bp), respectively, and an additional pair of chromosomes based on hybridization to ISR-2a over approximately half the chromosome, were readily identified. The other chromosomes in the

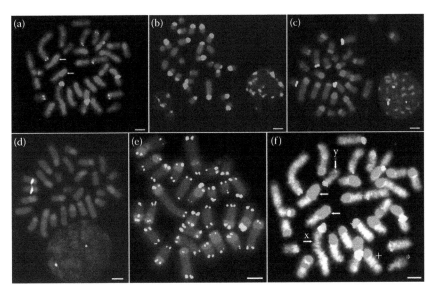

FIGURE 12.1 (**See color insert.**) Fluorescence in situ hybridization of probes containing tandemly repetitive DNA to mitotic *Ixodes scapularis* chromosomes and cell nuclei (see panels b through d) prepared from cell line ISE18. Chromatin was stained with DAPI (a through e, blue; f, gray): (a) nucleolar regions (NORs, ribosomal DNA) (red); *I. scapularis* repeat-1 (ISR-1, 90 bp tandem repeat) (green); (b) ISR-2a (95 bp tandem repeat) (red); (c) ISR-2b (96 bp tandem repeat) (green); (d) ISR-3 (385 bp tandem repeat) (yellow); (e) (TTAGG)$_n$ telomere-localizing tandem repeat (green); and (f) preliminary *I. scapularis* karyotype based on the relative localization pattern of probes hybridizing to the NORs (yellow), ISR-1 (light blue), ISR-2a (red), ISR-2b (green), and ISR-3 (purple). The putative X and Y chromosomes are marked accordingly and shown with arrows. The asterisk indicates a chromosome fragment. The two arrows in the center of the panel show a chromosome pair consistently identified based on a strong signal for ISR-2a. The plus symbol shows an extra chromosome in this spread, commonly observed in ISE18. Scale bars = 5 μm. (With kind permission from Springer Science+Business Media: *Chromosome Res.*, Genome organization of major tandem repeats in the hard tick, *Ixodes scapularis*, 18, 2010, 357–70, Meyer, J. M., Kurtti, T. J., Van Zee, J. P., and Hill, C. A., Figures 2 through 4.)

karyotype that could not be reliably paired or distinguished were grouped according to their hybridization signals to these markers. These groups include the chromosomes that show signals for ISR-2a + NOR (four chromosomes), ISR-1 + ISR-2a (four chromosomes), and ISR-2a only (10 chromosomes). Several probes comprising gDNA fractions resistant to digestion with the methylation-sensitive restriction enzyme *Hpa*II were developed to potentially identify methylated centromeric DNA. These probes hybridized to the terminal heterochromatic regions of the chromosomes presumably associated with peri-centromeric DNA. The C_0t-1 DNA fraction, presumably enriched for highly repetitive, fold-back DNA, showed strong hybridization to the termini of nearly all chromosomes (data not shown) supporting a correlation with the distribution of highly repetitive and presumably heterochromatic regions (i.e., ISR-1–3).

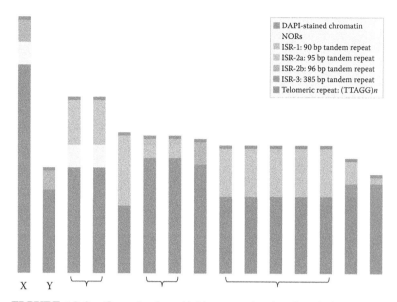

FIGURE 12.2 (**See color insert.**) Ideogram showing the relative arrangement of tandemly repetitive DNA based on hybridization of *Ixodes scapularis* ISE18 cell line chromosomes. The X and Y sex-determining chromosomes are labeled, and groups of chromosomes sharing similar hybridization patterns are shown with brackets. The individual chromosomes within these groups could not be readily distinguished from one another based on their relative sizes or distribution of the tandemly repetitive DNA markers examined. Chromosomes are drawn to scale based on the representative example provided by Meyer et al. (2010). The considerable amount of variability observed in the relative sizes of ISE18 chromosomes among different chromosome spreads did not permit generation of a true karyotype where chromosomes are assigned numbers based on size and FISH marker distribution.

A total of 45 completely sequenced and assembled BAC clones were also hybridized to ISE18 chromosomes (data not shown). A nonspecific hybridization pattern was observed with 42 BAC clones that are thought to reflect repeats dispersed among euchromatic regions of the chromosomes. Only three BAC clone hybridizations resulted in specific signals; these patterns matched that of hybridizations with markers for either the NORs or the ISR-3 tandem repeat family (*Ixodes scapularis* Genome Consortium, unpublished data).

12.2.7.2 *Rhipicephalus microplus* Karyotype

Chromosome morphology during meiosis suggested that the *R. microplus* chromosomes were holocentric instead of metacentric as previously reported (Oliver 1977). Ten bivalents of similar size and a single X chromosome were observed that fits with the previous observation of 20 autosomes and an XX–XO sex-determination system. The preliminary FISH-based karyotype was constructed for *R. microplus* based on the hybridization patterns of the tandem repeats RMR-1 and 2 and the NOR (Figure 12.3).

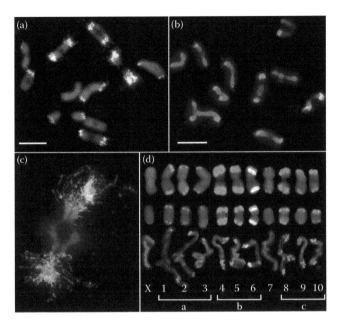

FIGURE 12.3 (**See color insert.**) Fluorescence in situ hybridization of probes containing tandemly repetitive DNA to meiotic chromosomes of *Rhipicephalus* (*Boophilus*) *microplus*. Chromatin was stained with DAPI (blue); (a) co-localization of two probes (red + green = yellow signal) containing the *R. microplus* repeat-1 (RMR-1) (149 bp); (b) localization of a probe containing RMR-2 (178–216 bp) (red) and a probe for the 28S rDNA (green); (c) cohybridization of RMR-1 (green) and RMR-2 (red) to a bivalent; and (d) preliminary *R. microplus* karyotype based on the relative localization of probes hybridizing to the 28S rDNA (green) and RMR-1 (red). Chromosomes were separated into three groups (a through c) and ordered according to relative descending length and RMR-1 DNA quantity (red). Bivalent 6 also shows hybridization to an rDNA probe (green), and the X chromosome is identified. Scale bars = 10 μm. (With kind permission from Springer Science+Business Media: *Chromosome Res.*, The position of repetitive DNA sequence in the southern cattle tick genome permits chromosome identification, 17, 2009, 77–89, Hill, C.A. et al., Figures 4 and 5.)

Both RMR-1 and 2 hybridized to sub-telomeric heterochromatin. RMR-1 hybridized to all but one bivalent and the X chromosome. The relative lengths of the bivalents combined with the relative intensity of RMR-1 hybridization enabled the determination of three groups of chromosomes. The three longest bivalents, each of which contained a relatively small amount of RMR-1 DNA were placed in the first group (group A). The second group (group B) was composed of three chromosomes that contained the greatest relative amount of RMR-1 DNA. The third group (group C) was composed of the shortest three bivalents each of which contained a moderate quantity of RMR-1 DNA. The relative amount of RMR-1 DNA on one of these bivalents (bivalent 8) made its recognition almost unequivocal. An absence of RMR-1 hybridization permitted the identification of the X chromosome and one bivalent

(bivalent 7). RMR-2 hybridized to each of the bivalents and the X chromosome.

12.2.8   Troubleshooting

12.2.8.1   High Background Signal Associated with Heterochromatic Regions

The $C_{o}t$-1 fraction prepared from *I. scapularis* gDNA was used in an approach to block signal from repetitive sequence and enhance the detection of higher complexity sequence. The fraction localized exclusively to the heterochromatic ends of the *I. scapularis* chromosomes and is expected to have limited utility for blocking repetitive sequence distributed throughout the euchromatin. Opportunities exist for further work to develop methods to improve the visualization of DNA sequences on the repeat-rich chromosomes of ixodid ticks.

12.2.8.2   Gray/Particulate Haze Surrounding Chromosomes

In some cases, a gray or particulate haze may be visible surrounding the chromosomes. This phenomenon is frequently observed when chromosomes are prepared using tissues harvested from ticks. Presumably, this is due to the presence of excess cytoplasmic material and cell debris that can interfere with chromosome and signal visualization. This issue can be overcome by pretreatment of slides using either a higher concentration (10–30 μL) or a longer pepsin treatment (3–5 minutes) than described in the protocol in Section 12.2.4. Note that to avoid precipitation, the pepsin should first be added to a clean beaker, rather than directly to the acid solution.

12.2.8.3   Old Chromosome Preparations

Chromosome preparations can be held at −20°C in fixative for up to 6 months. If older slides will be used, it is recommended to pretreat the slides with pepsin either at a higher concentration (10–30 μL) or for a longer duration (3–5 minutes) than described in the protocol in Section 12.2.4.

12.2.8.4   Chromosome Spreads Prepared from Cell Lines

Where possible, karyotypes should be based on chromosomes prepared from whole tissue. Cell lines that are available for several species of ticks (Bell-Sakyi et al. 2007) can also be used for this purpose, although it should be noted that the chromosomes of cells maintained in continuous culture typically accumulate repetitive DNA and chromosomal aberrations (aneuploidy) as noted by Chen et al. (1994) in chromosomes prepared from the ISE18 cell line. Thus, it is important to evaluate multiple spreads to develop a representative karyotype.

12.3 APPLICATIONS OF THE MAPS

Physical, genetic, and sequence maps are indispensable tools that can be used to address many of the questions commonly posed by tick and tick-borne disease researchers. The IscaW1 assembly and corresponding annotation have proved useful for the identification of genes and gene products associated with tick processes such as host location, blood feeding and digestion, developmental regulation, and pathogen acquisition and transmission. In addition, the genome sequence has been used to find classes of repetitive DNA and non-coding features such as microRNAs that may have an impact on gene and genome regulation (*Ixodes scapularis* Genome Consortium, unpublished data). The assignment and orientation of assembled genomic scaffolds or supercontigs is one of the most obvious applications of physical mapping. Beyond this, physical mapping can facilitate studies of chromosome biology by permitting analyses of the arrangement of classes of DNA on the chromosomes, including the identification of features of interest such as inversions, deletions, and insertions, the study of chromosome synteny between species, as well as genome evolution. For species of ixodid ticks, high-resolution linkage maps are expected to prove invaluable in applications such as genome scaffolding, population genetics and taxonomic studies, and mapping of loci associated with quantitative traits such as host preference, competence to transmit pathogens and parasites, and resistance to pesticides (i.e., quantitative trait loci [QTL] mapping).

12.3.1 Integration of Linkage Maps, Chromosomes, and Genomic Sequences

Genetic and physical maps are an important component of most genome sequencing efforts, because sequence-linked genetic markers associated with linkage groups and physical maps can assist the assembly and assignment of genome sequence along the chromosomes. Currently, the integration of genetic, sequence, and physical map data is underway for one ixodid tick, *I. scapularis*. The RADseq method (Baird et al. 2008; Hohenlohe et al. 2010; Pfender et al. 2011) has gained popularity for linkage map construction (Davey and Blaxter 2011) and has facilitated genetic variant discovery by sequencing of short stretches of DNA flanking restriction enzyme sites (i.e., RADtags), allowing orthologous sequences to be targeted in multiple individuals (Chutimanitsakun et al. 2011). The availability of the *I. scapularis* draft genome has provided an opportunity to integrate linkage, chromosome, and genome maps for this species via the RADseq approach. In preliminary work, five RADtag libraries prepared from five individual Wikel strain *I. scapularis* females were pooled and sequenced on the Illumina platform, generating more than 30,000 SNPs from millions of short (~100 bp) reads. Some of these markers will be used in genotyping of *I. scapularis* parents and F1 progeny to generate the first high-density linkage map for this species. Efforts are ongoing to integrate the genetic, sequence, and physical maps by localization of IscaW1 scaffold-linked BAC clones to the *I. scapularis* chromosomes using the above-described FISH mapping protocols (Gulia-Nuss, pers. comm.). Technically, map integration is feasible for any tick species assuming the availability of genomic resources such as sequence data, genetic markers, tick mapping populations, and BAC clones. With increasing demands for application of the RAD method in organisms that lack a reference genome, simpler library preparation protocols such as 2b-RAD (type IIB restriction enzyme-based RAD) (Wang et al. 2012) and ddRAD (double digest RAD) (Peterson et al. 2012) will have utility for linkage mapping in ixodid ticks.

12.3.2   Chromosome and Genome Organization and Function

Beyond the published work described earlier, little is currently known regarding chromosomal organization among acarine species. Analyses of *I. scapularis* chromosomes prepared by several methods revealed that this species has $2n = 28$ chromosomes, an XX–XY sex-determination system, and acro- or telo-centric chromosomes (Oliver 1977; Chen et al. 1994, Munderloh et al. 1994; Meyer et al. 2010). Studies of meiotic chromosome spreads prepared from *R. microplus* testes revealed an XX–XO sex-determination system with 22 diploid chromosomes in females and 21 in males (Hill et al. 2009). Previous studies have reported acro-centric chromosomes in *R. microplus* (Oliver and Bremner 1968; Hilburn et al. 1989; Garcia et al. 2002) with the presumed centromere at the tip of the chromosome and an indistinguishable short arm (Hilburn et al. 1989). More recently, results obtained using FISH and high-resolution digital images instead suggest holocentric chromosomes in this tick (Hill et al. 2009).

Studies have investigated the arrangement of coding and noncoding DNA on some of the larger scaffolds (>1 Mb) comprising the *I. scapularis* IscaW1 genome assembly, and a handful of sequenced and assembled *I. scapularis* and *R. microplus* BAC clones. Physical mapping suggests an accumulation of TRs in heterochromatic regions associated with the presumed telomeres and centromeres of these two tick species (Hill et al. 2009; Meyer et al. 2010), and various sequencing and genetic studies support low gene density and the accumulation of multiple families of TEs, many of which have also been identified in other species of arthropods (*Ixodes scapularis* Genome Consortium, unpublished data).

12.4 CONCLUSIONS

The studies of several research groups have provided important insights into the chromosome biology of species of ticks and mites (Oliver 1977, Hilburn et al. 1989, Chen et al. 1994, Munderloh et al. 1994). These studies have been followed by directed efforts to advance genomics research in two species of ixodid ticks, namely the prostriate tick *I. scapularis* and the metastriate tick *R. microplus*. The development of preliminary FISH-based karyotypes and physical maps for both species provides a foundation for advanced cytogenetic research. Now, progress in the field of tick cytogenetics requires the development of karyotypes for multiple species of ixodid and argasid ticks. Success with physical mapping studies in ticks has been restricted to the localization of repetitive DNA; however, the development of protocols to detect and study the arrangement of genes and other euchromatic sequences on chromosomes is desperately needed. Also required are physical mapping protocols that can accommodate the high repeat content associated with many tick genomes. These advances, when coupled with broader support for genome sequencing initiatives in species of the Ixodida, will ultimately provide much-needed insights into the genetic basis of many biological processes of relevance to tick and tick-borne disease research.

REFERENCES

13th International Conference on Lyme Borreliosis and Other Tick-Borne Diseases, August 18, 2013. http://iclb2013.com/overview.htm.

Baird, N. A., P. D. Etter, T. S. Atwood, M. C. Currey, A. L. Shiver, Z. A. Lewis, E. U. Selker, W. A. Cresko, and E. A. Johnson. 2008. Rapid SNP discovery and genetic mapping using sequenced RAD markers. *PLoS One* 3:e3376.

Barker, S. C. and A. Murrell. 2002. Phylogeny, evolution and historical zoogeography of ticks: A review of recent progress. *Exp Appl Acarol* 28:55–68.

Bell-Sakyi, L., E. Zweygarth, E. F. Blouin, E. A. Gould, and F. Jongejan. 2007. Tick cell lines: Tools for tick and tick-borne disease research. *Trends Parasitol* 23:450–57.

Benson, G. 1999. Tandem repeats finder: A program to analyze DNA sequences. *Nucleic Acid Res* 27:573–80.

Camicas, J.-L., J.-E. Hervy, F. Adam, and P. C. Morel. 1998. *The Ticks of the World (Acarida, Ixodida). Nomenclature, Described Stages, Hosts, Distribution.* Orstom Editions, Paris, France.

Centers for Disease Control and Prevention. 2012. Notices to readers: Final 2011 reports of nationally notifiable infectious diseases. *Morb Mortal Wkly Rep* (MMWR) 61:624–37. http://www.cdc.gov/mmwr/preview/mmwrhtml/mm6132a8.htm?s_cid = mm6132a8_w.

Chen, C., U. G. Munderloh, and T. J. Kurtti. 1994. Cytogenetic characteristics of cell lines from *Ixodes scapularis* (Acari: Ixodidae). *J Med Entomol* 31:425–34.

Chutimanitsakun, Y., R. Nipper, A. Cuesta-Marcos, L. Cistué, A. Corey, T. Filichkina, E. A. Johnson, and P. M. Hayes. 2011. Construction and application for QTL analysis of a restriction site associated DNA (RAD) linkage map in barley. *BMC Genomics* 12:4.

Cornman, R. S., M. C. Schatz, J. S. Johnston, Y. P. Chen, J. Pettis, G. Hunt, L. Bourgeois et al. 2010. Genomic survey of the ectoparasitic mite *Varroa destructor*, a major pest of the honeybee *Apis mellifera*. *BMC Genomics* 11:602.

Davey, J. W. and M. L. Blaxter. 2011. RADSeq: Next-generation population genetics. *Brief Funct Genomics* 9:416–23.

Delaye, C., L. Beati, A. Aeschlimann, F. Renaud, and T. de Meeus. 1997. Population genetic structure of *Ixodes ricinus* in Switzerland from allozymic data: No evidence of divergence between nearby sites. *Int J Parasitol* 27:769–73.

Dutt, M. K. 1954. Chromosome studies on *Rhipicephalus sanguineus* Laterille and *Hyalomma aegyptium* Neumann (Acarina: Ixodidae). *Curr Sci* 23:194–6.

Foil, L. D., P. Coleman, M. Eisler, H. Fragoso-Sanchez, Z. Garcia-Vazquez, F. D. Guerrero, N. N. Jonsson et al. 2004. Factors that influence the prevalence of acaricide resistance and tick-borne diseases. *Vet Parasitol* 125:163–81.

Garcia, R. N., C. Garcia-Fernandez, S. M. L. Garcia, and V. L. S Valente. 2002. Meiotic chromosomes of a southern Brazilian population of *Boophilus microplus* (Acari, Ixodidae). *Iheringia Sér Zool Porto Alegre* 92:63–70.

Geraci, N. S., J. S. Johnston, J. P. Robinson, S. K. Wikel, and C. A. Hill. 2007. Variation in genome size of argasid and ixodid ticks. *Insect Biochem Mol Biol* 37:399–408.

Goodman, J. L, D. T. Dennis, and D. E. Sonenshine. *Tick Borne Diseases of Humans*. 2005. Edited by Goodman, J. L., D. T. Dennis, and D. E. Sonenshine. ASM Press, Washington, DC. Pp. xiii+401.

Goroschenko, Y. L. 1962. The karyotypes of argasid ticks of the USSR fauna in connection with their taxonomy. *Tsitologiya* 4:137–49.

Grbić, M., T. Van Leeuwen, R. M. Clark, S. Rombauts, P. Rouzé, V. Grbić, E. J. Osborne et al. 2011. The genome of *Tetranychus urticae* reveals herbivorous pest adaptations. *Nature* 479:487–92.

Guerrero, F. D., P. Moolhuijzen, D. G. Peterson, S. Bidwell, E. Caler, M. Bellgard, V. M. Nene, and A. Djikeng. 2010. Reassociation kinetics-based approach for partial genome sequencing of the cattle tick, *Rhipicephalus (Boophilus) microplus*. *BMC Genomics* 11:374.

Guerrero, F. D., V. M. Nene, J. E. George, S. C. Barker, and P. Willadsen. 2006. Sequencing a new target genome: The *Boophilus microplus* (Acari: Ixodidae) genome project. *J Med Entomol* 43:9–16.

Hilburn, L. R., S. J. Gunn, and R. B. Davey. 1989. The genetics of new world *Boophilus microplus* (Canestrini) and *Boophilus annulatus* (Say) in their possible control. *Bull Soc Vector Ecol* 14:222–31.

Hill, C. A. Genome analysis of major tick and mite vectors of human pathogens. December 15, 2010. Submitted to the NIH-NIAID-NHGRI Pathogens and Vectors Working Group.

Hill, C. A., F. D. Guerrero, J. P. Van Zee, N. S. Geraci, J. G. Walling, J. J. Stuart. 2009. The position of repetitive DNA sequence in the southern cattle tick genome permits chromosome identification. *Chromosome Res* 17:77–89.

Hill, C. A., V. M. Nene, and S. K. Wikel. March 1, 2004. Proposal for Sequencing the Genome of the Tick, *Ixodes scapularis*. Submitted to National Institutes of Allergy and Infectious Disease.

Hill, C. A. and S. K. Wikel. 2005. The *Ixodes scapularis* genome project: An opportunity for advancing tick research. *Trends Parasitol* 21:151–3.

Hohenlohe, P. A., S. Bassham, P. D. Etter, N. Stiffler, E. A. Johnson, and W. A. Cresko. 2010. Population genomics of parallel adaptation in threespine stickleback using sequenced RAD tags. *PLoS Genet* 6:e1000862.

Hoogstraal, H. 1956. Notes on African *Haemaphysalis* ticks. III. The hyrax parasites, *H. bequaerti* sp. *nov.*, *H. orientalis* N. and W., 1915 (new combination), and *H. cooleyi* Bedford, 1929 (Ixodoidea, Ixodidae). *J Parasitol* 142:156–72.

Horak, I. G., J. L. Camicas, and J. E. Keirans. 2002. The Argasidae, Ixodidae and Nuttalliellidae (Acari: Ixodida): A world list of valid tick names. *Exp Appl Acarol* 28:27–54.Hoy M. A., F. Yu, J. M. Meyer, O. A. Tarazona, A. Jeyaprakash, K. Wu. 2013. Transcriptome sequencing and annotation of the predatory mite *Metaseiulus occidentalis* (Acari: Phytoseiidae): A cautionary tale about possible contamination by prey sequences. Exp Appl Acarol 59(3):283–96. doi: 10.1007/s10493-012-9603-4. Epub 2012 Aug 25.

Hoy M. A., F. Yu, J. M. Meyer, O. A. Tarazona, A. Jeyaprakash, K. Wu. 2013. Transcriptome sequencing and annotation of the predatory mite *Metaseiulus occidentalis* (Acari: Phytoseiidae): A cautionary tale about possible contamination by prey sequences. *Exp Appl Acarol* 59(3):283–96. doi: 10.1007/s10493-012-9603-4. Epub 2012 Aug 25.

Ijdo, J. W., R. A. Wells, A. Baldini, and S. T. Reeders. 1991. Improved telomere detection using a telomere repeat probe (TTAGGG)$_n$ generated by PCR. *Nucleic Acids Res* 19:4780.

Ixodes scapularis Genome Consortium. The genome sequence of the Lyme disease tick, *Ixodes scapularis.*Unpublished data.

Jiang, J. M. and B. S. Gill. 2006. Current status and the future of fluorescence *in situ* hybridization (FISH) in plant genome research. *Genome* 49:1057–68.

Jongejan, F. and G. Uilenberg. 2004. The global importance of ticks. *Parasitol* 129:S3–14.

Kahn, J. 1964. Cytotaxonomy of ticks. *Quart J Micr Sci* 105:123–37.

Klompen, J. S. H., W. Black, J. E. Keirans, and J. H. Oliver. 1996. Evolution of ticks. *Annu Rev Entomol* 41:141–61.

Lindgren, E. and T. G. T. Jaenson. 2006. *Lyme borreliosis in Europe: Influences of Climate and Climate Change, Epidemiology, Ecology and Adaptation Measures.* World Health Organization, Regional Office for Europe, Copenhagen, Denmark.

Lorite, P., J. A. Carrillo, and T. Palomeque. 2002. Conservation of (TTAGG) (n) telomeric sequences among ants (Hymenoptera, Formicidae). *J Hered* 93:282–5.

Luo, S., A. E. Hall, S. E. Hall, and D Preuss. 2004. Whole-genome fractionation rapidly purifies DNA from centromeric regions. *Nat Methods* 1(1):1–5.

Meyer, J. M. and C. A. Hill. 2014. Tick genetics, genomics and proteomics. In: *Biology of Ticks.* Edited by Roe, M. and D. Sonenshine. Second Edition. Oxford University Press, Oxford, United Kingdom, pp. 61–86.

Meyer, J. M., T. J. Kurtti, J. P. Van Zee, and C. A. Hill. 2010. Genome organization of major tandem repeats in the hard tick, *Ixodes scapularis. Chromosome Res* 18:357–70.

Meyne, J., H. Hirai, and H. T. Imai. 1995. FISH analysis of the telomere sequences of bulldog ants (Myrmecia: formicidae). *Chromosoma* 104:14–8.

Munderloh, U. G., Y. Liu, M. Wang, C. Chen, and T. J. Kurtti.1994. Establishment, maintenance and description of cell lines from the tick *Ixodes scapularis. J Parasitol* 80:533–43.

Okazaki, S., K. Tsuchida, H. Maekawa, H. Ishikawa, and H. Fujiwara. 1993. Identification of a pentanucleotide telomeric sequence, (TTAGG)$_n$, in the silkworm *Bombyx mori* and in other insects. *Mol Cell Biol* 13:1424–32.

Oliver, J. H. Jr. 1977. Cytogenetics of mites and ticks. *Annu Rev Entomol* 22:407–29.

Oliver, J. H. and K. C. Bremner. 1968. Cytogenetics of ticks. III. Chromosome and sex determination in some Australian hard ticks (Ixodidae). *Ann Entomol Soc Am* 61:837–44.

Oliver, J. H. Jr., M. R. Owsley, H. J. Hutcheson, A. M. James, C. Chen, W. S. Irby, E. M. Dotson, and D. K. McLain. 1993. Conspecificity of the ticks *Ixodes scapularis* and *I. dammini* (Acari: Ixodidae). *J Med Entomol* 30:54–63.

Palmer M. J., J. A. Bantle, X. Guo, W. S. Fargo. 1994. Genome size and organization in the ixodid tick *Amblyomma americanum* (L.). *Insect Mol Biol* 3(1):57–62.

Peterson, B. K., J. N. Weber, E. H. Kay, H. S. Fisher, and H. E. Hoekstra. 2012. Double digest RADseq: An inexpensive method for *de novo* SNP discovery and genotyping in model and non-model species. *PLoS One* 7:e37135.

Peterson, D. G., S. R. Schulze, E. B. Sciara, S. A. Lee, J. E. Bowers, A. Nagel, N. Jiang, D. C. Tibbitts, S. R. Wessler, and A. H. Paterson. 2002. Integration of C_0t analysis, DNA cloning, and high-throughput sequencing facilitates genome characterization and gene discovery. *Genome Res* 12(5):795–807.

Pfender, W. F., M. C Saha, E. A. Johnson, and M. B. Slabaugh. 2011. Mapping with RAD (restriction-site associated DNA) markers to rapidly identify QTL for stem rust resistance in *Lolium perenne*. *Theor Appl Genet* 122(8):1467–80.

Sahara, K., F. Marec, and W. Traut. 1999. TTAGG telomeric repeats in chromosomes of some insects and other arthropods. *Chromosome Res* 7:449–60.

Seraji-Bozorgzad, N. and A. C. Tselis. 2013. Non-lyme tick-borne diseases: A neurological perspective. *Curr Neurol Neurosci Rep* 13:388.

Sonenshine, D. E., S. M. Ceraul, W. E. Hynes, K. R. Macaluso, and A. F. Azad. 2002. Expression of defensin-like peptides in tick hemolymph and midgut in response to challenge with *Borrelia burgdorferi, Escherichia coli* and *Bacillus subtilis*. *Exp Appl Acaro* 28:127–34.

Spyropoulos, B. and P. B. Moens. 1994. *In situ* hybridization of meiotic prophase chromosomes. *Methods Mol Biol* 33:131–9.

Ullmann, A. J., C. M. Lima, F. D. Guerrero, J. Piesmann, and W. C. Black, 4th. 2005. Genome size and organization in the black-legged tick, *Ixodes scapularis* and the southern cattle tick, *Boophilus microplus*. *Insect Mol Biol* 14:217–22.

Ullmann, A. J., J. Piesman, M. C. Dolan, and W. C. Black, 4th. 2003. A preliminary linkage map of the hard tick, *Ixodes scapularis*. *Insect Mol Biol* 12:201–10.

Ullmann, A. J., J. J. Stuart, and C. A. Hill. 2008. Tick, Chapter 8. In: *Genome Mapping and Genomics in Animals, Vol. 1, Genome Mapping and Genomics in Arthropods*. Edited by Cole, C. and W. Hunter. Springer-Verlag, Heidelberg, Germany. Pp. 103–17.

Vítková, M., J. Král, W. Traut, J. Zrzavy, and F. Marec. 2005. The evolutionary origin of insect telomeric repeats, (TTAGG)n. *Chromosome Res* 13:145–56.

Walker, D. 1998. Tick-transmitted infectious diseases in the United States. *Annual Rev Public Health* 19:237–69.

Wang, S., E. Meyer, J. K. McKay, and M. V. Matz. 2012. 2b-RAD: A simple and flexible method for genome-wide genotyping. *Nat Methods* 9:808–10.

Zwick, M. S., R. E. Hanson, and M. N. Islam-Faridi. 1997. A rapid procedure for the isolation of C_0t-1 DNA from plants. *Genome* 40:138–42.

Index

Milton Keynes UK
Ingram Content Group UK Ltd.
UKHW051537141024
449569UK00028B/1515